PRINCIPLES OF TURBULENCE CONTROL

PRINCIPLES OF TURBULENCE CONTROL

Baochun Fan & Gang Dong

Nanjing University of Science and Technology
Nanjing, China

WILEY

國防工業出版社
National Defense Industry Press

This edition first published 2016

Published by John Wiley & Sons Singapore Pte Ltd, 1 Fusionopolis Walk, #07-01 Solaris South Tower, Singapore 138628, under exclusive license granted by National Defense Industry Press for all media and languages excluding Simplified and Traditional Chinese and throughout the world excluding Mainland China, and with non-exclusive license for electronic versions in Mainland China.

For details of our global editorial offices, for customer services, and for information about how to apply for permission to reuse the copyright material in this book please see our website at www.wiley.com.

Library of Congress Cataloging-in-Publication Data

Fan, Baochun, author.
Principles of turbulence control / Baochun Fan, Gang Dong.
pages cm
Includes bibliographical references and index.
ISBN 978-1-118-71801-8 (cloth)
1. Turbulence. 2. Transition flow. I. Dong, Gang, 1970- author. II. Title.
TA357.5.T87F36 2015
624.1'75 – dc23

2015019359

A catalogue record for this book is available from the British Library.

Set in 10/12pt, TimesLTStd by SPi Global, Chennai, India.
Printed and bound in Singapore by Markono Print Media Pte Ltd

1 2016

Contents

About the Authors

Baochun Fan is a professor at the Nanjing University of Science and Technology, China, in the State Key Laboratory of Transient Physics. His research areas are in fluid mechanics, flow control, combustion, detonation, and explosion dynamics. He has been carrying out fundamental and applied research in these areas for the past 30 years. He has served on many academic committees.

Gang Dong is a professor at the Nanjing University of Science and Technology, China, in the State Key Laboratory of Transient Physics. His research interests are in turbulent flows, flow control, combustion, and detonations. He received his PhD from Jiangsu University, China, in 1998. Over the past 20 years, he has published about 100 papers in peer-reviewed journals and conferences. He is a committee member of the Detonation Professional Group in the Chinese Society of Theoretical and Applied Mechanics.

Preface

Turbulence control suggests a taming of turbulence so as to eliminate some deleterious effects while enhancing others to achieve desired goals, such as drag reduction, mixing enhancement, noise reduction, suppression of separation and oscillation, reduction of pollutant emission and fluid structure interaction, and others. The potential benefits of realizing efficient flow-control systems for the economy and environment is immense undoubtedly.

In the past, controlling turbulent flow was believed artificially to be extremely difficult, until the discovery of turbulent coherent structures. Improved understanding of the underlying physics of turbulent flow makes it possible to realize efficient turbulence control. The turbulent control targeting specific coherent structures is most effective when the control input is introduced locally at a high receptivity region. Microelectromechanical systems (MEMS) developed recently have provided opportunities for targeting the small-scale coherent structures in the near-wall region. Furthermore, the best possible control method with lower consumed energy should be designed by solving the optimal closed-loop control problem, where the control input varies with the instantaneous flow fields continuously. There are three essential components in optimal control technique, i.e. sensors, controller and actuators, which are combined on a single programmable chip in MEMS. In the control process, the information about instantaneous flow fields, measured by means of sensors is inputted into the controller. Then the controller gives desired control variables to the actuators by using a control law based on an approach of the optimal flow control, and finally the actuators actualize the control, leading the variations of the flow field. The new measured information will be got for changed flow field, and the input control variables are adjusted inevitably. Therefore, turbulence control has developed into an interdisciplinary subject involving fluid dynamics, control theory, MEMS, material and numerical analysis, and others.

The interest in turbulence control continues to grow and the amount of related published papers has been increasing in recent years because of the potential applications. This book aims to offer the reader an introduction to the fundamental and theories of turbulence control, and written in a pedagogic style serving as a supplementary text for courses for undergraduate and graduate students in mechanical engineering.

The book contains three parts with two chapters in each part. Wall turbulence is presented in the first part, where the spectral method applied in the direct simulation of wall turbulence, and the statistical analysis and the spectral analysis, used to deal with and to analyze experimental and numerical results throughout the book are discussed in Chapter 1, and then the formations, evolutions, and characterizations of wall-turbulence coherent structures are described in Chapter 2. In the second part, the open-loop control of wall turbulence is introduced, where the flow control strategies are classified into two types, that is, controlled via altering the boundary conditions, such as the oscillating wall and deformed wall, and controlled via altering the governing equations directly, such as body force distributed in the flow field. The two control strategies are presented in Chapters 3 and 4, respectively. The optimal closed-loop control that is attracting much attention because of its large control effect with small control input is discussed in the third part, including linear and nonlinear optimal controls. Linear optimal control is presented in Chapter 5, and the discussions on nonlinear optimal control in Chapter 6 are concerned with the suboptimal controls based on the spectral method or neural network and the adjoint-based optimal control based on the flow field and its adjoint flow field, which are governed by Navier–Stokes (N-S) equations and the adjoint equations modeling a ghost flow, respectively.

This book is organized into six chapters, Chapter 1 and Chapters 3–6 were written by Fan Baochun, while Chapter 2 was written by Dong Gang. In writing this book, I am happy to acknowledge the research support from the National Science Foundation of China, under Grant No. 11172140. I also owe a great debt to my colleagues and former students in the Nanjing University of Science and Technology, who contributed to the completion of the book in various ways. Above all, I wish to thank my wife, Jia Chenyuan, for her love and patient support.

Fan Baochun
Nanjing, China

Part One
Wall Turbulence

1

Statistical Analysis and Spectral Method

Statistical and spectral analyses are commonly used to characterize random phenomena which are seemingly chaotic, irregular, and unpredictable. They can also be used as the effect tools to study and describe turbulent flow, since the turbulent flow is random.

Many different computational approaches have been developed to solve turbulence problem, which is a very difficult problem. The most widely used models include statistical models and pseudospectral methods that are based on statistical and spectral analysis, respectively. In the statistical models, such as turbulent-viscosity models, e.g., $k-\varepsilon$ model, Reynolds-stress models, probability-density function (PDF) methods, and large eddy simulation (LES), the turbulent flow is described in terms of some statistics. In the pseudospectral methods, the turbulent flow is described in terms of the spectral coefficients in the spectral space.

However, in order to extract the essential kinetic and dynamic characteristics of turbulence and interpret it properly from experimental and numerical results, the statistical and spectral analyses are also used to deal with and to analyze these data, amount of which has grown through application of new experimental and computational tools.

Throughout the book, the turbulent control problems are solved numerically by the pseudospectral methods, the experiment and numerical dada involved are dealt with by statistical or spectral analysis, and the discussions on turbulent flow are based on the statistical and spectral descriptions. Therefore, the statistical analysis and spectral method are introduced briefly in Chapter 1.

In Section 1.1, statistical analysis is presented, and in Section 1.2, the statistical representation of turbulent flow is discussed. Sections 1.3 and 1.4 are concerned with spectral series expansions for Fourier series and other orthogonal basis. The fundamental concepts and technical of spectral method, a numerical method for partial differential equations, are introduced in Section 1.5 and its spectacular applications to Navier–Stokes (N-S) equations to turbulent flows are discussed in Section 1.6.

1.1 Statistical Analysis and Spectral Method[1–3]

1.1.1 Average Value

The fluid velocity field in turbulent flow is random, which varies significantly and irregularily in both position and time, and described by random variables. Considering a random

Principles of Turbulence Control, First Edition. Baochun Fan and Gang Dong.

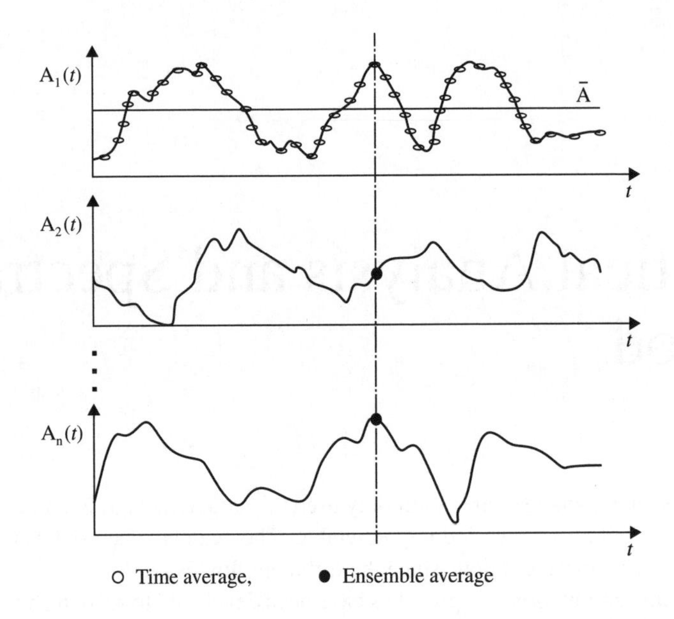

Figure 1.1 Time average and ensemble average of random variables

event expressed by a random variable $A(\mathbf{x}, t)$, which is statistically steady, the experimental measurements are taken on n repetitions at a specified position under the same set of conditions, and measured independent curves, $A_1(t), A_2(t), \ldots, A_n(t)$, are obtained. The curve is called a realization of random events $A(\mathbf{x}, t)$.

As shown in Figure 1.1, the time average for one realization is defined as

$$\bar{A} = \lim_{T\to\infty} \frac{\lim_{T\to\infty} \int_t^{t+T} A(t)dt}{T} \tag{1.1}$$

Therefore, a random variable A is decomposed into a mean $\bar{A}$ and a fluctuating part A', representing the deviation from the mean, such that

$$A = \bar{A} + A' \tag{1.2}$$

The mean value of a fluctuating quantity itself is zero.

Similarly, the space average is define as

$$\bar{A} = \lim_{V\to\infty} \frac{\int_V A(\mathbf{x})d\mathbf{x}}{V} \tag{1.3}$$

The ensemble average of the realizations under the same set of conditions is defined by

$$\langle A \rangle = \lim_{n\to\infty} \frac{A_1 + A_2 + \cdots + A_n}{n} \tag{1.4}$$

Hence, $A = \langle A \rangle + A'$ (1.5).

1.1.2 Probability Density and Statistical Moments

1.1.2.1 Probability Density

A realization of random events $A(\mathbf{x}, t)$ at a specified position is shown in Figure 1.2.

Define indicator function

$$\varphi(A,t) = \begin{cases} 1 & \text{if} \quad A(t) < A \\ 0 & \text{if} \quad A(t) \geq A \end{cases} \tag{1.6}$$

where A is a given value and $A(t)$ is an arbitrary value. Then we have

$$\lim_{T\to\infty} \int_t^{t+T} \varphi(A,t)dt = \sum \Delta t_i = T_L$$

T_L is a time during which $A(t) < A$.

$$\text{Let } \overline{\varphi} = \frac{T_L}{T} = P(A,t)$$

that is, a ratio between total duration of certain conditions satisfied and total duration of averaging, representing the percentage of time spent by $A(t)$ under the given level. P is called the cumulative distribution function (CDF) ($0 \leq P \leq 1$) and monotonically increases as A increase, as shown in Figure 1.3.

The probability density function (PDF) is defined to be the derivative of CDF:

$$P(A) = \frac{dP}{dA} \tag{1.7}$$

which represents the probability of events with $A < A(t) < A + \Delta A$. That is,

$$\Delta P = P\Delta A = P_{\text{rob}}\{A < A(t) < A + \Delta A\}$$

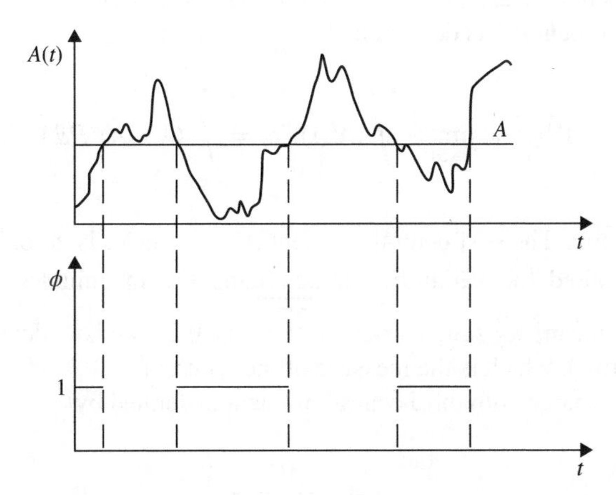

Figure 1.2 Indication function

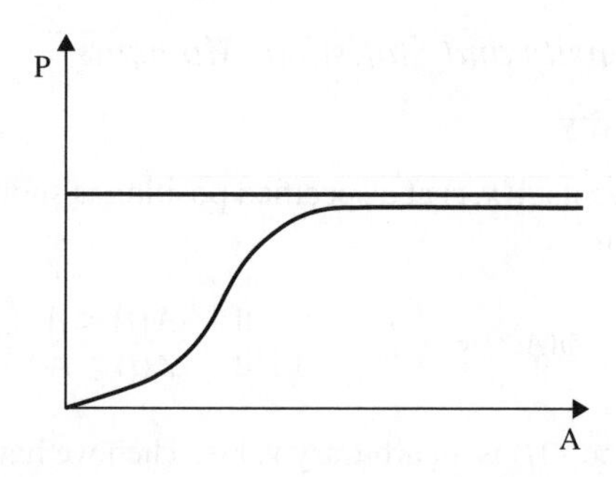

Figure 1.3 Cumulative distribution function

where P_{rob} denotes probability. It satisfies the normalized condition

$$\int_{-\infty}^{+\infty} P(A)dA = 1 \tag{1.8}$$

Obviously, the average (or mean) of the random variable can be obtained by PDF

$$\overline{A} = \int_{-\infty}^{+\infty} AP(A)dA \tag{1.9}$$

1.1.2.2 Statistical Moments

The mean values of the various powers of A are called moments. The nth moment $\overline{A^n}$ is defined by

$$\overline{A^n} = \lim_{T\to\infty} \frac{1}{T} \int_{t}^{t+T} A^n(t)dt = \int_{-\infty}^{+\infty} A^n(t)P(A)dA = \frac{\int_{-\infty}^{+\infty} A^n P(A)dA}{\int_{-\infty}^{+\infty} P(A)dA} \tag{1.10}$$

where the first moment $\overline{A}$, that is, $n = 1$, is the familiar mean value.

The nth central moment $\overline{A'^n}$ is defined by

$$\overline{A'^n(t)} = \lim_{T\to\infty} \frac{1}{T} \int_{t}^{t+T} (A'(t))^n dt = \int_{-\infty}^{\infty} (A - \overline{A})^n P dA \tag{1.11}$$

where A' is fluctuation. The first central moment $\overline{A'}$, of course, is zero. The second central moment $\overline{A'^2(t)}$ is called the variance, characterizing the magnitudes of the fluctuations with respect to its mean, its square root $\sqrt{\overline{A'^2(t)}}$ is the standard deviation, often called root-mean-square (rms), which is the measure of the width of P.

The skewness associated with third central moment is defined by

$$S = \frac{\overline{A'^3}}{\left(\overline{A'^2}\right)^{3/2}} \tag{1.12}$$

which gives an ideal of asymmetry in P about the origin. If the values of all odd moments are zero, P is symmetric about mean. If $S < 0$, the distribution cure of P shifts toward the negative fluctuation direction, the tail of the curve on the left side is longer. The negative fluctuation prevails.

The kurtosis associated with fourth central moment is defined by

$$K = \frac{\overline{A'^4}}{\left(\overline{A'^2}\right)^2} \tag{1.13}$$

which is a measure of whether the curve of P is peaked or flat relative to a normal distribution induced later. The curve of P with high kurtosis has a distinct peak near the mean, declines rather rapidly, and has heavy tails.

1.1.2.3 Characteristic Function

A random variable A can be written in the complex exponential form, that is, $e^{ikA(t)}$, where k is the wave number. The mean called the characteristic function is

$$f(k,t) = \overline{e^{ikA}} = \int_{-\infty}^{+\infty} e^{iAk} P(A,t) dA \tag{1.14}$$

Differentiating Eq. (1.14) with respect to k, we have

$$\frac{\partial^n f}{\partial k^n} = \int_{-\infty}^{+\infty} (iA)^n e^{ikA} P dA$$

$$\left(\frac{\partial^n f}{\partial k^n}\right)_{k=0} = (i)^n \int_{-\infty}^{+\infty} A^n P dA = (i)^n \overline{A^n}$$

$$\overline{A^n} = (i)^n \left(\frac{\partial^n f}{\partial k^n}\right)_{k=0} \tag{1.15}$$

This means moments are related to derivatives of characteristic function $f(k,t)$ at the origin $k = 0$.

The characteristic function can be written as a Taylor series of moments

$$f = \sum_{n=0}^{\infty} \frac{(ik)^n}{n!} \overline{A^n} \tag{1.16}$$

therefore the characteristic function in principle can be determined from all derivatives.

On substituting Eq. (1.16) into Eq. (1.14) yields

$$P(A,t) = \frac{1}{2\pi} \int_{-\infty}^{+\infty} f(k,A) e^{-ikA(t)} dk \tag{1.17}$$

Then, P is given from the determined characteristic function, f.

1.1.2.4 Normal Distribution

In many cases, there is a probability density function called the standard normal distribution (or standard Gaussian distribution) expressed by

$$P = \frac{1}{\sqrt{2\pi}} e^{-\frac{A^2}{2}} \tag{1.18}$$

Obviously, we have $\overline{A} = 0$ and $\overline{A^2} = 1$.

$$\text{If } P = \frac{1}{\sigma\sqrt{2\pi}} e^{-\frac{1}{2}\left(\frac{A-a}{\sigma}\right)^2} \tag{1.19}$$

then, it is said to be the normal distribution (or Gaussian distribution), in which the first central moment is a, rms is σ, all odd central moments are zero, and even moments B_{2n} are expressed by

$$B_{2n} = (2n-1)!!\sigma^{2n} \tag{1.20}$$

where "!!" denotes the double factorial. Also, skewness, $S = 0$ and kurtosis, $K = 3$.

The characteristic function is

$$f = \frac{1}{\sigma\sqrt{2\pi}} \int_{-\infty}^{+\infty} e^{ikA} e^{-\frac{1}{2}\left(\frac{A-a}{\sigma}\right)^2} dA = \frac{1}{\sigma\sqrt{2\pi}} \int_{-\infty}^{+\infty} \cos(kA) e^{-\frac{1}{2}\left(\frac{A-a}{\sigma}\right)^2} dA = e^{\left\{ika - \frac{k^2}{2}\sigma^2\right\}} \tag{1.21}$$

The shape of Gaussian distribution is symmetric about the peak at $A = a$, and there exist the grads near the peak, as shown in Figure 1.4. Skewness S and kurtosis K, redefined as $K = \frac{\overline{A'^4}}{\left(\overline{A'^2}\right)^2} - 3$ here, for any arbitrary PDF represent the deviation from the symmetric shape of Gaussian distribution.

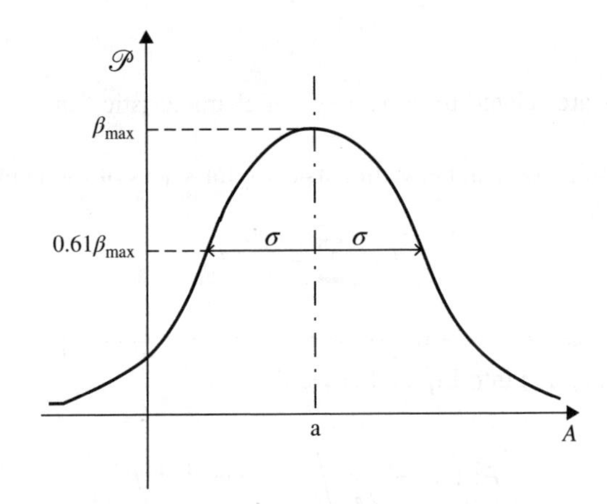

Figure 1.4 Normal distribution function

Let us consider N statistically independent random variables $A_i(t)$, where $A_i(t)$ are said to be independent if $\overline{A_i'A_j'} = 0$. We assume that all $A_i(t)$ have identical probability densities and that their mean values are zero. Then, the sum of all $A_i(t)$ has a Gaussian probability density, that is, A defined by $A = \sum_{i=1}^{N} A_i$ is a normal distribution function, which is called the central limit theorem.

1.1.3 Correlation Function

1.1.3.1 Auto-Correlation Function

Only the distributions of fluctuations at one point in time or space are discussed in previous section. Therefore, the relations between neighboring fluctuations will be discussed further here.

Using superscript i, denoting the measured position, the indicator function is defined by

$$\varphi(A_1, t_1)\varphi(A_2, t_2) = \begin{cases} 1 & A^{(i)}(t_1) < A_1 \text{ and } A^{(i)}(t_2) < A_2 \\ 0 & \text{otherwise} \end{cases} \tag{1.22}$$

where A_1 and A_2 are two given values and $A^{(i)}(t)$ is an arbitrary value at the ith measured position. t_1 and t_2 represent different times.

The cumulative probability distribution function is

$$P_{\text{rob}}\{A(t_1) < A_1; A(t_2) < A_2\} = \overline{\varphi(A_1, t_1)\varphi(A_2, t_2)} = P_2 \tag{1.23}$$

The correlated probability density function is

$$\mathsf{P}_2(A_1, A_2; t_1, t_2) = \frac{\partial^2 P_2(A_1, A_2; t_1, t_2)}{\partial A_1 \partial A_2} \tag{1.24}$$

It is the fraction time that the random variable $A(t)$ is between A_1 and $A_1 + dA_1$ at time t_1, as well as is between A_2 and $A_2 + dA_2$ at time t_2. It can also be written as

$$\Delta P_2 = \mathsf{P}_2 \Delta A_1 \Delta A_2 = P_{\text{rob}}\{A_1 < A(t_1) < A_1 + \Delta A_1, A_2 < A(t_2) < A_2 + \Delta A_2\} \tag{1.25}$$

which represents a statistical mass of a square area shown in Figure 1.5, if P_2 is regard as a density.

It also satisfies the following equations:

$$\int_{-\infty}^{+\infty}\!\!\int \mathsf{P}_2(A)dA_1 dA_2 = 1 \tag{1.26}$$

$$\int_{-\infty}^{+\infty} \mathsf{P}_2(A)dA_2 = \mathsf{P}(A_1, t) \tag{1.27}$$

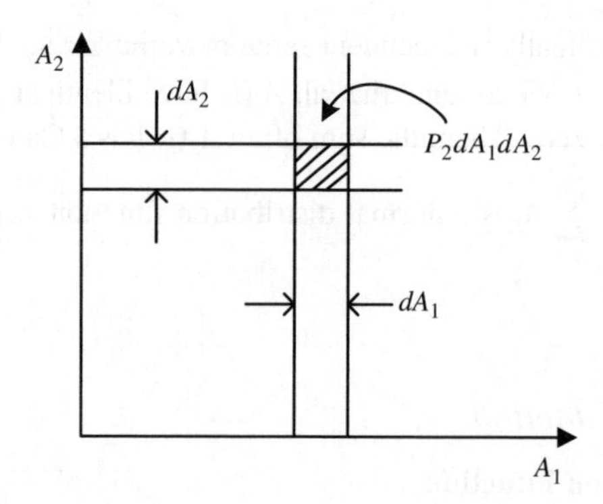

Figure 1.5 Correlated probability density

The central moment is

$$\overline{A^n(t_1)} = \int_{-\infty}^{+\infty}\int A_1^n P_2 dA_1 dA_2 \tag{1.28}$$

and the correlated moment is

$$\overline{A^m(t_1)A^n(t_2)} = \int_{-\infty}^{+\infty}\int A_1^m A_2^n P_2 dA_1 dA_2 \tag{1.29}$$

The most important correlated moment is $\overline{A(t_1)A(t_2)}$, which is defined as

$$\overline{A(t_1)A(t_2)} = \int_{-\infty}^{+\infty}\int A_1 A_2 P_2 dA_1 dA_2 \tag{1.30}$$

This is called the auto-correlation function, which expresses a survival capability of the random event, that is, how many traces remained at the time t_1 can be found at the time t_2 as random event developing. It will vanish as $\Delta t = t_2 - t_1 \rightarrow \infty$.

1.1.3.2 Joint Correlation Function

Considering the probability density for two random variables A and B, the indicator function is defined by

$$\varphi_A(A_1, t, \mathbf{x})\varphi_B(B_1, t, \mathbf{x}) = \begin{cases} 1 & A < A_1, B < B_1 \\ 0 & \text{otherwise} \end{cases} \tag{1.31}$$

where A_1 and B_1 are two given values. The probability for both $A < A_1$ and $B < B_1$ is

$$P_{\text{rob}}\{A < A_1 \text{ and B} < B_1\} = \overline{\varphi_A \varphi_B} = P_2.$$

The joint probability density is

$$P_2 = \frac{\partial^2 P_2}{\partial A \partial B} \tag{1.32}$$

Also, it is written as $\Delta P_2 = P_2 \Delta A \Delta B = P_{\text{rob}}\{A_1 < A < A_1 + \Delta A_1, B_1 < B < B_1 + \Delta B_1\}$.

The most important correlated moment is $\overline{AB}$, which is defined as

$$\overline{AB} = \int_{-\infty}^{+\infty}\int ABP_2 dAdB \tag{1.33}$$

This is called joint correlation function between A and B. When $\overline{AB} = 0$, the two random events A and B are said to be uncorrelated or independent.

1.2 Statistical Analysis of Turbulence

1.2.1 Reynolds Stress and Turbulent Kinetic Energy

Turbulence consists of random velocity fluctuations, so that it must be treated with statistical methods. All fluctuating quantities in turbulence would be decomposed into mean values and fluctuations, representing the mean character of the turbulent flow field and the deviation from the mean at moments, which is referred to as Reynolds decomposition.

In turbulence, a description of the flow at all points in time and space by solving the N-S equations is not feasible. If the Reynolds decomposition is applied to the N-S equations, then we take the average of all terms in the resulting equations, the correlation functions for fluctuating quantities will yield, such as $R_{ij} = \overline{u_i' u_j'}$. In general, the equations for a set statistics at a level (e.g., product of two parameters) contain additional statistics at a high level (e.g., triple products), with more unknowns than equations. Because of these unknown functions, the Reynolds average N-S equations are not closed, unless the additional statistics are modeled.

Reynolds stresses τ_{ij}^t is defined as

$$\tau_{ij}^t = -\rho \overline{u_i' u_j'} \tag{1.34}$$

which plays a crucial role in the equations for the mean velocity field and results in the difference between the instantaneous velocity field and the mean field, by transporting momentums between the turbulence and the mean flow. Therefore, it can be perceived as an agent producing stress in the mean flow.

The second central moment of fluctuating velocities $R_{ii} = \overline{u_i' u_i'} = \overline{u_1'^2 + u_2'^2 + u_3'^2}$, where repeated indices indicate a summation over all three values of the index, is written as

$$K = \frac{1}{2} R_{ii} = \overline{\frac{u_1'^2}{2} + \frac{u_2'^2}{2} + \frac{u_3'^2}{2}} \tag{1.35}$$

It is called the kinetic energy of turbulence, representing the energy of turbulent fluctuation.

$$\text{Let } I = \frac{\sqrt{\frac{1}{3}\left(\overline{u_1'^2} + \overline{u_2'^2} + \overline{u_3'^2}\right)}}{\overline{u}} \tag{1.36}$$

It is called turbulent strength, where $\overline{u} = \sqrt{\overline{u_1^2} + \overline{u_2^2} + \overline{u_3^2}}$. It can also be defined by

$$I_i = \frac{u'_{irms}}{\overline{u}}$$

In order to discuss the correlated relations between neighboring velocity fluctuations, we define the spatially auto-correlation coefficient by

$$R(r) = \frac{\overline{u'_A u'_B}}{u'_{A.rms} u'_{B.rms}} \tag{1.37}$$

where subscripts A and B represent two neighboring points with the distance of r. Since $\overline{u'_A u'_B} \leq \sqrt{\overline{u'^2_A}} \cdot \sqrt{\overline{u'^2_B}}$, we have $R(r) \leq 1$. As shown in Figure 1.6, the correlativity becomes stronger with the decrease of r, that is, $\lim_{r\to 0} R(r) \to 1$, as $B \to A$, whereas $\lim_{r\to\infty} R(r) \to 0$.

Let

$$l_T = \int_0^\infty R(r)dr \tag{1.38}$$

where l_T is called the characteristic scale of vortex, represented by a square area shown in Figure 1.6. It can be perceived as a scale of an imaginary fluid parcel, which moves in turbulence as a whole, that is, with the same fluctuating frequency and amplitude.

In order to discuss correlated relations between velocity fluctuations at two different times, we define the temporally auto-correlation coefficient by

$$R^*(t) = \frac{\overline{u'(t)u'(t+\Delta t)}}{u'_{t,\mathrm{rms}} u'_{t+\Delta t,\mathrm{rms}}} \tag{1.39}$$

Let $t^* = \int_0^\infty R^*(t)dt$

$$\text{we have } l_T^* = u'_{\mathrm{rms}} t^* \tag{1.40}$$

where t^* represents life of an imaginary fluid parcel. l_T^* with the same order of the magnitude with l_T denotes the moving path of the fluid parcel before the disaggregation and the loss of its integral character.

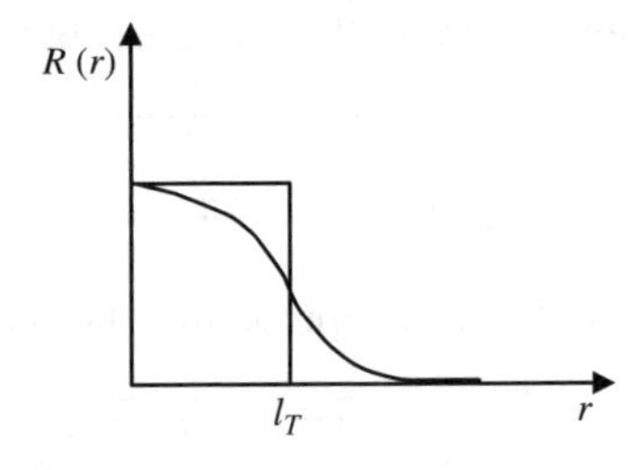

Figure 1.6 Characteristic scale of turbulence

1.2.2 Variable-Interval Time Average Method

Despite the fact that the turbulent flow is seemingly random and chaotic, the near wall turbulence possesses characteristic coherent structures, generally consisting of streaks and streamwise vortices (to be discuss in Chapter 2). The turbulence is formed intermittently and periodically via bursting events related to a violent breakup of a lifted steak. Bursting can be commonly detected by the detection function, then the ensemble average of the flow field around the bursting over all bursting events can be calculated, which may reveal the characteristic flow pattern as the turbulent onset.[4, 5]

1.2.2.1 Variable-Interval Time Average and Variable-Interval Space Average[6, 7]

The variable-interval time average of a fluctuating quantity $A(x_i, t)$, called VITA, is defined as

$$\widehat{A}(x_i, t, T) = \frac{1}{T} \int_{t-\frac{1}{2}T}^{t+\frac{1}{2}T} A(x_i, s)ds \tag{1.41}$$

where T is the averaging time. If one wants to obtain a local average of some particular phenomenon, the averaging time must be of the order of the time scale of the phenomenon under study. Note that

$$\overline{A}(x_i) = \lim_{T\to\infty} \widehat{A}(x_i, t, T)$$

where the bar indicates the conventional time average. The time average value will be fixed, if it is statistically steady.

Similarly, the variable-interval space average of a fluctuating quantity $A(x_i, t)$, called VISA, is defined as

$$\widehat{A}(x_i, t, L) = \frac{1}{L} \int_{x_i-L/2}^{x_i+L/2} A(s, t)ds \tag{1.42}$$

where L is the width of the spatial averaging.

1.2.2.2 Detection Function of Turbulent Bursting[8, 9]

"Bursting" is referred to the production of turbulence in boundary layer, which can be commonly detected by the detection function based on a threshold. For example, the detection criterion is completed using the VITA variance; the detection function $D(t)$ is defined as

$$D(t) = \begin{cases} 1 & \text{if} \quad \overset{\wedge}{\text{Var}} > k \cdot u'^2_{\text{rms}} \text{ and } \partial u'/\partial t > 0 \\ 0 & \text{otherwise} \end{cases} \tag{1.43}$$

where u' is the fluctuating component of the streamwise velocity. The localized mean of u' is

$$\widehat{u'}(x_i, t, T) = \frac{1}{T} \int_{t-T/2}^{t+T/2} u'(x_i, s)ds$$

and the localized second moment of u' is defined as

$$\widehat{u'^2}(x_i, t, T) = \frac{1}{T} \int_{t-T/2}^{t+T/2} u'^2(x_i, s)ds$$

Finally, the localized variance is defined as

$$\widehat{\mathrm{Var}}(x_i, t, T) = \widehat{u'^2}(x_i, t, T) - [\widehat{u'}(x_i, t, T)]^2 \tag{1.44}$$

Note that, $u'^2_{\mathrm{rms}} = \lim_{T \to \infty} \mathrm{Var}$

k is the threshold level and u'^2_{rms} is the rms of the fluctuating streamwise velocity.

The process of bursting detection from the measured fluctuating streamwise velocity by VITA is shown in Figure 1.7. The measured velocity at a position near the wall is shown in the upper figure, based on which the localized variance is obtained by Eq. (1.44), as shown in the middle. Finally, the distributions of detection function $D(t)$ can be drawn in the bottom one, from which it is clear that only one bursting is detected.

Figure 1.8 shows the variations of the fluctuating velocity with time at some positions with different y^+, where superscript "+" indicates variables normalized by wall-unit scales, to be introduced in Chapter 2. The impulse vertical lines at the horizontal axis shown in Figure 1.8(b) represent the detection functions detected from the streamwise velocity at $y^+ = 15$. It can be seen the bursting events are observed for four times, when the streamwise velocity fluctuate dramatically.

If we define $D(x)$ in the form associated with VISA

$$D(x) = \begin{cases} 1 & \text{if } \mathrm{Var} > k \cdot u'^2_{\mathrm{rms}} \text{ and } \partial u'/\partial x < 0 \\ 0 & \text{otherwise} \end{cases}$$

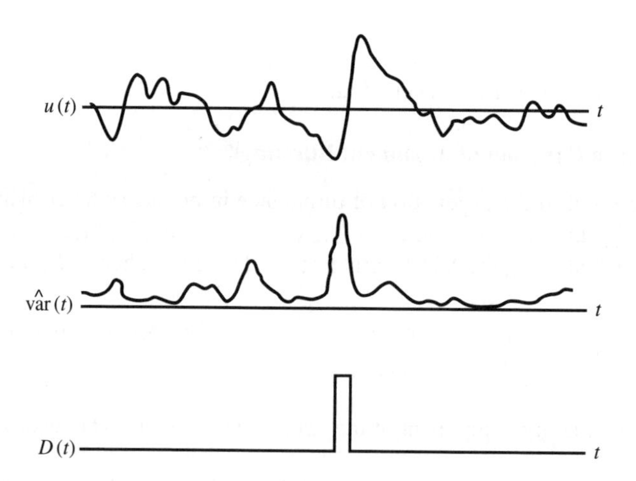

Figure 1.7 Detection process of bursting events.[4] *Source*: Blackwelder 1976. Reproduced with permission of Cambridge University Press

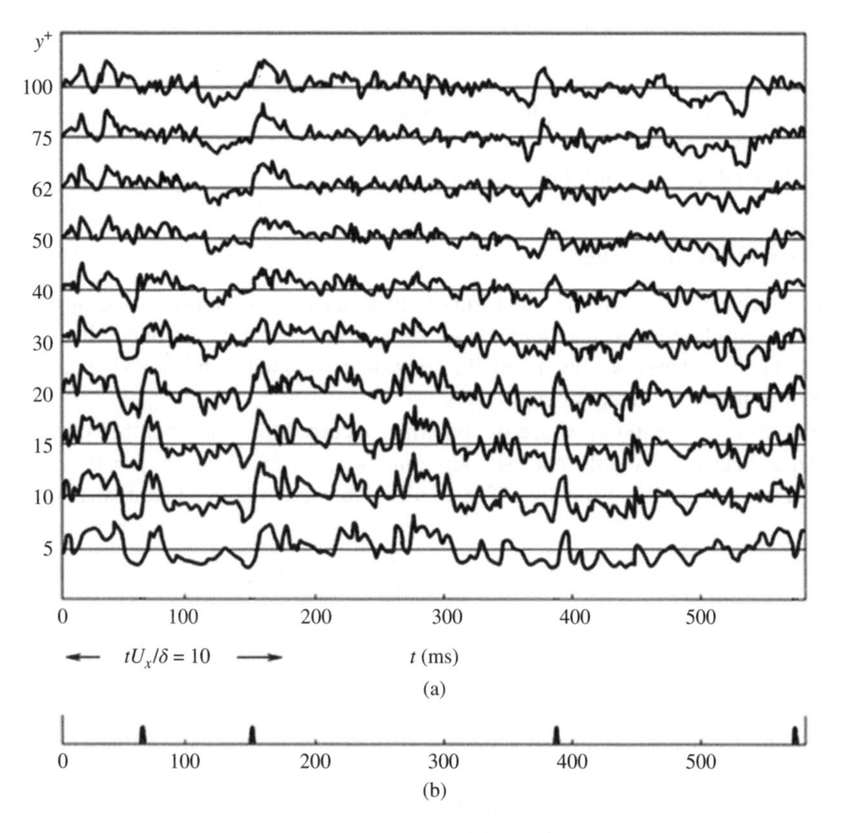

Figure 1.8 Instantaneous streamwise velocities (a) at different y^+ locations and detector function (b) obtained at $y^+ = 15$.[4] *Source*: Blackwelder 1976. Reproduced with permission of Cambridge University Press

where $\widehat{u}'(x_i, t, L) = \frac{1}{L} \int_{x_i - L/2}^{x_i + L/2} u'(s, t) ds$

$$\widehat{u}'^2(x_i, t, L) = \frac{1}{L} \int_{x_i - L/2}^{x_i + L/2} u'^2(s, t) ds$$

$$\widehat{\text{Var}}(x_i, t, L) = \widehat{u}'^2(x_i, t, L) - [\widehat{u}'(x_i, t, L)]^2$$

$D(x) = 1$ denotes the turbulent bursting then the turbulent bursting generated in the instantaneous flow field can be singled out, therefore both the positions, at which the bursting events occurred and the total number of the events detected at a moment are determined.

The detection of bursting phenomenon is very depending on the choice of detection function and threshold values. Since a burst occurs commonly in conjunction with violent outward ejections, the burst can be detected using the quadrant analysis. This technique prescribes the occurrence of an ejection when $u' < 0$ and $v' > 0$ and the $|uv|$ product exceed a set threshold.

Then the detection function $D(x)$ is defined as

$$D(x) = \begin{cases} 1 & \text{if} \quad |u'v'| > hu'_{\text{rms}}v'_{\text{rms}} \\ 0 & \text{otherwise} \end{cases}$$

Where h is a threshold, and || denotes the absolute value.

1.2.2.3 Conditional Average

Bursting is regard as one of the most important processes in wall-bounded turbulence. By means of the conditional average techniques, it is possible to study the burst phenomenon via the detailed flow pattern associated with the bursting, which will conduce to the understanding of the wall-bounded turbulent phenomenon.

For a reference position x_i where the bursting occurred, the total number of the bursting events detected at x_i, N, and the corresponding moment for each burst, $t_j, j = 1, \dots, N$, can be obtained using the detection function. The conditional averaging of a quantity A is defined by

$$\langle A(x_i, \tau) \rangle_{y^+} = \frac{1}{N}\sum_{j=1}^{N} A(x_i, t_j + \tau) \tag{1.45}$$

where y^+ indicates the position of detection probe and t_j is the time as the bursting event occurs. The positive or negative τ is used to determine the temporal behavior of A before and after burst.

The variations of the conditional averages of streamwise velocity at different y^+ with the time τ are shown in Figure 1.9. It is evident that, when the bursting event occurs, the velocity varies dramatically for $y^+ \leq 25$, whereas it does not vary considerably for $y > 25$.

In order to obtain a spatial structure rather than the temporal structure, the conditional averaging process is modified in the following way:

$$\langle A(\xi, t_i) \rangle_{y^+} = \frac{1}{N}\sum_{j=1}^{N} A(x_j + \xi, t_i)$$

where $x_j, j = 1, \dots, N$, denotes the detection point where the bursting event occurs and N is the total number of the bursting events detected at time t_i. The positive or negative ξ is used to determine the spatial behavior of A in the vicinity of the detection points in the $(x - z)$ planes at y^+.

The profiles of conditional averages of streamwise velocity at different y^+ locations are shown in Figure 1.10. The peaks of the profiles occur just upstream of the detected point $\xi = 0$, whereas it occurs after the burst as shown in Figure 1.9.

1.3 Fourier Transform and Spectrum

1.3.1 Harmonic Wave

Harmonic vibration around an undeflected position is described by

$$x(t) = a\cos\left(\frac{2\pi}{T}t + \alpha\right) \tag{1.46}$$

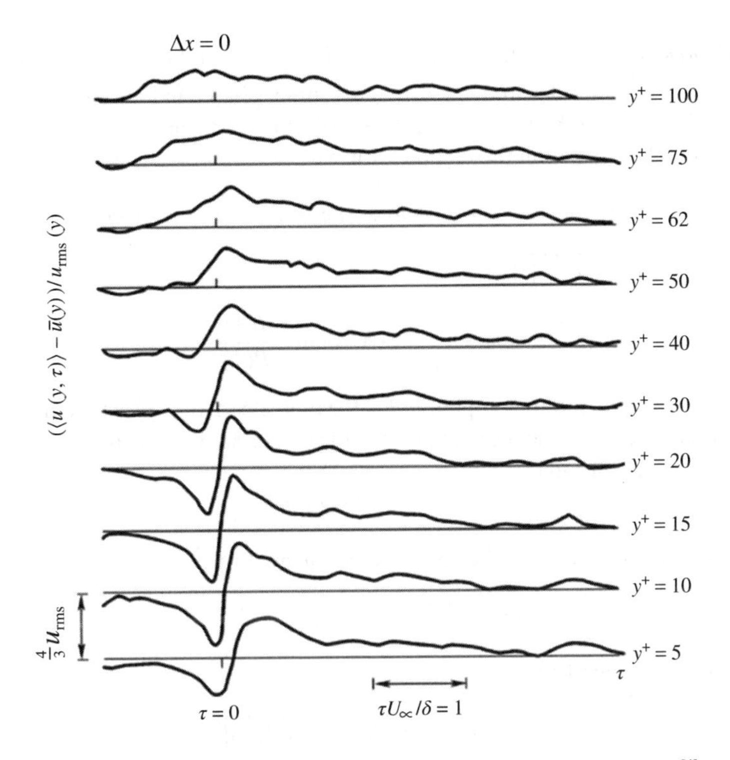

Figure 1.9 Conditionally averaged streamwise velocities at different y^+ locations.[4] *Source*: Blackwelder 1976. Reproduced with permission of Cambridge University Press

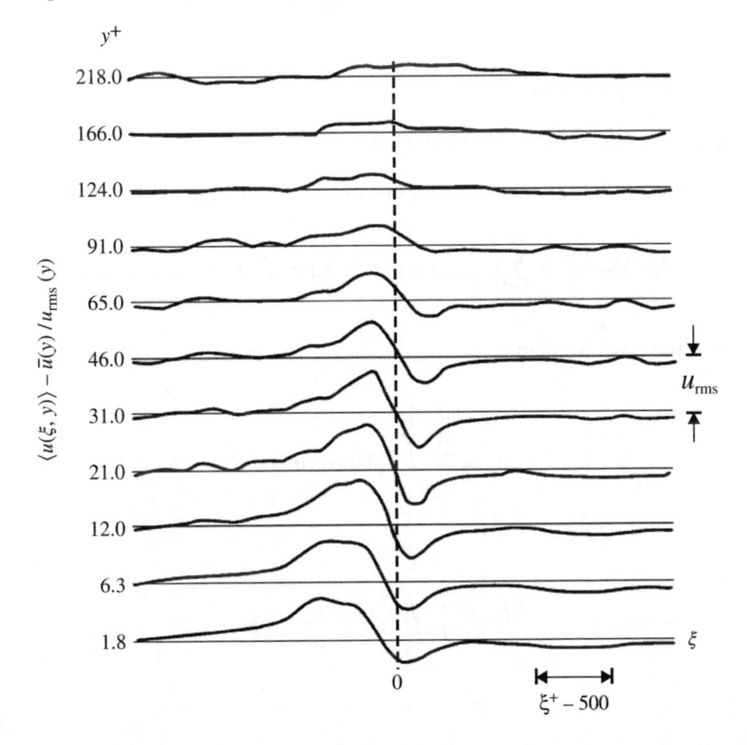

Figure 1.10 Conditionally averaged streamwise velocities at different y^+ locations as a function of streamwise coordinate.[6] *Source*: Kim J 1983. Reproduced with permission of AIP Publishing LLC

where $x(t)$ denotes the displacement, a is the amplitude of vibration, T is the period, $f = \frac{1}{T}$ is the frequency, $\omega = \frac{2\pi}{T}$ circular frequency, α is the initial phase, and $\omega t + \alpha$ is the phase.

Harmonic vibration can be represented by a projection of a circularly moving point on the horizontal axis and is written in the complex form as

$$A = a\cos(\omega t + \alpha) + ia\sin(\omega t + \alpha) = R(A) + iI(A) \tag{1.47}$$

where $a = |A| = \sqrt{[R(A)]^2 + [I(A)]^2}$

$$\omega t + \alpha = \arg A = \tan^{-1}\frac{I(A)}{R(A)}$$

where R denotes the real part of complex A, I is the image part, and a is the radius of the circle, that is, amplitude. At $t = 0$, the moving point is located at A_0 with the initial phase angle α.

Based on the Euler formula

$$e^{i\theta} = \cos\theta + i\sin\theta$$

Equation (1.47) can be written in the complex exponential form as

$$A = ae^{i(\omega t+\alpha)} = ae^{i\alpha}e^{i\omega t} = Fe^{i\omega t} \tag{1.48}$$

where F is a complex number:

$$a = |F| = \sqrt{[R(F)]^2 + [I(F)]^2}$$

$$\alpha = \arg F = \tan^{-1}\frac{I(F)}{R(F)}$$

1.3.2 *Fourier Transform*[10]

1.3.2.1 Fourier Series of Periodic Function

Fourier series associated with an arbitrary periodic function $f_T(t)$, with the period T, is

$$f_T(t) = \frac{a_0}{2} + \sum_{n=1}^{\infty}(a_n\cos n\lambda t + b_n\sin n\lambda t) \quad n = 1, 2, 3, \ldots, \tag{1.49}$$

where $\lambda = \frac{2\pi}{T}$ is the base frequency:

$$a_n = \frac{2}{T}\int_{-\frac{T}{2}}^{\frac{T}{2}} f_T(t)\cos n\lambda t\, dt$$

$$b_n = \frac{2}{T}\int_{-\frac{T}{2}}^{\frac{T}{2}} f_T(t)\sin n\lambda t\, dt$$

Writing in the complex exponential form, it becomes

$$f_T(t) = \sum_{-\infty}^{+\infty} F(n\lambda)e^{in\lambda t} \tag{1.50}$$

where the sum is over the infinite discrete frequencies. The complex Fourier coefficients are

$$F(n\lambda) = \frac{1}{T}\int_{-\frac{T}{2}}^{\frac{T}{2}} f_T(t)e^{in\lambda t}dt = \frac{\lambda}{2\pi}\int_{-\frac{T}{2}}^{\frac{T}{2}} f_T(t)e^{in\lambda t}dt$$

which satisfy conjugate symmetry:

$$F(n\lambda) = F^*(-n\lambda)$$

where an asterisk denotes the complex conjugate. $F(0)$ is the zero frequency component with respect to $n = 0$:

$$F(n\lambda)e^{in\lambda t} + F(-n\lambda)e^{-in\lambda t} = 2|F(n\lambda)|\cos(n\lambda t + \arg F(n\lambda)) \tag{1.51}$$

where

$$|F(n\lambda)| = \sqrt{[R(F(n\lambda))]^2 + [I(F(n\lambda))]^2}$$

$$\alpha = \arg F(n\lambda) = \tan^{-1}\frac{I(F(n\lambda))}{R(F(n\lambda))}$$

Hence,

$$f_T(t) = \sum_{-\infty}^{+\infty} 2|F(n\lambda)|\cos(n\lambda t + \arg F(n\lambda)) \tag{1.52}$$

$$F(n\lambda) = \frac{\lambda}{2\pi}\int_{-\frac{T}{2}}^{\frac{T}{2}} f_T(t)e^{in\lambda t}dt \tag{1.53}$$

It can be concluded that an arbitrary periodic function $f_T(t)$ can be regarded as a superimposition of infinite harmonic waves, in which discrete circular frequencies are integer multiple of base frequency, amplitudes are $2|F(n\lambda)|$, and initial phases are $\arg(F(n\lambda))$.

1.3.2.2 Fourier Transform of Nonperiodic Function

Let $\omega = \frac{n2\pi}{T}$, $\Delta\omega = \omega_{n+1} - \omega_n = \lambda$. As $T \to \infty$, that is, nonperiodic function, $d\omega = \frac{2\pi}{T}$, so that the integration is adapted in series expansion defined by Eq. (1.50) instead of the sum. We have

$$f(t) = \int_{-\infty}^{+\infty} \frac{d\omega}{2\pi}\int_{-\infty}^{+\infty} f(\tau)e^{-i\omega t}d\tau e^{i\omega t} = \frac{1}{2\pi}\int_{-\infty}^{+\infty} F(\omega)e^{i\omega t}d\omega = F^{-1}[F(\omega)] \tag{1.54}$$

or

$$F(\omega) = \int_{-\infty}^{+\infty} f(t)e^{-i\omega t}dt = F(f(t)) \tag{1.55}$$

where $F(\omega)$ is a function of ω. $F(t)$ is defined as Fourier transform and $F^{-1}(\omega)$ is Fourier inverse transform. $f(t)$ and $F(\omega)$ are a transform pair.

Writing Eq. (1.54) in the cosine transform, we obtain

$$f(t) = \frac{1}{\pi} \int_{0}^{+\infty} \left[\int_{-\infty}^{+\infty} f(\tau) \cos \omega(t - \tau) d\tau \right] d\omega \tag{1.56}$$

then

$$f(t) = \int_{-\infty}^{+\infty} \{R(F(\omega)) \cos \omega t + I(F(\omega)) \sin \omega t\} d\omega$$

$$= \int_{-\infty}^{+\infty} |F(\omega)| e^{i(\omega t - \varphi(\omega))} d\omega \tag{1.57}$$

where

$$|F(\omega)| = \sqrt{[R(F(\omega))]^2 + [I(F(\omega))]^2}$$

$$\alpha = \tan^{-1} \frac{I(F(\omega))}{R(F(\omega))}$$

$R(F(\omega))$ is the even function and $I(F(\omega))$ is the odd function.

1.3.2.3 Fourier Transform of Periodic Function

In order to discuss the Fourier transform of periodic functions, in which the frequencies of harmonic waves are discrete, impulse function δ is induced, defined by

$$\delta(t) = \begin{cases} 1 & t = 0 \\ 0 & t \neq 0 \end{cases} \tag{1.58}$$

We have $\int_{-\infty}^{+\infty} \delta(t)dt = 1$

When $t \neq a$, $\delta(t - a) = 0$

$$\int_{-\infty}^{+\infty} \delta(t - a)\varphi(t)dt = \varphi(a) \tag{1.59}$$

Hence, $F[\delta(t)] = \int_{-\infty}^{+\infty} \delta(t)e^{-i\omega t}dt = e^0 = 1$

$$F^{-1}[2\pi\delta(\omega)] = \frac{1}{2\pi} \int_{-\infty}^{+\infty} 2\pi\delta(\omega)e^{-i\omega t}d\omega = 1$$

and

$$F^{-1}[2\pi\delta(\omega - a)] = \frac{1}{2\pi}\int_{-\infty}^{+\infty} 2\pi\delta(\omega - a)e^{-i\omega t}d\omega = e^{iat} \quad (1.60)$$

Therefore, $\delta(t)$ and 1 are a transform pair, 1 and $2\pi\delta(\omega)$ are a transform pair as well as e^{iat} and $2\pi\delta(\omega - a)$ are also a transform pair.

By substituting Eq. (1.50) into Eq. (1.55), we obtain

$$F((f(t)) = \int_{-\infty}^{+\infty}\sum_{-\infty}^{+\infty} F(n\lambda)e^{in\lambda t}e^{-i\omega t}dt = \sum_{-\infty}^{+\infty} F(n\lambda)\int_{-\infty}^{+\infty} e^{in\lambda t}e^{-i\omega t}dt = \sum_{-\infty}^{+\infty} F(n\lambda)F(e^{in\lambda t}) \quad (1.61)$$

By substituting Eq. (1.50) into Eq. (1.61), we obtain

$$F((f(t)) = \sum_{-\infty}^{+\infty} F(n\lambda)F(e^{in\lambda t}) = \sum_{-\infty}^{+\infty} 2\pi F(n\lambda)\delta(\omega - n\lambda) = F(\omega) \quad (1.62)$$

In the same way, we obtain

$$F^{-1}[F(\omega)] = \frac{1}{2\pi}\int_{-\infty}^{+\infty}\sum_{-\infty}^{+\infty} 2\pi F(n\lambda)\delta(\omega - n\lambda)e^{i\omega t}d\omega = \sum_{-\infty}^{+\infty} F(n\lambda)e^{in\lambda t} = f(t) \quad (1.63)$$

It is proved that $f(t)$ and $F(\omega) = \sum_{-\infty}^{+\infty} 2\pi F(n\lambda)\delta(\omega - n\lambda)$ are a transform pair for periodic functions, that is,

$$F(f(t)) = F(\omega) = \int_{-\infty}^{+\infty} f(t)e^{-i\omega t}dt$$

$$F^{-1}(F(\omega)) = f(t) = \frac{1}{2\pi}\int_{-\infty}^{+\infty} F(\omega)e^{i\omega t}d\omega$$

with $F(\omega) = \sum_{-\infty}^{+\infty} 2\pi F(n\lambda)\delta(\omega - n\lambda)$.

$F(\omega)$ is regarded as a spectrum of the waveform function $f(t)$ as defined in electrical engineering.

If the signal $f(t)$ is applied to a filter, regarded as a physical system having an input and an output, then the output $f_{\text{out}}(t)$ will be produced. This filtering procedure can be described as follows: first, one analyzes $f(t)$ into its spectrum $F(\omega)$, then multiplies each spectral component by the corresponding transfer factor $T(\omega)$ to obtain the spectrum of $f_{\text{out}}(t)$, that is, $F_{\text{out}}(\omega)$, and finally synthesizes $f_{\text{out}}(t)$ from its spectrum. Thus,

$$F_{\text{out}}(\omega) = T(\omega)F(\omega)$$

and then $f_{\text{out}}(t) = \frac{1}{2\pi}\int_{-\infty}^{\infty} T(\omega)F(\omega)e^{i\omega t}d\omega$

Since multiplication of transform corresponds to convolution of original functions, that is,

$$f_{\text{out}}(t) = G(t) * f(t) = \int G(\tau, t) f(t-\tau) d\tau$$

where $G(t)$ is the Fourier transform of $T(\omega)$, which would be the characteristic of the filter.

1.3.3 Energy Spectrum

The squared modulus of a transform is defined as energy spectrum, which has the character of energy density measured per unit of ω; that is, $|F(\omega)|^2$ is the energy spectrum of $f(t)$.

Based on Parseval's theorem for Fourier series

$$\int_{-\infty}^{+\infty} f^2(t)dt = \frac{1}{2\pi}\int_{-\infty}^{+\infty} |F(\omega)|^2 d\omega = \frac{1}{2\pi}\int_{-\infty}^{+\infty} S(\omega)d\omega \tag{1.64}$$

where $S(\omega) = |F(\omega)|^2$. The integral, representing the amount of energy in the system, has usually a finite value for nonperiodic function, whereas it would have to become infinite, if the function $f(t)$ is periodic. Now we can consider an average power defined as

$$P = \frac{1}{T}\int_{-\frac{T}{2}}^{+\frac{T}{2}} f^2(t)dt = \sum_{-\infty}^{+\infty} |F(n\lambda)|^2$$

where $S(n\lambda) = |F(n\lambda)|^2$ is called the discrete power spectrum.

It can be written in the form of Fourier transform as

$$P = \int_{-\infty}^{+\infty} \left|\frac{1}{2\pi}F(\omega)\right|^2 d\omega \tag{1.65}$$

where $F(\omega) = \sum_{-\infty}^{+\infty} 2\pi F(n\lambda)\delta(\omega - n\lambda)$.

1.4 Spectral Series Expansion of Function

1.4.1 Orthogonal Basis

Consider an infinite sequence formed by basis functions

$$\phi_1(x), \phi_2(x), \ \ldots \ , \phi_k(x), \ \ldots \tag{1.66}$$

This series is the so-called linear independence as any one basis function in the series is independent linearly with others and is said to be orthogonal with respect to the weight function $w(x)$ further, if

$$(\phi_i, \phi_j) = \delta_{ij} v_i^2$$

where the inner product $(a,b) = \int_{\alpha}^{\beta} abwdx$, and a and b are two functions defined on the interval $[\alpha,\beta]$. δ_{ij} is the delta function defined by

$$\delta_{ij} = \begin{cases} 1 & i = j \\ 0 & i \neq j \end{cases}$$

and v_i is the normalized constant.

Let us consider an unknown function u that can be expressed by a sum of the orthogonal basis:

$$u = \sum_{k=-\infty}^{\infty} \tilde{u}_k \phi_k$$

where $\tilde{u}_k$ is a spectral coefficient. Actually, this expansion gives a mapping between the function u in the physical space and the spectral coefficients $\tilde{u}_k$ in the spectral space. u can be obtained by $\tilde{u}_k$,vice versa.

Fourier series and Chebyshev polynomials are in common use as the basis functions in the spectral method, the former is appropriate for the periodic problems and the latter for nonperiodic.

1.4.2 Fourier Series

As discussed in Section 1.3.2, a periodic function $u(x)$ defined on the interval $(0, 2\pi)$ can be expanded into the Fourier series in the complex form as

$$u(x) = \sum_{k=-\infty}^{\infty} \tilde{u}_k \phi_k(x) \tag{1.67}$$

where $\phi_k(x) = e^{ikx}$ and k is wave number. Then

$$\int_0^{2\pi} u(x)e^{-ilx}dx = \int_0^{2\pi} \sum_{k=-\infty}^{\infty} \tilde{u}_k \phi_k(x)\phi^*{}_l(x)dx = 2\pi\tilde{u}_k$$

Hence, the spectral coefficients are

$$\tilde{u}_k = \frac{1}{2\pi}\int_0^{2\pi} u(x)e^{-ikx}dx \qquad k = 0, \pm 1, \pm 2, \ldots . \tag{1.68}$$

Since

$$\int_0^{2\pi} \phi_k(x)\phi_l^*(x)dx = 2\pi\delta_{kl} \tag{1.69}$$

where $\{\phi_k\}$ is orthogonal over the interval $(0, 2\pi)$. If $u(x)$ is a real function, its spectral coefficients are complex.

By differentiating Eq. (1.67), we obtain

$$u'(x) = \sum_{k=-\infty}^{\infty} ik\tilde{u}_k e^{ikx}$$

$$u^{(l)}(x) = \sum_{k=-\infty}^{\infty} (ik)^l \tilde{u}_k e^{ikx} \tag{1.70}$$

where superscript "′" represents first-order derivative, and "l" represents lth derivative. It means that lth derivative in spectral (Fourier transform) space consists of multiplying each spectral coefficient by $(ik)^l$.

1.4.3 Chebyshev Polynomials

The Chebyshev polynomial of the first kind $\{T_k(x)\}$ is the polynomial of degree k defined on $-1 \leq x \leq 1$ by

$$T_k(x) = \cos(k\cos^{-1}x), \quad k = 0, 1, 2, \ldots, \tag{1.71}$$

Therefore, $-1 \leq T_K \leq 1$. By setting $x = \cos\theta$, we have

$$T_k = \cos k\theta$$

The differentiation of T_k gives

$$T'_k = k \sin\, k\theta / \sin\theta$$

By the application of trigonometrical formulas, we obtain a recurrence relation on the derivative

$$2T_k(x) = \frac{1}{k+1}T'_{k+1}(x) - \frac{1}{k-1}T'_{k-1}(x) \quad k \geq 1 \tag{1.72}$$

or

$$T'_k(x) = 2k\sum_{n=0}^{K} \frac{1}{c_{k-1-2n}} T_{k-1-2n}(x) \tag{1.73}$$

where $K = (k-1)/2$. A similar formula for the lth derivative is obtained by successive differentiation of Eq. (1.72).

The orthogonality property is

$$\int_{-1}^{1} T_m(x)T_n(x)\frac{dx}{\sqrt{1-x^2}} = \int_0^{\pi} \cos n\theta \cos m\theta d\theta = \begin{cases} 0, & m \neq n \\ c_m\frac{\pi}{2} & m = n \end{cases} \tag{1.74}$$

where $c_m = \begin{cases} 2 & m = 0 \\ 1 & m \geq 1 \end{cases}$.

Therefore, the Chebyshev polynomials are orthogonal over the interval $[-1, 1]$ with the weight $w(x) = (1 - x^2)^{-\frac{1}{2}}$.

From the trigonometrical identity

$$\cos(k+1)\theta + \cos(k-1)\theta = 2\cos\theta\cos k\theta,$$

the recurrence relationship of $\{T_k(x)\}$ is deduced as

$$T_{k+1}(x) = 2xT_k(x) - T_{k-1}(x) \tag{1.75}$$

That allows us to deduce the expression of the polynomials T_k, $k \geq 2$ from $T_0(x) = 1$ and $T_1(x) = x$.

Let $|u| < 1$, then

$$\sum_{k=0}^{\infty} u^k e^{ik\theta} = \frac{1}{1 - ue^{i\theta}}$$

The real part is

$$\sum_{k=0}^{\infty} u^k \cos k\theta = \frac{1 - u\cos\theta}{1 - 2u\cos\theta + u^2}$$

that is,

$$F(u, x) = \frac{1 - ux}{1 - 2ux + u^2} = \sum_{k=0}^{\infty} T_k(x)u^k \quad x \in [-1,\ 1] \tag{1.76}$$

It is shown that the polynomials $T_k(x)$ are coefficients of the expansion of $F(u, x)$ on the variable u in the exponential series. $F(u, x)$ is called the generation function of $T_k(x)$.

Chebyshev series expansion of the function $u(x)$ defined in the interval $[-1, 1]$ is

$$u(x) = \sum_{k=0}^{\infty} \tilde{u}_k T_k(x) \tag{1.77}$$

Based on the orthogonality property of Eq. (1.74), we have

$$\int_{-1}^{1} uT_k(x)\frac{dx}{\sqrt{1-x^2}} = \int_{-1}^{1} \sum_{m=0}^{N} \tilde{u}_m T_m(x) T_k(x) \frac{dx}{\sqrt{1-x^2}} = \tilde{u}_k c_k \frac{\pi}{2}$$

That is,

$$\tilde{u}_k = \frac{2}{c_k \pi} \int_{-1}^{1} u(x) T_k(x) \frac{dx}{\sqrt{1-x^2}} \tag{1.78}$$

Then the expansion coefficients $\tilde{u}_k$ can be determined from known $u(x)$. Equations (1.77) and (1.78) are a transform pair between the physical space and the spectral space.

From Eqs. (1.77) and (1.72), we obtain

$$u' = \sum_{m=0}^{\infty} \tilde{u}_m T'_m = \sum_{m=0}^{\infty} \tilde{u}_m^{(1)} T_m(x) \tag{1.79}$$

$$u'' = \sum_{m=0}^{\infty} \tilde{u}_m T''_m = \sum_{m=0}^{\infty} \tilde{u}_m^{(2)} T_m(x) \tag{1.80}$$

where superscript "$'$" represents first-order derivative and "$''$" represents second-order derivative:

$$\tilde{u}_m^{(1)} = \frac{2}{c_m} \sum_{p=m+1}^{\infty} p\tilde{u}_p \tag{1.81}$$

$$\tilde{u}_m^{(2)} = \frac{1}{c_m} \sum_{p=m+2}^{\infty} p\,(p^2 - m^2)\,\tilde{u}_p \tag{1.82}$$

Obviously, both $\tilde{u}_m^{(1)}$ and $\tilde{u}_m^{(2)}$ are polynomials of $\tilde{u}_k$. The derivative in spectral (Chebyshev transform) space can be represented in a sum of each spectral coefficient multiplying with corresponding weight.

1.5 Fundamentals of Spectral Methods[11–14]

1.5.1 Fundamental Concepts

1.5.1.1 Weighted Residual Method

Considering a differential equation written here as

$$L(u) = f \quad \alpha < x < \beta \tag{1.83}$$

with the following boundary conditions:

$$B_{-}u = g_{-} \text{ at } x = \alpha, \quad B_{+}u = g_{+} \text{ at } x = \beta \tag{1.84}$$

where L is an operator, B_- and B_+ correspond to Dirichlet, Neumann, or mixed conditions.

If the solution of the equation, that is, function $u(x)$, is approximated by u^N, the truncated series expansion after N terms is defined as

$$u^N = \sum_{k=0}^{N} \tilde{u}_k \phi_k \quad u^N \in X_N$$

where ϕ_k is trial (or basis) function. The choice of the trial functions is a key issue in the spectral method. The chosen trial functions are orthogonal. If the solution is periodic, use Fourier series, and if the domain is finite, but the solution is not periodic, Chebyshev polynomials are best.

Now we introduce the residual $R_N(x)$ defined by

$$R_N(x) = L_N(u^N) - f(u^N)$$

Spectral methods belong to the general class of weighted residual methods, based on the residual-minimizing conditions, defined through the inner product

$$(u, v) = \int_{\alpha}^{\beta} u\, v\, \omega\, dx = 0$$

where $u(x)$ and $v(x)$ are two functions defined on $[\alpha, \beta]$ and $\omega(x)$ is some given weight function. Therefore, we have

$$(R_N, \psi)_N = \int_{\alpha}^{\beta} R_N\ \psi\ \omega_* dx = 0 \quad \psi \in Y_N \tag{1.85}$$

where ψ is the test function and ω_* is the weight function. The dimension of the discrete set Y_N depends on the problem under consideration.

1.5.1.2 Collocation Method

The choice of the test functions and of the weights is associated with the spectral method. If the inner product in Eq. (1.85) is evaluated by a type of numerical quadrature known as "Gaussian integration," a set of points $\{x_i\}$, called the collocation points, might be obtained. In turn, the collocation method, one of the formulations in the spectral method, corresponds to the choice

$$\psi_i = \delta(x - x_i) \quad \text{and} \quad \omega_* = 1$$

From Eq. (1.85), we simply obtain

$$R_N(x_i) = 0$$

or

$$(L_N u^N)_{x_i} = f(x_i) \tag{1.86}$$

Therefore, in the collocation method, the residual is exactly zero at the collocation points.

Now, we will present a way to determine the collocation points. As we know, the main idea of the numerical integration is to fit a polynomial of degree N, $P_N(x)$ to the integrand $f(x)$ and then integrate $P_N(x)$, by summation of $P_N(x_i)$ with weight $\{w_i\}$ over the set of points $\{x_i\}$, which is formularized with

$$\int_{a}^{b} f(x)dx \approx \sum_{i=0}^{N} w_i P_N(x_i)$$

If we allow $\{x_i\}$ as well as $\{w_i\}$ to be unknowns, we can maximize the accuracy of the numerical integration by the choice of $\{x_i\}$ and $\{w_i\}$.

Based on Gauss–Jacobi integration theorem,[15] for the set of orthogonal polynomials $\{p_j(x)\}$ on $x \in [-1, 1]$ with respect to the weight function $\rho(x)$, the collocation points $\{x_i\}$ are chosen to be the solutions of the following equation:

$$p_{N+1}(x) = 0 \tag{1.87}$$

And the points $\{w_i\}$ are the solutions of the following set of linear equations:

$$\sum_{i=0}^{N} x_i^k w_i = \int_{-1}^{1} x^k \rho(x) dx \quad k = 0, \ldots, N \tag{1.88}$$

Then the Gauss–Jacobi integration

$$\int_a^b p_j(x)\rho(x)dx = \sum_{i=0}^{N} w_i p_j(x_i) \tag{1.89}$$

is exact for all $p_j(x)$ which are polynomials of at most degree $(2N + 1)$. The positive numbers $\{w_i\}$ are called "weights."

However, the roots $\{x_i\}$, which correspond to the collocation points, are all in the interior of $[-1, 1]$, in order to impose boundary conditions at one or both end points; the points $\{x_i\}$ are chosen to be the solutions of the following equation:

$$g(x) = p_{N+1}(x) + ap_N(x) = 0 \tag{1.90}$$

where $a = -p_{N+1}(-1)/p_N(-1)$, thus $g(-1) = 0$, we have $x_0 = -1$, the first point x_0 is fixed on the left boundary. Let the points $\{w_i\}$ be the solutions of Eq. (1.88), the quadrature formula

$$\int_{-1}^{1} p_j(x)\rho(x)dx = \sum_{i=0}^{N} w_i p_j(x_i) \tag{1.91}$$

is called Gauss–Radau integration. It can be proved that Gauss–Radau integration is exact for all $p_j(x)$ which are polynomials of at most degree $2N$.

If the points $\{x_i\}$ are chosen to be the solutions of the following equation:

$$g(x) = p_{N+1}(x) + ap_N(x) + bp_{N-1}(x) = 0 \tag{1.92}$$

where the coefficients a and b are determined by equations $g(\pm 1) = 0$, we have

$$x_0 = -1 \text{ and } x_N = 1$$

the first and final points x_0 and x_N are chosen at the two boundaries.

Let the points $\{w_i\}$ be the solutions of Eq. (1.88), then the quadrature formula

$$\int_{-1}^{1} p_j(x)\rho(x)dx = \sum_{i=0}^{N} w_i p_j(x_i) \tag{1.93}$$

is called Gauss–Lobatto integration, where all $p_j(x)$ are polynomials of at most degree $(2N - 1)$. The proof of this result is similar to the previous one.

These Gauss-type quadratures show that the integration with the integrand $p_j(x)\rho(x)$, where $p_j(x)$ and $\rho(x)$ are orthogonal polynomial and its weight function respectively, can be exactly displaced by a summation of $w_i p_j(x_i)$, where the collocation points $\{x_i\}$ and the weight points $\{w_i\}$ are determined on the special equations.

Considering the Chebyshev polynomials, e.g., the collocation points as well as the weight points are determined by Gauss-type quadratures, that is, Gauss–Jacobi integration, in terms of Eqs. (1.87) and (1.88),

$$x_j = \cos\frac{(2j+1)\pi}{2N+2} \quad w_j = \frac{\pi}{N+1} \quad j = 0, \ldots, N \tag{1.94}$$

Gauss–Radau integration, in terms of Eqs. (1.90) and (1.88),

$$x_j = \cos\frac{2\pi j}{2N+1} \quad w_j = \begin{cases} \dfrac{\pi}{2N+1} & j = 0 \\ \dfrac{2\pi}{2N+2} & 1 \le j \le N-1 \end{cases} \tag{1.95}$$

Gauss–Lobatto integration, in terms of Eqs. (1.92) and (1.88)

$$x_j = \cos\frac{\pi j}{N} \quad w_j = \begin{cases} \dfrac{\pi}{2N} & j = 0, N \\ \dfrac{\pi}{N} & 1 \le j \le N-1 \end{cases} \tag{1.96}$$

which is used most commonly.

1.5.2 Fourier–Galerkin Method

1.5.2.1 Discrete Fourier Series

The more familiar spectral method is the Fourier–Galerkin method where the trial functions are trigonometric functions $\{e^{ikx}\}$. Therefore, the solution of Eq. (1.83) is sought in the form of the truncated Fourier series

$$u(x) \approx u^N(x) = \sum_{k=-N/2}^{N/2-1} \tilde{u}_k e^{ikx}$$

where $u^N(x)$ is a polynomial of degree N.

When the technique of collocation is applied, where the collocation points $\{x_i\}$ associated with the Fourier series are defined by

$$x_i = \frac{2\pi}{N} i \qquad i = 0, 1, \ldots, N-1 \tag{1.97}$$

we have

$$u(x_i) = \sum_{k=-N/2}^{N/2-1} \tilde{u}_k e^{ikx_i} \tag{1.98}$$

Consequently,

$$\tilde{u}_k = \frac{1}{N} \sum_{i=0}^{N-1} u(x_i) e^{-ikx_i} \tag{1.99}$$

If the function $u(x)$ is assumed to be periodic, it satisfies

$$u(x_0) = u(x_n)$$

The solutions at the noncollocation points can be obtained by the interpolation among the collocation points in the form

$$u(x) = \sum_{i=0}^{N-1} u(x_i) g_i(x) \tag{1.100}$$

where the function $g_i(x)$ is reduced to

$$g_i(x) = \frac{1}{N} \sum_{p=-\frac{N}{2}}^{\frac{N}{2}-1} e^{ip(x-x_i)} = \frac{1}{N} \sin\left[\frac{N}{2}\left(x - x_i\right)\right] \cot \frac{x - x_i}{2}$$

is a trigonometric polynomial and satisfies

$$g_i(x_j) = \delta_{ij}.$$

1.5.2.2 Differentiation at Collocation Point

Then from Eq. (1.98), the lth derivative at a given collocation point x_j is expressed as

$$u^{(l)}(x_j) = \sum_{k=-\frac{N}{2}}^{\frac{N}{2}-1} (ik)^l \tilde{u}_k e^{ikx_j} \tag{1.101}$$

By differentiating Eq. (1.100), we can construct differentiation formulas expressing the derivative, of any order, in physical space at a given collocation point, in terms of the values of the function itself at all collocation points, that is,

$$(u(x_j))^{(l)} = \sum_{s=0}^{\frac{N}{2}-1} u(x_s) g_s^{(l)}(x_j) = \sum_{s=0}^{\frac{N}{2}-1} (D_l)_{js} u(x_s) \tag{1.102}$$

where

$$(D_1)_{js} = \begin{cases} \frac{1}{2}(-1)^{j+s} \cot \dfrac{x_j - x_s}{2}, & s \neq j \\ 0, & s = j \end{cases}$$

$$(D_2)_{js} = \begin{cases} \frac{1}{2}(-1)^{j+s+1} \dfrac{1}{\sin^2((x_j - x_s)/2)}, & s \neq j \\ -\dfrac{2N^2+1}{6}, & s = j \end{cases}$$

1.5.2.3 Fourier–Galerkin Method

In the Fourier–Galerkin method, the trial functions are trigonometric functions, $\{e^{ikx}\}$, which satisfy the boundary conditions, and the collocation points $\{x_i\}$ associated with the text function $\psi_i = \delta(x - x_i)$ are defined by

$$x_i = \frac{2\pi}{N} i \quad i = 0, 1, \ldots, N-1 \tag{1.103}$$

We have

$$(L_N u^N)_{x_j} = f(x_j) \tag{1.104}$$

Using the discrete orthogonality property of the trial function $\{e^{ikx}\}$ and the derivative character expressed by Eq. (1.101), a closed system furnishing $N+1$ equations can be obtained for determining the $N+1$ spectral coefficients $\tilde{u}_k$ from Eq. (1.86). This is because boundary conditions are satisfied by the trial functions and we have $X_N = Y_N$.

1.5.3 Chebyshev–Tau Method[16]

1.5.3.1 Discrete Chebyshev Series

When the solution is not periodic, taking Chebyshev polynomials $\{T_k(x)\}$ as a trial function is more suitable in the spectral method, called Chebyshev–Tau method. Here, the solution of Eq. (1.83) is sought in the form of the truncated Chebyshev series

$$u(x) \approx u^N(x) = \sum_{k=0}^{N} \tilde{u}_k T_k(x) \tag{1.105}$$

where $u^N(x)$ is a polynomial of degree N. When the technique of collocation is applied, we have

$$u(x_i) = \sum_{k=0}^{N} \tilde{u}_k T_k(x_i)$$

Consequently,

$$\tilde{u}_k = \frac{1}{\gamma_k} \sum_{j=0}^{N} u(x_j) T_k(x_j) w_j \tag{1.106}$$

where $\gamma_k = \frac{\pi}{2} c_k \;\; k < N$

$$\gamma_N = \begin{cases} \frac{\pi}{2} & \text{for Gauss and Gauss–Radau} \\ \pi & \text{for Gauss–Lobatto} \end{cases}$$

The solutions at the noncollocation points can be obtained by the interpolation among the collocation points in the form

$$u(x) = \sum_{i=0}^{N-1} u(x_i) \psi_i(x) \tag{1.107}$$

where the function $\psi_i(x)$ is Lagrange interpolated polynomial with $\psi_i(x_j) = \delta_{ij}$.

For Gauss–Lobatto integration,

$$\psi_j(x) = \frac{(-1)^{j+1}(1 - x^2) T'_N(x)}{\bar{c}_j N^2 (x - x_j)} \quad j = 0, 1, \ldots, N \tag{1.108}$$

where $\bar{c}_0 = \bar{c}_N = 2, \bar{c}_j = 1, 1 \leq j \leq N - 1$.

1.5.3.2 Differentiation at Collocation Point

At a given collocation point, Eqs. (1.79) and (1.80) are written as

$$u'(x_j) = \sum_{m=0}^{N} \tilde{u}_m T'_m(x_j) = \sum_{m=0}^{N} \tilde{u}_m^{(1)} T_m(x_j) \tag{1.109}$$

$$u''(x_j) = \sum_{m=0}^{N} \tilde{u}_m T''_m(x_j) = \sum_{m=0}^{N} \tilde{u}_m^{(2)} T_m(x_j) \tag{1.110}$$

where

$$\tilde{u}_m^{(1)} = \frac{2}{c_m} \sum_{\substack{p=m+1 \\ p+m \text{ odd}}}^{N} p \tilde{u}_p \quad m = 0, \ldots, N - 1 \tag{1.111}$$

$$\tilde{u}_N^{(1)} = 0, \;\; \tilde{u}_{N-1}^{(1)} = 2N \tilde{u}_N \tag{1.112}$$

$$\tilde{u}_m^{(2)} = \frac{1}{c_m} \sum_{\substack{p=m+2 \\ p+m \text{ even}}}^{N} p(p^2 - m^2)\tilde{u}_p \quad m = 0, \ldots, N-2 \tag{1.113}$$

$$\tilde{u}_{N-1}^{(2)} = \tilde{u}_N^{(2)} = 0 \tag{1.114}$$

where $c_m = \begin{cases} 2 & m = 0 \\ 1 & m \geq 1 \end{cases}$.

For lth derivative, we have

$$u^l(x_j) = \sum_{m=0}^{N} \tilde{u}_m^{(l)} T_m(x_j)$$

In terms of Eqs. (1.72) and (1.73), we obtain

$$2k\tilde{u}_k = c_{k-1}\tilde{u}_{k-1}^{(1)} - \tilde{u}_{k+1}^{(1)} \quad k \geq 1 \tag{1.115}$$

Since $u_k^{(1)} = 0$ at $k \geq N$, hence

$$c_k\tilde{u}_k^{(1)} = \tilde{u}_{k+2}^{(1)} + 2(k+1)\tilde{u}_{k+1} \quad 0 \leq k \leq N-1 \tag{1.116}$$

By successive differentiations, the general recurrence formula for $\tilde{u}_k^{(l)}$ is obtained

$$c_k\tilde{u}_k^{(l)} = \tilde{u}_{k+2}^{(l)} + 2(k+1)\tilde{u}_{k+1}^{(l-1)} \quad 0 \leq k \leq N-1 \tag{1.117}$$

By differentiating Eq. (1.107), the derivative in physical space at a given collocation point is given by

$$(u(x_j))' = \sum_{s=0}^{N} u(x_s)\psi_s'(x_j) = \sum_{s=0}^{N} (D_N)_{js} u(x_s) \tag{1.118}$$

For Gauss–Lobatto integration, in terms of Eq. (1.108), we have

$$(D_N)_{lj} = \begin{cases} \dfrac{\bar{c}_l}{\bar{c}_j} \dfrac{(-1)^{l+j}}{x_l - x_j} & l \neq j \\ \dfrac{-x_j}{2(1 - x_j^2)} & 1 \leq l = j \leq N-1 \\ \dfrac{2N^2 + 1}{6} & l = j = 1 \\ -\dfrac{2N^2 + 1}{6} & l = j = N \end{cases} \tag{1.119}$$

1.5.3.3 Chebyshev–Tau Method

In the Chebyshev–Tau method, the trial functions are Chebyshev polynomials $\{T_k(x)\}$, which cannot satisfy the boundary conditions so that $Y_N < X_N$, and the collocation points $\{x_i\}$ associated with the text function $\psi_i = \delta(x - x_i)$ might be given by Eqs. (1.94)–(1.96), depended on which type of Gaussian integrations is chosen. Then, we have

$$(L_N u^N)_{x_j} = f(x_j) \tag{1.120}$$

By means of discrete Chebyshev series (Eq. (1.106)) and the derivative character expressed by Eqs. (1.111)–(1.114), a system only furnishing $N - 1$ equations related to $\tilde{u}_k$ can be obtained, which is not closed for determining the $N + 1$ spectral coefficients $\tilde{u}_k$. The complement equations should be derived from boundary conditions. Considering a channel flow, on upper and lower walls, we have

$$u(-1) = u(1) = 0$$

Since $T_k(1) = 1$ and $T_k(-1) = (-1)^k$, then

$$\sum_{k=0}^{N} \tilde{u}_k = 0 \tag{1.121}$$

$$\sum_{k=0}^{N} (-1)^k \tilde{u}_k = 0 \tag{1.122}$$

Now, we obtain a closed system to determine all spectral coefficients, $\tilde{u}_k$. This so-called Tau method is a modification of the Galerkin method allowing the use of trial functions not satisfying the boundary conditions.

1.5.4 *Helmholtz Equation*

Helmholtz equation is a one-dimensional ordinary differential equation written as

$$\frac{d^2u}{dy^2} - \lambda u = f \tag{1.123}$$

Where u is unknown, f is a function of y, and λ is a known constant.

Based on the collocation method, the Chebyshev–Tau approximation of equation (1.123) is

$$\tilde{u}_n^{(2)} - \lambda \tilde{u}_n = \tilde{f}_n \; n = 0, 1, \ldots, N - 2 \tag{1.124}$$

When combined with Eq. (1.113), this yields

$$\sum_{\substack{p=n+2 \\ p+n \text{ even}}}^{N} p(p^2 - n^2)\tilde{u}_p - \lambda \tilde{u}_n = \tilde{f}_n \quad n = 0, 1, \ldots, N - 2 \tag{1.125}$$

Using the recursion equation (1.116), that is, $2n\tilde{u}_n^{(1)} = c_{n-1}\tilde{u}_{n-1}^{(2)} - \tilde{u}_{n+1}^{(2)}$, Eq. (1.124) is also written as

$$2n\tilde{u}_n^{(1)} = c_{n-1}(\tilde{f}_{n-1} + \lambda\tilde{u}_{n-1}) - (\tilde{f}_{n+1} + \lambda\tilde{u}_n) \quad n = 1, \ldots, N-3$$

Institution of Eq. (1.115), that is, $c_k\tilde{u}_k^{(1)} = \tilde{u}_{k+2}^{(1)} + 2(k+1)\tilde{u}_{k+1}$, we have

$$\begin{aligned} 2n\tilde{u}_n^{(1)} &= \frac{c_{n-1}}{2(n-1)}[c_{n-2}(\tilde{f}_{n-2} + \lambda\tilde{u}_{n-2}) - (\tilde{f}_n + \lambda\tilde{u}_n)] \\ &\quad - \frac{1}{2(n+1)}[c_n(\tilde{f}_n + \lambda\tilde{u}_n) - (\tilde{f}_{n+2} + \lambda\tilde{u}_{n+2})] \\ &\quad n = 2, \ldots, N-4 \end{aligned}$$

This simplifies to

$$\begin{aligned} &\frac{c_{n-2}}{4n(n-1)}\lambda\tilde{u}_{n-2} + \left(1 - \frac{\lambda}{2\left(n^2-1\right)}\right)\tilde{u}_n + \frac{\lambda}{4n(n+1)}\tilde{u}_{n+2} \\ &= \frac{c_{n-2}}{4n(n-1)}\tilde{f}_{n-2} - \frac{1}{2(n^2-1)}\tilde{f}_n + \frac{1}{4n(n+1)}\tilde{f}_{n+2} \\ &\quad n = 2, \ldots, N-4 \end{aligned} \tag{1.126}$$

Noting that the four equations are dropped in going from Eqs. (1.124) to (1.126), we can write Eq. (1.124) as

$$\begin{aligned} &\frac{c_{n-2}}{4n(n-1)}\lambda\tilde{u}_{n-2} + \left(1 - \frac{\lambda\beta_n}{2\left(n^2-1\right)}\right)\tilde{u}_n + \frac{\lambda\beta_{n+2}}{4n(n+1)}\tilde{u}_{n+2} \\ &= \frac{c_{n-2}}{4n(n-1)}\tilde{f}_{n-2} - \frac{\beta_n}{2(n^2-1)}\tilde{f}_n + \frac{\beta_{n+2}}{4n(n+1)}\tilde{f}_{n+2} \\ &\quad n = 2, \ldots, N \end{aligned} \tag{1.127}$$

where $\beta_n = \begin{cases} 1 & 0 \le n \le N-2 \\ 0 & n > N-2 \end{cases}$

There are three types of boundary conditions for Eq. (1.123):

1. *Dirichlet boundary conditions:*

$$u(\pm 1) = u_\pm$$

which may also be written as

$$\sum_{n=0}^{N} \tilde{u}_n = u_+$$

$$\sum_{n=0}^{N} (-1)^n \tilde{u}_n = u_-$$

or

$$\sum_{\substack{n=0\\ n\ \text{even}}}^{N} \tilde{u}_n = \frac{u_+ + u_-}{2} = c_{\text{even}} \tag{1.128}$$

and

$$\sum_{\substack{n=0\\ n\ \text{odd}}}^{N} \tilde{u}_n = \frac{u_+ - u_-}{2} = c_{\text{odd}} \tag{1.129}$$

Since the even and odd coefficients are uncoupled in Eqs. (1.127)–(1.129), two complete linear systems are obtained, in which the structure for the coefficients is quasi-tridiagonal. For example, the linear algebraic system for the even coefficients is

$$\begin{pmatrix} 1 & 1 & 1 & \cdots & & & & 1 \\ A & A & A & & & & & \\ & A & A & A & & & & \\ & & A & A & A & & & \\ & & & & \vdots & & & \\ & & & & A & A & A & \\ & & & & & A & A & \\ & & & & & & A & A \end{pmatrix} \begin{pmatrix} \tilde{u}_0 \\ \tilde{u}_2 \\ \tilde{u}_4 \\ \\ \vdots \\ \tilde{u}_{N-4} \\ \tilde{u}_{N-2} \\ \tilde{u}_N \end{pmatrix} = \begin{pmatrix} c_{\text{even}} \\ B_0 \\ B_2 \\ \\ \vdots \\ B_{N-6} \\ B_{N-4} \\ B_{N-2} \end{pmatrix} \tag{1.130}$$

where A denotes the nonzero coefficient from Eq. (1.127) and B is the right-hand side of Eq. (1.127).

Helmholtz equation with the Dirichlet boundary conditions may be solved by solving the quasi-tridiagonal systems for the even and odd coefficients, respectively, for $\tilde{u}_n$, and then reversing the Chebyshev transform on $\tilde{u}_n$ to produce u_n.

2. *Neumann boundary conditions:*

$$\frac{du}{dy}(\pm 1) = a_\pm \tag{1.131}$$

The Chebyshev–Tau approximation may be written as

$$\sum_{n=0}^{N} \tilde{u}_n^{(1)} = a_+$$

$$\sum_{n=0}^{N} (-1)^n \tilde{u}_n^{(1)} = a_-$$

or

$$\sum_{\substack{n=0\\ n\ \text{even}}}^{N} \tilde{u}_n^{(1)} = \frac{a_+ + a_-}{2} = a_{\text{even}} \tag{1.132}$$

$$\sum_{\substack{n=0\\ n\ \text{odd}}}^{N} \tilde{u}_n^{(1)} = \frac{a_+ - a_-}{2} = a_{\text{odd}} \tag{1.133}$$

Since

$$\tilde{u}_n^{(1)} = \frac{2}{c_n} \sum_{\substack{p=n+1 \\ p+\ n\text{odd}}}^{N} p\tilde{u}_p \quad n = 0, \ldots, N-1 \tag{1.111}$$

We have

$$\sum_{\substack{n=0 \\ n\ \text{odd}}}^{N} \tilde{u}_n^{(1)} = \sum_{\substack{n=0 \\ n\ \text{even}}}^{N} n^2 \tilde{u}_n = a_{\text{odd}} \tag{1.134}$$

$$\sum_{\substack{n=0 \\ n\ \text{even}}}^{N} \tilde{u}_n^{(1)} = \sum_{\substack{n=0 \\ n\ \text{odd}}}^{N} n^2 \tilde{u}_n = a_{\text{even}} \tag{1.135}$$

Since the even and odd coefficients decouple in Eqs. (1.127), (1.134), and (1.135), two complete linear systems are obtained, which is the same as that of the Dirichlet problem.

If $\lambda = 0$, the compatibility condition for Eqs. (1.123) and (1.131) is

$$\int_{-1}^{1} f dy = a_+ - a_-$$

Written in the discrete form

$$\sum_{n=0}^{N-2} \frac{-2}{n^2 - 1} \tilde{f}_n = a_+ - a_- \tag{1.136}$$

3. *Robin boundary conditions:*

$$\frac{du}{dy}(\pm 1) + b_\pm u(\pm 1) = B_\pm \tag{1.137}$$

which may also be written as

$$\sum_{n=0}^{N} \tilde{u}_n^{(1)} + b_+ \sum_{n=0}^{N} \tilde{u}_n = B_+$$

$$\sum_{n=0}^{N} (-1)^n \tilde{u}_n^{(1)} + b_- \sum_{n=0}^{N} (-1)^n \tilde{u}_n = B_-$$

or equivalently

$$\sum_{n=0}^{N} (n^2 + b_+) \tilde{u}_n = B_+ \tag{1.138}$$

$$\sum_{n=0}^{N} (-1)^{n+1} (n^2 - b_-) \tilde{u}_n = B_- \tag{1.139}$$

Equations (1.127), (1.138), and (1.139) form a complete linear system, which solution process is more costly since the even and odd modes do not decoupled.

1.6 Spectral Method of Navier–Stokes Equations

The incompressible N-S equations are written as

$$\nabla \cdot \mathbf{u} = 0 \tag{1.140}$$

$$\frac{\partial \mathbf{u}}{\partial t} + \mathbf{u} \cdot \nabla \mathbf{u} = -\nabla p + \nu \nabla^2 \mathbf{u} \tag{1.141}$$

where all variables are normalized with respect to the channel half-width and center line velocity for a channel flow. Equation (1.140) is the mass conservation equation, and Eq. (1.141) is the momentum conservation equation. $\nu = \frac{1}{\mathrm{Re}}$, Re is the Reynolds number, $\mathbf{u}$ is velocity vector, and p is the pressure.

Writing in a general form

$$\frac{\partial \mathbf{u}}{\partial t} = \mathbf{F}(\mathbf{u}, \mathbf{x}, t) \tag{1.142}$$

For numerical calculations, it is in general most efficient to apply spectral methods only to the spatial dependence, since the time dependence can be marched forward from one time level to another. Therefore, the discretization of time derivative in Eq. (1.142) should be discussed first.

1.6.1 Time Integration Method [17, 18]

1.6.1.1 Time-Marching Method

For a time-dependent equation, time marching means that the solution at time level $n+1$ can calculated from the known values at old time level, where n is the running index in the t direction. After we have known values at time level $n+1$, then the same procedure is used to calculate values at time level $n+2$. In this fashion, the solution is progressively obtained by marching in steps of time. There are three type approaches used commonly for time advancement, they are explicit, implicit, and semi-implicit.

Explicit is the simpler numerical scheme, the unknown values at time level $n+1$ can be obtained from the known values at the time levels directly, that is,

$$\mathbf{u}^{n+1} = \mathbf{G}(\mathbf{u}^n, \mathbf{u}^{n-1}, \ldots)$$

where schemes can be distinguished according to the number of involved time levels, such as the one-step (e.g., Runge–Kutta schemes) and the multistep methods (e.g., Adams–Bushforth schemes).

The second-order and third-order Adams–Bushforth schemes denoted by AB2 and AB3, respectively, are expressed, respectively, by

$$\mathbf{u}^{n+1} = \mathbf{u}^n + \Delta t \left\{ \frac{3}{2}\mathbf{F}\left(\mathbf{u}^n, \mathbf{x}, t^n\right) - \frac{1}{2}\mathbf{F}\left(\mathbf{u}^{n-1}, \mathbf{x}, t^{n-1}\right) \right\} \tag{1.143}$$

and

$$\mathbf{u}^{n+1} = \mathbf{u}^n + \Delta t \left\{ \frac{23}{12}\mathbf{F}(\mathbf{u}^n, \mathbf{x}, t^n) - \frac{4}{3}\mathbf{F}\left(\mathbf{u}^{n-1}, \mathbf{x}, t^{n-1}\right) + \frac{5}{12}\mathbf{F}\left(\mathbf{u}^{n-2}, \mathbf{x}, t^{n-2}\right) \right\} \tag{1.144}$$

where Δt is the interval of a time step. The stringent time step restrictions should be considered in the explicit approach for stability.

The implicit scheme is somewhat more complicated comparing with the explicit scheme, where the unknowns are not only expressed on the left-hand side but also occur in a function on the right-hand side. Such as

$$\frac{\mathbf{u}^{n+1} - \mathbf{u}^n}{\Delta t} = \mathbf{F}\left(\mathbf{u}^{n+1}, t^{n+1}\right) \quad \text{Backward Euler scheme (BE)} \tag{1.145}$$

$$\frac{(3/2)\mathbf{u}^{n+1} - 2\mathbf{u}^n + (1/2)\mathbf{u}^{n-1}}{\Delta t} = \mathbf{F}\left(\mathbf{u}^{n+1}, t^{n+1}\right)$$

Second-order backward finite difference (BFD2) (1.146)

$$\frac{(11/6)\mathbf{u}^{n+1} - 3\mathbf{u}^n + (3/2)\mathbf{u}^{n-1} - (1/3)\mathbf{u}^{n-2}}{\Delta t} = \mathbf{F}\left(\mathbf{u}^{n+1}, t^{n+1}\right)$$

Third-order backward finite difference (BFD3) (1.147)

$$\frac{(25/12)\mathbf{u}^{n+1} - 4\mathbf{u}^n + 3\mathbf{u}^{n-1} - (4/3)\mathbf{u}^{n-2} + (1/4)\mathbf{u}^{n-3}}{\Delta t} = \mathbf{F}\left(\mathbf{u}^{n+1}, t^{n+1}\right)$$

Fourth-order backward finite difference (BFD4) (1.148) and

$$\frac{\mathbf{u}^{n+1} - \mathbf{u}^n}{\Delta t} = \frac{\mathbf{F}(\mathbf{u}^{n+1}, t^{n+1}) + \mathbf{F}(\mathbf{u}^n, t^n)}{2} \quad \text{Crank–Nichoson scheme (CN)} \tag{1.149}$$

In the semi-implicit scheme, the explicit scheme is applied for some terms, as well as the implicit scheme is applied for others. A nonlinear equation system is expressed by

$$\frac{\partial \mathbf{u}}{\partial t} = \mathbf{F}(\mathbf{u}, \mathbf{x}, t) + L(\mathbf{u}, \mathbf{x}, t) \tag{1.150}$$

where **F** and **L** represent nonlinear and linear operators, respectively. If the explicit scheme AB3 is applied to the nonlinear part and the implicit scheme CN to the linear part, then we obtain a semi-implicit scheme, called AB3CN, expressed as

$$\mathbf{u}^{n+1} = \mathbf{u}^n + \Delta t\left\{\frac{23}{12}\mathbf{F}(\mathbf{u}^n, \mathbf{x}, t^n) - \frac{4}{3}\mathbf{F}\left(\mathbf{u}^{n-1}, \mathbf{x}, t^{n-1}\right) + \frac{5}{12}\mathbf{F}\left(\mathbf{u}^{n-2}, \mathbf{x}, t^{n-2}\right)\right\}$$
$$+\frac{\Delta t}{2}\{\mathbf{L}\left(\mathbf{u}^{n+1}, t^{n+1}\right) + \mathbf{L}(\mathbf{u}^n, t^n)\} \tag{1.151}$$

Semi-implicit backward finite differences with different accuracies for N-S equation (1.141), denoted as SBFD, can be expressed by

$$\frac{1}{\Delta t}\sum_{j=0}^{k} a_j \mathbf{u}^{n+1-j} = \sum_{j=0}^{k-1} b_j \mathbf{F}\left(\mathbf{u}^{n-j}\right) \tag{1.152}$$

where superscript n denotes the time level, and a_j and b_j are constant given in Table 1.1.

Thus, the scheme SBFD3 can expressed as

$$\frac{1}{\Delta t}\left(\frac{11}{6}\mathbf{u}^{n+1} - 3\mathbf{u}^n + \frac{3}{2}\mathbf{u}^{n-1} - \frac{1}{3}\mathbf{u}^{n-2}\right) - 3\mathbf{F}(\mathbf{u}^n) + 3\mathbf{F}\left(\mathbf{u}^{n-1}\right) - \mathbf{F}\left(\mathbf{u}^{n-2}\right) = 0 \tag{1.153}$$

or

$$\frac{11}{6}\mathbf{u}^{n+1} = -\left(-3\mathbf{u}^n + \frac{3}{2}\mathbf{u}^{n-1} - \frac{1}{3}\mathbf{u}^{n-2}\right) + \Delta t[3\mathbf{F}(\mathbf{u}^n) - 3\mathbf{F}\left(\mathbf{u}^{n-1}\right) + \mathbf{F}\left(\mathbf{u}^{n-2}\right)] \tag{1.154}$$

1.6.1.2 Time Splitting Method[19, 20]

The terms contained in N-S equations represent the different physical process, such as, $\frac{\partial \mathbf{u}}{\partial t}$ is a local derivative describing the velocity changing with time at an instantaneous position, flow is steady if $\frac{\partial \mathbf{u}}{\partial t} = 0$; $\mathbf{u} \cdot \nabla \mathbf{u}$, called inertia or a convective term, denotes the effect of flow on the velocity field; ∇p, called pressure adjustment, represents the effect of pressure difference; $\nu \nabla^2 u$, called diffusion or a dissipation term, represents the effect of viscosity.

The key idea of time-splitting method is the replacement of simultaneous processes by sequential steps. For example, the split can be by physics: advection on one fractional step, pressure adjustment on another, and diffusion/viscosity on a third, described as follows and

Table 1.1 Coefficients of SBDF

Scheme	Order	a_0	a_1	a_2	a_3	a_4	b_0	b_1	b_2	b_3
SBDF1	1	1	−1				1			
SBDF2	2	3/2	−2	1/2			2	−1		
SBDF3	3	11/6	−3	3/2	−1/3		3	−3	1	
SBDF4	4	25/12	−4	3	−4/3	1/4	4	−6	4	−1

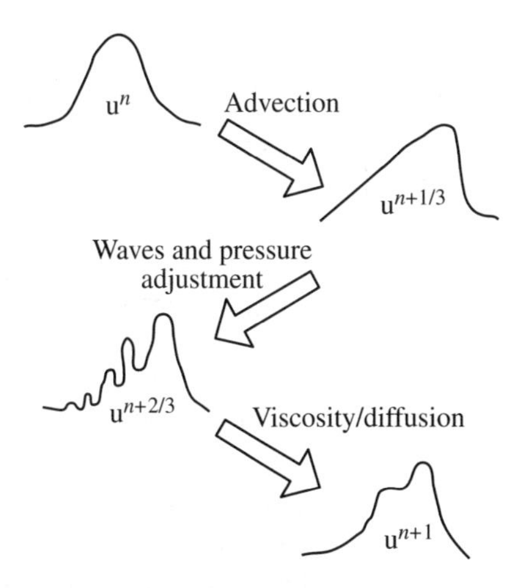

Figure 1.11 Splitting by a physical process.[14] *Source*: Boyd J P 2001. Reproduced with permission of Dover Publication

shown in Figure 1.11:

$$\frac{D\mathbf{u}^{n+1/3}}{Dt} = 0 \tag{1.155}$$

$$\frac{\partial \mathbf{u}^{n+2/3}}{\partial t} = -\nabla p^{n+2/3} \tag{1.156}$$

$$\frac{\partial \mathbf{u}^{n+1}}{\partial t} = -\nu \nabla^2 \mathbf{u}^{n+1} \tag{1.157}$$

where Du/Dt denotes the total derivative

$$\frac{D}{Dt} = \frac{\partial}{\partial t} + \mathbf{u} \cdot \nabla$$

Each fractional equation can be solved individually using a different reasonable numerical method.

1.6.2 Spectral Method based on Time Marching Algorithms (1)

1.6.2.1 Semi-Implicit Time Marching Algorithm

Consider an incompressible fluid flow in a rectangular channel as shown in Figure 1.12. For a turbulent channel flow, the velocity and pressure fields are decomposed into base and fluctuating parts, called base-fluctuation decomposition:

$$\mathbf{u}(\mathrm{x}, t) = U(y)\mathbf{e}_x + \mathbf{u}'(\mathrm{x}, t) \tag{1.158}$$

$$p(\mathbf{x}, t) = \Pi_x(t)x + p'(\mathbf{x}, t) \tag{1.159}$$

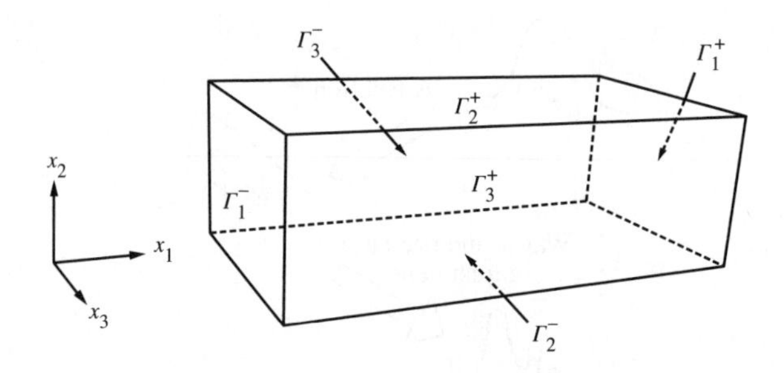

Figure 1.12 Schematic of channel flow

where $\mathbf{x} = (x, y, z)$ or $\mathbf{x} = (x_1, x_2, x_3)$ represents the streamwise, normal, and spanwise directions, respectively. $U(y)$ is the base flow varying only in the y direction, and $\mathbf{e}_x$ is a unit vector in the x direction. The base pressure $\Pi_x(t)x$ is linear in the x direction, where $\Pi_x(t)$ varies linearly with an average drag coefficient on the wall. Then the pressure gradient is

$$\nabla p(\mathbf{x}, t) = \Pi_x(t)\mathbf{e}_x + \nabla p'(\mathbf{x}, t) \tag{1.160}$$

Substituting Eqs. (1.158) and (1.160) into Eqs. (1.140) and (1.141), respectively, gives

$$\nabla \cdot \mathbf{u}' = 0 \tag{1.161}$$

$$\frac{\partial \mathbf{u}'}{\partial t} + \nabla p' - \nu\nabla^2\mathbf{u}' = -\mathrm{N}(\mathbf{u}') + \mathbf{C} \tag{1.162}$$

where nonlinear term $\mathbf{N}(\mathbf{u}') = \mathbf{u}' \cdot \nabla u'$ and spatially constant term $\mathbf{C} = \left[\nu\frac{\partial^2 U}{\partial y^2} - \Pi_x\right] \mathrm{e}^x$.

Substituting Eq. (1.161) into the divergence of Eq. (1.171) leads to

$$\nabla^2 p' = \nabla \cdot \mathbf{C} \tag{1.163}$$

Equations (1.161)–(1.163) constitute the equations for fluctuating quantities.

Taking the SBDF3 scheme, that is, Eq. (1.154) in temporal discretization, Eq. (1.162) can then be written as

$$\nu\nabla^2\mathbf{u}' - \frac{11}{6\Delta t}\mathbf{u}' - \nabla p' = -\mathbf{R} \tag{1.164}$$

where superscript "$n + 1$" is suppressed to convenience writing:

$$\mathbf{R} = -\frac{1}{\Delta t}\left[-3\mathbf{u}'^{n} + \frac{3}{2}\mathbf{u}'^{n-1} - \frac{1}{3}\mathbf{u}'^{n-2}\right] - [3\mathbf{N}(\mathbf{u}'^{n}) - 3\mathbf{N}(\mathbf{u}'^{n-1}) + \mathbf{N}(\mathbf{u}'^{n-2})] + \mathbf{C}$$

Obviously, $\nabla \cdot \mathbf{R} = \nabla \cdot \mathbf{C}$.

Suppressing the time superscripts gives

$$\nu\nabla^2\mathbf{u} - \varepsilon\mathbf{u} - \nabla p = -\mathbf{R} \tag{1.165}$$

$$\nabla \cdot \mathbf{u} = 0 \tag{1.166}$$

$$\nabla^2 p = \nabla \cdot \mathbf{R} \tag{1.167}$$

Considering the coherent structures of wall turbulence, to be presented in the next chapter, the boundary conditions in the x and z directions are periodic, while the upper and lower walls give rise to no-slip boundary conditions. Thus, we have

$$\begin{cases} \mathbf{u}(x,y,z,t) = W(x,\pm 1,z,t) \\ \mathbf{u}(x+L_x,y,z,t) = \mathbf{u}(x,y,z,t) \\ \mathbf{u}(x,y,z+L_z,t) = \mathbf{u}(x,y,z,t) \end{cases} \tag{1.168}$$

where L_x and L_z are the periodic intervals in the x and z directions, respectively, and W is the wall velocity.

1.6.2.2 Fourier–Galerkin Method

One-dimensional Fourier transform as described by Eq. (1.58) can easily be extended into a two-dimensional system $(x,\ z)$, expressed as

$$\tilde{u}_{\mathbf{k}} = \frac{1}{4\pi^2}\int_0^{2\pi}\int_0^{2\pi} u(x,z)e^{-i\mathbf{k}\cdot\mathbf{x}}d\mathbf{x}$$

where $\mathbf{x} = 2\pi\left(\frac{x}{L_x}\frac{z}{L_z}\right)^T$, $\mathbf{k} = \begin{pmatrix} k_x \\ k_z \end{pmatrix}$
and its inverse transform

$$u(x,z) = \sum_{k_x=-\infty}^{\infty}\sum_{k_z=-\infty}^{\infty}\tilde{u}_{\mathbf{k}}e^{i\mathbf{k}\cdot\mathbf{x}} \tag{1.169}$$

By differentiating Eq. (1.169), we have

$$\frac{\partial u}{\partial x_j} = \frac{2\pi i k_j}{L_j}\sum_{k_x=-\infty}^{\infty}\sum_{k_z=-\infty}^{\infty}\tilde{u}_{\mathbf{k}}e^{i\mathbf{k}\cdot\mathbf{x}}$$

$$\frac{\partial^2 u}{\partial x_j^2} = -4\pi^2\left(\frac{k_j}{L_j}\right)^2\sum_{k_x=-\infty}^{\infty}\sum_{k_z=-\infty}^{\infty}\tilde{u}_{\mathbf{k}}e^{i\mathbf{k}\cdot\mathbf{x}}$$

where subscript j represents the component in the j direction.

Hence, from Eqs. (1.165) and (1.167), we have

$$\nu\frac{d^2\tilde{u}_{\mathbf{k}}}{dy^2} - \lambda\tilde{u}_{\mathbf{k}} - i2\pi\frac{k_x}{L_x}\tilde{p}_{\mathbf{k}} = -\tilde{R}_{x\mathbf{k}} \tag{1.170}$$

$$\nu\frac{d^2\tilde{v}_{\mathbf{k}}}{dy^2} - \lambda\tilde{v}_{\mathbf{k}} - \frac{d\tilde{p}_{\mathbf{k}}}{dy} = -\tilde{R}_{y\mathbf{k}} \tag{1.171}$$

$$\nu \frac{d^2 \tilde{w}_{\mathbf{k}}}{dy^2} - \lambda \tilde{w}_{\mathbf{k}} - i2\pi \frac{k_z}{L_z} \tilde{p}_{\mathbf{k}} = -\tilde{R}_{z\mathbf{k}} \quad (1.172)$$

$$\frac{d^2 \tilde{p}_{\mathbf{k}}}{dy^2} - 4\pi^2 \left(\frac{k_x^2}{L_x^2} + \frac{k_z^2}{L_z^2} \right) \tilde{p}_{\mathbf{k}} = i2\pi \frac{k_x}{L_x} \tilde{R}_{x\mathbf{k}} + i2\pi \frac{k_z}{L_z} \tilde{R}_{z\mathbf{k}} + \frac{d\tilde{R}_{y\mathbf{k}}}{dy} \quad (1.173)$$

where $\lambda = 4\pi^2 \nu \left(\frac{k_x^2}{L_x^2} + \frac{k_z^2}{L_z^2} \right) + \varepsilon$. Equations (1.170)–(1.173) constitute a complete system of equations, which can be written in a brief form as

$$\tilde{p}'' - \boldsymbol{\kappa}^2 \tilde{p} = \tilde{F} \quad (1.174)$$

$$\nu \tilde{v}'' - \lambda \tilde{v} - \tilde{p}' = -\tilde{R}_y \quad (1.175)$$

$$\nu \tilde{u}'' - \lambda \tilde{u} - i\kappa_x \tilde{p} = -\tilde{R}_x \quad (1.176)$$

$$\nu \tilde{w}'' - \lambda \tilde{w} - i\kappa_z \tilde{p} = -\tilde{R}_z \quad (1.177)$$

where superscript "′" and "″" represent first-order and second-order derivatives individually and $\boldsymbol{\kappa} = 2\pi \left(\frac{k_x}{L_x} \frac{k_z}{L_z} \right)^T$.

1.6.2.3 Aliasing Removal (3/2 Rule)

Since $\mathbf{N}(\mathbf{u}) = \mathbf{u} \cdot \nabla \mathbf{u}$ in Eq. (1.162) is a nonlinear quadratic term, the Fourier–Galerkin treatment of this term should be considered when the equation is solved by the spectral method. Consider the quadratic term

$$w(x) = u(x)v(x)$$

In case of an infinite series expansion, the Fourier–Galerkin approximation takes the form of convolution sum:

$$\tilde{w}_k = \sum_{p+q=k} \tilde{u}_p \tilde{v}_q \quad (1.178)$$

Where $u(x) = \sum_{p=-\infty}^{\infty} \tilde{u}_p e^{ipx}$

$$v(x) = \sum_{q=-\infty}^{\infty} \tilde{v}_q e^{iqx}$$

$$\tilde{w}_k = \frac{1}{2\pi} \int_0^{2\pi} w(x) e^{-ikx} dx$$

In the collocation method, where collocation points $x_j = 2\pi j/N$, the Fourier–Galerkin approximation takes the form

$$u_j^N = \sum_{k=-N/2}^{N/2-1} \tilde{u}_k \exp(ikx_j) \quad j = 0, \ldots, N-1$$

$$v_j^N = \sum_{k=-N/2}^{N/2-1} \tilde{v}_k \exp(ikx_j) \quad j = 0, \ldots, N-1$$

where the series expansion is truncated after N terms, and the maximal wave number K in the spectra is $K = N/2 - 1$.

Since $w_j^N = u_j^N v_j^N \quad j = 0, 1, \ldots, N-1$

$$\hat{w}_k = \frac{1}{N}\sum_{j=0}^{N-1} w_j^N e^{-ikx_j} = \sum_{\substack{p+q=k \\ |p|,|q| \le N/2}} \tilde{u}_p \tilde{v}_q \tag{1.179}$$

thus, Eq. (1.179) is replaced with

$$\begin{aligned}\hat{w}_k &= \frac{1}{N}\sum_{j=0}^{N-1}\left(\sum_{p=-N/2}^{N/2-1} \tilde{u}_p \exp(ipx_j)\right)\left(\sum_{q=-N/2}^{N/2-1} \tilde{v}_q \exp(iqx_j)\right) \\ &= \sum_{p=-N/2}^{N/2-1}\sum_{q=-N/2}^{N/2-1} \tilde{u}_p \tilde{v}_q \left(\frac{1}{N}\sum_{j=0}^{N-1} \exp\left[i(p+q-k)x_j\right]\right)\end{aligned} \tag{1.180}$$

Using the orthogonality relation

$$\frac{1}{N}\sum_{j=0}^{N-1} \exp[i(p+q-k)x_j] = \begin{cases} 1 & p+q-k = 0, \pm N, \pm 2N, \ldots \\ 0 & p+q-k = \text{other} \end{cases} \tag{1.181}$$

Since $-N/2 \le p+q-k \le N/2-1$, $\hat{w}_k$ is nonzero only if

$$p+q-k = 0 \quad \text{or} \quad p+q-k = \pm N$$

We have

$$\hat{w}_k = \sum_{p+q=k} \tilde{u}_p \tilde{v}_q + \sum_{p+q=k\pm N} \tilde{u}_p \tilde{v}_q = \tilde{w}_k + \sum_{p+q=k\pm N} \tilde{u}_p \tilde{v}_q \tag{1.182}$$

The first term on the right-hand side is the convolution sum defined by Eq. (1.178), and the second term is the aliasing error produced by truncation of an infinite series expansion.

The aliasing error can be removed by extension of spectra from K to K', where $K' = M/2 - 1$. M will be found later to be equal to $(3/2)N$. The extended spectra is defined as

$$\tilde{U}_k = \begin{cases} \tilde{u}_k & |k| \le N/2 \\ 0 & \text{otherwise} \end{cases} \quad \text{and} \quad \tilde{V}_k = \begin{cases} \tilde{v}_k & |k| \le N/2 \\ 0 & \text{otherwise} \end{cases} \tag{1.183}$$

Thus, the $\tilde{U}_k$ (or $\tilde{V}_k$) coefficients are the $\tilde{u}_k$ (or $\tilde{v}_k$) coefficients padded with zeros for the additional wave numbers.

Let

$$y_j = 2\pi j/M$$

$$W_j = U_j V_j$$

$$U_j = \sum_{k=-M/2}^{M/2-1} \tilde{U}_k e^{iky_j} \quad j = 0, 1, \ldots, M-1$$

$$V_j = \sum_{k=-M/2}^{M/2-1} \tilde{V}_k e^{iky_j}$$

Likewise, let

$$\widehat{W}_k = \frac{1}{M} \sum_{j=0}^{M-1} W_j e^{-iky_j} \quad k = -M/2, \ldots, M/2-1$$

Then

$$\widehat{W}_k = \sum_{p+q=k} \tilde{u}_p \tilde{v}_q + \sum_{p+q=k\pm M} \tilde{U}_p \tilde{V}_q \tag{1.184}$$

If $M = 3N/2$, when $-N/2 \le k \le N/2 - 1$, the second sum of Eq. (1.184) is taken on

$$-2N \le p + q \le -N - 1 \quad \text{or} \quad N \le p + q \le 2N - 1$$

Since $-M/2 \le p \le M/2 - 1$ and $-M/2 \le q \le M/2 - 1$, the above-mentioned constraint conditions are simplified further as

$$p < -N/2 \quad \text{or} \quad p > N/2 - 1$$

$$q < -N/2 \quad \text{or} \quad q > N/2 - 1$$

which leads to $\tilde{U}_p$ and $\tilde{V}_q$ being zero, such that the second sum vanishes. This de-aliasing technique is called the 3/2 rule.

1.6.2.4 Influence Matrix Technique

The pressure equation (1.174) and the normal velocity equation (1.175) form a complete set for $\tilde{p}$ and $\tilde{v}$:

$$\begin{cases} \tilde{p}'' - \boldsymbol{\kappa}^2 \tilde{p} = \tilde{F} & \tilde{v}'(\pm 1) = h_\pm \\ \nu \tilde{v}'' - \lambda \tilde{v} - \tilde{p}' = -\tilde{R}_y & \tilde{v}(\pm 1) = g_\pm \end{cases} \tag{1.185}$$

It is difficult to solve, because the pressure $\tilde{p}_{\pm}$ at the wall is unknown a priori while $\tilde{p}$ appears in the $\tilde{v}$ differential equation. A method called the influence matrix technique is employed usually to obtain the solutions of these equations.

$$\text{Let } \begin{pmatrix} \tilde{p} \\ \tilde{v} \end{pmatrix} = \begin{pmatrix} \tilde{p}_1 \\ \tilde{v}_1 \end{pmatrix} + \begin{pmatrix} \tilde{p}_2 \\ \tilde{v}_2 \end{pmatrix} \tag{1.186}$$

where $(\tilde{p}_1, \tilde{v}_1)$ and $(\tilde{p}_2, \tilde{v}_2)$ satisfy the following equations, respectively:

$$\begin{cases} \tilde{p}_1'' - \boldsymbol{\kappa}^2 \tilde{p}_1' = \tilde{F} & \tilde{p}_1(\pm 1) = 0 \\ \nu \tilde{v}_1'' - \lambda \tilde{v}_1 - \tilde{p}_1' = -\tilde{R}_y & \tilde{v}_1(\pm 1) = g_{\pm} \end{cases} \tag{1.187}$$

and

$$\begin{cases} \tilde{p}_2'' - \boldsymbol{\kappa}^2 \tilde{p}_2' = 0 & \tilde{p}_2(\pm 1) = \delta_{\pm} \\ \nu \tilde{v}_2'' - \lambda \tilde{v}_2 - \tilde{p}_2' = 0 & \tilde{v}_2(\pm 1) = 0 \end{cases} \tag{1.188}$$

The solution of Eq. (1.187) can be obtained by a method solving Helmholtz equation presented in Section 1.6.4. However, δ_+ and δ_- in Eq. (1.188) are unknown, therefore, $(\tilde{p}_2, \tilde{v}_2)$ should be decomposed further:

$$\begin{pmatrix} \tilde{p}_2 \\ \tilde{v}_2 \end{pmatrix} = \delta_+ \begin{pmatrix} \tilde{p}_+ \\ \tilde{v}_+ \end{pmatrix} + \delta_- \begin{pmatrix} \tilde{p}_- \\ \tilde{v}_- \end{pmatrix} \tag{1.189}$$

where $(\tilde{p}_+, \tilde{v}_+)$ and $(\tilde{p}_-, \tilde{v}_-)$ satisfy the following equations, respectively:

$$\begin{cases} \tilde{p}_+'' - \boldsymbol{\kappa}^2 \tilde{p}_+ = 0 & \tilde{p}_+(+1) = 1, \quad \tilde{p}_+(-1) = 0 \\ \nu \tilde{v}_+'' - \lambda \tilde{v}_+ - \tilde{p}_+' = 0 & \tilde{v}_+(\pm 1) = 0 \end{cases} \tag{1.190}$$

and

$$\begin{cases} \tilde{p}_-'' - \boldsymbol{\kappa}^2 \tilde{p}_- = 0 & \tilde{p}_-(+1) = 0, \quad \tilde{p}_-(-1) = 1 \\ \nu \tilde{v}_-'' - \lambda \tilde{v}_- - \tilde{p}_-' = 0 & \tilde{v}_-(\pm 1) = 0 \end{cases} \tag{1.191}$$

$(\tilde{p}_+, \tilde{v}_+)$ and $(\tilde{p}_-, \tilde{v}_-)$ as well as $\tilde{v}_{\pm}'(\pm 1)$ and $\tilde{v}_1'(\pm 1)$ are obtained by solving the Helmholtz equations, that is, Eqs. (1.191) and (1.199), meanwhile h_+ and h_- can be determined by iteration. Then δ_+ and δ_- are determined further by solving the following linear algebra equations:

$$\mathbf{A} \begin{bmatrix} \delta_+ \\ \delta_- \end{bmatrix} = \begin{bmatrix} \tilde{v}_+'(+1) & \tilde{v}_-'(+1) \\ \tilde{v}_+'(-1) & \tilde{v}_-'(-1) \end{bmatrix} \begin{bmatrix} \delta_+ \\ \delta_- \end{bmatrix} = \begin{bmatrix} h_+ - \tilde{v}_1'(+1) \\ h_- - \tilde{v}_1'(-1) \end{bmatrix} \tag{1.192}$$

which is produced from the requirement of the boundary conditions of pressure equation. The matrix $\mathbf{A}$ is called the influence matrix.

1.6.2.5 Tau Correction

The pressure equation (1.167) is derived from the continuity equation (1.166) and the momentum equation (1.165) based on the properties of the continuous differentiation operators. Errors will be introduced into the pressure equation when discretized, such that the solutions do not satisfy the continuity equation (1.166). In order to correct the errors, additional terms, called tau term, are introduced into the right-hand side of the equations, which yields

$$\nu\tilde{u}_m^{(2)} - \lambda\tilde{u}_m - i\kappa_x\tilde{p}_m = -\tilde{R}_{x,m} - \tilde{\tau}_{x,m} \tag{1.193}$$

$$\nu\tilde{v}_m^{(2)} - \lambda\tilde{v}_m - \tilde{p}_m^{(1)} = -\tilde{R}_{y,m} - \tilde{\tau}_{y,m} \tag{1.194}$$

$$\nu\tilde{w}_m^{(2)} - \lambda\tilde{w}_m - i\kappa_z\tilde{p}_m = -\tilde{R}_{z,m} - \tilde{\tau}_{z,m} \tag{1.195}$$

$$\tilde{d}_m = i\kappa_x\tilde{u}_m + i\kappa_z\tilde{w}_m + \tilde{v}_m^{(1)} = 0 \tag{1.196}$$

$$m = 0, 1, \ldots, N$$

The application of the discrete divergence to the above-mentioned complete system yields

$$\tilde{p}_m^{(2)} - \boldsymbol{\kappa}^2\tilde{p}_m = \tilde{F}_m + \left(i\kappa_x\tilde{\tau}_{x,m} + i\kappa_z\tilde{\tau}_{z,m} + \tilde{\tau}_{y,m}^{(1)}\right)$$

Since $\tilde{\tau}_{x,m}$, $\tilde{\tau}_{y,m}$, and $\tilde{\tau}_{z,m}$ vanish for $0 \le m \le N-2$, we have

$$\tilde{p}_m^{(2)} - \boldsymbol{\kappa}^2\tilde{p}_m = \tilde{F}_m + \tilde{\tau}_{y,m}^{(1)} \quad m = 0, 1, \ldots, N-2 \tag{1.197}$$

Hence, the complete set for $\tilde{p}$ and $\tilde{v}$ is

$$\begin{cases} \tilde{p}_m^{(2)} - \boldsymbol{\kappa}^2\tilde{p}_m = \tilde{F}_m + \tilde{\tau}_{y,m}^{(1)} & \tilde{v}^{(1)}(\pm 1) = \tilde{h}_\pm \quad m = 0, \ldots, N-2 \\ \nu\tilde{v}_m^{(2)} - \lambda\tilde{v}_m - \tilde{p}_m^{(1)} = -\tilde{R}_{y,m} - \tilde{\tau}_{y,m} & \tilde{v}(\pm 1) = \tilde{g}_\pm \quad m = 0, \ldots, N \end{cases} \tag{1.198}$$

Where $\tilde{\tau}_y = (0, \ldots, 0, \tilde{\tau}_{y,N-1}, \tilde{\tau}_{y,N})^T$.
According to the recurrence relation (1.116),

$$\tilde{\tau}_y^{(1)} = ((N-1)\tilde{\tau}_{y,N-1}, 2N\tilde{\tau}_{y,N}, 2(N-1)\tilde{\tau}_{y,N-1}, 2N\tilde{\tau}_{y,N}, \ldots, 2(N-1)\tilde{\tau}_{y,N-1}, 2N\tilde{\tau}_{y,N}, 0)^T,$$

all spectral coefficients, even if $0 \le m \le N-2$, are disturbed by the truncation errors. Hence, Eq. (1.198) is modified further, using $\tilde{\sigma}_m$ to displace $\tilde{\tau}_{y,m}$

$$\begin{cases} \tilde{p}_m^{(2)} - \boldsymbol{\kappa}^2\tilde{p}_m = \tilde{F}_m + \tilde{\sigma}_m^{(1)} & \tilde{v}^{(1)}(\pm 1) = \tilde{h}_\pm \quad m = 0, \ldots, N-2 \\ \nu\tilde{v}_m^{(2)} - \lambda\tilde{v}_m - \tilde{p}_m^{(1)} = -\tilde{R}_{y,m} - \tilde{\sigma}_m & \tilde{v}(\pm 1) = \tilde{g}_\pm \quad m = 0, \ldots, N \end{cases} \tag{1.199}$$

where $\tilde{\sigma}^{(1)} = ((N-1)\tilde{\sigma}_{N-1}, 2N\tilde{\sigma}_N, 2(N-1)\tilde{\sigma}_{N-1}, 2N\sigma_N, \ldots, 2(N-1)\tilde{\sigma}_{N-1}, 2N\tilde{\sigma}_N, 0)^T$.
Obviously, for $m \le N-1$,

$$\tilde{\sigma}_m^{(1)} = \frac{2}{c_m} m' \tilde{\sigma}_{m'} \tag{1.200}$$

where

$$m' = \begin{cases} N-1 & m \text{ even} \\ N & m \text{ odd} \end{cases} \tag{1.201}$$

To solve Eq. (1.199), consider the B_1 problem

$$\begin{cases} \tilde{p}_{1,m}^{(2)} - \kappa^2 \tilde{p}_{1,m} = \tilde{F}_m & \tilde{v}^{(1)}(\pm 1) = \tilde{h}_{\pm} \quad m = 0, \ldots, N-2 \\ \nu \tilde{v}_{1,m}^{(2)} - \lambda \tilde{v}_{1,m} - \tilde{p}_{1,m}^{(1)} = -\tilde{R}_{y,m} & \tilde{v}(\pm 1) = \tilde{g}_{\pm} \quad m = 0, \ldots, N-2 \end{cases} \tag{1.202}$$

and the B_0 problem

$$\begin{cases} \tilde{p}_{0,m}^{(2)} - \kappa^2 \tilde{p}_{0,m} = \frac{2}{c_m} m' & \tilde{v}^{(1)}(\pm 1) = \tilde{h}_{\pm} \quad m = 0, \ldots, N-2 \\ \nu \tilde{v}_{0,m}^{(2)} - \lambda \tilde{v}_{0,m} - \tilde{p}_{0,m}^{(1)} = 0 & \tilde{v}(\pm 1) = \tilde{g}_{\pm} \quad m = 0, \ldots, N-2 \end{cases} \tag{1.203}$$

m' is determined by Eq. (1.201).

Equation (1.202) identifying with Eq. (1.185) is a complete set for unmodified $\tilde{p}_1$ and $\tilde{v}_1$. Equation (1.203) embodied by the tau terms is a supplementary equation for the tau correction. Both problems for $\tilde{p}_{0,m}, \tilde{v}_{0,m}$ and $\tilde{p}_{1,m}, \tilde{v}_{1,m}$ can be solved by the influence matrix technique.

Substituting Eqs. (1.113) and (1.114) into the v equations in Eqs. (1.192) and (1.193), respectively, yields

$$\begin{cases} \tilde{\sigma}_{1,N-1} = \lambda \tilde{v}_{1,N-1} + 2N p_{1,N} - \tilde{R}_{y,N-1} \\ \tilde{\sigma}_{1,N} = \lambda \tilde{v}_{1,N} - \tilde{R}_{y,N} \end{cases} \tag{1.204}$$

and

$$\begin{cases} \tilde{\sigma}_{0,N-1} = \lambda \tilde{v}_{0,N-1} + 2N p_{0,N} \\ \tilde{\sigma}_{0,N} = \lambda \tilde{v}_{0,N} \end{cases} \tag{1.205}$$

The unknown variables in the above-mentioned equations, that is, $\tilde{\sigma}_{1,N}$, $\tilde{\sigma}_{1,N-1}$, $\tilde{\sigma}_{0,N}$, and $\tilde{\sigma}_{0,N-1}$ would be obtained straightforwardly based on the solutions for Eqs. (1.202) and (1.203).

Let $\tilde{p}_m = \tilde{p}_{1,m} + \beta \tilde{p}_{0,m}$

Substituting into the p equations in Eqs. (1.199), (1.202), and (1.203) gives

$$\tilde{\sigma}_m^{(1)} = \frac{2}{c_m} m' \beta \tag{1.206}$$

Compared with Eq. (1.200),

$$\beta = \tilde{\sigma}_{m'} \tag{1.207}$$

Let $\tilde{v}_m = \tilde{v}_{1,m} + \beta \tilde{v}_{0,m}$

Substituting into the v equations in Eqs. (1.199), (1.202), and (1.203)

$$\tilde{\sigma}_m = \tilde{\sigma}_{1,m} + \beta \tilde{\sigma}_{0,m} \tag{1.208}$$

From Eqs. (1.207) and (1.208),

$$\tilde{\sigma}_{m'} = \tilde{\sigma}_{1,m'} / (1 - \tilde{\sigma}_{0,m'}) \quad (m' = N-1, N) \tag{1.209}$$

Then we can show that

$$\begin{cases} \tilde{p}_m = \tilde{p}_{1,m} + \tilde{\sigma}_{m'}\tilde{p}_{0,m} & m = 0, \ldots, N \\ \tilde{v}_m = \tilde{v}_{1,m} + \sigma_{(m+1)'}\tilde{v}_{0,m} \end{cases} \tag{1.210}$$

Now, the spectral coefficients resulting from a combined Fourier transform in $x - z$ and a Chebyshev transform in y, such as $\tilde{\mathbf{u}}_{k_x,n_y,k_z}$ and $\tilde{p}_{k_x,n_y,k_z}$, can be computed numerically.

Finally, the primary functions in N-S equations, that is, Eqs. (1.140) and (1.141), will be achieved by the inverse transform, from the mathematical form, such as,

$$\mathbf{u}(\mathbf{x}) = \sum_{k_x=-N_x/2+1}^{N_x/2} \sum_{n_y=0}^{N_y-1} \sum_{k_z=-N_z/2+1}^{N_z/2} \tilde{\mathbf{u}}_{k_x,n_y,k_z} T_{n_y} e^{2\pi i(k_x x/L_x + k_z z/L_z)}$$

1.6.3 *Spectral Method based on Time Marching Algorithms (2)*

For incompressible flow, there is no connection between pressure and density, and pressure satisfies the Poisson equation that is usually solved numerically by the finite difference method. In order to solve N-S equations by the spectral method, Eqs. (1.161) and (1.162) can be written as

$$\frac{\partial u_i}{\partial x_i} = 0 \tag{1.211}$$

$$\frac{\partial u_i}{\partial t} = -\frac{\partial p}{\partial x_i} + H_i + \nu\frac{\partial^2 u_i}{\partial x_j^2} \tag{1.212}$$

where H_i denotes a sum of convective term and mean pressure gradient.

Introducing two quantities ϕ and g as dependent variables, defined by

$$\phi = \frac{\partial^2 v}{\partial x_j^2} \tag{1.213}$$

$$g = \frac{\partial u}{\partial z} - \frac{\partial w}{\partial x} \tag{1.214}$$

thus

$$\frac{\partial \phi}{\partial t} = h_v + \nu\frac{\partial^2 \phi}{\partial x_j^2} \tag{1.215}$$

$$\frac{\partial g}{\partial t} = h_g + \nu\frac{\partial^2 g}{\partial x_j^2} \tag{1.216}$$

where

$$h_v = -\frac{\partial}{\partial y}\left(\frac{\partial H_1}{\partial x} + \frac{\partial H_3}{\partial z}\right) + \left(\frac{\partial^2}{\partial x^2} + \frac{\partial^2}{\partial z^2}\right) H_2$$

$$h_g = \frac{\partial H_1}{\partial z} - \frac{\partial H_3}{\partial x}$$

Utilizing AB2CN to Eqs. (1.215) and (1.216), we have

$$\begin{cases} \left(1 - \frac{\nu\Delta t}{2}\nabla^2\right) g^{n+1} = \frac{\Delta t}{2}\left(3h_g^n - h_g^{n-1}\right) + \left(1 + \frac{\nu\Delta t}{2}\nabla^2\right) g^n \\ g(\pm 1) = 0 \end{cases} \tag{1.217}$$

and

$$\begin{cases} \left(1 - \frac{\nu\Delta t}{2}\nabla^2\right) \phi^{n+1} = \frac{\Delta t}{2}\left(3h_v^n - h_v^{n-1}\right) + \left(1 + \frac{\nu\Delta t}{2}\nabla^2\right) \phi^n \\ \nabla^2 v^{n+1} = \phi^{n+1} \\ v^{n+1}(\pm 1) = a_\pm \\ \frac{\partial v^{n+1}}{\partial y}(\pm 1) = b_\pm \end{cases} \tag{1.218}$$

Note that Eq. (1.218) is analogous to Eq. (1.185) formally, then let

$$v^{n+1} = v_1^{n+1} + c_+ v_+^{n+1} + c_- v_-^{n+1} \tag{1.219}$$

where v_1,v_+ and v_- satisfy the following equations, respectively:

$$\begin{cases} \left(1 - \frac{\nu\Delta t}{2}\nabla^2\right) \phi_1^{n+1} = \frac{\nabla t}{2}\left(3h_v^n - h_v^{n-1}\right) + \left(1 + \frac{\nu\Delta t}{2}\nabla^2\right) \phi_1^n & \phi_1^{n+1}(\pm 1) = 0 \\ \nabla^2 v_1^{n+1} = \phi_1^{n+1} & v_1^{n+1}(\pm 1) = a_\pm \end{cases} \tag{1.220}$$

$$\begin{cases} \left(1 - \frac{\nu\Delta t}{2}\nabla^2\right) \phi_+^{n+1} = 0 & \phi_+^{n+1}(1) = 1, \quad \phi_+^{n+1}(-1) = 0 \\ \nabla^2 v_+^{n+1} = \phi_+^{n+1} & v_+^{n+1}(\pm 1) = 0 \end{cases} \tag{1.221}$$

$$\begin{cases} \left(1 - \frac{\nu\Delta t}{2}\nabla^2\right) \phi_-^{n+1} = 0 & \phi_-^{n+1}(1) = 0, \quad \phi_-^{n+1}(-1) = 1 \\ \nabla^2 v_-^{n+1} = \phi_-^{n+1} & v_-^{n+1}(\pm 1) = 0 \end{cases} \tag{1.222}$$

Those equations can be solved numerically by spectral method, that is, the Fourier–Galerkin method in the streamwise and spanwise directions, and the Chebyshev–Tau in the wall normal direction, and finally we obtain v_1^{n+1}, v_+^{n+1}, and v_-^{n+1}。

Based on the influence matrix method, the algebraic system is written as

$$\begin{pmatrix} \tilde{v}'_+(+1) & \tilde{v}'_-(+1) \\ \tilde{v}'_+(-1) & \tilde{v}'_-(-1) \end{pmatrix} \begin{pmatrix} \tilde{c}_+ \\ \tilde{c}_- \end{pmatrix} = \begin{pmatrix} \tilde{b}_+ - \tilde{v}'_1(+1) \\ \tilde{b}_- - \tilde{v}'_1(-1) \end{pmatrix} \tag{1.223}$$

by which the constants $\tilde{c}_+$ and $\tilde{c}_-$ can be determined, then $\tilde{v}^{n+1}$ will be obtained from Eq. (1.219). Likewise, the normal vorticity g is computed from Eq. (1.215) by the spectral method. Finally, the streamwise velocity u and the spanwise velocity w are obtained from Eqs. (1.211) and (1.214).

1.6.4 Spectral Method based on Time-Split Method[21, 22]

Time-split method is in general a method of approximation of the N-S equations, replacing simultaneous processes by sequential steps. Here it is split into two fractional steps:

$$\frac{\partial u_i}{\partial t} = -\frac{\partial}{\partial x_j}u_i u_j + \nu\frac{\partial}{\partial x_j}\frac{\partial}{\partial x_j}u_i \quad \text{(convective and diffusive step)} \tag{1.224}$$

and

$$\frac{\partial u_i}{\partial t} = -\frac{\partial \phi}{\partial x_i} \quad \text{(pressure adjustment step)} \tag{1.225}$$

To ensure the velocity field to be divergence-free, a scale ϕ to be determined is induced in Eq. (1.225) instead of original pressure p. If the semi-implicit scheme AB2CN is applied, then

$$\frac{\hat{u}_i - u_i^n}{\Delta t} = \frac{1}{2}\left(3H_i^n - H_i^{n-1}\right) + \frac{1}{2}\nu\left(\frac{\partial^2}{\partial x_1^2} + \frac{\partial^2}{\partial x_2^2} + \frac{\partial^2}{\partial x_3^2}\right)\left(\hat{u}_i + u_i^n\right) \tag{1.226}$$

where $H_i = -(\partial/\partial x_j)u_i u_j$ is the convective term.

And Eq. (1.225) is written as

$$\frac{u_i^{n+1} - \hat{u}_i}{\Delta t} = -\frac{\partial \phi^{n+1}}{\partial x_i} \tag{1.227}$$

In order to ensure

$$\frac{\partial u_1^{n+1}}{\partial x_1} + \frac{\partial u_2^{n+1}}{\partial x_2} + \frac{\partial u_3^{n+1}}{\partial x_3} = 0$$

we have

$$\left(\frac{\partial^2}{\partial x_1^2} + \frac{\partial^2}{\partial x_2^2} + \frac{\partial^2}{\partial x_3^2}\right)\phi^{n+1} = \frac{1}{\Delta t}\left(\frac{\partial \hat{u}_1}{\partial x_1} + \frac{\partial \hat{u}_2}{\partial x_2} + \frac{\partial \hat{u}_3}{\partial x_3}\right) \tag{1.228}$$

and

$$p = \phi + (\Delta t/2\nu)\nabla^2\phi \tag{1.229}$$

So, Eq. (1.226) can be written as

$$\nu\nabla^2\hat{\mathbf{u}} - \frac{2}{\Delta t}\hat{\mathbf{u}} = -\mathbf{R} \tag{1.230}$$

which is the governing equation about the intermediate velocity fields in time splitting method. Only the boundary conditions for the velocity field are given, and those of the intermediate velocity field are unknown. Here, the periodic conditions are still imposed for both streamwise and spanwise directions in intermediate velocity field, but the boundary conditions at the upper and lower walls would be obtained by the following derivatives.

Let $\mathbf{u}_i^*(\mathbf{x}, t_n + \Delta t)$ satisfy the differential equation

$$\begin{cases} \dfrac{\partial u_i^*}{\partial t} = H_i^* + v\dfrac{\partial}{\partial x_j}\dfrac{\partial}{\partial x_j}u_i^* \\ u_i^*\left(\mathbf{x}, t_n\right) = u_i(\mathbf{x}, t_n) \end{cases} \tag{1.231}$$

where $\mathbf{u}_i(\mathbf{x}, t_n)$ is the solution of the N-S equation (1.141). Hence,

$$\begin{aligned} \hat{u}_i &\approx u_i^*(\mathbf{x},\ \ t_n + \Delta t) \\ &= u_i^*(\mathbf{x},\ \ t_n) + \Delta t\frac{\partial u_i^*}{\partial t} + O(\Delta t^2) \\ &= u_i^*(\mathbf{x},\ \ t_n) + \Delta t(H_i^* + v\nabla^2 u_i^*) + O(\Delta t^2) \end{aligned}$$

Since $u_i^*(\mathbf{x}, t_n) = u_i(\mathbf{x}, t_n)$,

$$\begin{aligned} \hat{u}_i &= u_i^*(\mathbf{x},\ \ t_n) + \Delta t(H_i^* + v\nabla^2 u_i^*) + O(\Delta t^2) \\ &= u_i(\mathbf{x},\ \ t_n) + \Delta t(H_i + v\nabla^2 u_i) + O(\Delta t^2) \\ &= u_i(\mathbf{x},\ \ t_{n+1}) + \Delta t\frac{\partial p}{\partial x_i} + O(\Delta t^2) \end{aligned}$$

From Eq. (1.229), $\hat{u}_i = u_i^{n+1} + \Delta t\frac{\partial \phi^n}{\partial x_i}$ Also, it is written as

$$\hat{\mathbf{u}}|_w = \mathbf{u}^{n+1}|_w + \Delta t\left.\frac{\partial \phi^n}{\partial \mathbf{x}}\right|_w \tag{1.232}$$

It is the boundary condition at the upper and lower walls for Eq. (1.230).
Equation (1.228) can be written in the form as

$$\nabla^2 \phi = f \tag{1.233}$$

The boundary conditions at streamwise and spanwise directions are periodic, and the conditions at the upper and lower walls are written as

$$\left.\frac{\partial \phi^{n+1}}{\partial \mathbf{x}}\right|_w = \left.\frac{\partial \phi^n}{\partial \mathbf{x}}\right|_w \tag{1.234}$$

Equations (1.230) and (1.233) would appear in the same form with that of Eqs. (1.165) and (1.167), hence $\hat{\mathbf{u}}$ and ϕ^{n+1} can be calculated by the spectral method, in succession, the solutions $\mathbf{u}^{n+1}$ are obtained by Eq. (1.227).

1.7 Closed Remarks

The realization of a random process appears unpredictable, but some properties, that is, its statistical properties, are quite reproducible. The probability density function (PDF) representing the probability of the event involved in the process is commonly used to describe the statistical properties, which can be constructed by the infinite statistical moments possessed in random variables. Several statistical moments, such as mean value, rms, skewness, kurtosis, Reynolds stress, turbulent kinetic energy, etc., are traditionally used in discussing the turbulent problem. The turbulence is formed intermittently via bursting events that can be detected by the detection function. By means of the conditional average the flow field around the bursting, the characteristic flow pattern associated with the bursting will be revealed.

By decomposition of state variables of flow fields, such as Reynolds decomposition, LES decomposition (low-pass filtering) etc, a closed equation describing the statistical properties of turbulence can be induced based on the statistical models, this widely used approach equations can be solved numerically.

A periodic function $f(\mathbf{x})$ can be regarded as a superimposition of infinite harmonic waves, or Fourier series in the complex form, by Fourier transform, where the weight function $F(\mathbf{k})$ is called the spectrum of $f(\mathbf{x})$. Actually, this transform gives a mapping between the function $f(\mathbf{x})$ in the physical space $\mathbf{x}$ and the spectrum $F(\mathbf{k})$ in the spectral space $\mathbf{k}$. Likewise, a function $f(\mathbf{x})$ can also be expressed by a sum of orthogonal polynomials, such as Legendre polynomials, Chebyshev polynomials, etc. The mapping between the physical space and the spectral space can also be performed by the transform.

Since the derivative in spectral space can be represented in terms of all spectral coefficients, the partial differential equations in physical space are then transformed into a set of algebraic equations in spectral space, which conduces to the numerical computation. Subsequently, the resulting spectral coefficients are transformed back to physical space. The method solving the equations in spectral space via the discrete technique of collocation is called the pseudospectral method.

The choice of the basis (or trial) function and the test function is essentially associated with the spectral method. In the collocation method, the test function is taken as a delta function at the collocation points given by special formulas, such as Gaussian integrations. Based on this method, the choice of the basis function is then considered. For example, the basis functions are trigonometric functions in the Fourier–Galerkin method for the periodic problems, which satisfy the boundary conditions, so that a closed system can be obtained. Furthermore, the basis functions are Chebyshev polynomials in the Chebyshev–Tau method for the nonperiodic problems, which cannot satisfy the boundary conditions, so that the complement equations should be derived from boundary conditions.

The near-wall turbulence governed by the N-S equations can be simulated numerically by the spectral methods. In calculations, it is in general most efficient to apply spectral methods only to the spatial dependence, since the time-dependence can be marched forward from one time level to another. Both a Chebyshev–Tau method in the wall normal direction and a dealiazed Fourier method in the homogeneous directions, that is, streamwise and spanwise directions, are used for the spatial derivatives. In order to ensure that the computed solutions satisfy both the incompressibility constraint and the momentum equation when the time advancement is carried out using a semi-implicit time marching schemes, a Chebyshev–Tau influence-matrix method, including a Tau-correction step, is employed for the linear term and the pressure term. Aliasing errors in the streamwise and spanwise directions are removed by the spectral truncation method referred to as 3/2 rule.

References

[1] Tennekes H and Lumley D L. A First Course in Turbulence, Academic Press, Inc., New York, 1972.
[2] Panchev S. Random Functions and Turbulence, Pergamon Press, Oxford, 1971.
[3] Pope S B. Turbulent Flows, Cambridge University Press, Cambridge, 2000.
[4] Blackwelder R F and Kaplan R E. On the wall structure of the turbulent boundary layer. J. Fluid Mech., 1976, 76:89–112.
[5] Blackwelder R F and Eckelmann H. Streamwise vortices associated with the bursting phenomenon. J. Fluid Mech., 1979, 94:577–594.
[6] Kim J. On the structure of wall-bounded turbulent flows. Phys. Fluids, 1983, 26(8):2088–2097.
[7] Kim J. Turbulence structures associated with the bursting event. Phys. Fluids, 1985, 26(1):52–58.
[8] Bogard D G and Tiederman W G. Burst detection with single point velocity measurements. J. Fluid Mech., 1986, 162:389–413.
[9] Lu S S and Willmarth W W. Measurements of the structure of the Reynolds stress in a turbulent boundary layer. J. Fluid Mech., 1973, 60:481–511.
[10] Bracewell R N. The Fourier Transform and its Applications, McGraw-Hill Companies, Inc., New York, 2000.
[11] Adrian R J, Christensen K T and Liu Z C. Analysis and interpretation of instantaneous turbulent velocity fields. Exp. Fluid, 2000, 29:275–290.
[12] Canuto C M, Hussaini Y, Quarteroni A and Zang T A. Spectral Methods in Fluid Dynamics, Springer-Verlag, Berlin, 1988.
[13] Peyret R. Spectral Methods for Incompressible Viscous Flow, Springer-Verlag, Inc., 2002.
[14] Boyd J P. Chebyshev and Fourier Spectral Methods, Dover Publication, Inc., 2001.
[15] Davis P J and Rabinowitz P. Methods of Numerical Integration, Academic Press, 1984.
[16] Kleiser L and Schumann U. Treatment of incompressibility and boundary conditions in 3-D numerical spectral simulations of plane channel flows. In: Hirschel EH (ed.), Third GAMM Conference Numerical Methods in Fluid Mechanics. Vieweg, Braunschweigh, 1980, pp. 165–173.
[17] Chorin A J. On the convergence of discrete approximation to the Navier–Stokes equations. Math. Comp., 1969, 23:341–353.
[18] Teman R. Navier–Stokes Equations Theory and Numerical Analysis, 3rd ed. North-Holland, Amsterdam, 1987, pp. 395–426.
[19] Kim J, Moin P and Moser R. Turbulence statistics in fully developed channel flow at low Reynolds number. J. Fluid Mech., 1987, 177:133–166.
[20] Kim J and Moin P. Application of a fractional step method to incompressible Navier–Stokes equations. J. Comput. Phys, 1985, 59:308–323.
[21] Choi H, Moin P and Kim J. Active turbulence control for drag reduction in wall-bounded flows. J. Fluid Mech., 1994, 262:75–110.
[22] LeVeque R L and Oliger J. Numerical Analysis Project. Manuscript NA-81-16, Computer Science Department, Stanford University, Stanford, CA, 1981.

the pressure [illegible] adjusting errors in the streamwise and spanwise directions are [illegible] by the spectral truncation method referred to as 3/2 rule.

References

[1] [illegible] A First Course in Turbulence, [illegible], 1972.
[2] [illegible] Press, [illegible].
[3] [illegible] University Press, Cambridge, [illegible].
[4] [illegible] F. and [illegible] R. On the wall [illegible] Fluid Mech. [illegible]
[5] [illegible] structures associated with the [illegible] Fluid Mech. [illegible]
[6] Kim [illegible] Phys. Fluids, [illegible]
[7] [illegible] structures associated with the [illegible] Phys. Fluids [illegible]
[8] [illegible] Phys. [illegible]
[9] [illegible] Reynolds stress in [illegible] Fluid [illegible]
[10] [illegible] Transform and its Applications, [illegible], New York, [illegible]
[11] [illegible]
[12] [illegible] Methods in Fluid Dynamics, Springer-Verlag, [illegible]
[13] [illegible] Methods for [illegible] Dover Publications, [illegible]
[14] [illegible]
[15] [illegible] Methods, [illegible] Press, 1984.
[16] [illegible]
[17] [illegible]
[18] [illegible]
[19] [illegible]
[20] [illegible]
[21] [illegible]
[22] [illegible] Numerical analysis [illegible] Computer Science Department, Stanford University, Stanford, CA, 198[illegible]

2

Wall Turbulence and Its Coherent Structure

Since the intermittent of the shear turbulence has been found in 1950s, it is recognized that there exists a large-scale coherent structure, except for the small-scale random motion, for the turbulent vortex motions. In the 1960–1970s, with the developments and applications of the flow visualization technique and the hot-line measurement, the coherent structure of the turbulent boundary layer was further studied. It is found that the coherent structure exhibits the distinctive properties in the viscous sublayer, buffer layer, and outer layer of the turbulent boundary layer. Statistically, the variance of the structure is significantly larger than the mean of the structure, this leads to the measurability of the coherent motion. Therefore, the physical understanding of the shear turbulence, especially wall turbulence, was greatly deepened. After 1980s, with the help of the direct numerical simulation (DNS), the coherent structure of the turbulent boundary layer was further identified.

The coherent structure of wall turbulence has a significant influence on the generation of skin-friction drag and heat and mass transfers of the fluid, therefore, understanding the coherent structure of wall turbulence is the necessary prerequisite to control the fluid motion in the boundary layer.

As a fundamental, boundary layer flow associated with wall turbulence and flow instability is briefly introduced in Section 2.1. The instability of fluid motion comes from the external disturbance (refer to receptivity). When a small disturbance is added, the flow instability can lead to the formation of wall turbulence and its coherent structure; this process is called as the transition from laminar flow to turbulent flow, which undergoes the linear instability stage and nonlinear instability stage in boundary layer flow. The transition to turbulence is closely related to the generation of skin-friction drag in wall turbulence; it is important to understand the process for controlling or delaying the transition. Section 2.2 describes the basic process of the flow transition, that is, the formation of wall turbulence. In this section, the receptivity of the flow transition, the linear instability including transient growth, and the nonlinear instability including a turbulent spot are introduced. Note that the larger amplitude of external disturbance can rapidly and directly lead to the formation of the turbulence, this is called as the "bypass" transition which is also introduced in Section 2.2.

Principles of Turbulence Control, First Edition. Baochun Fan and Gang Dong.

Once wall turbulence is generated, the coherent structure can be sustained by itself in the near-wall region (inner layer, $y^+ < 200$, the superscript + represents wall unit); the self-sustaining of the structure plays an important role in macroscopic properties in wall turbulence, e.g., skin-friction drag. Once the coherent motion and its self-sustaining are understood, we can effectively control the flow behaviors of wall turbulence. Section 2.3 introduces the structure of the turbulent boundary layer and the key elements of the coherent structure (streaky structure and streamwise vortices) of wall turbulence. The formations and evolutions of the structures are given in Section 2.4. Particularly, a coherent motion based on the soliton and its relevant vortex structures is also included in this section. Finally, the bursting and self-sustaining processes, associated with the turbulence production and the generation of skin-friction drag in wall turbulence, are discussed in Section 2.5.

2.1 Boundary Layer Flow and Flow Stability

2.1.1 Boundary Layer Flow

The concept of boundary layer was firstly proposed by Ludwig Prandtl in 1904. He postulated that the effect of fluid viscosity on the flow properties is limited within a thin layer near the wall when a fluid passes through a surface of flat plate, and that the fluid outside the layer is independent of fluid viscosity and regarded as an inviscid potential flow. The inner layer is called as the boundary layer, as shown in Figure 2.1. The uniform free-stream (U_0) passes through the surface of a plate; due to the fluid viscosity, the velocity of the fluid attached on the plate surface is zero, thus, a region (boundary layer) from the plate surface to the free-stream (U_0) can be formed. In the boundary layer of Figure 2.1, the incoming flow velocity along the normal direction of the plate surface varies and forms a velocity gradient; we can use the boundary layer thickness, δ, to characterize the boundary layer. Usually, δ can be defined as the distance from the plate surface at which the viscous flow velocity is 99% of the free-stream velocity (the surface velocity of an inviscid flow).

The boundary layer thickness is dependent of interaction between the inertial forces and viscous forces, that is, dependent on a Reynolds number ($\mathrm{Re} = U_0 L/\nu$, where U_0 is the free-stream velocity, L is the characteristic length of boundary layer, e.g., plate length, and ν is the kinematic viscosity). The larger the Re, the stronger the inertial effect and thus thinner the boundary layer. According to different descriptions of size and thickness of the boundary layer, Reynolds numbers used to characterize the flow include

$$\mathrm{Re}_x = \frac{U_0 x}{\nu} \tag{2.1}$$

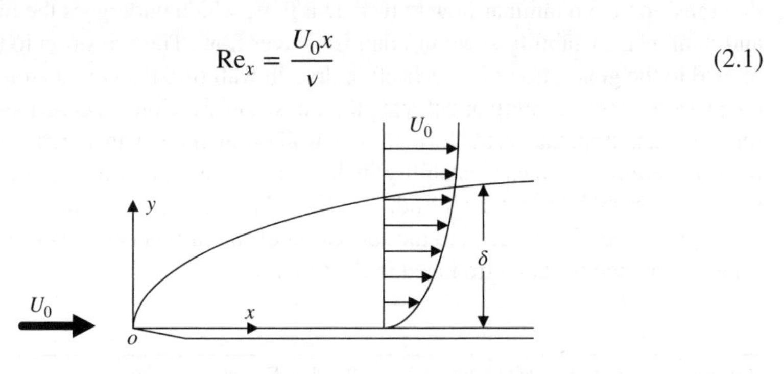

Figure 2.1 Sketch of boundary layer flow

where x is the distance from the leading edge of the flat plate,

$$\mathrm{Re}_\delta = \frac{U_0\delta}{\nu} \tag{2.2}$$

where δ is the thickness of the boundary layer,

$$\mathrm{Re}_{\delta^*} = \frac{U_0\delta^*}{\nu} \tag{2.3}$$

where δ^* is the displacement thickness of the boundary layer, which represents a distance by which the flat plate would have to be displaced in the inviscid case to give the same total mass flow as the viscous case, and

$$\mathrm{Re}_\theta = \frac{U_0\theta}{\nu} \tag{2.4}$$

where θ is the momentum thickness of the boundary layer, which represents a distance by which the flat plate would have to be displaced in the inviscid case to give the same total momentum as the viscous case.

Many types of fluid flow can be regarded as boundary layer flow, such as external flow around the flat plate, channel flow, pipe flow, etc. Figure 2.2 gives the two simplest types of boundary layer flow, that is, Poiseuille flow and Blasius flow.

In the Poiseuille flow, it is assumed that both upper and lower flat plates are motionless, fluid within the plates flows along the x direction, thus, the velocity profile is parabolic, that is,

$$u = -\frac{1}{2\mu}\frac{dp}{dx}\left(h^2 - y^2\right) \tag{2.5}$$

where μ is the dynamic viscosity. Equation (2.5) shows that the velocity profile in Poiseuille flow is associated with the pressure gradient along with the streamwise direction (dp/dx).

The Blasius flow describes the steady two-dimensional boundary layer that forms on a semi-infinite plate which is held parallel to an incoming flow U_0. The velocity profile of the Blasius flow can be obtained by similarity solution of steady, incompressible, two-dimensional boundary layer equations for continuity and momentum.

2.1.2 Flow Stability

External disturbances may change the flow state of a fluid, which is associated with the flow stability. The stability of fluid flow can be defined as the "immunity" ability of the fluid flow

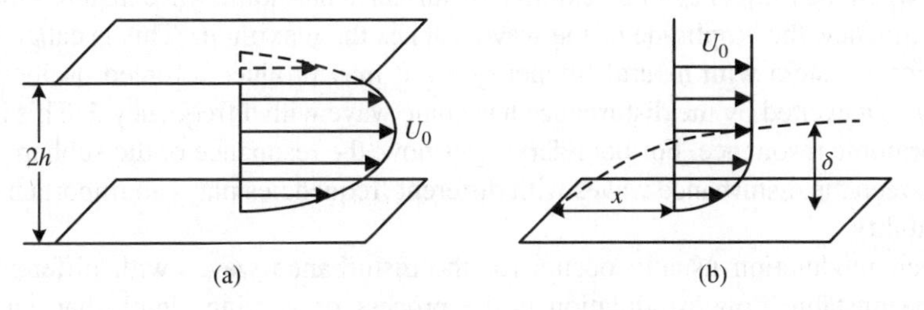

Figure 2.2 (a) Poiseuille flow and (b) Blasius flow

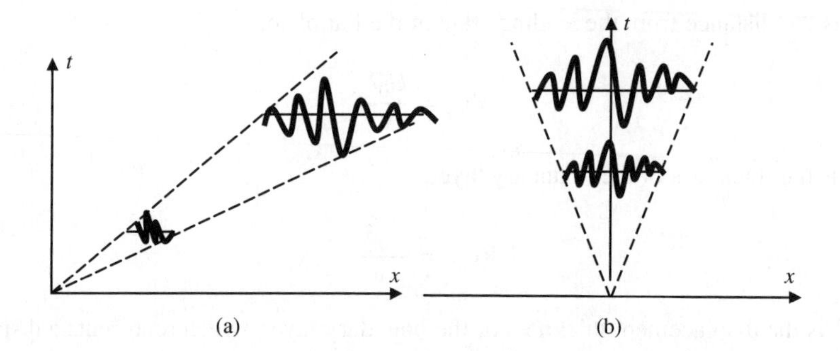

Figure 2.3 (a) Convective instability and (b) absolute instability for fluid flow

state to external or internal disturbances. If flow of a fluid transforms from one state to another state (the transformation may occur either gradually or abruptly), we say that fluid flow is unstable, otherwise fluid flow is stable. The instability process of fluid flow is either temporal or spatial; however, it is important that the unstable flow can develop in both temporal and spatial modes. This development of the instability can be divided into the convective instability and the absolute instability, as shown in Figure 2.3.

If a small amplitude disturbance wave is convected away from the disturbed source location and ultimately leaves the flow undisturbed then the flow is convectively unstable (Figure 2.3a). However, if the disturbed wave exponentially grows in time at every locations of space and flow field cannot return to its initial state after a period of time then the flow is said to be absolutely unstable (Figure 2.3b).

Flow instability relates to the disturbance wave. Various disturbances firstly enter the boundary layer and convert into the disturbance waves in the fluid. Then the disturbance waves are further unstabilized and broken-down through linear and nonlinear growth and finally form turbulence. Theoretically, a disturbance wave can be regarded as the superposition of harmonic (or quasi-harmonic) waves. Also, the disturbance wave is also produced in a harmonic form by a well-controlled experiment.

The simple harmonic motion of the disturbance wave represents a basic and periodic motion. A periodic disturbance wave $f(t)$ can be expressed as the superposition of a series of simple harmonic wave (also known as disturbance harmonic wave). A simple harmonic wave with a fundamental frequency ($n = 1$) is called as fundamental wave, while the disturbance harmonic wave with higher frequency ($n > 1$) is called as nth harmonic wave.

The disturbance harmonic waves with different frequencies can interact with each other. For example, when the frequency of an external disturbance harmonic wave matches the system's natural frequency, the amplitude of the wave reaches the maximum. This is called as the resonance. For a system with natural frequency $n\lambda$, it may produce a forced oscillation with a frequency λ/n excited by the disturbance harmonic wave with a frequency λ. This is called as the subharmonic resonance. For boundary layer flow, the resonance or the subharmonic resonance between the disturbance waves with different frequencies plays an important role in the flow instability.

However, modulation usually occurs for the disturbance waves with different frequencies in the unstable flow. Modulation is the process of varying simple harmonic motion (signal), controlled by another simple harmonic motion (carrier wave). The modulation includes frequency modulation (FM), amplitude modulation (AM), and phase modulation (PM). For the FM process, given a transmitted signal $x_m(t)$ and a carrier wave single $x_c(t) = A\exp(i2\pi f_c t), f_c = 1/T_c$, the frequency modulated signal is

$$x_c(t) = A\exp\left[i2\pi\int_0^t f(\tau)\,d\tau\right] \tag{2.6}$$

where $f(t) = f_c + f_\Delta x_m(t)$ and f_Δ represents the maximum frequency shift of f_c.

For the disturbance harmonic waves in boundary layer flow, the oscillation parameters (frequency, amplitude, and phase) of these waves may vary by the modulation process. The variations of the wave parameters have the significant influence on the flow stability, which is further discussed in the next section.

2.1.3 Linear Stability Theory of Flow

The object of linear stability theory in fluid flow is to obtain the stability condition of the flow excited by a small amplitude disturbance. In the 1900s, Orr and Sommerfeld proposed the two-dimensional linear stability theory. They assumed that the disturbance is infinitesimal, and then the equations governing fluid flow can be linearized by extracting the solutions of base flow and neglecting the nonlinear term. In the linearized equations, if the amplitude of disturbances superimposed on the base flow decay in time, then the base flow is stable, if the amplitude grows to a large level, the base flow is unstable.

For Navier–Stokes (N-S) equations, as shown in Eqs. (1.140) and (1.141), we can decompose the total flow field as the base flow and the disturbance

$$u = U + u', \quad v = V + v', \quad w = W + w', \quad p = P + p' \tag{2.7}$$

where, U, V, W, P are the velocity components and pressure of the base flow, u', v', w', and p' are the velocity components and pressure of the disturbance.

Considered a two-dimensional, incompressible flow, the base flow and the disturbance are

$$U = U(y), \quad V = W = 0, \quad P = P(x, y) \tag{2.8}$$

and

$$u' = u'(x, y, t), v' = v'(x, y, t), w' = 0, p' = p'(x, y, t) \tag{2.9}$$

then Eq. (2.7) can be expressed as

$$u = U + u', v = v', p = P + p' \tag{2.10}$$

Inserting Eqs. (2.10) into the N-S equations, we have

$$\frac{\partial u'}{\partial x} + \frac{\partial v'}{\partial y} = 0 \tag{2.11}$$

$$\frac{\partial u'}{\partial t} + (U + u')\frac{\partial u'}{\partial x} + v'\frac{\partial (U + u')}{\partial y} = -\frac{\partial (P + p')}{\partial x} + \nu\left[\frac{\partial^2 u'}{\partial x^2} + \frac{\partial^2 (U + u')}{\partial y^2}\right] \tag{2.12a}$$

$$\frac{\partial v'}{\partial t} + (U + u')\frac{\partial v'}{\partial x} + v'\frac{\partial v'}{\partial y} = -\frac{\partial (P + p')}{\partial y} + \nu\left(\frac{\partial^2 v'}{\partial x^2} + \frac{\partial^2 v'}{\partial y^2}\right) \tag{2.12b}$$

After subtracting the N-S equations for the base flow and neglecting the quadratic terms in the disturbance variable, one obtains

$$\frac{\partial u'}{\partial t} + U\frac{\partial u'}{\partial x} + v'\frac{\partial U}{\partial y} = -\frac{\partial p'}{\partial x} + \nu\left(\frac{\partial^2 u'}{\partial x^2} + \frac{\partial^2 u'}{\partial y^2}\right) \tag{2.13a}$$

$$\frac{\partial v'}{\partial t} + U\frac{\partial v'}{\partial x} = -\frac{\partial p'}{\partial y} + \nu\left(\frac{\partial^2 v'}{\partial x^2} + \frac{\partial^2 v'}{\partial y^2}\right) \tag{2.13b}$$

Because Eq. (2.13) are linear and the solutions can be superimposed, the small disturbance can be decomposed in Fourier series. Each term of the disturbance in Fourier series represents a fluctuation that is regarded as a wave propagating along the x direction. We introduce the disturbance stream function $\psi(x, y, t)$, then the single fluctuation of disturbance can be expressed as

$$\psi(x, y, t) = \phi(y)e^{i(\alpha x - \omega t)} \tag{2.14}$$

where $\phi(y)$ is the amplitude of the disturbance that consists of real part ϕ_r and imaginary part ϕ_i, $\phi = \phi_r + i\phi_i$. α and ω are the complex numbers. α represents the disturbance parameter growing in space, $\alpha = \alpha_r + i\alpha_i$, where α_r is the wave number of the disturbance in the x direction, α_i is the growth or decay factor of the disturbance amplitude in the x direction. ω corresponds to the disturbance parameter growing in time, $\omega = \omega_r + i\omega_i$, where ω_r is the oscillation frequency of the disturbance and ω_i is the disturbance parameter varying in time.

Appling the relationship between the velocity and the stream function to Eq. (2.14), we have

$$u' = \frac{\partial \psi}{\partial y} = \phi'(y)e^{i(\alpha x - \omega t)} \tag{2.15a}$$

$$v' = -\frac{\partial \psi}{\partial x} = -i\alpha\phi(y)e^{i(\alpha x - \omega t)} \tag{2.15b}$$

Equations (2.13a) and (2.13b), derivations of x and y, respectively, can eliminate term p', and then inserting Eq. (2.15) yields a forth-order differential equation:

$$(U - c)(\phi'' - \alpha^2\phi) - U''\phi = -\frac{i}{\alpha \mathrm{Re}}(\phi^{(4)} - 2\alpha^2\phi'' + \alpha^4\phi) \tag{2.16}$$

where $c = \alpha/\omega$, $\mathrm{Re} = 1/\nu$, and ν is the kinematic viscosity.

Equation (2.16) is the basic equation of linear stability for fluid flow, also called Orr–Sommerfeld equation (OSE). The boundary conditions of OSE can be given as

$$\begin{aligned} \phi(y_1) = \phi'(y_1) = 0 \quad y = y_1 \\ \phi(y_2) = \phi'(y_2) = 0 \quad y = y_2 \end{aligned} \tag{2.17}$$

Thus, the linear stability problem for a two-dimensional laminar flow becomes the eigenvalue problem for solving Eq. (2.16) and its boundary conditions (2.17), that is, for the parallel mean flow, the nonzero solution of small disturbance can be obtained only if the eigenvalue relationship between the parameters $\alpha, c(\text{or}\,\omega)$, and Re must be satisfied. The nonzero solution represents the disturbance harmonic waves. Tollmien (in 1929) and Schlichting (in 1933) independently obtained the solution that is named as Tollmien–Schlichting (TS) wave.

2.2 Transition of Boundary Layer Flow

Fluid flow includes laminar flow and turbulent flow. When the velocity of fluid flow is small, fluid flows in parallel layers with no disruption between the layers. This is called laminar flow. However, when the velocity of fluid flow is high, the transfers of mass, momentum, and energy between the layers occur, and the vortical motion arises in fluid with the variations of velocity and pressure in time and space. This is called turbulent flow. After Osborne Reynolds' experiment on fluid dynamics in a pipe in 1883, it was found that when the Reynolds number of fluid flow gradually increases, the laminar fluid initially flows in wavy form and then becomes the turbulence with the rapid increases of amplitude and frequency of the wavy motion. The process is called transition. The transition of boundary layer flow exists widely in natural and man-made phenomena, such as, flows on the surface of various vehicles and blades. This transition can greatly promote the formation of skin-friction drag, noise, and the vibration of the body; therefore, it is significantly important to study and control the transition in boundary layer flow. For example, if we can postpone the transition from laminar flow to turbulent flow or maintain the laminar state on the surface of the vehicles using a certain flow control method, skin-friction drag on the surface of vehicles will be greatly decreased and therefore leads to the increases of the movement speed and the energy-saving efficiency of vehicles.

From the point of view of physical mechanism, the transition from the laminar to turbulent state is a complicated process, in which some key segments are still open. Up to now, no general theories or mathematical formulae can be used to describe the whole transition process. However, we can confirm that the transition is closely related to the flow instability and depends on the flow type and circumstance. The effect of flow instability on the transition in the boundary layer is discussed in this section.

2.2.1 Basic Process

Usually, unstable laminar flow does not become turbulent flow immediately but undergoes a successive process that depends on the Reynolds number. The fluid maintains the laminar state for a small Reynolds number flow. When the Reynolds number increases, the instantaneous velocity of the flow varies in time and space, and the flow sometimes fluctuates and sometimes decays. After the Reynolds number reaches a certain value, the flow becomes a fully developed turbulent state. We can define two Reynolds numbers, Re_{cr} and Re_{tr}, the former is defined as the critical Reynolds number when the laminar flow just begins to be unstable and the latter is defined as the transitional Reynolds number when the fully developed turbulence just forms. Usually, the range of Re_{cr} is 2000–2300, while Re_{tr} is larger than Re_{cr} and depends on the flow state and the circumstance.

Maybe the first complete study on the transition of boundary layer flow was performed in 1947[1]; the study laid the experimental foundation of the flow instability. After that, a large number of studies have addressed to the transition of boundary layer flow. It has been well known that the transition can be completed through the different paths that are dependent of the initial disturbance amplitude. These possible paths of the transition to turbulence are shown in Figure 2.4. Along with the increase of the amplitude of forcing environmental disturbance, the transition path changes from A to E. Path A represents a regular transition path (also called a natural transition) with a small amplitude of initial external disturbance. Figure 2.5 shows the physical picture of path A transition from laminar to turbulent flow. In this path, the

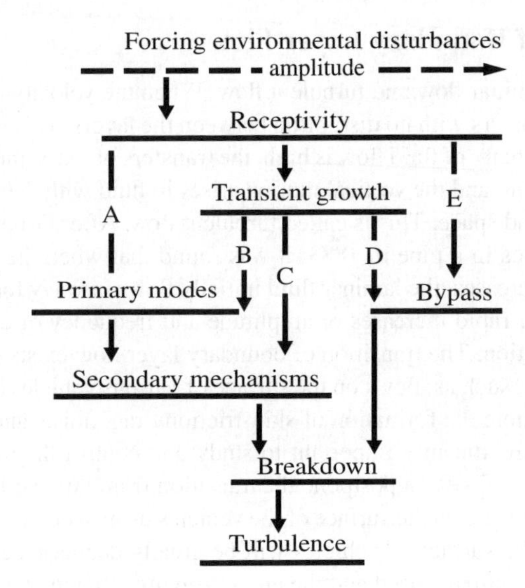

Figure 2.4 The paths of a transition from receptivity to turbulence.[2] *Source*: SaricWS 2002. Reproduced with permission of Annual Reviews

external disturbance firstly enters into the fluid (corresponding to receptivity) and forms the two-dimensional linear disturbance waves (TS waves) propagating toward right. The amplitude of the TS waves linearly grows during the linear stage (corresponding to the primary mode in Figure 2.4). When the wave amplitude grows to a certain level, a nonlinear development occurs. The TS waves gradually develop into the three-dimensional hairpin vortex (also called Λ vortex) structure. This process is completed through the secondary instability mechanisms. Further, local regions with nonlinear fluctuation (turbulent spot) appear in the flow field, and finally, these regions unite and lead to the fully developed turbulent flow. Path E in Figure 2.4 represents a bypass transition to turbulence. In this path, an initial disturbance with enough large amplitude from external free-stream turbulence (FST) enters the fluid in receptivity stage and then rapidly leads to the turbulence without undergoing the primary and secondary modes of path A. This means that a regular path of transition is bypassed and hence path E is termed as the bypass transition. Paths B, C, and D fall somewhere between the regular and bypass transitions. A common feature of three paths is that a transient growth process plays a role in early stage of a transition, although the late transition may develop in the different paths that are depended on the amplitude of the initial disturbance.

2.2.2 *Receptivity Stage*

The receptivity of flow is the first stage of the transition to turbulence. The aim of studying the stage is to understand what the external origin for the formation of the disturbances in boundary layer flow is and how the external disturbances transform into the disturbance waves of fluid. Morkovin[4] firstly proposed the abovementioned questions which now is called receptivity problem.

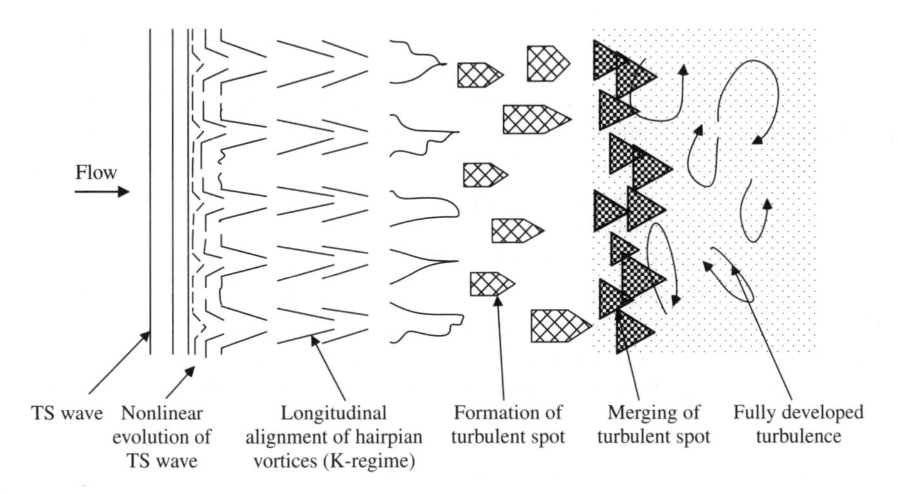

Figure 2.5 Regular transition of boundary layer flow.[3] *Source*: Alfredsson P H 1996. Reproduced with permission of Springer

The external disturbances to the boundary layer involve various factors and events that enter into the fluid and then lead to the flow instability. These external disturbances include acoustics/vorticity/turbulence of free stream, pressure gradient, wall roughness, wall temperature, vibration, suction, etc., and can be imposed artificially or naturally.

From the viewpoint of disturbance type of free stream, the receptivity can be divided into acoustics and vorticity. The former corresponds the receptivity of flow to the external acoustical wave (e.g., noise), the latter corresponds the receptivity to the vortex structure and turbulence of the free stream.

From the viewpoint of excited location of the external disturbances, the receptivity can be divided into the leading-edge and the local receptivities. In the leading-edge receptivity, the local flow varies when incoming flow meets the leading edge of a flat plate, and the variation transforms the external disturbance into the internal one. The receptive location includes the leading edge and the limited region downstream the leading edge. In the latter case, the flow varies at the downstream location far from the leading edge. The surface roughness and the abruptness (e.g., bulge, gap, ditch, etc.) of the flat plate play an important role in the local variation of the flow.

In addition, from the viewpoint of the exerted method of external disturbances, the receptivity can be divided into artificial and natural methods. For artificial receptivity, some man-made approaches, such as vibrating bar, suction, and heating of the walling, are used to produce the disturbance to the boundary layer. For the natural receptivity, an environmental factor, such as, acoustical noise, vortices/turbulence of the free stream, excites the disturbance.

2.2.2.1 Acoustics Receptivity

The acoustical wave with a given wavelength and frequency cannot be transformed into the disturbance wave immediately when it enters the boundary layer, this is because the properties of wavelength and frequency of the acoustical wave are different from the natural properties of the disturbance wave. When the flow velocity of the boundary layer is small, the propagating speed of the acoustical wave is larger than that of flow; the wavelength of the acoustical wave is also larger than that of disturbance wave, thus, the wavelength undergoes a decreasing process. Generally, the external disturbance including the acoustical wave firstly affects the locations where the local flow varies rapidly, and then forms the disturbance wave by inducing local periodical variations of flow in the boundary layer. The common local characteristics include the leading edge of the flat plate and the surface roughness.

The amplitude of the disturbance wave (e.g., TS wave) induced by the acoustical wave at the leading edge of the flat plate firstly grows and then decays along the flow direction, the process continues until the flow instability occurs (corresponding to the lower branch of neutral curve in linear stability theory, see Section 2.2.3). A ratio of amplitude of the disturbance wave in the unstable flow, $|u'_{TS}|_{I}$, to the amplitude of the acoustical wave of the free stream, $|u'_{ac}|_{LE}$, can be used to define the receptive coefficient of the leading edge:

$$K_I = |u'_{TS}|_I / |u'_{ac}|_{LE} \tag{2.18}$$

In Eq. (2.18), the subscripts I, TS, ac, and LE represent the lower branch of the neural curve, TS wave, acoustical wave, and leading edge of the flat plate, respectively. Figure 2.6 gives the variations of the receptive coefficient of the leading edge along with the frequency of the disturbance wave F, where F is the dimensionless frequency, $F = 2\pi f \nu / U_0{}^2 \times 10^6$ (f is the frequency of the disturbance wave, ν is the fluid viscosity, and U_0 is the velocity of incoming flow). Figure 2.6 also shows that different shapes of the leading edge have a different receptivity to the acoustical wave.

The roughness or local bulge of the flat plate in the downstream region can lead to the local receptivity to the acoustical wave. For example, the TS wave excited by the acoustical wave can be measured by setting a small bulge object at the downstream region of the flat plate. When the bulge object is higher than the viscous sublayer of the boundary layer, the receptivity can show the linear relationship with the amplitude of the acoustical wave and the height of the bulge object.[6] An asymptotic analysis based on the triple-deck structure theory[7] shows that the transient modulation between the acoustical wave and the local variation of the mean flow induced by the bulge object within the boundary layer leads to the formation of the TS wave.

2.2.2.2 Vorticity Receptivity

Unlike the acoustical disturbance, the propagating speed of the external vorticity disturbance is comparable to that of the free stream, the wavelength of the vorticity disturbance is usually three times that of the internal disturbance wave (TS wave) in the boundary layer; hence, the vorticity disturbance cannot be directly transformed into the internal disturbance wave. Therefore, an interaction of the vorticity disturbance with the rapid variation of the mean flow is needed to reduce the wavelength in generation of the internal disturbance wave that satisfies the dispersion relationship of the TS wave in the boundary layer.

The external vorticity disturbance takes effect mainly in the upper part of the boundary layer, since the disturbance can rapidly decay in the viscous sublayer of the boundary layer. The mechanism is completely different from that of the acoustical receptivity. The acoustical receptivity is due to the interaction between the acoustical wave and the rapid variation of the mean flow within the viscous sublayer, while the vorticity receptivity is due to the interaction

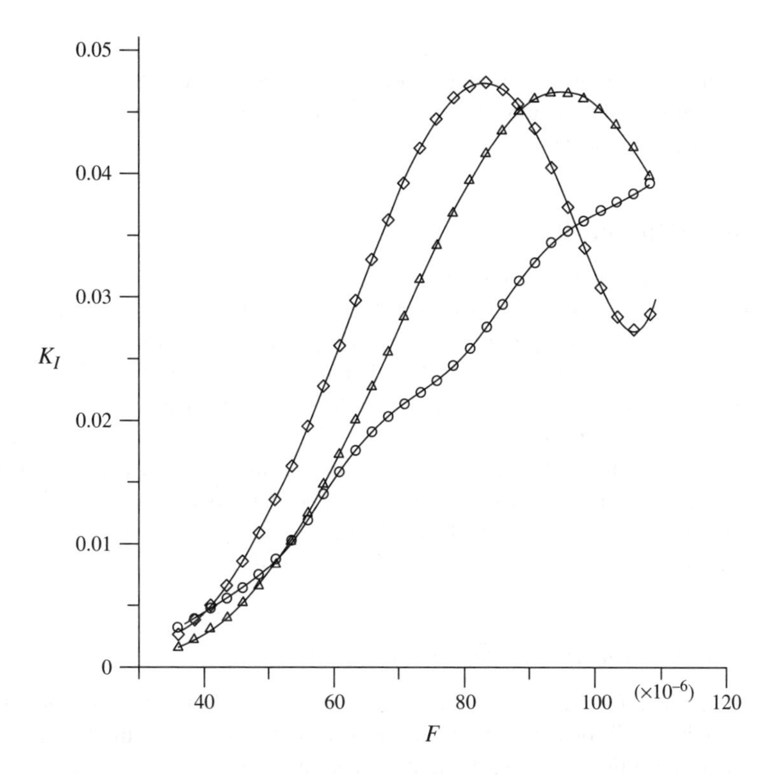

Figure 2.6 Receptive coefficient versus dimensionless frequency of the disturbance wave, the different symbols represent the different shapes of leading edge of the flat plate.[5] *Source*: Alfredsson P H 1996. Reproduced with permission of Springer

between the vortices and the local variation of the mean flow with a small scale within the upper layer of the boundary layer.[8] The theoretical studies suggested that the vorticity receptivity is one third of the acoustical receptivity at the same geometry of the boundary layer and the same free-stream conditions.

In a sense, the vorticity disturbance can be regarded as the one of the representations of FST; one can use turbulence intensity of the incoming flow, Tu, to characterize the strength of the vorticity disturbance in the free stream:

$$Tu = \frac{\sqrt{\frac{1}{3}(\overline{u'^2} + \overline{v'^2} + \overline{w_{\prime}^2})}}{U_0} \tag{2.19}$$

where u', v', and w' are the velocity fluctuation components of the incoming flow.

Increasing the Tu can enhance the amplitude of the disturbance wave. Kendall[9] experimentally showed that the three type motions can be excited by FST in the boundary layer: (1) vortex stretching motion; (2) vibration of fluid with slow growth of the amplitude at the outer layer of the boundary layer; and (3) TS wave with the rapid growth of the amplitude. The results are the same as the results of the vorticity disturbance. Yang and Voke[10] suggested that the TS wave excited by the FST is the result from the normal velocity fluctuation v'. With the action of the leading edge of the flat plate, the normal velocity fluctuation can transform into the streamwise velocity fluctuation and further lead to the formation of the local disturbance vorticity in the boundary layer.

2.2.2.3 Artificial Receptivity

Both the acoustical and vorticity receptivities mentioned above can be taken as the natural receptivity. However, artificial approaches, such as vibrating bar, suction, wall heating or cooling, can also lead to the receptivity to the disturbance. For the natural receptivity, the transformation from the external disturbance to the internal disturbance must satisfy the wavelength criterion; however, the artificial excitation can produce the disturbances with a wide range of wave number in the boundary layer, including the TS wave that satisfies the dispersion relationship of the disturbance. Therefore, it is easy for the artificial receptivity to produce the disturbance. In the experiments, the artificial approaches have been widely used to excite the disturbance wave within the boundary layer in order to study the receptivity, the transition from laminar to turbulent flow, and the approaches controlling flow.

2.2.3 *Linear Instability and Transient Growth*

Laminar fluid can be affected by surrounding circumstance and thus produce the disturbance wave when certain conditions are satisfied (that is, receptivity). When the Reynolds number is enough small, the fluid can resist the disturbance by itself (or the fluid is immune to the disturbance) and keeps laminar. After the Reynolds number exceeds the critical level, the fluid will lose immunity to the disturbance and become unstable.

Linear unstable flow is the primary stage of the fluid motion, which can be analyzed using the linear stability theory described in Section 2.1.3. Different eigenvalues of Eq. (2.14) determine different stabilities of the disturbance wave. When α is a real number and ω is an imaginary number, Eq. (2.14) characterizes the spatially periodic disturbance whose amplitude varies along with time. This is called the temporally developing mode of the disturbance. In this case, c in Eq. (2.16) is also expressed as

$$c = \frac{\omega}{\alpha} = c_r + ic_i \tag{2.20}$$

where c_r is the propagating velocity of the disturbance wave in the x direction and is called the phase velocity; c_i is the growing or decaying coefficient of the disturbance and is called the developing coefficient. If $c_i < 0$ (that is, $\omega_i < 0$), the disturbance decays in time and the flow is stable, while if $c_i > 0$ (that is, $\omega_i > 0$), the disturbance grows in time and the flow is unstable. Therefore, the OSE described by Eqs. (2.16) and (2.17) actually represents the linear stability analysis in time.

When α is an imaginary number and ω is a real number, Eq. (2.14) represents the case that the disturbance varies along the x direction with a given disturbance frequency. This is called the spatially developing mode of the disturbance. When $\alpha_i < 0$, the disturbance grows in the x direction, and when $\alpha_i > 0$, the disturbance decays in the x direction.

When both α and ω are real numbers, Eq. (2.14) means that the amplitude and frequency of the disturbance vary neither in time nor in space. In this case, the disturbance is neutral.

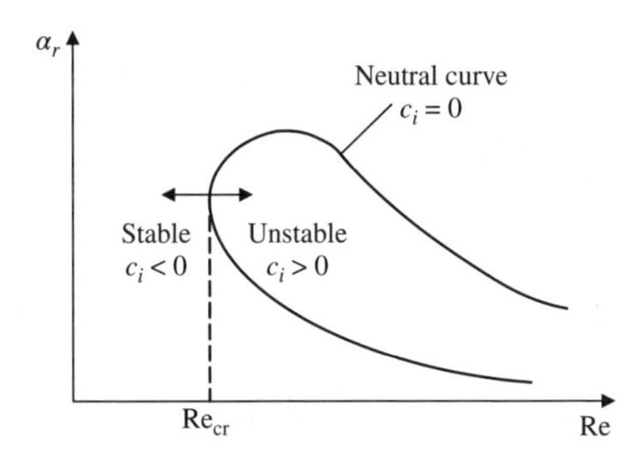

Figure 2.7 Neutral curve of flow for a temporally developing mode

Solving differential equations of the disturbance can be used to find the critical condition of the flow instability. Usually, one can obtain solutions of the eigenvalue with the given $U(y)$, that is, obtain the eigenvalue relationship, $f(\alpha, \omega, \text{Re}) = 0$. For a temporally developing mode, given the Reynolds number of the mean flow and the wave number α_r of the disturbance, one can solve Eq. (2.16) with the boundary condition (2.17) and yield eigenfunction $\phi(y)$ and complex number $c_r + ic_i$. Figure 2.7 shows the schematic of the neutral curve for fluid flow. Each point determined by a couple of α and Re corresponds a pair of c_r and c_i. In Figure 2.7, only the curve for $c_i = 0$ is drawn. The curve separates the stable region (disturbance decays, $c_i < 0$) from the unstable region (disturbance grows, $c_i > 0$). Plotting a tangent to the curve that parallels to the vertical coordinate, the Reynolds number corresponding to point of the tangent is the minimum Reynolds number on the neutral curve, that is, the critical Reynolds number, Re_{cr}, or, stable limit. When a Reynolds number of the flow is smaller than Re_{cr}, flow is stable and the disturbance decays in time, while when a Reynolds number of flow is larger than Re_{cr}, the flow may be damped, neutral, or growing. Based on the point corresponding to Re_{cr} on the neutral curve, the curve can be divided into two parts, the bottom part is called the bottom branch curve (also called I branch curve) and the upper part is called the upper branch curve (also called II branch curve).

It can be proved that there exists inflection point for mean velocity profile $U(y)$ in the boundary layer when instability occurs for laminar flow with neglecting the viscosity. This inviscid instability (also called inflection point instability) takes a role in the streak instability during the evolution of the coherent structure in wall turbulence and will be further discussed in Section 2.4. Note that the flow instability can also occur for viscous flow when no inflection point is present in the mean velocity profile. This is due to the disturbance amplification by fluid viscosity.

Aforementioned linear stability analysis is based on the linearized N-S equations. For the temporally developing mode of the disturbance, a positive or negative sign of ω_i determines the damped ($\omega_i < 0$), growing ($\omega_i > 0$), or neutral ($\omega_i = 0$) state of the disturbance development in time. In this analysis, the mode of eigenvalue (eigenvector) is orthogonal to each other; thus, when these eigenvectors decay in time ($\omega_i < 0$), the resultant vector also decays. The operator of linearized N-S equations, however, is intrinsically nonorthogonal; this means that although each mode of the disturbance decays in time, the superposition of the modes may initially grow followed by exponential decay. The phenomenon is called transient growth

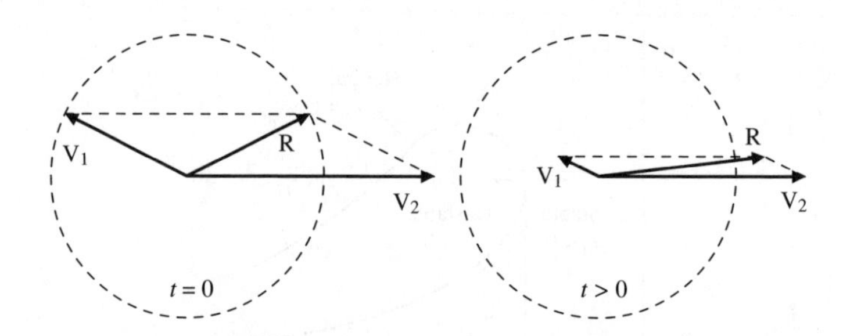

Figure 2.8 Transient growth of the resultant (R) of two nonorthogonal vectors (V_1 and V_2)

(also called algebraic growth due to the growth in algebraic form) and can be simply illustrated in Figure 2.8 in geometrical schematic.

Transient growth phenomenon was firstly noted by Ellingsen and Palm in 1975[11] and was later investigated by other researchers in earlier studies (e.g., Refs. [12, 13]). Butler and Farrell[14] proposed a method for determining optimal disturbance parameters for maximum transient growth in plane Couette and Poiseuille flows and Blasius boundary layer flows. The determination of optimal disturbance has been further extended by many studies (e.g., Refs. [15–18]).

Mathematically, the transient growth is an initial problem of a dynamic system. If an appropriate initial condition can result in the largest energy growth of system, the subsequent growth might be thought as a measure for the risk of transition and breakdown mechanisms of fluid flow. Thus, we can define a maximum growth rate of initial disturbance in terms of relative changes in energy at time t:

$$G(t) = \max_{u_0} \frac{\|u(t)\|^2}{\|u_0\|^2} = \max_{u_0} \frac{\langle u(t), u(t) \rangle}{\langle u_0, u_0 \rangle} \tag{2.21}$$

Here, u_0 denotes the initial conditions and $u(t)$ denotes the subsequent flow at time t. For the linearized N-S equations, we defined an operator $\mathcal{H}(t)$; thus, the $u(t)$ for a dynamic system can be expressed as follows:

$$u(t) = \mathcal{H}(t)u_0 \tag{2.22}$$

Substituting Eq. (2.22) into Eq. (2.21), we obtain

$$G(t) = \max_{u_0} \frac{\langle \mathcal{H}(t)u_0, \mathcal{H}(t)u_0 \rangle}{\langle u_0, u_0 \rangle} = \max_{u_0} \frac{\langle \mathcal{H}(t)^*\mathcal{H}(t)u_0, u_0 \rangle}{\langle u_0, u_0 \rangle} \tag{2.23}$$

where $\mathcal{H}(t)^*$ is the adjoint operator of $\mathcal{H}(t)$. Computing $G(t)$ numerically transforms the problem to that of finding the maximum eigenvalue, $\lambda_{\max}$, of $\mathcal{H}(t)^*\mathcal{H}(t)$, that is,

$$G(t) = \max_{u_0} \frac{\langle \lambda_{\max} u_0, u_0 \rangle}{\langle u_0, u_0 \rangle} = \lambda_{\max} \tag{2.24}$$

The details of the numerical procedure can refer to Schmid and Henningson.[18]

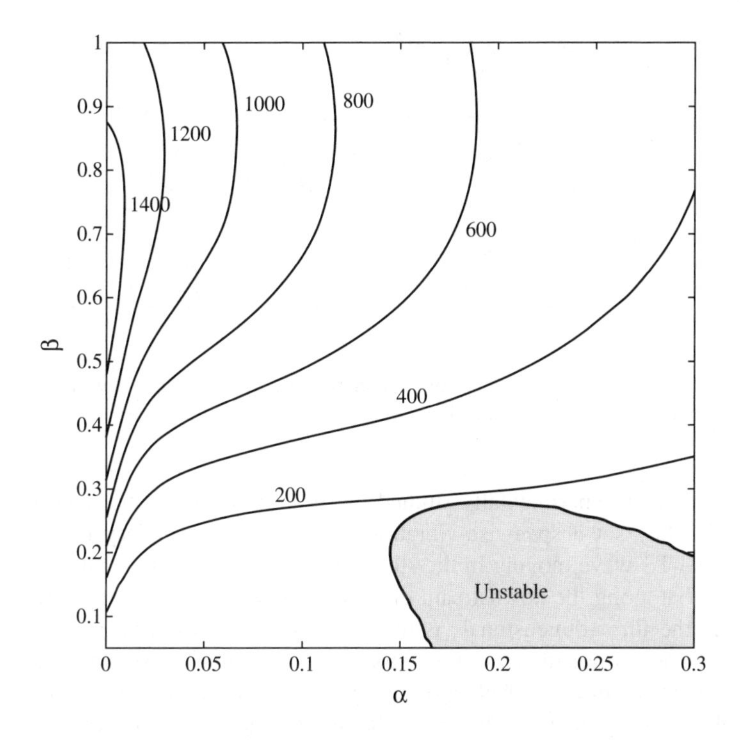

Figure 2.9 Maximum transient energy growth for the Blasius boundary layer in wave number space, $\mathbf{Re}_{\delta^*} = \mathbf{1000}$[17]

For an initial disturbance with form of $u_0 = \hat{u}_0 e^{i(\alpha x + \beta z)}$, the $G(t)$ can be computed for each pair of streamwise and spanwise wave numbers (α, β) at all times, as shown in Figure 2.9. For this transient growth map of Blasius boundary layer flow, maximum $G(t)$ falls into the location of zero of streamwise number that implies the worst initial disturbance corresponds to streamwise elongated vortices. The vortices induce the streamwise streaks with alternating high and low streamwise velocity (also called lift-up mechanism[12]). In some cases, the transient growth of streaks can be large enough to trigger nonlinear interactions and a subsequent transition. This is especially true in the bypass transition that will be discussed in Section 2.2.5.

2.2.4 Nonlinear Instability and Turbulent Spot

When the amplitude of the disturbance wave grows to a certain level, e.g., the velocity of disturbance is larger than 1% of the incoming velocity, the transition will show a nonlinear process. In such process, the three-dimensional disturbance appears and develops periodically along the spanwise direction of flat-plate boundary layer. Simultaneously, the two-dimensional TS wave also varies periodically in the spanwise direction. Due to the complexity of the nonlinear instability, it is difficult to describe the entire process theoretically. However, a better understanding to the nonlinear instability is possible, with the help of the progress in experimental techniques and theoretical analyses.

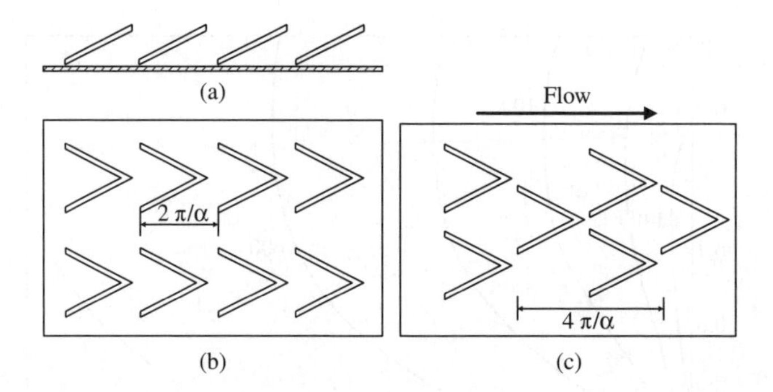

Figure 2.10 Alignments of three-dimensional Λ-shape vortices. (a) and (b) are the K-regime of the transition, (c) is the N-regime of the transition

Klebanoff *et al.*[19] experimentally studied the flow transition in the flat-plate boundary layer in 1962. They set a spanwise-vibrating ribbon in the boundary layer to excite the two-dimensional TS wave moving in the streamwise direction. In this classical experiment, they observed that along the downstream direction the two-dimensional TS wave gradually translates into the three-dimensional Λ-shape vortex structures whose head tilts upward. Their experiment also found that these Λ-shape vortices align in the streamwise direction and that the streamwise-disturbed wavelength of the vortices is the same as that of the TS wave (fundamental wave), as shown in Figure 2.10a and b. Later experiments further confirmed this transition type which is called the K-regime of the transition (after Klebanoff *et al.*[19]). In 1968, Knapp and Roache[20] found by a smoke visualization technique that another type of transition exists, in which the Λ-shape vortices align staggered in the streamwise direction, as shown in Figure 2.10c. In this type, the streamwise-disturbed wavelength of the vortices is twice that of the TS wave. The alignment of the Λ-shape vortices involved in subharmonic resonance of the disturbance waves with low-frequency and therefore the type of transition is called the subharmonic transition, also called the N-regime (New type) or H-regime (after Herbert[21]) of transition. The aforementioned studies suggested that the two-dimensional TS waves induce the three-dimensional, nonlinear disturbances that dominate the later evolutions of the flow transition. Figure 2.11 shows the Λ-shape vortices patterns of two types of transitions by DNS computations.[22]

The K-regime of the transition usually occurs when initial amplitude of the TS wave is enough large; it is a consequence of development of the TS wave with a fundamental frequency f_1 induced by variation of local mean flow. The induction can promote the spanwise modulations of the disturbance wave with a frequency of an integral number of fundamental frequency, and thus it leads to the growth of disturbance amplitude along the streamwise direction. Compared with the K-transition, the subharmonic transition is understood more clearly, although discovery of the latter is later than that of the former. Many studies showed that the subharmonic transition occurs when initial amplitude of the TS wave is small. The feature of the subharmonic transition is associated with the appearance and the nonlinear effect of the three-dimensional subharmonic disturbance wave. The detailed description and analysis of both transitions can be further referred to Kachanov.[23]

In order to study the nonlinear instability of fluid flow, some theoretical approaches were proposed to reveal the mechanisms and conditions of nonlinear developments of

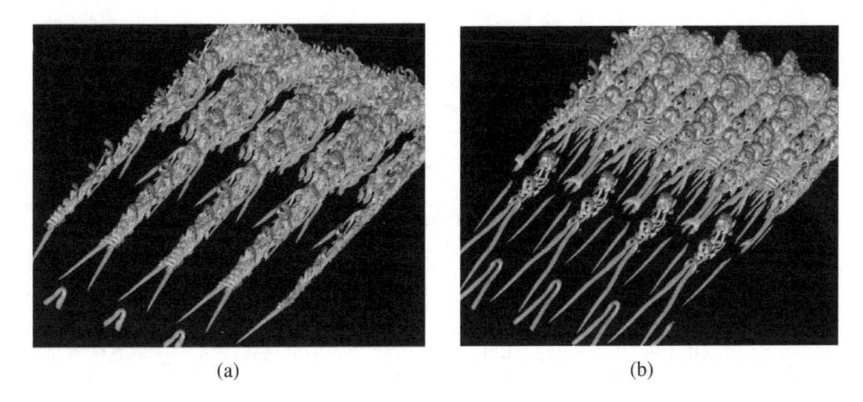

(a) (b)

Figure 2.11 The Λ-shape vortices patterns of (a) K-regime and (b) N-regime of the transition.[22] The flow is from the lower left to the upper right. *Source*: Sayadi T 2012. Reproduced with permission of AIP Publishing LLC

three-dimensional disturbance. The first approach was the weak nonlinear instability theory which was given by Landau,[24] a Russian physicist, in 1944. Later, some more reasonable theories, such as, second instability theory,[21, 25, 26] resonant triad model,[27] etc. were proposed to interpret the mechanism of nonlinear instability.

The later stage of the nonlinear transition in the boundary layer is the appearance of the turbulent spot in local flow field. The turbulent spots can evolve and merge with each other, and finally develop into turbulent flow in the entire flow field, accompanied by the turbulent intermittent. In the flow downstream, flow sometimes is turbulent and sometimes is laminar, the alternation of the flow state at the same spatial region is called as intermittent, and the local region of the turbulent fluctuation in the laminar region is called as turbulent spot.

The concept of a turbulent spot was first suggested by Emmons in 1953.[28] He found in his experiment that the violent nonlinear instability in fluid flow does not immediately lead to the turbulence in the whole flow space but forms the local turbulent fluctuation, that is, turbulent spot. Figure 2.12a shows photo of a turbulent spot in the top view; the head of the turbulent spot (right part) shows a rhombus-like or heart-like shape. Generally, the formation

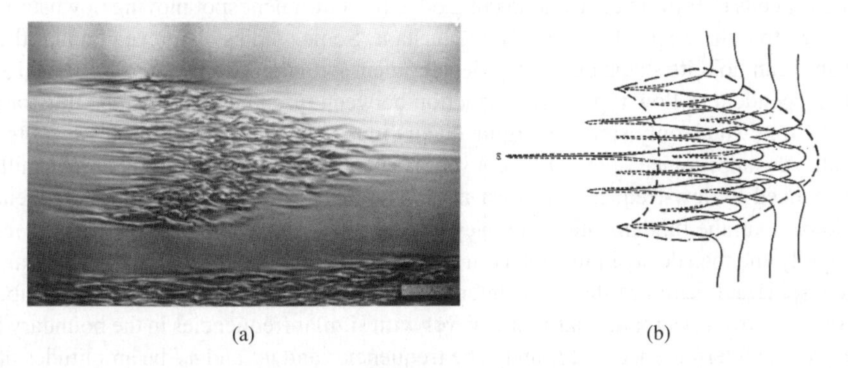

(a) (b)

Figure 2.12 Turbulent spot. (a) Experimental photo in the top view ($Re = 2 \times 10^5$), (b) schematic in the top view.[29] Flow is from left to right

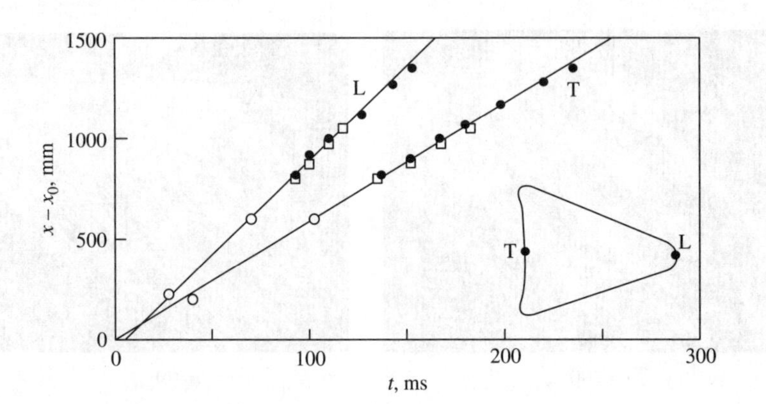

Figure 2.13 Propagating speed of trailing edge (*T*) and leading edge (*L*) of the turbulent spot, $x_0 = 300\,\mathrm{m/s}$, $U_0 = 10\,\mathrm{m/s}$. Open symbols are the experimental data of Grek *et al.*,[31] solid symbols are the experimental data of Wygnanski *et al.*[32]

of the turbulent spot is associated with the flow instability, but there is still lack of a unified description. Kachanov[23] suggested that the necessary condition of the turbulent spot formation is an enough strong temporal modulation of initial amplitude of the disturbance waves. Once the condition is satisfied, the disturbance waves can spontaneously modulate in the spanwise direction and ultimately developed into the local turbulent region with violent disturbances. The study by Delo and Smits[29] revealed that a turbulent spot is formed by the twine of several Λ-shape vortices generated by the nonlinear disturbances (Figure 2.12b). Recently, Lee and Wu[30] presented that the turbulent spot is actually composed of several soliton-like coherent structures (SCS) and their bounded vortices, whose structures will be further discussed in Section 2.4.3.

In the later stage of the transition, the turbulent spot can grow in time and space. Different turbulent spots may merge (instead of modulate) each other and finally lead to a fully developed turbulent flow downstream. For a single turbulent spot, its growth is linear with time, the propagating speed of its trailing edge is apparently smaller than that of its leading edge; however, both speeds are smaller than local streamwise velocity of fluid. Figure 2.13 shows the measured speeds of trailing edge and leading edge for a turbulent spot moving downstream. It can be seen that these speeds are basically constant. Some studies also revealed that the turbulent spot can hold its shape during the development with the leading edge angle of 18–24° that is independent of flow type. The interaction between turbulent spots can be illustrated in Figure 2.14, in which only simple merging occurs in the development of turbulent spots.

Usually, the appearance of the turbulent spot is accompanied by the turbulent intermittence that is induced by low-frequency disturbances. The disturbance waves, whose frequency is much lower than the fundamental frequency and its subharmonic frequency of TS wave, can grow rapidly and then decay immediately in the region of the K-regime of the transition and the turbulent spot, and therefore shows the intermittence in time and space. One can use superposition between two disturbance harmonic waves with similar frequencies in the boundary layer to interpret the intermittence. Let f_1 and f_2 be frequencies and u_1' and u_2' be amplitudes of two waves, the superposed quasi-harmonic wave frequency is $f \sim f_1, f_2$, and the superposed amplitude ranges from maximum $u_1' + u_2'$ to minimum $|u_1' - u_2'|$. If $\Delta f = |f_1 - f_2| \ll f_1$ or f_2, then the

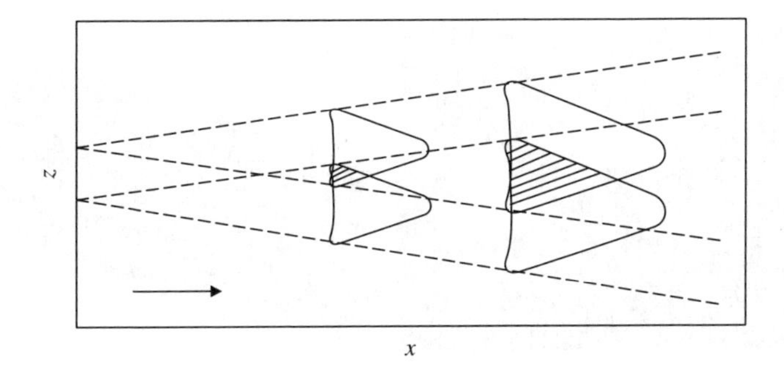

Figure 2.14 Development and merging of two turbulent spots[31]

period of the quasi-harmonic wave is $T_\Delta = 1/\Delta f > T_0 = 1/f$. It means that the quasi-harmonic wave produces an alternating process of wave amplitude with a long period and a frequency of $\Delta f = 1/T_\Delta$. The process can be regarded as the turbulent intermittence.

Based on the concept of the intermittence, an intermittence function can be defined as

$$I(x, y, z, t) = \begin{cases} 0 & \text{laminar} \\ 0 \sim 1 & \text{transition} \\ 1 & \text{turbulent} \end{cases} \tag{2.25}$$

Thus, we can use an intermittence factor to characterize the turbulence intensity in fluid flow:

$$\gamma(x, y, z, t) = \frac{1}{T}\int_0^T I(x, y, z, t)dt \tag{2.26}$$

Equation (2.26) gives the time-averaged value of turbulent intermittence function within the time domain $[0, T]$.

Based on the experimental results, one can obtain the intermittence factor as a function of x (streamwise direction) by the fitting method[33]

$$\gamma = \begin{cases} 1 - \exp\left[-\left(x - x_{tr}\right)^2 n\sigma/U\right] & x > x_{tr} \\ 0 & x < x_{tr} \end{cases} \tag{2.27}$$

where x_{tr} is the streamwise position of the flow transition, n is the growth rate of the turbulent spot, and σ is the propagating speed of the turbulent spot.

Based on the abovementioned discussions, the turbulent intermittence is associated with the turbulent spot. Equation (2.27) is common for different pressure gradients and turbulent intensities of incoming flow, which implies the inherent physical property of the flow transition.

2.2.5 *Bypass Transition*

The bypass transition is a different path from the regular transition, in which the initial stage in the regular transition is bypassed. In this case, external disturbances with large amplitude can

(a) (b)

Figure 2.15 Bypass transition subjected to free-stream turbulence. (a) Visualized measurement,[36] (b) direct numerical simulation, the upper, middle, and lower sheets are located in free-stream, at $y = \delta$ and $y = \delta/3$, respectively, δ is the thickness of the boundary layer at inlet.[37] *Source*: Matsurbara M 2007. Reproduced with permission of Cambridge University Press; Annual Reviews

directly and rapidly lead to turbulence in boundary layer flow instead of undergoing the linear TS wave mode. The external disturbances giving rise to the bypass transition include FST, acoustic disturbance, vorticity disturbance, etc. The transition to turbulence subjected to FST has been widely studied theoretically, experimentally, and numerically, due to its important role in practical applications.

Like the regular transition, our understanding of the bypass transition is still superficial at present. The early stage of the bypass transition involves receptivity and generation of the unsteady streaky structures with spanwise-alternating high and low streamwise velocities of fluid. The streaky structure was first noted by Klebanoff in 1971[34] and then was called as the Klebanoff mode by Kendall.[35] The streamwise elongated streaks can rapidly break down and result in the formation of the turbulent spot in the later stage of the transition. Figure 2.15 shows the process of the bypass transition subjected to FST. The streamwise streaks followed by the appearance of the turbulent spot are obviously observed.

The generation of streamwise streaky structure is the result of receptivity and growth of external free-stream vortical disturbances with the low-frequency parts. Some studies (e.g., Refs. [14,16,36,38]) suggested that the transient growth theory (see Section 2.2.3) can well interpret the bypass transition process, in which the optimal disturbances produce streamwise counter-rotating vortices that in turn generate elongated streamwise streaks by the lift-up effect. On the contrary, Durbin and Wu[37] invoked that the low-frequency free-stream vortical disturbances enter the boundary layer by shear filtering mechanism and produce even lower frequency disturbance modes. These very low-frequency disturbances are then amplified and elongated in the streamwise direction by the shear and finally result in the streaky structure (rearward jet of streamwise component of velocity fluctuations termed by these authors). In addition, an interaction between a pair of oblique waves can also lead to the formation of streaky structure.[39] The experimental observations in plane Poiseuille flow[40] showed the nonlinear generation of streamwise vortices by a pair of oblique waves giving rise to streaks that are in turn amplified by a transient growth mechanism.

The later stage of the bypass transition is the formation and consequent merging of turbulent spots. A secondary instability of the streaks can be thought of responding for the breakdown

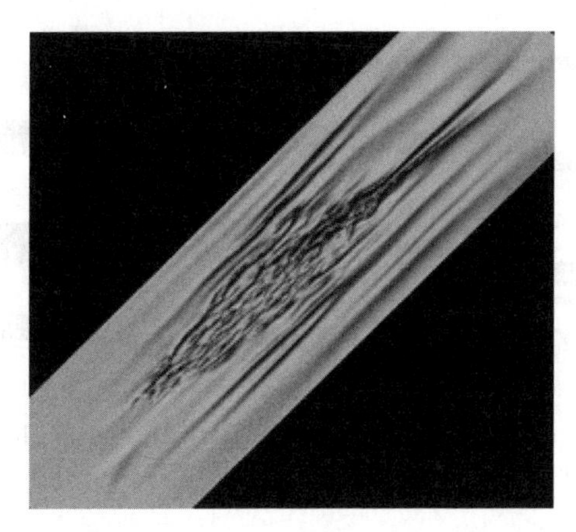

Figure 2.16 Bypass transition subjected to passing wakes, flow is from lower left to upper right.[42]
Source: Wu X 1999. Reproduced with permission of Cambridge University Press

to turbulence. Due to the inflection point instability of the velocity profile in spanwise and normal directions, $U(y, z)$, for streaks, the secondary instability mechanism will take the place of the transient growth mechanism and then excite the turbulent spot in the laminar boundary layer when streak amplitude exceeds the threshold value. The details of streaks instability will be further discussed in Section 2.4.1. On the contrary, Jocobs and Durbin[41] found that no evidence of streak instability can be used to support the secondary instability mechanism. Because the disturbances with small scale and high frequency in FST cannot enter the boundary layer due to the shear filtering, the streaks in the boundary layer lack the source of disturbance and consequently are relatively stable. The formation of the turbulent spot is due to the interaction between the lift-up rearward jet (streaks) and the free-stream vortices via Kelvin–Helmholtz instability. The turbulent spot originates near the top of the boundary layer. It is different from that described in Section 2.2.4, in which the turbulent spot is usually stimulated by forcing at the wall.

The passing wakes also lead to the bypass transition, due to their particular applications in turbomachinery. However, the process is more complex because wakes prior to the transition usually are anisotropic turbulence, accomplished by the intermittence and the large-scale organized structures. It is believed that last stage the wake-induced transition is also associated with the appearance and growth of turbulent spots.[37, 42] Figure 2.16 shows a matured turbulent spot structure in the bypass transition subjected to passing wakes. As can be seen that the formation of a turbulent spot in the wake-induced transition is similar to that in the FST-induced transition, as long as there is no flow separation in the boundary layer.

2.3 Coherent Structure of Wall Turbulence

The universal definition of a coherent structure of wall turbulence does not still reach an agreement up to now. According to the definition by Robinson,[43] a coherent structure is that of the flow over which at least one fundamental flow variable (velocity component, density, temperature, etc.) exhibits significant correlation with itself or with another variable over a range of

Figure 2.17 Top view of a coherent structure in the near-wall region. The black part represents the low-speed streak ($y^+ = 20$) and the gray part represents streamwise vortices ($0 < y^+ < 60$).[44] Flow is from left to right

space and/or time that is significantly larger than the smallest local scales of the flow. Although a complete scenario of the turbulent coherent structure is still need to be revealed, a common understanding of basic feathers for the structure has been obtained with the help of experimental and numerical techniques in recent decades. Robinson[43] summarized the following useful framework for understanding the coherent structure in wall turbulence: (1) low-speed streaks in the viscous sublayer; (2) ejections of low-speed fluid outward from the wall, including lifting low-speed streaks; (3) sweeps of high-speed fluid inward toward the wall, including inrushes from the outer region; (4) vortical structures of various forms; (5) sloping near-wall shear layers, exhibiting local concentrations of spanwise vorticity and $\partial u'/\partial x$; (6) near-wall "pockets" visible in the laboratory as regions swept clean of marked near-wall fluid; (7) large (δ-scale) motions capped by three-dimensional bulges in the outer turbulent/potential interface; and (8) shear-layer "backs" of large-scale outer-region motions, consisting of sloping (δ-scale) discontinuities in the streamwise velocity.

Although there exist various moderate-scale vortices and their intermittent motions in the logarithmic layer and outer layer, the distinctive structure deposits in the near-wall region ($y^+ < 100, y^+$ represents height in wall unit normal to the wall) that includes viscous layer, buffer layer, and part of the logarithmic layer. Contrary to the typical vortices structure of the outer layer, the complex structures of small-scale vortices in the near-wall region contribute to skin-friction drag. By studying the structures and their evolutions that occurred in the near-wall region, we can artificially control the flow pattern that aims at drag reduction.

It is now clear that fluid motion in the near-wall region is the coherent motion of the small-scale structure whose basic elements are "streak" and "vortex," as shown in Figure 2.17. An interaction of low-speed and high-speed streaks with vortex structures of various scales leads to a quasi-periodically dynamical evolution (that is, coherent motion). In this section, the features and identification of the streak and the vortex structure are discussed. The formation and the evolution of these structures will be introduced in the next section.

2.3.1 Statistical Properties of Near-Wall Turbulence

Preceding the discussion of the coherent structure, we firstly give the structure and the statistical properties of the boundary layer. For a laminar flow, a single structure along the direction

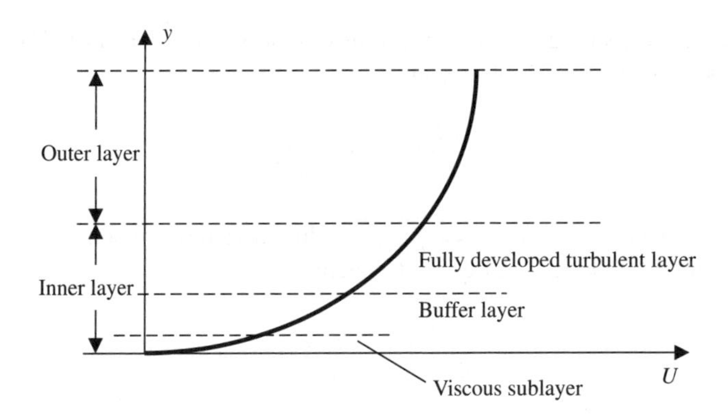

Figure 2.18 Structure of the turbulent boundary layer

normal to the wall is shown in the boundary layer. For a turbulent flow, however, a complex multiple-layer structure can be found, as shown in Figure 2.18.

Traditionally, the turbulent boundary layer is composed of the inner layer ($y/\delta \leq 0.2$) and the outer layer ($y/\delta > 0.2$); further, the inner layer includes the viscous sublayer ($0 < y^+ < 5$), the logarithmic layer ($30 < y^+ < 200$), and the buffer layer ($5 < y^+ < 30$) between the sublayer and the logarithmic layer. Some types of the turbulent boundary layer flow, such as pipe flow, channel flow, and turbulent flow on the flat-plate boundary layer, are regarded as wall turbulence. Without loss of generality, we can analyze the statistical properties of the inner layer in the two-dimensional turbulent boundary layer.

It is assumed that flow has only streamwise (x direction) mean velocity, and that the velocity has only normal (y direction) profile, that is, $U(y)$. The velocity fluctuation is statistically uniform in streamwise (x) and spanwise (z) directions. The streamwise pressure gradient is zero. A no-slip boundary condition is satisfied on the wall, that is, the mean velocity and the velocity fluctuation on the wall are vanished. Based on the abovementioned assumptions, the velocities and the pressure for two-dimensional incompressible flow in the boundary layer are expressed as in Eq. (2.10). Substituting Eq. (2.10) into the N-S equations (see Eqs. (1.140) and (1.141)) and taking the time averaging for both hands of the N–S equations, the momentum equation of U can be expressed as

$$\frac{\mu}{\rho}\frac{\partial^2 U}{\partial y^2} - \frac{\partial \overline{u'v'}}{\partial y} = 0 \tag{2.28}$$

here a relationship of $\nu = \mu/\rho$ is used. Integrating Eq. (2.28), one obtains

$$\mu\frac{\partial U}{\partial y} - \rho\overline{u'v'} = \tau_w \tag{2.29}$$

where $-\rho\overline{u'v'}$ is called Reynolds stress and the constant τ_w is the wall shear stress.

The left-hand side of Eq. (2.29) represents total shear stress that is sum of the molecular viscous stress and the Reynolds stress. According to the Boussinesq assumption that relates the second-order term of turbulent fluctuation to the mean term, the local Reynolds stress is in proportion to the velocity gradient of mean flow, that is, $-\rho\overline{u'v'} = \mu_t \partial U/\partial y$, here μ_t is the turbulent viscosity. Thus, Eq. (2.29) becomes

$$\frac{\partial U}{\partial y}(\mu + \mu_t) = \tau_w \quad (2.30)$$

Defining a friction velocity, u_τ, as $u_\tau = \sqrt{\tau_w/\rho}$, a dimensionless velocity, $U^+ = U/u_\tau$, and a dimensionless length, $y^+ = yu_\tau/\nu$, Eq. (2.30) becomes

$$\frac{\partial U^+}{\partial y^+}\left(1 + \frac{\mu_t}{\mu}\right) = 1 \quad (2.31)$$

For the viscous sublayer ($0 < y^+ < 5$), the molecular viscosity dominates the flow and the turbulent fluctuation may be neglected, thus $\mu_t/\mu \ll 1$, Eq. (2.31) is written as $\partial U^+/\partial y^+ = 1$, that is,

$$U^+ = y^+ \quad (2.32)$$

Equation (2.32) is the mean velocity profile of the viscous sublayer. Due to the linear profile property, the layer is also called the linear sublayer.

For the logarithmic layer ($30 < y^+ < 200$), the turbulent fluctuation exceeds the molecular viscosity, thus $\mu_t/\mu \gg 1$, then Eq. (2.31) becomes

$$\frac{\mu_t}{\mu}\frac{\partial U^+}{\partial y^+} = 1 \quad (2.33)$$

Due to the eddy viscosity coefficient, $\nu_t = \kappa y u_\tau$, Eq. (2.33) is rewritten as

$$\kappa y^+ \frac{\partial U^+}{\partial y^+} = 1 \quad (2.34)$$

Integrating Eq. (2.34) yields

$$U^+ = \frac{1}{\kappa}\ln y^+ + B \quad (2.35)$$

Equation (2.35) is the logarithmic profile of mean velocity in the logarithmic layer. Based on the experimental results, the constants in Eq. (2.35) are $\kappa = 0.4$ and $B = 5.5$.

For the buffer layer ($5 < y^+ < 30$) between the viscous and logarithmic layers, both molecular viscosity and turbulent fluctuation dominate the flow and lead to the complicated flow behavior. No accurate expression can be applied to describe the layer. An approximation of mean velocity profile can be written as

$$U^+ = -3.05 + 5\ln y^+ \quad (2.36)$$

Figure 2.19 shows the computed (solid line) and measured (symbol) mean velocity profiles in the inner layer of the boundary layer, these profiles agree well with the curves (dashed lines) from Eqs. (2.32) and (2.35). Figure 2.20 gives the distributions of turbulent intensity (defined by the root-mean-square of the turbulent velocity fluctuation components) and Reynolds stress. In the viscous sublayer ($0 < y^+ < 5$), turbulent behavior is very weak (but still exists). Along

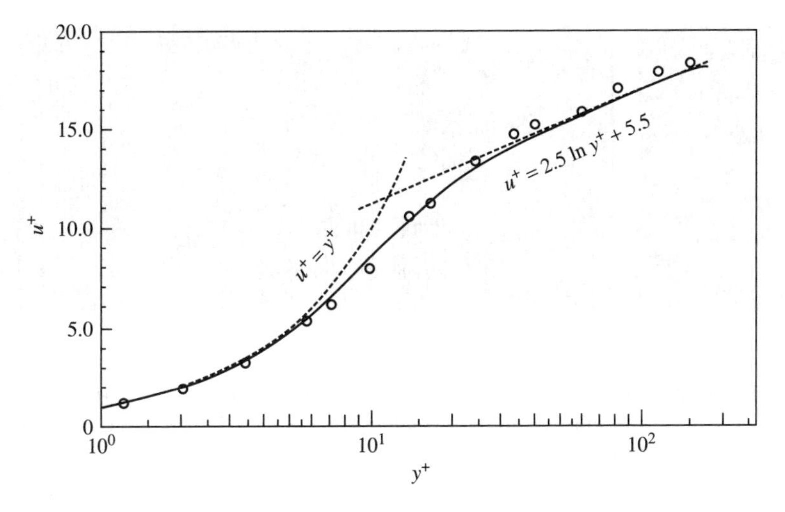

Figure 2.19 Mean velocity profiles of the turbulent boundary layer. Solid line is the computed result,[45] dashed lines are the curves from Eqs. (2.32) and (2.35), and symbols are the experimentally measured results.[46] *Source*: Kim J 1987. Reproduced with permission of Cambridge University Press

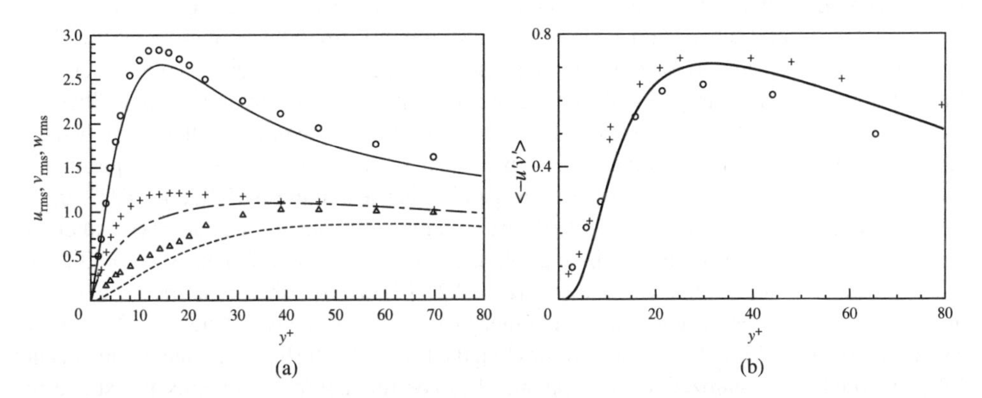

Figure 2.20 Profiles of (a) turbulent intensity and (b) Reynolds stress in the turbulent boundary layer. Lines are the computed results,[45] symbols are the experimentally measured results from Ref. [47] in (a) and from Ref. [46] in (b). u_{rms} : —, o; v_{rms} : ---, Δ; w_{rms} : — - —, + in (a). *Source*: Kim J 1979. Reproduced with permission of Cambridge University Press

with the increase of the normal distance (y^+), the turbulent intensity and the Reynolds stress grow rapidly. The root-mean-square of the streamwise velocity fluctuation component (u_{rms}) reaches its maximum in the buffer layer ($5 < y^+ < 30$), and the root-mean-squares of normal and spanwise velocity fluctuation components (v_{rms} and w_{rms}) reach their maximums in the logarithmic layer ($30 < y^+ < 200$). The Reynolds stress also reaches its maximum in the logarithmic layer. After their maximums, the turbulent intensity and the Reynolds stress decrease slowly with the further increase of the normal distance.

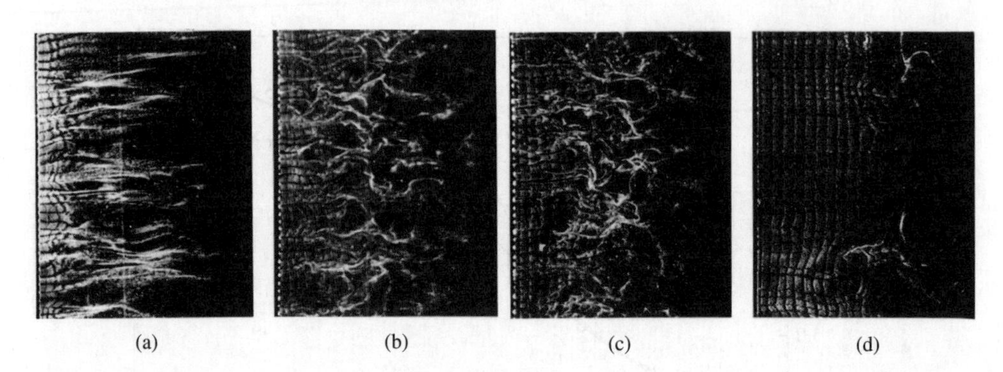

Figure 2.21 Streak structure in the turbulent boundary layer.[48] (a) $y^+ = 2.7$, (b) $y^+ = 38$, (c) $y^+ = 101$, (d) $y^+ = 407$. Flow is from left to right. *Source*: Kline S J 1967. Reproduced with permission of Cambridge University Press

2.3.2 *Structural Features and Identification of Streak*

2.3.2.1 Streak Features

In 1967, Kline *et al.*[48] in Stanford University found, using the hydrogen-bubble flow visualization technique, that streaky lines of fluid flow described by hydrogen bubbles in the viscous sublayer of the boundary layer do not move along straight lines parallel to the streamwise direction, but gathers laterally to form streamwise elongated regions which now is called as streaks. Later, further experiments revealed that the regions with different streamwise velocities can alternately appear in the spanwise direction. The high velocity region is called as high-speed streak, while the low one is called as low-speed streak. Figure 2.21 shows the experimental pictures of streaky structure using the hydrogen-bubble technique by Kline *et al.*[48] In the viscous sublayer ($y^+ = 2.7$, Figure 2.21a), the streamwise-elongated and spanwise-alternate streaks are obviously observed. In the logarithmic region near the buffer layer ($y^+ = 38$), the streaks are short and curly but still visible (Figure 2.21b). However, the streaks become obscure and invisible in the logarithmic layer and outer layer far from the wall (Figure 2.21c and d). The observed behavior of the streaks implies that the flow in the turbulent boundary layer is not fully random but is organized to some extent. The experimental measurements also suggested that the shape of the streaks is not completely same each other, even these streaks locate in the same layer. Therefore, the streaky structure is quasi-regular or coherent.

Earlier studies[48–50] indicated that low-speed streaks can undergo a series of events such as the lift-up, oscillation, and breakdown, and finally lead to the turbulence. Usually, these low-speed streaks stay at the viscous sublayer and the buffer layer and are nearly not observed clearly at $y^+ > 30$. Statistically, low-speed streaks align periodically in the spanwise direction with the spanwise spacing of $\Delta z^+ \approx 100$ ($\Delta z^+ \approx \Delta z u_\tau / \nu$) and the streamwise length of $\Delta x^+ \approx 1000$ ($\Delta x^+ \approx \Delta x u_\tau / \nu$).

2.3.2.2 Identification of Streak

From the experimental point of view, the hydrogen-bubble technique can well be used to identify the streak structure. In this visualization technique, a platinum wire cathode is mounted vertically upstream of the test section of flume. A graphite rod is used as an anode. Due to the pulsed electrolysis of water, the hydrogen-bubble cluster (time line) can be produced at intervals. Because the hydrogen bubble is very light and can move along the flow, a cluster of the hydrogen bubbles produced by a pulsed process represents the trajectory of the fluid particle. By measuring the streamwise spacing between neighboring hydrogen-bubble time lines and the pulsed time interval, one can compute local velocity of fluid particle. Figure 2.21 is the typical streak structure identified by the hydrogen-bubble visualization technique. Of the data obtained by DNS, we can use streamwise velocity fluctuation u' to characterize the streaks, where $u' > 0$ denotes regions of high-speed streak, whereas $u' < 0$ denotes regions of low-speed streak. The dark part in Figure 2.17 presents the computed low-speed streaks ($u' < 0$) of channel turbulent flow. The statistically mean spanwise spacing and streamwise length of low-speed streaks computed by DNS also qualitatively agree with those by experimental measurements. The agreement between computed and measured results also implies that the streak structure is a common element in near-wall turbulent flow.

2.3.3 Structural Features and Identification of Vortex

There exist various vortices with different spatial and temporal scales in wall turbulence. Vortex is the "muscle" of the viscous fluid, by which mass and momentum can be transferred and pumped between the fluid layers. Identifying and describing the three-dimensional vortex structures are necessary to understand the coherent motion in wall turbulence. In wall-bounded flow, streamwise vortices, including quasi-streamwise vortices, hairpin vortices, and vortex packet, are one of the most important structures in coherent motion, and the main reason of the turbulent bursting and the drag production.

2.3.3.1 Vorticity Dynamics of Fluid Motion

In fluid mechanics, vorticity is an important quantity for describing the local rotating motion of the fluid. We first define a curl in the velocity filed as

$$\boldsymbol{\omega} = \nabla \times \mathbf{u} \tag{2.37}$$

here $\boldsymbol{\omega}$ is the vorticity vector and $\mathbf{u}$ is the velocity vector.

Since the vorticity is a vector, one can define the following nomenclatures:

1. A curve in a flow field tangent to the vorticity $\boldsymbol{\omega}$ at every point is called a *vorticity line (filament)*.
2. A curved surface in a flow field tangent to the vorticity ω at every point is called a *vorticity surface (sheet)*.
3. A tube-like vortex surface formed by all vorticity lines passing through a closed curve, which itself is not a vorticity line and can shrink to a point inside the fluid, is called a *vorticity tube*.

Diverging the momentum equation (Eq. 1.141), one obtains the transport equation of vorticity as

$$\frac{\partial \boldsymbol{\omega}}{\partial t} + \mathbf{u} \cdot \nabla \boldsymbol{\omega} = \boldsymbol{\omega} \cdot \nabla \mathbf{u} + \nu \nabla^2 \boldsymbol{\omega} \tag{2.38}$$

The first term on the right-hand side of Eq. (2.38), $\boldsymbol{\omega} \cdot \nabla \mathbf{u}$, denotes the vorticity production due to the vortex stretching induced by strain rate of velocity gradient, hence the term is also called the stretching term; the velocity gradient component, $\partial u_i/\partial x_j$, in $\boldsymbol{\omega} \cdot \nabla \mathbf{u}$ can be expressed as

$$\frac{\partial u_i}{\partial x_j} = S_{ij} + \Omega_{ij} \tag{2.39}$$

where S_{ij} and Ω_{ij} represent the symmetric part (called strain rate tensor) and asymmetric parts (called rotating rate tensor) of $\partial u_i/\partial x_j$, respectively, and are expressed as

$$S_{ij} = \frac{1}{2}\left(\frac{\partial u_i}{\partial x_j} + \frac{\partial u_j}{\partial x_i}\right) \tag{2.40}$$

$$\Omega_{ij} = \frac{1}{2}\left(\frac{\partial u_i}{\partial x_j} - \frac{\partial u_j}{\partial x_i}\right) \tag{2.41}$$

The three components of Eq. (2.41) represent the following vorticity components:

$$\omega_x = -\Omega_{yz} = -\frac{1}{2}\left(\frac{\partial v}{\partial z} - \frac{\partial w}{\partial y}\right) \tag{2.42a}$$

$$\omega_y = -\Omega_{zx} = -\frac{1}{2}\left(\frac{\partial w}{\partial x} - \frac{\partial u}{\partial z}\right) \tag{2.42b}$$

$$\omega_z = -\Omega_{xy} = -\frac{1}{2}\left(\frac{\partial u}{\partial y} - \frac{\partial v}{\partial x}\right) \tag{2.42c}$$

here ω_x, ω_y, and ω_z are the streamwise (x), normal (y), and spanwise (z) vorticities, respectively.

2.3.3.2 Streamwise Vortex Feature

Streamwise vortex is the vortical element structure that is dominated mainly by streamwise vorticity, ω_x. Generally, streamwise vortices locate at the region ranged in $20 < y^+ < 40$ (buffer layer), whose head and tail can be extended to the logarithmic layer and the viscous sublayer, respectively. Streamwise vortices include hairpin vortex (also called horseshoe vortex), arch vortex, Λ vortex, quasi-streamwise vortex, etc.

The concept of the hairpin vortex was firstly proposed by Theodorsen in 1952,[51] however, the concept was not confirmed at that time. With the developments of the experimental technique, the hairpin-like or horseshoe-like vortices existed in the turbulent boundary layer were gradually proven. Figure 2.22a illustrates the structure of a hairpin vortex which is composed of a vortex head (dominated by ω_z), a vortex neck (dominated by ω_y), and a vortex leg (dominated by ω_x). The vortex head tilts upward in 45° to the wall and develops downstream, and the end of vortex legs with streamwise vorticity attaches to the wall. Because of the viscous attachment of vortex legs on the wall, motion speed of the head is larger than that of the leg. This leads to a stretching of the hairpin vortex whose vorticity increases based on Eq. (2.38). The stretched vortex promotes the violent ejection motion upward of the fluid inside the head and legs and the violent sweep motion downward of the fluid outside the head and legs.

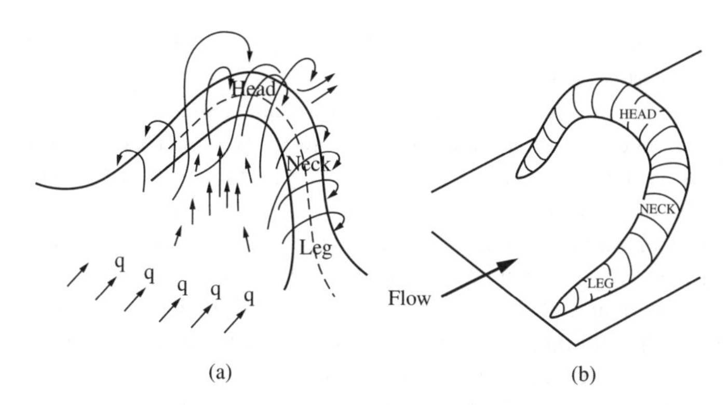

Figure 2.22 Schematic of the hairpin vortex. (a) Symmetric structure,[51] (b) Asymmetric structure.[43] *Source*: Theodorsen 1991. Reproduced with permission of Annual reviews; Cambridge University Press

The ejection and sweep motions produce a strong stress layer in the vicinity of the hairpin vortex and thus redistribute the mean velocity profile of fluid flow. Due to the inviscid (inflection) instability and other instabilities, a small-scale turbulent bursting occurs. Robinson[43] believed that the hairpin vortex whose two legs extend to the wall actually is seldom, the asymmetric hairpin vortex whose two legs have different lengths is the main vortex structure near the wall, as shown in Figure 2.22b.

In near-wall turbulence, a single hairpin vortex can also generate a vortex packet structure that includes a series of hairpin vortices aligned in the streamwise direction. Figure 2.23 shows the structure of a vortex packet. Two main features can be seen from the figure: (1) the head of hairpin vortices constitutes an interface with an angle of 12–20° to the streamwise direction, and (2) a low-speed streak exists beneath the head and between the legs of streamwise vortices. Because the streamwise length of the low-speed streak is far larger than that of the single

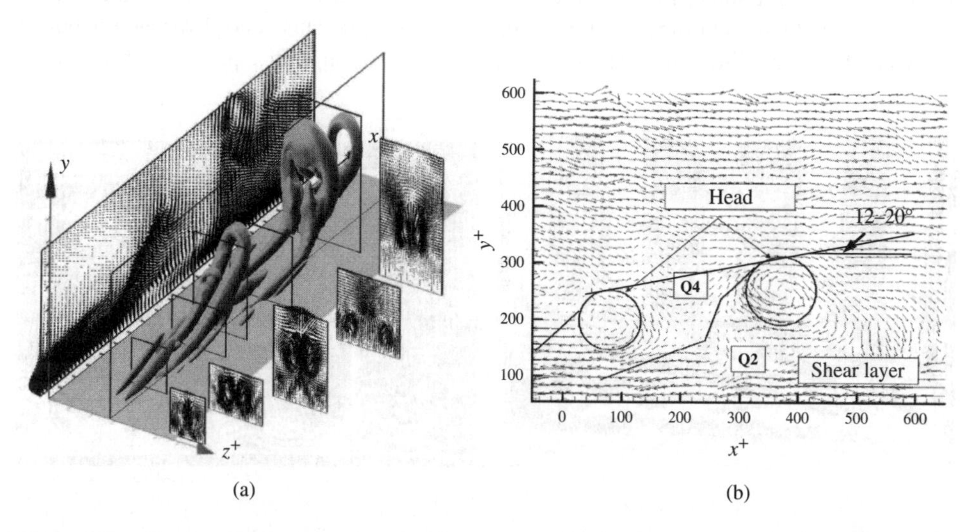

Figure 2.23 Vortex packet. (a) Structure, (b) velocity vector, $Re_\theta = 934$.[52] *Source*: Zhou J 1999. Reproduced with permission of Cambridge University Press

streamwise vortex, therefore, one can regard the low-speed streak as the footprint left by the streamwise-going vortex packet.

Another type of streamwise vortex structure is the quasi-streamwise vortex, which refers to any vortical element with a predominately streamwise (x) orientation. Usually, the quasi-streamwise vortex stays at the viscous sublayer and its downstream end gradually tilts upward with the streamwise motion. In addition, two legs of a symmetric hairpin (horseshoe) vortex or single leg of an asymmetric one can be regarded as a pair of or a single quasi-streamwise vortex, respectively. Because of the destruction by turbulent fluctuation and the viscous dissipation of fluid, hairpin vortices are difficult to survive for a long time, thus, Robinson[43] thought that a large number of the quasi-streamwise vortex constructs the main vortex structure in the near-wall region. These vortices induce the ejection of low-speed fluid or sweep of high-speed fluid by pumping action, and therefore contribute to the Reynolds stress in the vicinity of the wall. Very recently, Wu and Moin[53] performed a DNS computation for the zero-pressure-gradient flat-plate boundary layer. They found the striking preponderance of hairpin vortices in near-wall turbulence during the bypass transition. Their results imply that the main vortex structure in near-wall turbulence still needs to be examined.

2.3.3.3 Identification of a Streamwise Vortex

Due to the three-dimensional topology structure of the streamwise vortex, it is difficult to see the overall perspective of the vortex using measured or visual techniques at present. However, people can reveal the structure and feature of the streamwise vortex from a snapshot of flow field by means of the hydrogen-bubble method or the modern PIV technique. Moreover, a deduction from the data obtained by DNS computations can also help us to cognize the structure of the streamwise vortex.

The hydrogen-bubble method can be used not only to observe the streak structure (e.g., Figure 2.21), but also to identify the part of the streamwise vortex structure. Considering the effect of vortex on bubble, Lian and Su[54] proposed a criterion for identifying the streamwise vortex, including (1) superposition of the time lines (the clusters of hydrogen bubbles), (2) twine of the time lines, and (3) concentration of bubble on vortex core. According to the

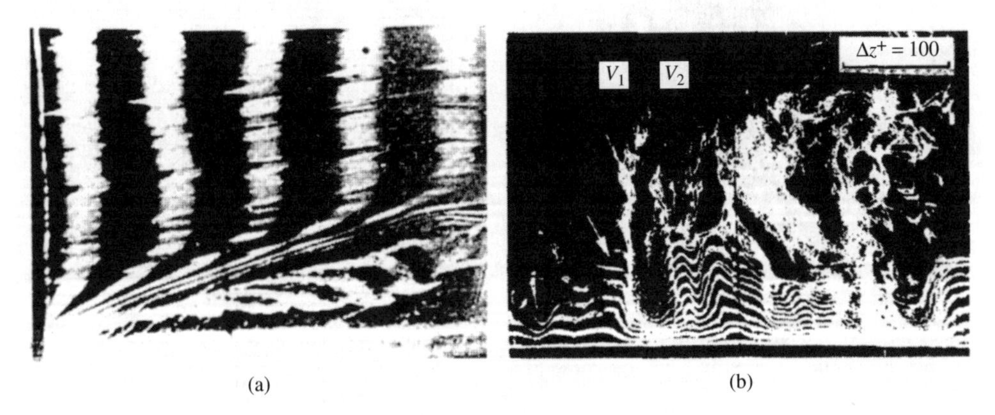

Figure 2.24 Streamwise vortex structures shown by the hydrogen-bubble method.[54] (a) Side view, flow is from left to right; (b) top view, flow is from down to up

criterion, Figure 2.24 gives the snapshots of the streamwise vortex structure. A superposition of time lines is apparently observed in the vicinity of the wall in Figure 2.24a and implies a streamwise vortex tilted upward in the streamwise direction, while V_1 and V_2 in Figure 2.24b represent two streamwise vortices, respectively.

The dataset from DNS computation provides plenty of information for turbulence. Based on the data of DNS, we can examine the vortex structure by introducing the following educing methods.

1. *Vorticity*:
   Vorticity can be used to characterize the rotating intensity of fluid in the flow field. In near-wall turbulence, however, the vorticity does not distinguish a vortex structure from the strong shear layer; thus, vorticity is not an appropriate tool for identifying the vortex structure, especially for the streamwise vortex.
2. *Invariants of the velocity gradient tensor*:
   For an arbitrary point O in the flow field, a Taylor series expansion of each velocity component u_i can be performed in terms of the space coordinates with the origin in O, that is,

$$u_i = \dot{x}_i = A_i + A_{ij}x_j + A_{ijk}x_jx_k \tag{2.43}$$

A first-order linear approximation of Eq. (2.43) is

$$u_i = \dot{x}_i = A_i + A_{ij}x_j \tag{2.44}$$

If point O is located at a critical point where all the components of the velocity vector vanish, we have $A_i = 0$ and the following characteristic equation of A_{ij} for an incompressible flow:

$$\det(\mathbf{A} - \lambda\mathbf{I}) = 0 \tag{2.45}$$

here $\mathbf{A}$ is the velocity gradient tensor ($A_{ij} = \partial u_i/\partial u_j$), λ is the eigenvalue of tensor matrix $\mathbf{A}$, and $\mathbf{I}$ is the unit matrix. Eq. (2.45) is also rewritten as

$$\lambda^3 + P\lambda^2 + Q\lambda + R = 0 \tag{2.46}$$

where P, Q, and R are called as first, second, and third invariants of a velocity gradient tensor, respectively, $P = -\mathrm{tr}(\mathbf{A})$, $Q = [\mathrm{tr}(\mathbf{A})]^2/2 - \mathrm{tr}(\mathbf{A}^2)$, $R = -\det(\mathbf{A})$. These invariants are independent of the reference frame. From continuity of the incompressible flow, $P = 0$, thus, Eq. (2.46) becomes

$$\lambda^3 + Q\lambda + R = 0 \tag{2.47}$$

The characteristic equation (2.47) has three roots, so that we can determine the topology of the local flow pattern based on the following discriminant D:

$$D = (R/2)^2 + (Q/3)^3 \tag{2.48}$$

When $D > 0$, Eq. (2.48) admits one real solution and two conjugate complex solutions (focus topology), whereas when $D < 0$, Eq. (2.48) admits three real solutions (node–saddle–saddle topology).

Perry and Chong[55] suggested that a vortex structure can be defined by the spatial region in which complex eigenvalues of the velocity gradient tensor occur ($D > 0$). The identification of the vortex structure can display the main characteristics of the hairpin vortex but is difficult to reveal the fine structure of the vortex.

The vortex structure is also defined as the region that the second invariant of velocity gradient tensor is positive ($Q > 0$). The second invariant Q of A_{ij} can be written as

$$Q_{ij} = \frac{\Omega_{ij}\Omega_{ij} - S_{ij}S_{ij}}{2} \tag{2.49}$$

The meanings of the S_{ij} and Ω_{ij} can be seen in Eqs. (2.40) and (2.41). In fact, the vortex structure embodied by Eq. (2.49) represents those regions in which rotating motion and vorticity concentration simultaneously occurs.

Another method for identifying the vortex structure is that based on the imaginary part of the complex eigenvalue of the velocity gradient tensor. Three roots of Eq. (2.47) can be expressed as

$$\lambda_1 = J + K,\ \ \lambda_{2,3} = -\frac{J+K}{2} \pm \frac{J-K}{2}\sqrt{-3} \tag{2.50}$$

where $J = \left(-\frac{R}{2} + \sqrt{\frac{R^2}{4} + \frac{Q^3}{27}}\right)^{1/3}$, and $K = -\left(\frac{R}{2} + \sqrt{\frac{R^2}{4} + \frac{Q^3}{27}}\right)^{1/3}$. The method is based on the fact that the imaginary part of the complex eigenvalue only appears in the region, in which local annular or spiral streamline exists. The definition suppresses the region that has vorticity but no local spiral motion, such as the shear layer. Figure 2.23a is an example of the hairpin vortex packet plotted by the method.

3. *Analysis of the Hessian of the pressure*:

Jeong and Hussain[56] proposed that a vortex structure in a turbulent flow is always accompanied by a local pressure minimum in a plane perpendicular to the vortex axis; thus, a vortex core is expressed by a connected region with two negative eigenvalues of tensor B_{ij}:

$$B_{ij} = (S_{ij}S_{ij} + \Omega_{ij}\Omega_{ij}) \tag{2.51}$$

Because the information on local pressure extrema is contained in the Hessian of the pressure, the gradient of the N–S equations is considered and then is decomposed into its symmetric and asymmetric parts. The transport equation of the former is

$$\frac{\partial S_{ij}}{\partial t} + u_i\frac{\partial S_{ij}}{\partial x_j} + B_{ij} = \nu\frac{\partial S_{ij}}{\partial x_k^2} - \frac{1}{\rho}\frac{\partial p}{\partial x_i \partial x_j} \tag{2.52}$$

The existence of a local pressure minimum requires two positive eigenvalues for the tensor B_{ij}. By neglecting the contribution of the first two terms of the left-hand side of Eq. (2.52), only the tensor B_{ij} is considered to determine the existence of a local pressure minimum due to a vortical motion. Because of the symmetric feature of B_{ij}, all its eigenvalues are real and can be ordered $\lambda_1 > \lambda_2 > \lambda_3$; therefore, one can define

$$\lambda_2 < 0 \tag{2.53}$$

as a connected region of the vortex so that B_{ij} has two negative eigenvalues.

2.4 Formation and Evolution of a Coherent Structure

It is well known that the basic elements of the coherent structure, that is, longitudinally elongated streaks and streamwise dominated vortices, play the important roles in both transitional (Section 2.2) and developed (Section 2.3) turbulent boundary layers. Either transition to turbulence or self-sustaining turbulence in near-wall flow is involved in the formation and evolution of the elements and the interaction between them, which are discussed in this section.

2.4.1 Formation and Instability of Streak

Longitudinally elongated streaks are prevalent in transitional and developed turbulent boundary layer flows. For transitional boundary layer flow, it is generally accepted that the streak is the dominated structure in the early stage of the bypass transition, although the mechanisms of the formation of a streaky structure have still not reached agreement as discussed in Section 2.2.5. For the developed wall turbulence, however, the formation of a streak is the result from streamwise vortices (hairpin or quasi-streamwise vortices). Streamwise vortices transfer surrounding fluid and produce a normal velocity that advects low-speed fluid near the wall upward, and hence leads to the formation of streak. Figure 2.25a shows the velocity vectors of a pair of counter-rotating vortices in the cross section of the flow direction. Vortices induce an upward normal velocity of fluid between them. This induction forms a low-speed region represented by contours of streamwise velocity, as shown in Figure 2.25b. The statistically mean streamwise length of low-speed streaks is about 1000 wall units (see Section 2.3.2) that is much larger than that of streamwise vortices; thus, streamwise vortices convecting downstream not only induce the formation of low-speed streaks but also leave them to remain where they are. Therefore, a packet of streamwise vortices can lead to a long low-speed streak.

However, the instability mechanism for a low-speed streak in both transitional and developed turbulent boundary layers seems similar to each other. The low-speed streak is stable in the viscous sublayer; however, due to the turbulent coherent (vortical) motion and other factors, it can lift upward in the streamwise direction and further destabilize and break down. Figure 2.26 illustrates the velocity distributions of fluid induced by a pair of counter-rotating

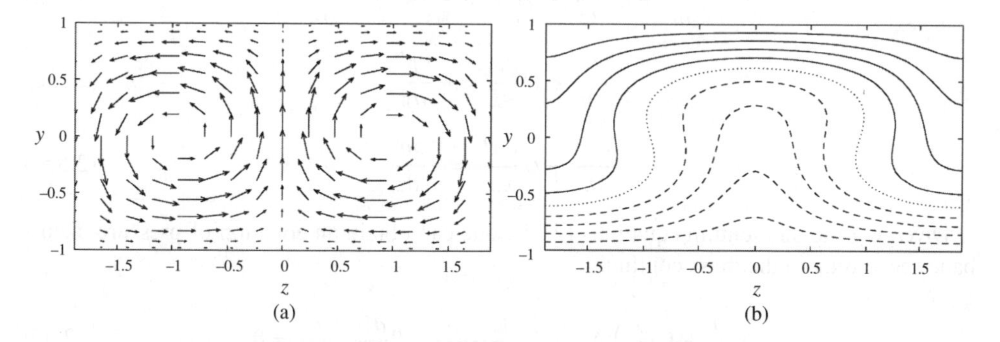

Figure 2.25 Formation of low-speed streaks in the cross section of the flow direction.[57] (a) Velocity vectors, (b) contours of streamwise velocity (dashed lines are the low-speed region). *Source*: Waleffe F 1997. Reproduced with permission of AIP Publishing LLC

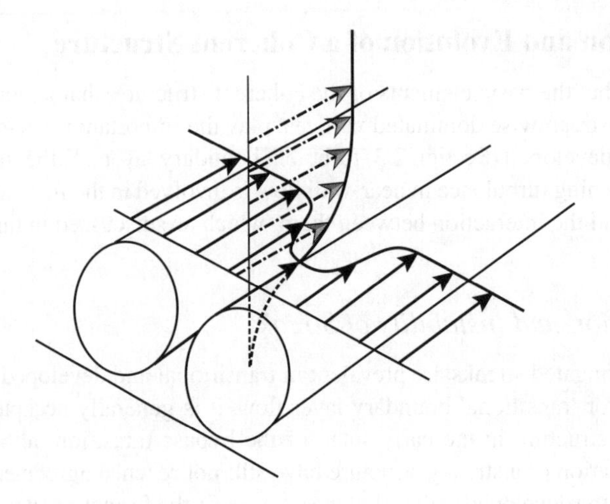

Figure 2.26 Distributions of streamwise and spanwise velocity profiles induced by a pair of counter-rotating streamwise vortices[58]

vortices. One can see that the streamwise vortical rolls result in the spanwise and normal distributions of streamwise velocity and therefore redistribute the shear stress. Because of the existence of the inflection points in these distributions, the instability of the low-speed streak occurs based on the inviscid instability theory.

A further analysis can be performed for the streak instability by linear stability theory. Considering a mean velocity profile whose distribution is only associated with spanwise and normal directions, $U(y, z)$ (also see Figure 2.25b), assuming a periodical distribution of $U(y, z)$ in the spanwise direction, and neglecting the viscous effect, we can obtain the following linearized equations of an infinitesimal disturbance by adopting the approach of Section 2.1.3:

$$\frac{\partial u'}{\partial x} + \frac{\partial v'}{\partial y} + \frac{\partial w'}{\partial z} = 0 \tag{2.54}$$

$$\frac{\partial u'}{\partial t} + U\frac{\partial u'}{\partial x} + \frac{\partial U}{\partial y}v' + \frac{\partial U}{\partial z}w' = -\frac{\partial p'}{\partial x} \tag{2.55a}$$

$$\frac{\partial v'}{\partial t} + U\frac{\partial v'}{\partial x} = -\frac{\partial p'}{\partial y} \tag{2.55b}$$

$$\frac{\partial w'}{\partial t} + U\frac{\partial w'}{\partial x} = -\frac{\partial p'}{\partial z} \tag{2.55c}$$

Diverging the momentum equations (2.55), one can obtain an equation of pressure disturbance according to the mass continuity:

$$\left(\frac{\partial}{\partial t} + U\frac{\partial}{\partial x}\right)\Delta p' - 2\frac{\partial U}{\partial y}\frac{\partial^2 p'}{\partial x \partial y} - 2\frac{\partial U}{\partial z}\frac{\partial^2 p'}{\partial x \partial z} = 0 \tag{2.56}$$

where $\Delta = \frac{\partial^2}{\partial x^2} + \frac{\partial^2}{\partial y^2} + \frac{\partial^2}{\partial z^2}$.

The pressure disturbance in the streamwise (x) direction is expressed as

$$p'(x, y, z, t) = \text{Real}(\widetilde{p}'(y, z)e^{i(\alpha x - ct)}) \tag{2.57}$$

here Real means the real part of pressure disturbance; α is the streamwise wave number; c is the phase speed, $c = c_r + ic_i$. Substituting Eq. (2.57) into Eq. (2.56), we have

$$(U - c)\left(\frac{\partial^2}{\partial x^2} + \frac{\partial^2}{\partial z^2} - \alpha^2\right)\widetilde{p}' - 2\frac{\partial U}{\partial y}\frac{\partial \widetilde{p}'}{\partial y} - 2\frac{\partial U}{\partial z}\frac{\partial \widetilde{p}'}{\partial z} = 0 \tag{2.58}$$

Equation (2.58) constitutes an eigenvalue problem of pressure disturbance. Once the eigenvalue relationship of the pressure disturbance is solved, one can obtain the corresponding relationship of the velocity disturbance by the following expressions:

$$i\alpha(U - c)\widetilde{v}' = -\partial\widetilde{p}'/\partial y \tag{2.59a}$$

$$i\alpha(U - c)\widetilde{w}' = -\partial\widetilde{p}'/\partial z \tag{2.59b}$$

$$i\alpha(U - c)\widetilde{u}' + \widetilde{v}'\partial U/\partial y + \widetilde{w}'\partial U/\partial z = -i\alpha\widetilde{p}' \tag{2.59c}$$

A Fourier series expansion for the pressure disturbance yields

$$\widetilde{p}'(y, z) = \sum_{k=-\infty}^{\infty} \widehat{p}_k'(y)e^{i(k+\gamma)\beta z} \tag{2.60}$$

where β is the spanwise wave number. On one hand, it is noted that γ and $\gamma \pm n$ (n is any integer) in Eq. (2.60) have the same eigenvalue relationship; on the other hand, Eq. (2.58) is symmetric about $z = 0$ in the spanwise (z) direction. Thus, the symmetries can categorize the eigenvalue relationship of Eq. (2.60) into two modes: a fundamental mode with $\gamma = 0$ and a subharmonic mode with $\gamma = 1/2$.

For the fundamental mode ($\gamma = 0$), Eq. (2.60) can be further decomposed based on the odd or even features of a function, as follows:

$$\widetilde{p}'(y, z) = \sum_{k=-\infty}^{\infty} \widehat{p}_k'(y)\cos(k\beta z) \quad \text{(even)} \tag{2.61a}$$

$$\widetilde{p}'(y, z) = \sum_{k=-\infty}^{\infty} \widehat{p}_k'(y)\sin(k\beta z) \quad \text{(odd)} \tag{2.61b}$$

Similarly, for the subharmonic mode ($\gamma = 1/2$), one can have

$$\widetilde{p}'(y, z) = \sum_{k=-\infty}^{\infty} \widehat{p}_k'(y)\cos\left(\frac{2k+1}{2}\beta z\right) \quad \text{(even)} \tag{2.62a}$$

$$\widetilde{p}'(y, z) = \sum_{k=-\infty}^{\infty} \widehat{p}_k'(y)\sin\left(\frac{2k+1}{2}\beta z\right) \quad \text{(odd)} \tag{2.62b}$$

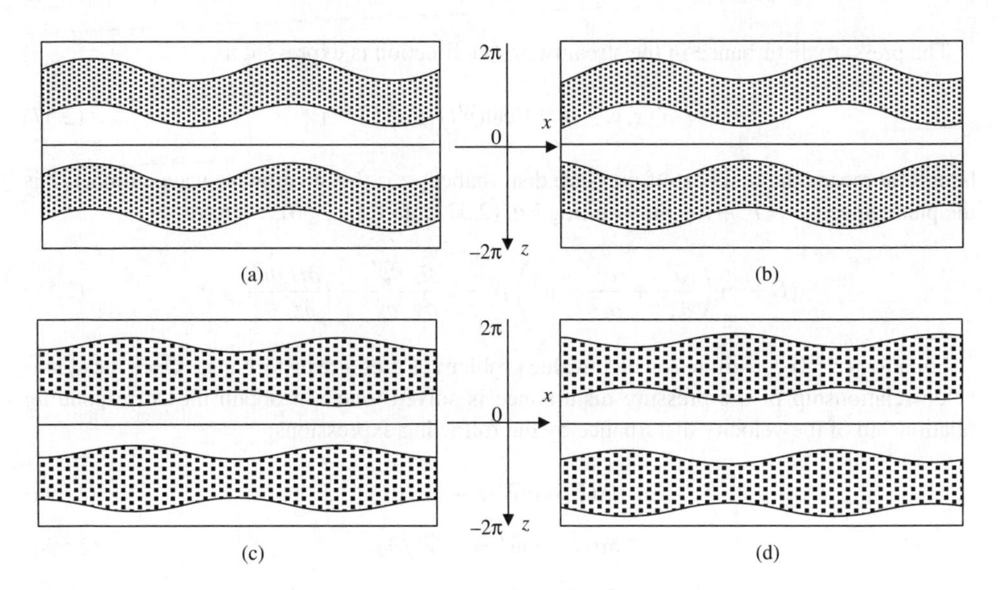

Figure 2.27 Instability modes of low-speed streaks. (a) Fundamental sinuous mode, (b) subharmonic sinuous mode, (c) fundamental varicose mode, (d) subharmonic varicose mode. *Source*: Andersson P 2001. Reproduced with permission of Cambridge University Press

Combining Eqs. (2.61b), (2.62a) with Eqs. (2.59) can result in the mathematical form of streak instability, we call it as "sinuous (or asymmetric)" instability, similarly, combining Eqs. (2.61a), (2.62b) with Eqs. (2.59) yields a mathematical form of a streak with a "varicose (or symmetric)" instability.

The abovementioned discussions can result in four types of the instability pattern induced by an infinitesimal disturbance for streaks in the $U(y, z)$ form with a spanwise periodically distribution, as illustrated in Figure 2.27. For these streak instabilities, Asai *et al.*[60, 61] experimentally suggested that the varicose instability is due to the normal inflection point instability of the mean velocity profile, while the sinuous instability is due to the spanwise inflection point instability of the mean velocity profile. They also concluded that the sinuous streaks are more unstable to disturbance than the varicose streaks are. The conclusion is then confirmed by the analysis of Liu *et al.*[62]

2.4.2 *Formation of a Vortex Structure*

Formation of a vortex structure in the near-wall region of both transitional and developed turbulent boundary layers is usually a complex process. In summary, there exist three paths to the formation of the vortex structure: (1) formation and instability of spanwise vortices, (2) formation of vortices due to the streak instability, and (3) regeneration of hairpin vortices.

2.4.2.1 Formation and Instability of Spanwise Vortices

Formation of a spanwise vortex starts from the amplification of a disturbance wave in the laminar flow. It is assumed that initially only the velocity profile in the normal (y) direction,

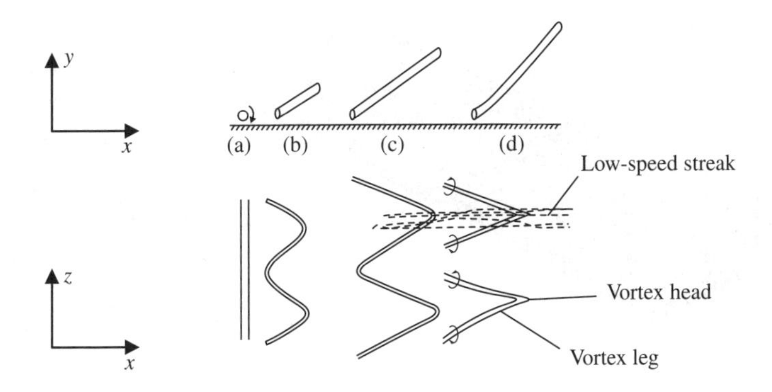

Figure 2.28 Instability of the spanwise vortex

(y), exists in the laminar boundary layer, then a spanwise vorticity (ω_z) layer forms in the near-wall region. Moreover, a two-dimensional disturbance (TS) wave from the flow is a vorticity wave that does not transport the mass but propagates and accumulates the vorticity.[63] When the disturbance wave grows, its vorticity is gradually concentrated. Once the disturbance wave exponentially grows to a critical level, the spanwise vorticity layer becomes roll up in a two-dimensional form through the Kelvin–Helmholtz instability. In this case, vorticity is further concentrated and developed into the individual spanwise vorticity region, that is, spanwise vorticity tube. Due to the three-dimensional nonlinear effect, the spanwise vorticity tube begins to deform slightly. As time goes on, vortical element far from the wall has quicker speed than that near the wall by the wall shear; hence, the vorticity tube connecting vortical elements is stretched and develops into the hairpin vortex. Figure 2.28 illustrates the process. One can see that the spanwise vorticity tube evolves in "W" shape in the spanwise direction. Simultaneously, the downstream part of the tube tilts upward, while the upstream part of the tube attaches on the wall. This leads to the stretching of the tube and therefore enhances the growth of vorticity. Because of the induction by the hairpin vortex, the low-speed streak is formed between the legs of the vortex.

For a flow transition process, formation and instability of the spanwise vortices are the starting point of the transform from laminar to turbulent flow. In the viscous sublayer of fully developed turbulent flow, laminar can be partly maintained because of the stability of the low-speed streak. This circumstance breeds spanwise vortices which in turn evolve into hairpin vortices with the nonlinear excitation of the disturbance waves. This excitation plays an important role in the formation and instability of vortices; therefore, one can artificially impose the disturbance waves in order to control the behavior of the coherent motion in the turbulent boundary layer.

2.4.2.2 Formation of Vortices due to the Streak Instability

It is well known that streamwise vortices produce low-speed streaks. However, streak profiles with the form of $U(y, z)$ are frequently observed without streamwise vortices. Although some researchers[64–67] suggested that new streamwise vortices can be develop from a parent vortex, Schoppa and Hussain[44] believed that an unstable streak is able to induce vortices that are dominated by streamwise vorticity.

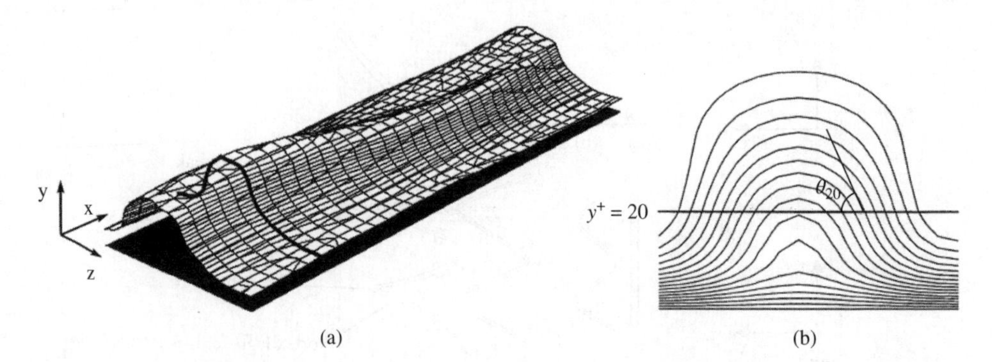

Figure 2.29 Lifted low-speed streak: (a) three-dimensional shape; (b) distribution of $u(y, z)$, θ_{20} defines a lifted intensity of the streak.[44] *Source*: Schoppa W 2002. Reproduced with permission of Cambridge University Press

By detecting the turbulent behavior in near-wall flow, Schoppa and Huassain invoked that the three-dimensional shape of a lifted streak can be illustrated in Figure 2.29a. The cross section of streaks has the form of a harmonic wave, as shown by contours of $U(y, z)$ in Figure 2.29b. Given an initially straight streak in the turbulent boundary layer, with the following form:

$$U(y, z) = U(y) + (\Delta u/2)\cos(\beta z) y \exp(-\eta y^2)$$
$$V = W = 0 \tag{2.63}$$

where $U(y)$ denotes turbulent base flow profile, Δu is the velocity deficit, β is the spanwise wave number, η is the factor used to give a plateau in normal vorticity at $y^+ = 10$–30, these authors imposed a nonnormal disturbance expressed as follows:

$$w'(x, y) = \varepsilon \sin(\alpha x) y \exp(-\eta y^2)$$
$$u' = v' = 0 \tag{2.64}$$

here ε is the disturbance amplitude and α is the streamwise wave number. Thus, a sinuous streak is produced in flow field without initial vortices. Due to nonnormal characteristics of the initial disturbance to the flow, the streak is rapidly destabilized via a transient growth (see Section 2.2.3) mechanism. The streak transient growth (STG) firstly produces a strong shear motion, and further promotes the streamwise vorticity sheets. In the subsequent nonlinear stage, the streamwise vorticity sheets are enhanced by the stretching mechanism and finally developed into quasi-streamwise vortices.

Figure 2.30 gives the top view and side view of streamwise vortices generated by low-speed streak instability via STG. In Figure 2.30a, opposite-signed streamwise vortices (SP and SN) on either side of a low-speed streak align in the streamwise (x) direction, the head of a streamwise vortex overlaps with the tail of another streamwise vortex in the top view. The streamwise length of a vortex is about 200 wall units which is much smaller than streamwise length of a low-speed streak. This allows an x-alignment of several streamwise vortices that in turn feedback the low-speed streak jointly. The side view in Figure 2.30b further reveals that downstream part of a quasi-streamwise vortex tilts upward. The tilting of the vortex

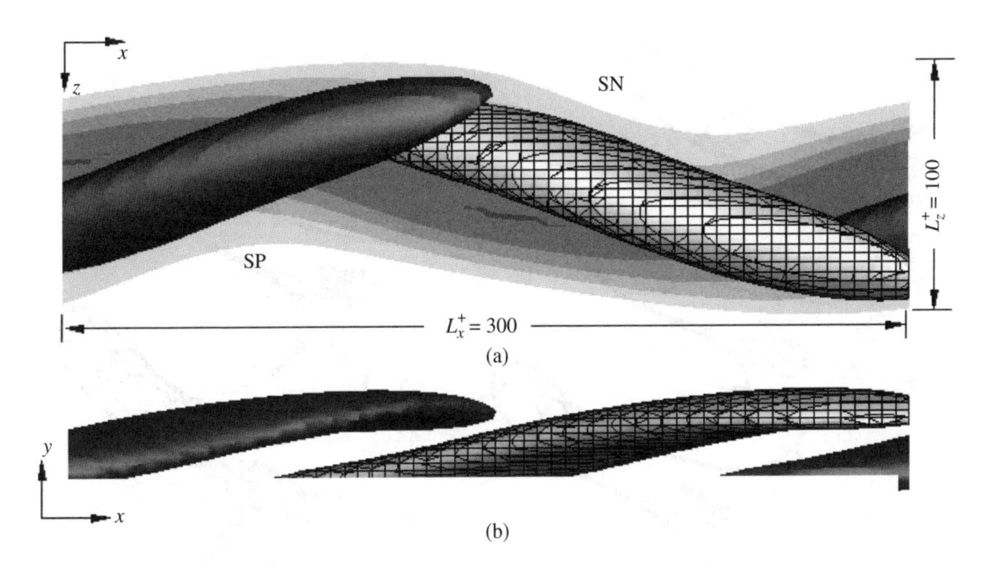

Figure 2.30 Quasi-streamwise vortices.[44] (a) Top view, (b) side view. The gray part in the bottom of (a) is the low-speed streak. *Source*: Schoppa W 2002. Reproduced with permission of Cambridge University Press

takes an important role in the formation and regeneration of hairpin vortices that is to be discussed below.

Experiments and computations of DNS have indicated that spanwise disturbances with the form of Eq. (2.64) are common in near-wall turbulence. The disturbances provide a source for destabilizing the low-speed streak. In addition, the ejection and sweep motions of fluid and the contribution to the Reynolds stress in the near-wall region induced by quasi-streamwise vortices generated from streak instability agree well with those observations in fully developed turbulence. These agreements imply the importance of the formation of vortices due to the streak instability.

2.4.2.3 Regeneration of Hairpin Vortices

As mentioned above, the hairpin vortex is one of the important vortex structures in near-wall turbulence and can be formed via nonlinear instability of a two-dimensional spanwise vorticity tube. However, the formation of hairpin vortices can also be achieved via a regeneration mechanism (or called Parent–Offspring mechanism). Smith and Walker[68, 69] pointed out that the formation of secondary hairpin vortices (offspring vortices) associates with the induction by the existing hairpin vortex (parent vortex). Figure 2.31 illustrates the regeneration process of the secondary vortices from a symmetric hairpin vortex. In this process, the pressure plays an important role in the vortex regeneration. Figure 2.31a shows the sharp, crescent-shaped ridges form and develop in the surface flow due to the action of adverse pressure gradient behind the vortex head and near the one vortex leg. Then, rapid outward movement of the erupting ridges concentrates the vorticity in Figure 2.31b. Next thing is that the erupting sheets gradually roll up (Figure 2.31c and d). Finally, the complete secondary hairpin vortices are generated behind the head and near the one leg of the original hairpin vortex.

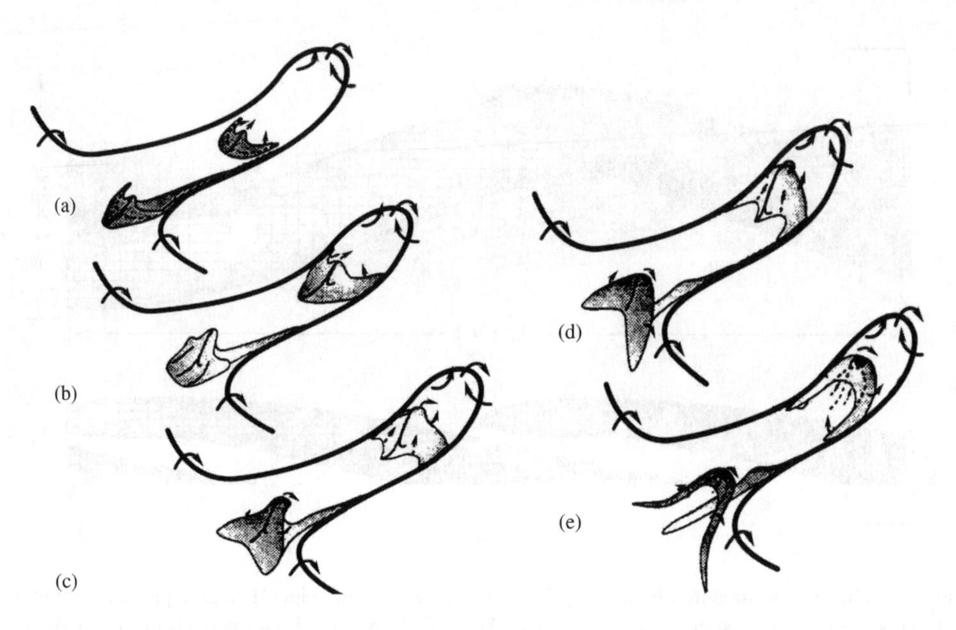

Figure 2.31 Regeneration of the secondary vortices[68]

However, Zhou *et al.*[52] found that regeneration of hairpin vortices is a complicated process that is accomplished by the formation of the hairpin vortex packet. The induction effect of the primary vortex plays an important role in the regeneration of the new vortices. Based on the Biot–Savart law, the velocity of fluid particle located at radius vector **x** can be formed induced by vorticity $\boldsymbol{\omega}(\mathbf{x}')$ at the $\mathbf{x}'$ that is in the vicinity of the **x**, that is,

$$\mathbf{u}(\mathbf{x}) = -\frac{1}{4\pi}\int_V \frac{(\mathbf{x}-\mathbf{x}')\times\boldsymbol{\omega}(\mathbf{x}')}{|\mathbf{x}-\mathbf{x}'|^3}dV(\mathbf{x}') \tag{2.65}$$

then a process involved in the formation and regeneration of hairpin vortices induced by a pair of counter-rotating quasi-streamwise vortices is illustrated in Figure 2.32. Initially, a pair of quasi-streamwise vortices locates near the wall, and the spacing between the downstream parts of both vortices is larger than that between the upstream parts of these vortices, as shown in Figure 2.32a. Thus, mutual induction is enhanced between the downstream parts and therefore produces a large normal velocity that lifts up the downstream parts of both vortices (see points A in Figure 2.32b). Due to the enhanced vorticity of downstream parts, fluid between two vortices is ejected upward and forms a shear layer in the mean flow that gives rise to the formation of spanwise vorticity. The spanwise vorticity rapidly rolls up the fluid and forms a spanwise vortex (see point B in Figure 2.32b). As time goes on, a viscous connection between the spanwise vortex and the downstream parts of quasi-streamwise vortices results in the formation of a hairpin vortex as shown in Figure 2.32c. In this case, the hairpin vortex becomes expand outward at the points C and D due to the larger curvatures of these points and further forms the Ω-like vortex as illustrated in Figure 2.32d. Because of the downward curvature of points E on the Ω-like vortex, the mutual induction makes both points E close to each other. This further enhances the mutual induction that lifts up the points E. Like the events occurred

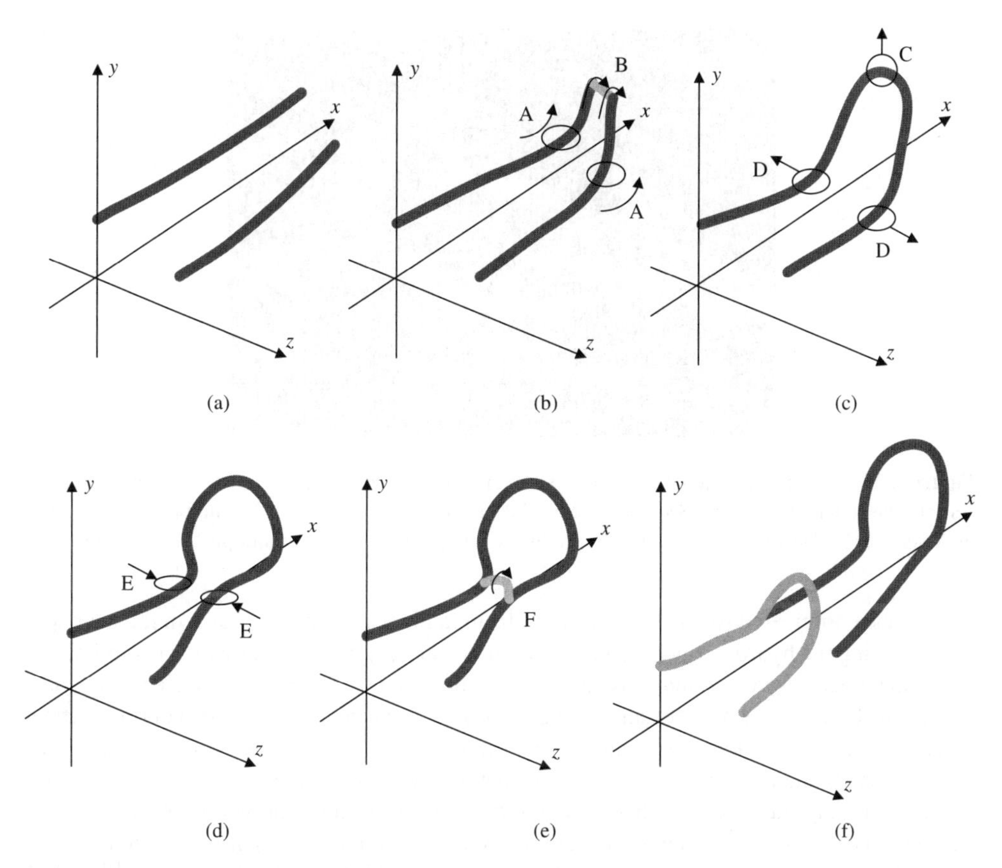

Figure 2.32 Formation and regeneration of the hairpin vortex and its off-spring vortex. *Source*: Zhou 1999. Reproduced with permission of Cambridge University Press

in Figure 2.32b, spanwise vorticity forms and subsequently rolls up a spanwise vortex, see point F in Figure 2.32e. The next thing is the breakup of the vortex in the vicinity of the point F. As a result, a new secondary hairpin vortex appears at the upstream of the primary hairpin vortex, as shown in Figure 2.32f.

The formation and regeneration of hairpin vortices can repeat and give rise to a hairpin vortex packet that is composed of primary, secondary, third, fourth hairpin vortices, etc. The details of the process can refer to Ref. [52]. Note that the hairpin vortex packet is the common structure in the near-wall region (low Reynolds number) of the flow with a high Reynolds number and that has been widely observed experimentally (e.g., see Figure 2.23b). In addition, the formation of the vortex packet is relatively insensitive to flow of the outer layer. Both features imply a universal property of the hairpin vortex packet in the near-wall coherent structure.

2.4.3 A Novel Coherent Motion: Soliton and Its Relevant Structures

Lee *et al.*[30,70,71] recently proposed a physical picture of coherent motion in the transitional boundary layer flow, based on their series of experimental investigations. In their studies, a

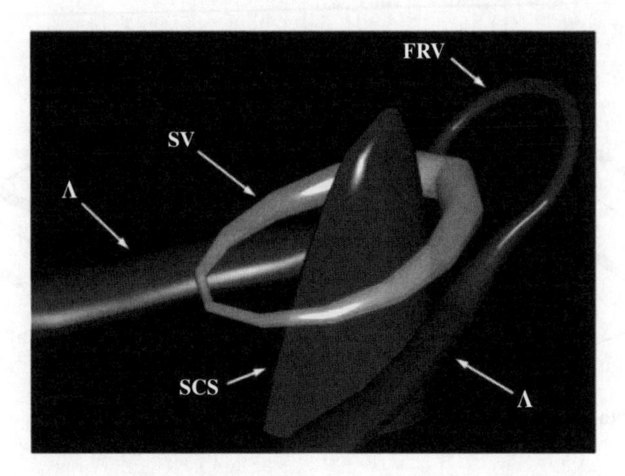

Figure 2.33 Three-dimensional topologic structure of SCS and relevant vortex structure. SCS – soliton-like coherent structure, SV – secondary closed vortex, Λ–Λ vortex, FRV – first ring vortex (head of Λ vortex).[30] *Source*: Lee C B 2008. Reproduced with permission of Cambridge University Press

SCS takes a dominated role in the transition from laminar to turbulent flows in the near-wall region. Using the hydrogen-bubble visualization technique, they found that the evolution of a SCS can induce a Λ-vortex and subsequently a secondary closed vortex (SCV). The sequential interaction between the Λ vortex and the SCV results in the formation of a chain of ring vortices that are in turn broken down to turbulence. Figure 2.33 illustrates the three-dimensional spatial pattern of SCS and relevant vortex structures before the formation of a chain of ring vortices.

By examining the large number of visual data taken by the hydrogen-bubble technique, Lee and Wu[30] found that the SCSs commonly exist in the transitional boundary layer. As claimed by these authors, a soliton can be characterized by the following features[70]: (1) *it is a wave with permanent form;* (2) *it is localized so that it decays rapidly or approaches a constant amplitude at infinity; and* (3) *it can strongly interact other solitons and remerge from the collision with unchanged shape apart from a phase shift*. These features imply that a soliton is a nonlinear traveling wave propagating with little loss of energy and shape instead of a vortex structure. Lee and Chen[71] speculated that the formation of the SCS is due to the nonlinear growth of interaction between two oblique disturbance waves. Once the SCS is formed, it can induce the generation of the Λ vortex. Figure 2.34 shows the formation process of the SCS and subsequent Λ vortex using the hydrogen-bubble visualization technique. A SCS has been formed in Figure 2.34b due to the interaction between the oblique disturbance waves, after that, a Λ vortex appears in Figure 2.34f–h. The generation of a Λ vortex can be interpreted using Figure 2.35. Locations A and B in Figure 2.35b and c represent an evolving SCS (nonlinear wave packet). At the right side of the SCS, a sweep motion of fluid occurs, thus the fluid above the SCS would also sweep down to the right side of the SCS due to the mass conservation. Moreover, the fluid at the left hand of SCS would also sweep down to the near-wall region. As a result, a strong shear layer along the border of the SCS forms and rolls up to a Λ vortex. These results clearly show that a Λ vortex appears after the SCS in time and below the SCS in space.

The next thing is the formation of the SCV which is an important event in the transitional boundary layer. Figure 2.36 shows the formation of a SCV between the legs of the Λ vortex. During the process, the head of the Λ vortex is stretched into a first ring vortex (FRV, see

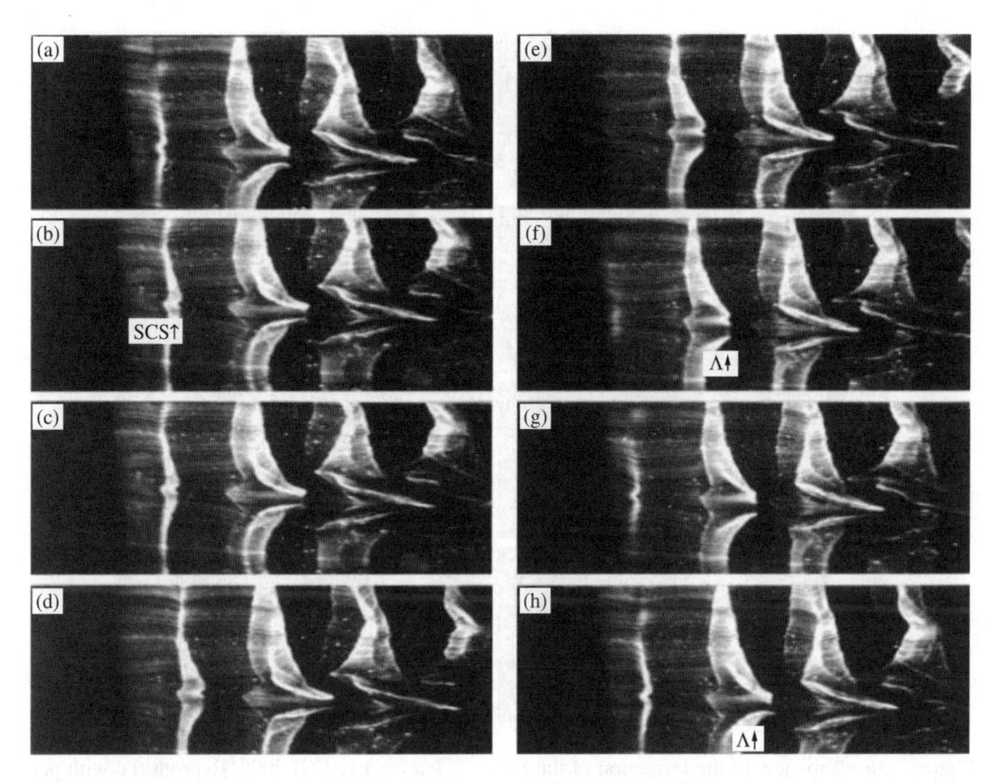

Figure 2.34 Top view of the formation of the soliton-like coherent structure (SCS) and Λ vortex by the hydrogen-bubble visualization technique.[71] *Source*: Lee C B 2008. Reproduced with permission of Cambridge University Press

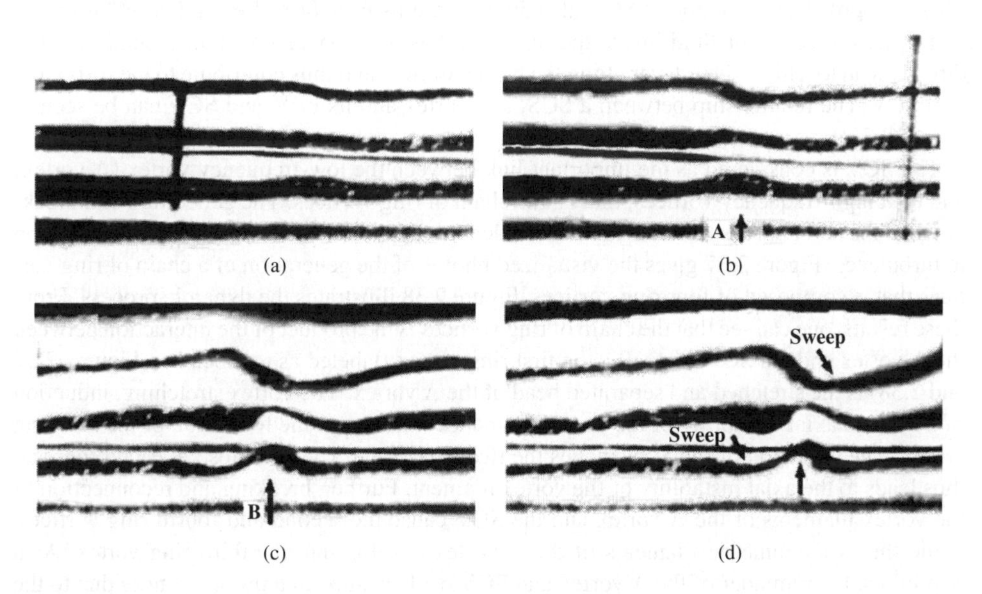

Figure 2.35 Side view of the formation of the soliton-like coherent structure (SCS) and Λ vortex. A and B represent an evolving SCS.[30] *Source*: Lee C B 2008. Reproduced with permission of Cambridge University Press

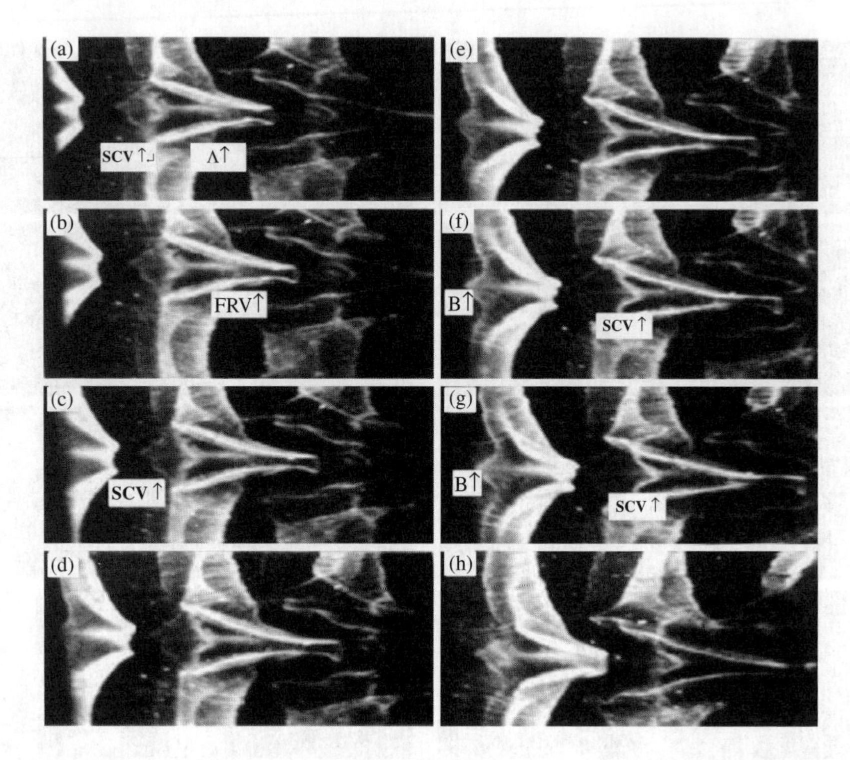

Figure 2.36 Top view of the formation of the SCV.[30] *Source*: Lee C B 2008. Reproduced with permission of Cambridge University Press

Figure 2.36b). The formation of SCV also directly results from SCS. Lee and Wu[30] revealed that upward ejection of fluid inside the SCS and downward sweep of fluid around the SCS directly lead to a high-shear layer along the border of SCS and thus contribute to the formation of a SCV. The relationship between a SCS, a Λ vortex and its FRV, and SCV can be seen in Figure 2.33.

The SCV is considered as the important link between the low-frequency vortex (Λ vortex) and next high-frequency vortices, that is, the chain of ring vortices. The generation and breakdown of the latter are associated with the turbulent production in the later stage of the transition to turbulence. Figure 2.37 gives the visualized photos of the generation of a chain of ring vortices that is composed of four ring vortices. Figure 2.38 illustrates the dynamic process. From these results, one can see that the chain of ring vortices is the product of the interaction between the Λ vortex and the SCV. Actually, the first ring vortex (labeled as number 1 in Figures 2.37 and 2.38) is the stretched and separated head of the Λ vortex. The vortex stretching, induction of SCV, and axial instability of the vortex filament contribute to the formation of the first ring vortex. The induction of the SCV makes the two legs of the Λ vortex close to each other and thus leads to the axial instability of the vortex filament. Further, breaking and reconnection of the vortex filaments of the Λ vortex and the SCV cause the second and fourth ring vortices, while the two symmetric filaments of the Λ vortex develop into the third ring vortex. As a byproduct, the reminder of the Λ vortex and SCV evolves into streamwise vortices due to the axial instability.

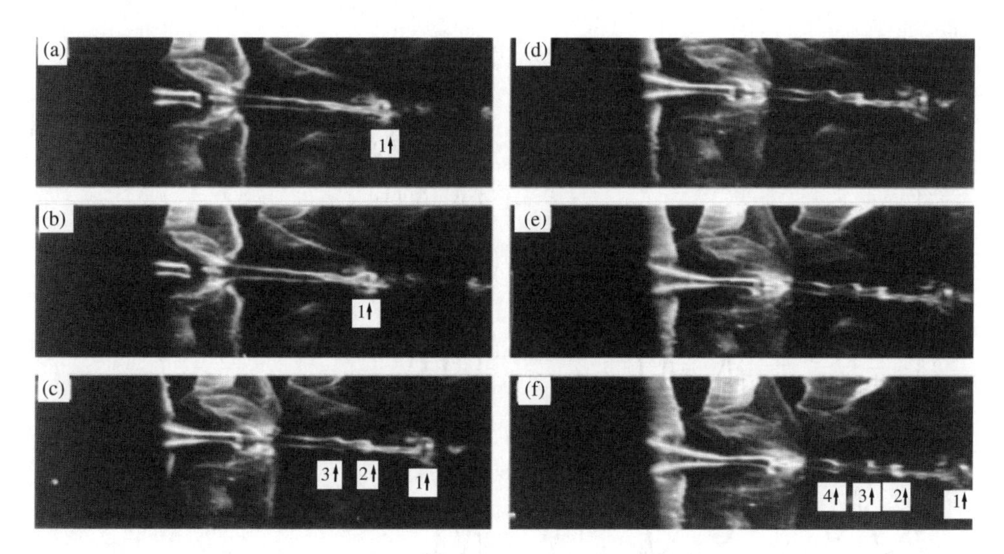

Figure 2.37 Top view of the formation of a chain of ring vortices. Numbers denotes the different ring vortices.[30] *Source*: Lee C B 2008. Reproduced with permission of Cambridge University Press

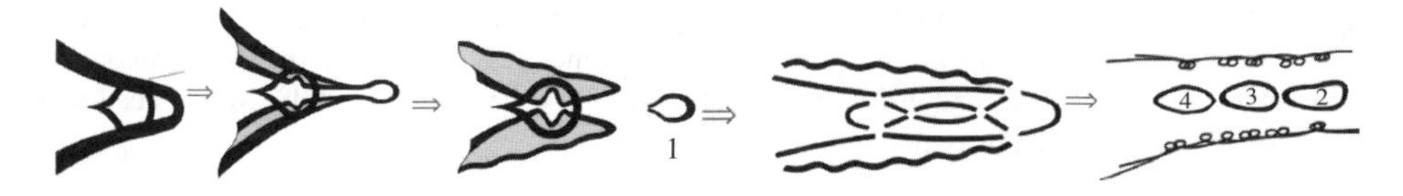

Figure 2.38 Schematic of the generation of a chain of ring vortices and streamwise vortices.[30] *Source*: Lee C B 2008. Reproduced with permission of Cambridge University Press

Lee and Wu[30] further pointed out that the breakdown of the chain of ring vortices plays an important role in the production of turbulence in the later stage of the transition to turbulence. These high-frequency vortices can be transformed into the even high-frequency fluctuations when the breakdown occurs. Figure 2.39 shows the hot-wire measurement results of development of the chain of ring vortices and their corresponding frequency spectra. One can see from Figure 2.39a that the four-spike velocities (corresponding four ring vortices) form periodically at upstream of $x = 500$ mm and gradually become random downstream. Figure 2.39b shows a wider frequency range with high-frequency component at downstream of $x = 700$ mm where the chain of ring vortices has broken down.

In addition, the SCSs-based coherent motion can also well interpret the formation of a turbulent spot in the later stage of the transition. The study by Lee and Wu[30] revealed that a turbulent spot is actually a complex of several SCSs and their bounded vortices, as shown in Figure 2.40. Obviously, the formation of the turbulent spot is the result from evolution of SCSs.

Although studies based on the soliton and relevant vortex structures mainly provide qualitative information for understanding the coherent motion of wall-bounded flow, the novel description tries to systematically draw the outline of the story occurred in boundary layer flow. In particular, Lee and his co-workers pointed out that there exists a universal scenario based on the SCSs frame to unify the various transitional paths as shown in Figure 2.4.

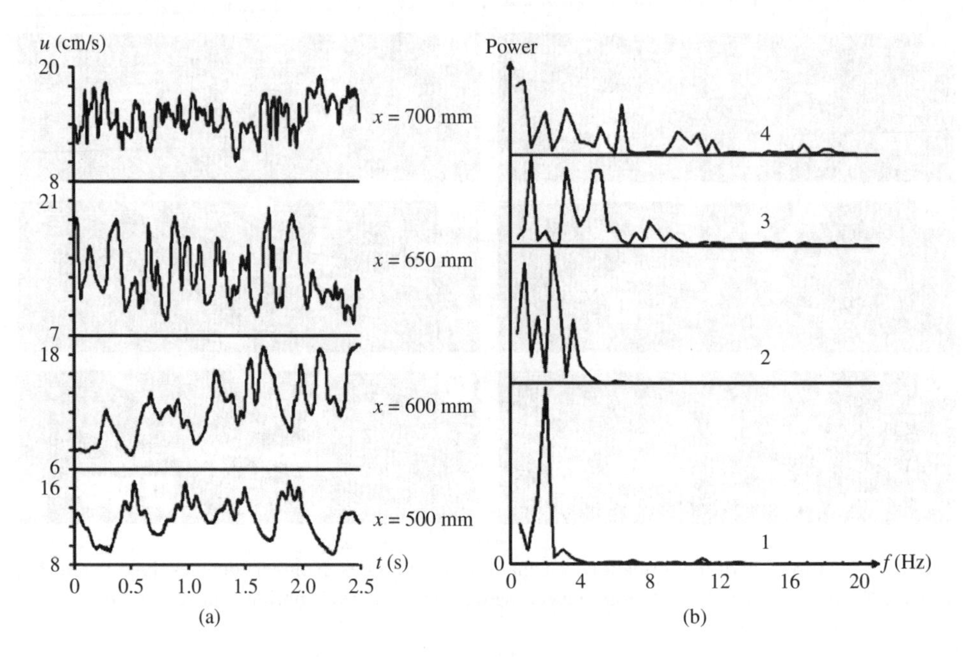

Figure 2.39 (a) Time traces starting from the four-spike stage to the multiple spike stage and (b) their corresponding spectra. The TS wave frequency was 2 Hz. Lines 1, 2, 3, and 4 in (b) are the frequency spectra at $x = 500$, 600, 650, and 700 mm, respectively.[30] *Source*: Lee C B 2008. Reproduced with permission of Cambridge University Press

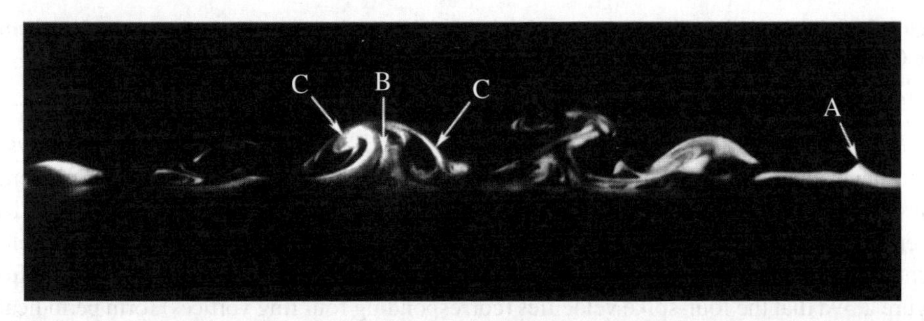

Figure 2.40 Cross section of a turbulent spot. A is a young SCS, B denotes several SCSs, and C denotes bounded vortices.[30] *Source*: Lee C B 2008. Reproduced with permission of Cambridge University Press

Moreover, the SCSs-based dynamics is not only common in transitional boundary layer flow, but is also possible in developed turbulent boundary layer flow.

2.5 Bursting and Self-Sustaining of Wall Turbulence

Based on the experimental, theoretical, and numerical studies, a global frame of the evolution of a coherent structure in wall turbulence has been constructed in the past decades. Actually, formation and development of the streaks and streamwise vortices are the segments of the evolution of the coherent structure in both transitional and developed turbulent boundary layers. Complete descriptions depend on the understanding of relationship between the streaks and vortices. Robinson[43] summarized various idealized models (also called conceptual models by the author) that described physical scenario of this relationship (coherent motion). These early works, however, are not conclusively accepted or abandoned, due to the limitation of the research techniques and available data. Along with the developments of experimental and numerical techniques, great progress has been made recently for understanding the evolution of a coherent structure in both boundary layer flows. For transitional boundary layer flow, the evolution of a coherent structure has been described in Section 2.2, by emphasizing, e.g., the three-dimensional development of Λ-like vortices in the K-regime of the transition or lift-up and the secondary instability of a streaky structure in the early stage of the bypass transition. A novel evolution of a coherent structure based on the soliton and its relevant vortices has also been systematically proposed[30] in transitional wall-bounded flows. In this section, we focus on the turbulent bursting event and the turbulent self-sustaining process. The former dominates the turbulent production in both transitional and developed boundary flows, while the latter is closely related to the production of skin-fraction drag in the developed turbulent boundary layer.

2.5.1 Bursting Event

Turbulent bursting is an important event in boundary layer flow and is a signal of the turbulence production. Since Runstadler *et al.*[72] proposed "bursting" for defining the turbulence production in 1963, turbulent bursting event has attracted much attention of researchers during past few decades. Earlier studies (e.g., Ref. [48]) indicated that the turbulent bursting event is involved in the whole process of lift-up, oscillation, and breakdown of low-speed streaks, especially the violent breakdown of the streaks. Simultaneously, several conditional sampling techniques have been developed to identify the bursting event, such as, quadrant analysis,[73] VITA (variable interval time average),[74] and VISA (variable interval space average).[75] Recent investigations by Lee *et al.*[30, 71] revealed that the coherent motions based on the soliton and its relevant vortex structures can well correspond to the turbulent bursting event that was described from the view of evolution of low-speed streaks. These authors found that the low-speed streak

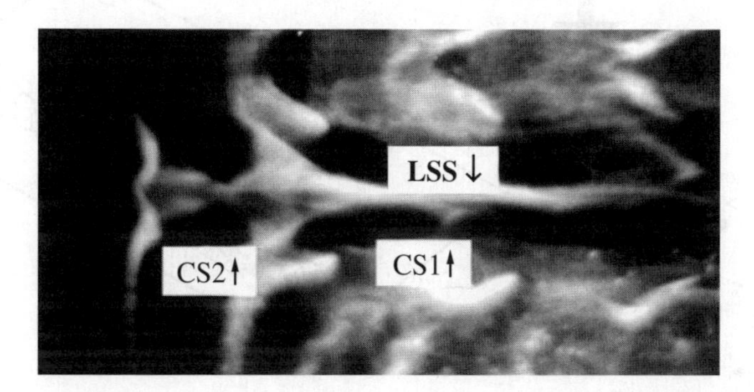

Figure 2.41 Top view of a low-speed streak (labeled as LSS) and SCSs (labeled as CS1 and CS2).[30] *Source*: Lee C B 2008. Reproduced with permission of Cambridge University Press

Table 2.1 Turbulent bursting event based on the classical and SCS-based interpretations[30]

	Classical interpretations	SCSs-based interpretations
Production position	Low-speed streak	Low-speed streak with several SCSs
Process 1	Rapid outward motion	Rapid outward motion of the SCSs
Process 2	Rapid growth of oscillatory motion	Formation of the SCV
Process 3	Chaotic flows	Formation of the chain of ring vortices

in a classical structure is actually composed of several CSCs, as shown in Figure 2.41. Because the streamwise velocity is higher at the front of the SCS than that at the rear of it, this leads to a gradual connection between the adjacent SCSs and forms a streaky shape in the streamwise direction. Lee and Wu believed that there exists a reasonable interpretation for a turbulent bursting event based on the SCSs frame. Table 2.1 gives the comparison of the bursting process between the classical and SCSs-based descriptions.

It is well known currently that the turbulent bursting is associated with motion behavior of the fluid near the wall. As mentioned above, there exist two predominant motions of fluid in the near-wall region of the boundary layer, that is, ejection and sweep motions which are associated with streamwise vortices as illustrated in Figure 2.42. The streamwise vortex can induce the low-speed fluid at one side of the vortex away from the wall and form a ejection activity ($u' < 0, v' > 0$, the superscript means the fluctuation). To fill the space in which fluid has been ejected, the high-speed fluid on the top of the vortex rapidly rushes toward

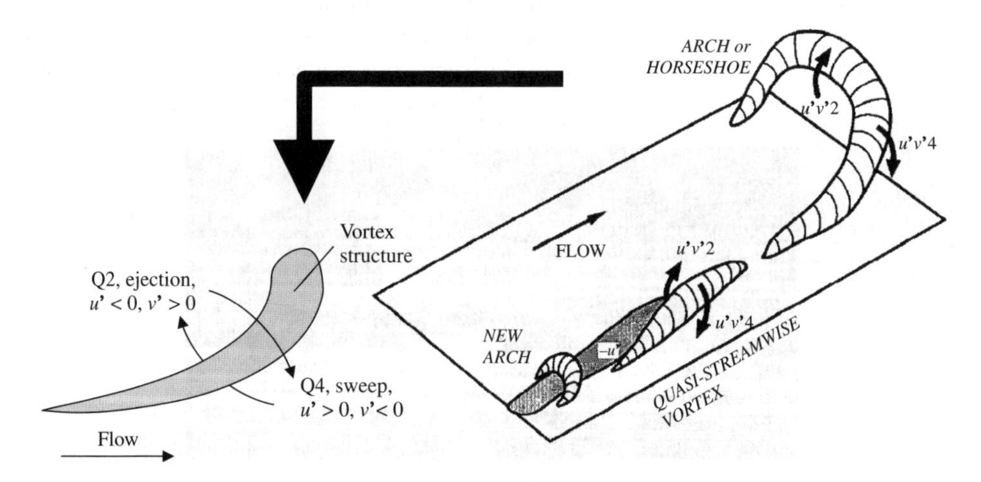

Figure 2.42 Relationship between the vortex structures and the fluid motions (ejection and sweep). *Source*: Robinson S K 1991. Reproduced with permission of Annual Reviews

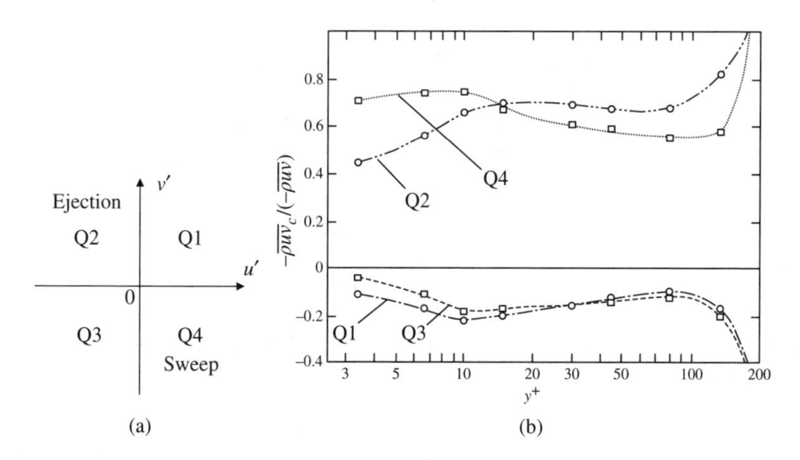

Figure 2.43 (a) Quadrants of the instantaneous $u'v'$-plane and (b) distributions of Reynolds stress of different quadrants.[73] *Source*: Wallace J M 1972. Reproduced with permission of Cambridge University Press

the wall and forms a sweep activity ($u' > 0, v' < 0$). Both of ejection and sweep activities produce the Reynolds stress ($-\overline{u'v'}$) that accounts for about 40–80% of the total Reynolds stress in near-wall turbulence,[73] and therefore contributes to the turbulent bursting and the skin-friction drag formation.

The quadrant analysis can be used to further identify the ejection and sweep activities, as shown in Figure 2.43. According to the sign of velocity fluctuation components, u' and v', the ejection and sweep activities correspond to the second (Q2) and fourth (Q4) quadrants, respectively, as shown in Figure 2.43a. A hot-film measurement result[73] in Figure 2.43b shows that Q2 and Q4 activities contribute to the most of the Reynolds stress compared with Q1 and Q3 activities. In the region near the wall ($y^+ < 15$), Reynolds stress produced by Q2 and Q4 activities grows, however, contribution of sweep (Q4) activity exceeds that of ejection (Q2) activity. This result implies that the formation of skin-friction drag is mainly due to the sweep of high-speed fluid toward the wall. Away from the wall, the Reynolds stresses by Q2 and Q4 reach their maxima and then gradually decay. Within the region of $y^+ > 15$, the contribution of Q2 to the Reynolds stress exceeds the contribution of Q4.

Abovementioned discussions indicate that the formation of skin-friction drag in near-wall turbulence mainly attributes to the violent sweep motion of high-speed fluid toward the wall that is induced by streamwise vortices. This implies that streamwise vortices play an important role in generating skin-friction drag.

2.5.2 *Self-Sustaining of a Coherent Structure*

Evolution of a coherent structure for the developed wall turbulence is closely associated with the self-sustaining of coherent motion. Studies by Hamilton *et al.*[76] and Waleffe[57] found that turbulence can be sustained in the form of an autonomous cycle in the near-wall region. During the cycle, the turbulence is sustained by the interaction between the streaks and streamwise vortices instead of any other inputs from outer flow, as illustrated in Figure 2.44. From

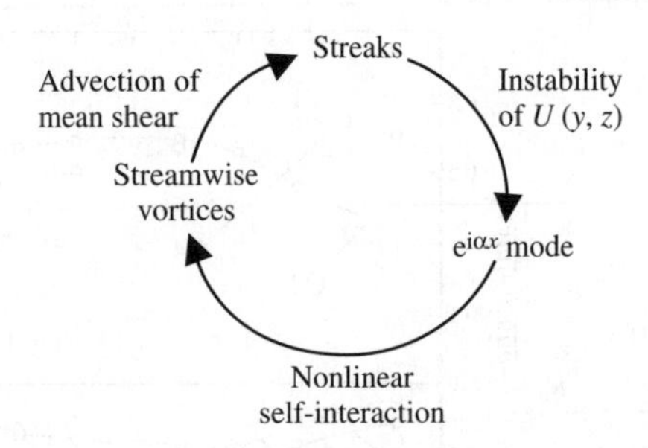

Figure 2.44 The self-sustaining process.[57] *Source*: Waleffe F 1997. Reproduced with permission of AIP Publishing LLC

the figure, it can be seen that the self-sustaining process consists of three phases, that is: (1) streamwise vortices redistribute the streamwise momentum and lead to the streaks with the velocity profile of $U(y, z)$, (2) the inflection (wake-like) instability of streaks then results in the growth of the disturbances with the form of $e^{i\alpha x}$, and (3) the primary nonlinear effect resulting from the development of the instability is to reenergize the original streamwise vortices and thus allowing the self-sustaining nonlinear process.

Jiménez and Moin[77] proposed a concept of "minimal channel unit" to identify the self-sustaining cycle in the near-wall region. They shrunk the turbulent channel size to the minimal level so that turbulence can survive without any outer inputs. In this two-wall minimal channel turbulent flow, the centerline of the channel locates within the inner region of either wall. Thus, the turbulent activity is isolated and not disturbed by motions of the outer region. Figure 2.45 shows the time histories of averaged wall shear for two sizes of minimal channel flow.[77] Here, τ_w is the mean wall shear, U_c is the mean flow velocity at

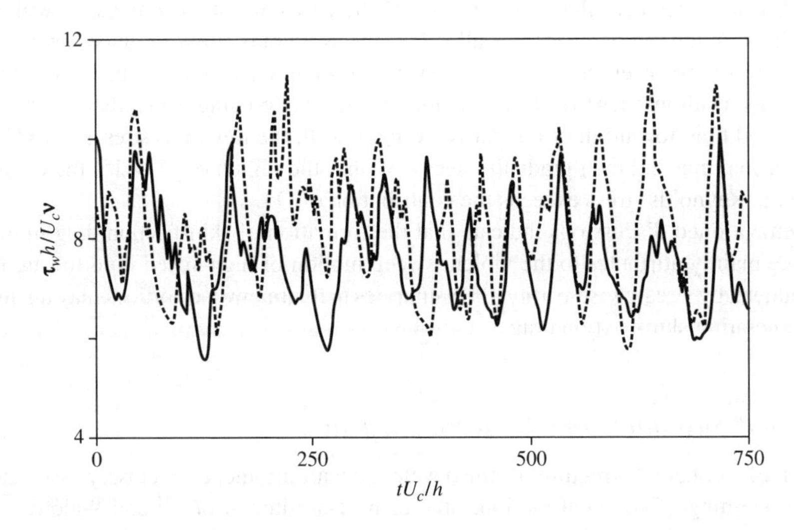

Figure 2.45 Instantaneous averaged wall shear history for different minimal turbulent channel sizes.[77] Only one wall is represented. Re = 5000. Solid line: $L_x \times L_z = \pi \times 0.18\pi$, dashed line: $L_x \times L_z = 0.5\pi \times 18\pi$. *Source*: Jiménez J 1991. Reproduced with permission of Cambridge University Press

the centerline, h is the half-height of channel, and ν is the viscosity. One can see the random but quasi-periodic behavior of turbulent flow. Each quasi-periodic cycle actually represents an intermittency cycle of turbulence that is a modulation of the bursting activities in time. The study also indicated that the time period of the cycle ranges from $t^+ = 600$ to 1500 that is depended on the Reynolds number. Furthermore, by selectively removing different links in the cycle for decaying the turbulence, Jiménez and Pinelli[78] confirmed that the cycle is able to sustain by itself in the near-wall region ranging from $y^+ = 20$ to 60 and that the turbulent statistical results of the minimal channel turbulence are the similar as those from fully developed turbulence.

The self-sustaining feature of the streaks-streamwise vortices cycle is the main mechanism of the persistent generation of the skin-friction drag in the near-wall region of boundary layer flow. If the cycle is broken off, we can decay the drag formation and even give rise to the laminarization of the turbulent boundary layer. Possible approaches that break off the cycle include the following:

1. *Elimination of streaks*. Streaks are the one of the sources for generating streamwise vortices. If one can eliminate the streaks stayed in the near-wall region of flow, the generation of streamwise vortices will be inhibited. This then leads to the inhibition of skin-friction drag.
2. *Suppression of instability of streaks*. Streamwise vortices can be generated only if streaks are unstable, although the streaks have existed. Therefore, if we stabilize the streaks in the viscous sublayer and prevent them from lift up and instability, the generation of streamwise vortices and subsequent skin-friction drag will be blocked.
3. *Elimination of streamwise vortices*. Streamwise vortices are the direct source of the formation of skin-friction drag. Although the streaks instability occurs, we can still reduce the drag by continuously eliminating the vortices (or a vortex packet) that are induced by unstable streaks or parent vortices.
4. *Decay or inhibition of a transition to turbulence*. A transition from laminar to turbulent flow is the process of growth and instability of the disturbance waves, which is involved in the linear and nonlinear growth of two-dimensional TS waves and the nonlinear instability of three-dimensional disturbance waves. Suppressing the growth of disturbance waves can result in decay or inhibition of the transition to turbulence and therefore skin-friction drag.

In the coherent motion of near-wall turbulence, streaks and streamwise vortices are indirect and direct sources of the formation for skin-friction drag, respectively. Statistically, streamwise vortices have a streamwise length of 200–300 wall units and a relatively short life time. Moreover, the vortex is easy to be disturbed by other factors of flow field. Therefore, controlling streamwise vortices is relatively difficult and usually requires sensitively close (feedback) loop strategies. However, the streaks usually have a streamwise length of about 1000 wall units and a relatively long life time. Thus, it is easier to control the streaks for the purpose of the drag reduction. Various control approaches for drag reduction will be introduced in the next chapters.

2.6 Closed Remarks

Wall turbulence is one of the complex flows of fluid and can be formed via transition from laminar to turbulent flow in the boundary layer excited by the external disturbances. Once the fully developed turbulence forms, it will evolve and sustain in a coherent structure form that is composed of velocity streaks and streamwise-dominated vortices. The transition in a boundary layer and the coherent structures for wall turbulence have the significant influences on the skin-friction drag formation and the heat and mass transfers. Understanding the transition and the coherent motion can be used well to predict and control the flow behavior in wall-bounded turbulence, and therefore attain the goal for drag reduction, vibration reduction, and transfer enhancement of mass and heat in practical applications.

For the process of the transition in the boundary layer, flow instability excited by the external disturbances is the main reason of the transition to a turbulent state of fluid. Due to the external disturbance, laminar fluid undergoes series of stages, including receptivity, linear instability (and/or transient growth), and nonlinear instability, and finally develops to turbulent fluid. In the receptivity stage, external disturbance, such as acoustical wave or vorticity, is transformed into the two-dimensional disturbance wave within the boundary layer. The initial growth of the disturbance wave can be linearly described by Orr–Sommerfeld equations whose eigen-solution is the TS wave. When the growth of the TS wave reaches the lower branch of the neutral curve of the linear stability theory, the instability of the TS wave is controlled by the nonlinear process. In this nonlinear stage, the two-dimensional TS wave is firstly transformed into the three-dimensional disturbance wave. The breakdown of the three-dimensional instability wave (also called secondary instability) can be divided into two types, that is, the K-regime transition and subharmonic transition. The former is associated with the interactions between instability wave and its high-order harmonic waves, while the latter is involved in the subharmonic resonance of the disturbance waves. The breakdown of the instability wave ultimately promotes the appearance of a turbulent spot. The development and merging of several turbulent spots lead to the formation of fully developed turbulence. Note that the two-dimensional disturbance wave (TS wave) may be bypassed when the amplitude of external disturbance is enough large. In this bypass transition, external disturbance subjected to free-stream turbulence or passing wake can induce the streaky structure in the near-wall region and then rapidly and directly lead to the generation of turbulence.

In the transitional and fully developed near-wall turbulence ($y^+ < 200$), coherent motion dominates the evolution of turbulent flow. Low-speed streaks in streamwise-elongated and spanwise-alternate fashion and streamwise vortices in the three-dimensional topological structure are the basic elements of the coherent structure. The streaks are formed through advecting the fluid by streamwise vortices, and then undergoing the lift-up and instability that in turn induce new streamwise vortices. On the other hand, streamwise vortices can generate off-spring vortices and a vortex packet that in turn promote the formation of the streaks. The streak–streamwise vortex interaction near the wall results in the relatively stable cycle (periodical) motion that is self-sustaining and independent of the outer layer structure of the boundary layer.

During the self-sustaining cycle of near-wall turbulence, streamwise vortices transfer fluid and thus lead to the eruptive ejection and sweep motions of fluid. These motions (especially sweep process) are the main reason of the turbulent bursting and skin-friction drag formation. According to the features of coherent motion of near-wall turbulence, people can pertinently control the boundary layer transition and coherent structure in order to achieve the goal of drag reduction.

References

[1] Schubauer G B and Skramstad H K. Laminar boundary-layer oscillations and transition on a fiat plate. J. Res. Nat. Bur. Stand., 1947, 38:251–292.
[2] Saric W S, Reed H L, and Kerschen E J. Boundary-layer receptivity to freestream disturbances. Annu. Rev. Fluid Mech., 2002, 34:291–319.
[3] Alfredsson P H, Bakchinov A A, Kozlov V V, and Matsubara M. Laminar turbulent transition at a high level of a free stream turbulence. In P W Duck and P Hall (Eds.), Nonlinear Instability and Transition in Three-Dimensional Boundary Layers. Kluwer, Dordrecht , 1996. pp. 423–436.
[4] Morkovin M V. On the many faces of transition. In C S Wells (Ed.), Viscous Drag Reduction. Plenum, London, 1969. pp. 1–31.
[5] Wanderly J B V and Corke T C. Boundary-layer receptivity to freesteam sound on elliptic leading edges of flat plates. J. Fluid Mech., 2001, 429:1–29.
[6] Aizin L B and Polyakov N F. Generation of Tollmien–Schlichting wave by sound at an isolated surface roughness. Novosibirsk: RAS. Sib. Branch Inst. Theoret. Appl. Mech., 1979, pp. 17–19.
[7] Goldstein M E. Scattering of acoustic waves into Tollmien–Schlichting waves by small streamwise variations in surface geometry. J. Fluid Mech., 1985, 154:509–529.
[8] Dietz A J. Local boundary-layer receptivity to a convected free-stream disturbance. J. Fluid Mech., 1999, 378:291–317.
[9] Kendall J M. Experiments on boundary layer receptivity to freestream turbulence. AIAA Paper 98-0530, 1998.
[10] Yang Z Y and Voke P R. Numerical simulation of transition under turbulence. Technical Report ME–FD/91.01, Department of Mechanical Engineering, University of Surrey, UK, 1991.
[11] Ellingsen T and Palm E. Stability of linear flow. Phys. Fluids, 1975, 18:487–488.
[12] Landahl M T. A note on an algebraic instability of inviscid parallel shear flows. J. Fluid Mech., 1980, 98:243–251.
[13] Hultgren L S and Gustavsson L H. Algebraic growth of disturbances in a laminar boundary layer. Phys. Fluids, 1981, 24:1000–1004.
[14] Butler K M and Farrell B F. Three-dimensional optimal perturbations in viscous shear flow. Phys. Fluids, 1992, A4:1637–1650.
[15] Trefethen L N, Trefethen A E, Reddy S C, and Driscoll T A. Hydrodynamic stability without eigenvalues. Science, 1993, 261:578–584.
[16] Andersson P, Berggren M, and Henningson D S. Optimal disturbances and bypass transition in boundary layers. Phys. Fluids, 1999, 11:134–150.
[17] Schmid P J. Linear stability theory and bypass transition in shear flow. Phys. Plasma, 2000, 7:1788–1794.
[18] Schmid P J and Henningson D S. Stability and Transition in Shear Flows. Springer, New York, 2001.
[19] Klebanoff P S, Tidstrom K D, and Sargent L M. The three-dimensional nature of boundary-layer instability. J. Fluid Mech., 1962, 12:1–34.
[20] Knapp C F and Roache P J. A combined visual and hot-wire anemometer investigation of boundary-layer transition. AIAA J., 1968, 6:29–36.
[21] Herbert T. Secondary instability of boundary layers. Annu. Rev. Fluid Mech., 1988, 20:487–526.
[22] Sayadi T, Hamman C W, and Moin P. Fundamental and subharmonic transition to turbulence in zero-pressure-gradient flat-plate boundary layers. Phys. Fluids, 2012, 24:091104.
[23] Kachanov Y S. Physical mechanisms of laminar-boundary-layer transition. Annu. Rev. Fluid Mech., 1994, 26:411–482.
[24] Landau L D. On the problem of turbulence. Dokl. Akad. Nauk SSSR, 1944, 44:311–314.
[25] Orszag S A and Patera A T. Secondary instability of wall-bounded shear flows. J. Fluid Mech., 1983, 128:347–385.
[26] Bayly B J and Orszag S A. Instability mechanisms in shear-flow transition. Annu. Rev. Fluid Mech., 1988, 20:359–391.
[27] Craik A D D. Nonlinear resonant instability in boundary layers. J. Fluid Mech., 1971, 50:393–413.
[28] Emmons H W. The laminar-turbulent transition in a boundary layer: Part 1. J. Aero. Science, 1951, 18:490–498.
[29] Delo C and Smits A J. Volumetric visualization of coherent structure in a low Reynolds number turbulent boundary layer. Int. J. Fluid Dyn., 1997, 1:1–22.
[30] Lee C B and Wu J Z. Transition in wall-bounded flows. Appl. Mech. Rev., 2008, 61:030802.

[31] Grek G R, Kozlov V V, and Ramazanov M P. Laminar-turbulent transition at high free stream turbulence. Novosibirsk: RAS. Sib. Branch Inst. Theoret. Appl. Mech., 1987, pp. 8–87.
[32] Wygnanski I J, Harritonidis J H, and Zilberman M. On the spreading of a turbulent spot in the absence of a pressure gradient. J. Fluid Mech., 1982, 123:69–90.
[33] Narasimha R. On the distribution of intermittency in the transition region of a boundary layer. J. Aero. Science, 1957, 24:711–712.
[34] Klebanoff P S. Effects of free-stream turbulence on a laminar boundary layer. Bull. Am. Phys. Soc., 1971, 10:1323.
[35] Kendall J M. Experimental study of disturbances produced in a pre-transitional laminar boundary layer by weak free stream turbulence. AIAA Paper 85-1695, 1985.
[36] Matsubara M and Alfredsson P H. Disturbance growth in boundary layers subjected to free-stream turbulence. J. Fluid Mech., 2001, 430:149–168.
[37] Durbin P and Wu X H. Transition beneath vortical disturbances. Annu. Rev. Fluid Mech., 2007, 39:107–128.
[38] Luchini P. Reynolds number independent instability of the boundary-layer over a flat surface: Part 2. Optimal perturbations. J. Fluid Mech., 2000, 404:289–309.
[39] Schmid P J and Henningson D S. A new mechanism for rapid transition involving a pair of oblique waves. Phys. Fluids, 1992, A4:1986–1989.
[40] Elofsson P A and Alfredsson P H. An experimental study of oblique transition in plane Poiseuille flow. J. Fluid Mech., 1998, 358:177–202.
[41] Jacobs R G and Durbin P A. Simulations of bypass transition. J. Fluid Mech., 2001, 428:185–212.
[42] Wu X, Jacobs R G, Hunt J C R and Durbin P A. Simulation of boundary layer transition induced by periodically passing wakes. J. Fluid Mech., 1999, 398:109–153.
[43] Robinson S K. Coherent motions in the turbulent boundary layer. Annu. Rev. Fluid Mech., 1991, 23:601–639.
[44] Schoppa W and Hussain F. Coherent structure generation in near-wall turbulence. J. Fluid Mech., 2002, 453:57–108.
[45] Kim J, Moin P and Moser R. Turbulence statistics in fully developed channel flow at low Reynolds number. J. Fluid Mech., 1987, 177:133–166.
[46] Eckelmann H. The structure of the viscous sublayer and the adjacent wall region in a turbulent channel flow. J. Fluid Mech., 1974, 65:439–459.
[47] Kreplin H and Eckelmann H. Behavior of the three fluctuating velocity components in the wall region of a turbulent channel flow. Phys. Fluids, 1979, 22:1233–1239.
[48] Kline S J, Reynolds W C, Schraub F A, and Runstadler P W. The structure of turbulent boundary layers. J. Fluid Mech., 1967, 30:741–773.
[49] Kim H T, Kline S J, and Reynolds W C. The production of turbulence near a smooth wall in a turbulent boundary layer. J. Fluid Mech., 1971, 50:133–160.
[50] Smith C R and Metzler S P. The characteristics of low-speed streaks in the near-wall region of a turbulent boundary layer. J. Fluid Mech., 1983, 129:27–54.
[51] Theodorsen T. Mechanism of turbulence. In Proc. 2nd Midwestern Conference on Fluid Mechanics, Columbus, Ohio State University, Ohio, 1952. pp. 1–19.
[52] Zhou J, Adrian R J, Balachandar S, and Kendall T M. Mechanisms for generating coherent packets of hairpin vortices in channel flow. J. Fluid Mech., 1999, 387:353–396.
[53] Wu X and Moin P. Direct numerical simulation of turbulence in a nominally zero-pressure-gradient flat-plate boundary layer. J. Fluid Mech., 2009, 630:5–41.
[54] Lian X Q and Su T C. The applications of the hydrogen bubble method to the investigations of complex flows. In The Visualization Society of Japan (Ed.), Atlas of Visualization II. CRC Press, Boca Raton, FL, 1996.
[55] Perry A E and Chong M S. A description of eddying motions and flow patterns using critical-point concepts. Annu. Rev. Fluid Mech., 1987, 19:125–155.
[56] Jeong J and Hussain F. On the definition of a vortex. J. Fluid Mech., 1995, 285:69–94.
[57] Waleffe F. On a self-sustaining process in shear flows. Phys. Fluids, 1997, 9:883–900.
[58] Holmes P, Lumley J L, Berkooz G, and Rowley C W. Turbulence, Coherent Structures, Dynamical Systems and Symmetry (2nd ed.). Cambridge University Press, Cambridge, 2012, p. 61.
[59] Andersson P, Brandt L, Bottaro A, and Henningson D S. On the breakdown of boundary layer streaks. J. Fluid Mech., 2001, 428:29–60.
[60] Asai M, Minadawa M, and Nishioka M. The instability and breakdown of a near-wall low-speed streak. J. Fluid Mech., 2002, 455:289–314.

[61] Konish F and Asai M. Experimental investigation of the instability of spanwise-periodic low-speed streaks. Fluid Dyn. Res., 2004, 34:299–315.
[62] Liu Y, Zaki T A, and Durbin P A. Floquet analysis of secondary instability of boundary layers distorted by Klebanoff streaks and Tollmien–Schlichting waves. Phys. Fluids, 2008, 20:124102.
[63] Wu J Z, Ma H Y, and Zhou M D. Vorticity and Vortex Dynamics. Springer, Berlin, 2006. pp. 535–538.
[64] Brooke J W and Hanratty T J. Origin of turbulence-producing eddies in a channel flow. Phys. Fluids A, 1993, 5:1011–1022.
[65] Hanratty T J and Papavassiliou D V. The role of wall vortices in producing turbulence, AIAA 98–2960, 1998.
[66] Heist D K, Hanratty T J, and Na Y. Observations of the formation of streamwise vortices by rotation of arch vortices. Phys. Fluids, 2000, 12:2965–2975.
[67] Bernard P S, Thomas J M, and Handler R A. Vortex dynamics and the production of Reynolds stress. J. Fluid Mech., 1993, 253:385–419.
[68] Smith C R, Walker J D A, Haidari A H, and Sobrun U. On the dynamics of near wall turbulence. Phil. Trans. Roy. Soc. London Ser. A, 1991, 336:131–175.
[69] Smith C R and Walker J D A. Turbulent wall-layer vortices. In S Green (Ed.), Fluid Vortices, Springer, Berlin, 1994.
[70] Lee C B. New features of CS solitons and the formation of vortices. Phys. Lett., 1998, A247:3971102.
[71] Lee C B and Chen S Y. Transition and Turbulence Control, In M Gad-el-Hak and H M Tsai (Eds.), Word Scientific, Singapore, 2005, pp. 38–85.
[72] Runstadler P G, Kline S J, and Reynolds W C. An experimental investigation of flow structure of the turbulent boundary layer. Report No. MD-8, Department of Mechanical Engineering, Stanford University, Stanford, CA, USA, 1963.
[73] Wallace J M, Eckelmann H, and Brodkey R S. The wall region in turbulent shear flow. J. Fluid Mech., 1972, 54:39–48.
[74] Blackwelder R F and Kaplan R E. On the wall structure of the turbulent boundary layer. J. Fluid Mech., 1976, 76:89–112.
[75] Kim J. Turbulence structures associated with the bursting event. Phys. Fluids, 1985, 28:52–58.
[76] Hamilton J M, Kim J, and Waleffe F. Regeneration mechanisms of near-wall turbulence structures. J. Fluid Mech., 1995, 287:317–348.
[77] Jiménez J and Moin P. The minimal flow unit in near-wall turbulence. J. Fluid Mech., 1991, 225:213–240.
[78] Jiménez J and Pinelli A. The autonomous cycle of near-wall turbulence. J. Fluid Mech., 1999, 389:335–359.

Part Two

Control of Wall Turbulence

3

Control of Turbulence with Active Wall Motion

A considerable degree of knowledge has been accumulated on the wall turbulence[1] as mentioned in Chapter 2. Although coherent motions in wall turbulence are grouped into many classes, the coherent structures in the near-wall region, whose basic elements are streaks and streamwise vortices including quasi-streamwise vortex (QSV) and hairpin vortex, are known to play a dominant role in the near-wall turbulent transport phenomena and the skin-friction drag on the wall.

The elongated QSVs, inclined in the vertical plane and tilted in the horizontal plane, overlap with alternating sign in x as staggered array. However, the legs of the hairpin vortex, which is closer to the wall than the head and the neck, can also be regarded as a pair of counter-rotating streamwise vortices. Therefore, QSVs populate the vicinity of the wall and construct main vortex structures. These vortices induce the wall-normal motion, ejecting low-speed fluid away from the wall and sweeping high-speed fluid toward the wall, which affects the wall shear stress strongly. The strongly energetic regions where the energy transfer occurs and the production and destruction of the Reynolds shear stress are mostly concentrated in the regions closed to QSV.[2, 3]

The streaks are recognized as the collection of the near-wall high- or low-speed fluids (negative or positive streamwise velocity fluctuations), and are formed due to the wall-normal motion associated with the streamwise vortices. A low-speed streak is created when low speed fluid at one side of the vortex is pumped away from the wall. To replace the ejected low momentum fluid, the high-speed fluid on the top of the vortex moves toward the wall, forming a high-speed streak. Therefore, the presence of streamwise vortices redistributes the longitudinal velocity into alternating high- and low-speed regions.

The low-speed streaks extend along the wall, migrate slowly and lift upward, then they begin to experience a violent oscillation or break-up as a sweep imposed by the outer flow field arrives. This process is named as "bursting." As a result of bursting events, the turbulent energy is produced due to transferring energy between an upstream high-speed fluid in outer regions and a downstream low-speed fluid in inner regions, which plays a key role in generating and maintaining the turbulent cycle. Moreover, the sweeps induced by streamwise vortices in

Principles of Turbulence Control, First Edition. Baochun Fan and Gang Dong.

the vicinity of the wall are strengthened through the inrush of high momentum fluid following the bursting, so that the sweeps are more important than ejection for the induction of high-skin friction in the turbulent boundary layer.

The skin-frictional drag of a turbulent wall-bounded flow is attributed to the existence of near-wall vortical structure and the associated ejection/sweep events. For channel flows with the same flux, a direct relation between the skin friction coefficient and Reynolds stress distribution can be derived as follows.[4]

Based on the base-fluctuation decomposition (see Section 1.6.2) with the base flow $U(y) = 1 - (1 - y)^2$, and the assumption of constant mass flux, periodic flow structures in the streamwise (x) and spanwise (z) directions, and no-slip boundary conditions, the Navier–Stokes (N-S) equation in the x direction averaged by integration in both the x and z directions, that is, the homogeneous directions, then becomes

$$\frac{\partial \overline{u'}}{\partial t} + \frac{\partial \overline{uv}}{\partial y} = -\Pi(t) + \text{Re}\frac{\partial^2 \overline{u}}{\partial^2 y} \tag{3.1}$$

where overbar "–" denotes an average over a y = constant plane.

Integration of Eq. (3.1) over y gives

$$\Pi(t) = -\frac{1}{\text{Re}}\left.\frac{\partial \overline{u}}{\partial y}\right|_w = -\tau_w = -\frac{C_f}{2} \tag{3.2}$$

where C_f is the skin friction coefficient, defined by $C_f = \tau_w / \left(\frac{1}{2}\rho U_c\right)$, U_c is the center line velocity.

Substitution of Eq. (3.2) into Eq. (3.1) results in

$$\frac{C_f}{2} = \frac{\partial}{\partial y}\left(\overline{uv} - \frac{1}{\text{Re}}\frac{\partial \overline{u}}{\partial y}\right) + \frac{\partial \overline{u'}}{\partial t} \tag{3.3}$$

By applying triple integration, that is, $\int_0^1 dy \int_0^y dy \int_0^y dy$, to Eq. (3.3), we have

$$C_f = \frac{4}{\text{Re}} + 6\int_0^1 (1 - y)(-\overline{uv})dy - 3\int_0^1 (1 - y)^2 \frac{\partial \overline{u'}}{\partial t} dy$$

For a stationary turbulence

$$\frac{\partial \overline{u'}}{\partial t} = 0$$

Hence

$$C_f = \frac{4}{\text{Re}} + 6\int_0^1 (1 - y)(-\overline{u'v'})dy \tag{3.4}$$

where all variables are normalized by the channel half-width and the center line velocity. $\overline{u'v'}$ is the Reynolds shear stress, produced directly by ejection and sweep activities ($v' > 0$ and $v' < 0$, respectively). This equation shows that skin-friction drag in a turbulent channel flow consists of the laminar drag and the y-weighted integral of $\overline{u'v'}$, which is regarded as a streamwise momentum transform in the wall-normal direction. The sweep activities are more significant than ejections in the vicinity of the wall due to the effects of bursting events, which creates the regions of high-skin friction and causes the y-weighted integral of $\overline{u'v'}$ to be always positive.

In the past, controlling turbulent flow artificially was believed to be extremely difficult, until the discovery of turbulent coherent structures. Improved understanding of the underlying physics of turbulent flow makes us possible to realize efficient turbulence control, especially with an objective of skin-friction drag reduction.[5, 6] Most of such attempts have focused on suppressing the near-wall coherent structures, only if the turbulence production cycle could be favorably altered by modifying or disturbing any part of the near-wall activities. Various control strategies have been proposed.[7, 8] Some control techniques are simple and more likely to be used in practice, such as longitudinal grooves or riblets, compliant coatings, and large-eddy breakup devices, and so on. It is known that turbulent control targeting toward specific coherent structures is most effective when the control input is introduced locally at a high receptivity region, but practical implementation of such control requires large number of microsensors and microactuators and will not be possible. However, microelectromechanical system (MEMS) developed recently,[9] in which electronic and mechanical components are combined on a single programmable chip having characteristic length of less than 1 mm, has provided opportunities for targeting the small-scale coherent structures in the near-wall region. Motivated by the quest to achieve efficient control of turbulent flow, many attempts have been made on the study of MEMS-based flow actuators, numerical devices for actuating a variety of flow fields have been advanced, whose design and build are more challenging than microsensors, such as blowing and/or suction embodied with various actuators, the synthetic jet actuator, piezoelectric flap actuator, electromagnetic actuator, and others.

The strategies of flow control are categorized in different ways[8, 10, 11] for example, based on energy expenditure (passive or active), the control loop (open-loop, closed-loop, or optimal control), the objectives of turbulent control (drag reduction, lift enhancement, and mixing augmentation, etc.) and control techniques applied at the wall or away from it (rigid-wall motion, compliant coating, and magneto- and electrohydrodynamic body forces, etc.), and others.

An incompressible fully developed turbulent flow can be described completely by the governing equations and corresponding initial and boundary conditions, thus the flow control strategies can also be classified into two types from the point of view of mathematics, that is controlled via altering the boundary conditions and controlled via altering the governing equations directly. The control via altering the boundary conditions includes: (1) surface protuberances (strips, bumps, grooves or riblets, and large-eddy breakup devices, etc.); (2) actively moving wall (oscillating wall and wavy wall, etc.) and compliant wall; (3) mass transfer through a porous wall or a wall with slots, as well as blowing and/or suction of primary fluid from the wall embodied with various actuators such as a resonance actuator, a speaker, a fan, a pump, and other devices; (5) energy transfer by heating and cooling of the surface, etc. Likewise, the control via altering the governing equations includes: (1) change in the fluid rheological properties, for example, viscosity, via introducing different additives, such as polymers, microbubbles, droplets, particles, dust or fiber, and surfactants, etc.; (2) body force distributing in the flow field, e.g., magneto- and electro-hydrodynamic body forces.

Before proceeding with closed-loop studies, we need to have a good understanding of open-loop dynamics, which is typically focused on improving understanding of flow control physics, and provides the foundation and motivation for closed-loop studies. Therefore, the open-loop control of wall turbulence is introduced in this section, and the optimal closed-loop control will be discussed in the next section. For the open-loop control in this section, two control strategies based on altering boundary conditions and the governing equations are presented in Chapters 3 and 4, respectively.

Considering the restriction on the length of the book, only turbulence control with actively moving walls is concerned in Chapter 3. Sections 3.1–3.3 are devoted to the turbulence control with oscillating or wavy plane walls. Where, flow induced by spanwise wall oscillation in a still viscous fluid, called Stokes second problem, is introduced in Section 3.1, as a preparative fundament for following discussions. Then, the experiments of near-wall turbulence flow controlled by lateral wall oscillations, performed in water channel and wind tunnel for incompressible and compressible flows, respectively, are presented in Section 3.2. Finally, in Section 3.3, numerical simulations of turbulence control with oscillating or wavy plane walls are provided and the detailed mechanisms are also discussed. Moreover, Sections 3.4–3.6 are devoted to the turbulence control with up-down oscillating or wavy deformed walls. The simple methods generating deformed walls are introduced in Section 3.4, and experiments of turbulence control with deformed walls in water channel and wind tunnel are presented in Section 3.5. Finally, numerical simulations of turbulence control with deformed walls are provided and the detailed mechanisms are also discussed in Section 3.6.

3.1 Stokes Second Problem

Investigation on flow induced by spanwise wall oscillation in a still viscous fluid is called Stokes second problem; moreover the boundary layer induced is referred to as Stokes layer.[12]

The wall moves sinusoidally according to

$$W = W_m \sin(2\pi t/T) \tag{3.5}$$

where, W is the spanwise velocity of the wall, W_m is the oscillation amplitude, and T is the oscillation period. Introducing the nondimensional parameters: $W^+ = W/W_m$, $T^+ = T{W_m}^2/\nu$, $t^+ = t{W_m}^2/\nu$, where ν is the kinematic viscosity, we have

$$W^+ = \sin(2\pi t^+/T^+) \tag{3.6}$$

Ignoring the higher order nonlinear terms, the governing equations for Stokes layer can be expressed as[13]

$$\frac{\partial w^+}{\partial t^+} = \frac{\partial^2 w^+}{\partial y^{+2}} \tag{3.7}$$

with the boundary conditions

$$w^+(y^+ = 0,\ \ t^+) = W^+$$

$$w^+(y^+ \to \infty,\ \ t^+) = 0$$

where $y^+ = yW_m/\nu$ is the nondimensional wall-normal distance.

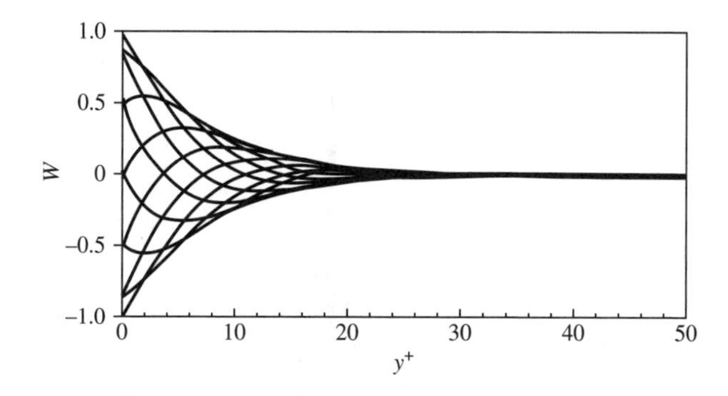

Figure 3.1 Spanwise velocity profiles and its envelop in Stokes layer[13]

The analytical solution of this equation is given as

$$w^{+}(y^{+},t^{+}) = e^{-\sqrt{\pi/T^{+}}y^{+}} \sin\left(\frac{2\pi t^{+}}{T^{+}} - \sqrt{\pi/T^{+}}y^{+}\right) \tag{3.8}$$

It indicates that the spanwise oscillation induced by the wall is damped away from the wall, and the Stokes layer at distance y^+ has a phase lag of $\sqrt{\pi/T^{+}}y^{+}$ with respect to the motion of the wall.

The envelope of these velocity profiles is expressed by

$$W_{\text{env}}^{+} = \exp(-\sqrt{\pi/T^{+}}y^{+}) \tag{3.9}$$

The different instantaneous spanwise velocity profiles and its envelope are shown in Figure 3.1, which are obtained from Eqs. (3.8) and (3.9). The effect of the oscillating wall decays exponentially from the wall, and is confined in the near-wall regions. The depth of the Stokes layer penetrating into the internal flow field is defined as

$$l^{+} = \sqrt{\frac{T^{+}}{\pi}} \ln\left(\frac{1}{W_{\text{th}}^{+}}\right) \tag{3.10}$$

where l^+ represents a thickness within the influence of the Stokes layer and W_{th}^{+} is the threshold value, usually set as $W_{\text{th}}^{+} = 0.5$.

According to Eq. (3.8),the maximal acceleration is

$$a_{m}^{+} = \frac{2\pi}{T^{+}} \exp\left(-\sqrt{\frac{\pi}{T^{+}}}y^{+}\right) \tag{3.11}$$

The wall moving back and forth in the spanwise direction can be regarded as a constant source of vorticity of alternate signs for the Stokes layer. Figure 3.2 shows the different instantaneous profiles of streamwise vorticity in a wall oscillation period, where y is normalized by the channel half-width δ. It can be observed that the streamwise vorticities are almost negative in time intervals $\left(\frac{7\pi}{4}, 2\pi\right)$ and $\left(0, \frac{3\pi}{4}\right)$, whereas they are positive in the time interval $\left(\frac{3\pi}{4}, \frac{7\pi}{4}\right)$.

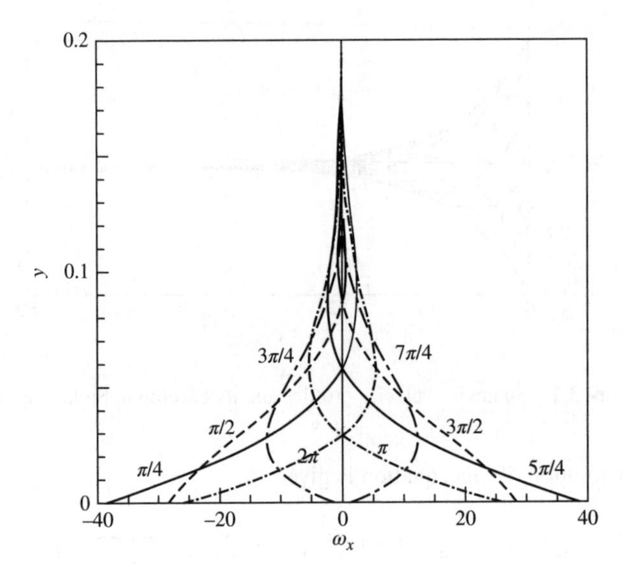

Figure 3.2 Streamwise vorticity profiles induced by wall oscillation in an oscillation period[14]

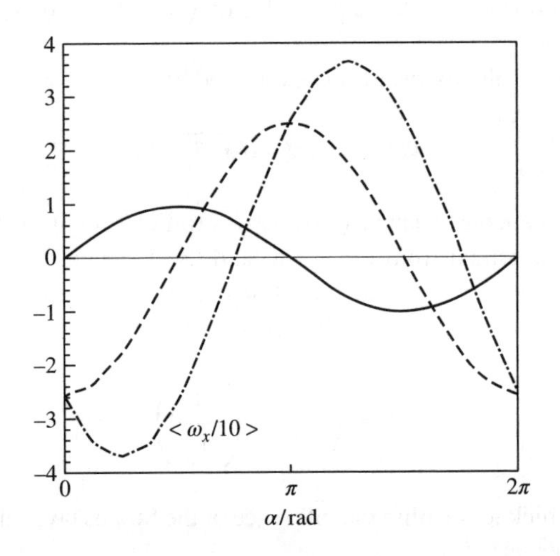

Figure 3.3 Variations of wall spanwise velocity, wall displacement, and streamwise vorticity with the wall phase angle α

The variations of wall spanwise velocity, wall displacement, and streamwise vorticity at $y^+ = 0.2$ with the wall phase angle, α during a cycle of wall oscillation are described by solid line, dashed line, and dash-dotted line, respectively, in Figure 3.3. For the spanwise velocity profile, the phase angle is lagged by $\phi = \pi/4$ relative to the wall displacement.

3.2 Experiments of Wall Turbulence with Spanwise Wall Oscillation

3.2.1 Incompressible Flow with Spanwise Wall Oscillation

For a channel flow, the spanwise oscillation in Stokes layer induced by laterally oscillating walls can modify the coherent structures of the near-wall turbulence, leading to suppression of the bursts and sweeps, and alternating high and low speed streaks.

3.2.1.1 Experimental Apparatus

The experiments of a near-wall turbulence flow controlled by lateral wall oscillations are usually performed in water channels and wind tunnels for incompressible and compressible flows, respectively. A schematic of water channel facility is shown in Figure 3.4.

The oscillating section is mounted on a pair of linear bearings; the motion of the surface is given by a crank slider mechanism connected to a computer-controlled stepping motor, whose oscillating wall mechanism is installed at a streamwise position of 3.5 m from the inlet of the test section, as shown in Figure 3.4. The spanwise wall displacement D followed a sinusoidal function of time

$$D = \frac{D_m}{2} \sin\left(\frac{2\pi t}{T}\right)$$

where $D_m/2$ is oscillation amplitude, D_m is peak-to-peak displacement of the wall oscillation, and T is the oscillation period.

When the freestream velocity is $U_c = 0.18$ m/s, then the measured boundary layer thickness is $\delta = 60\,\text{mm}$, the momentum thickness is $\theta = 7.5$ mm, the friction velocity is

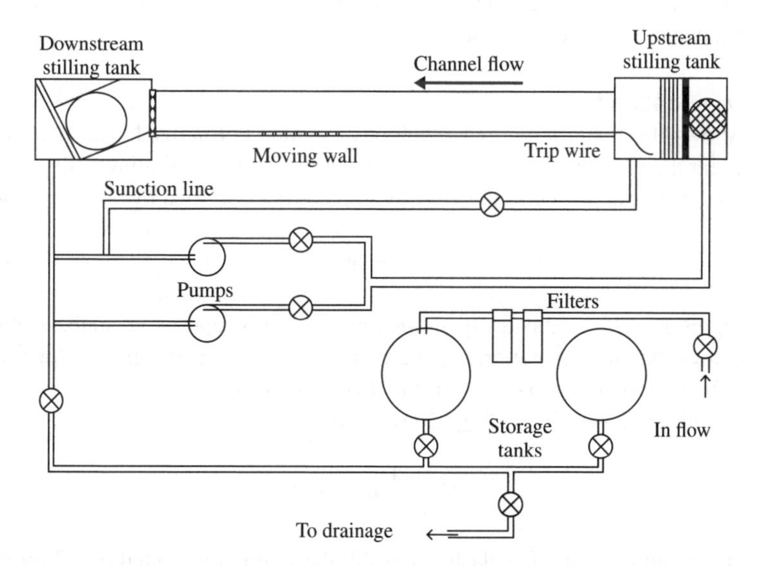

Figure 3.4 Schematic of water channel facility.[15] *Source*: Ricco P 2004. Reproduced with permission of Elsevier Limited

$u_\tau = 8.5$ mm/s, and $\nu/u_\tau = 0.1217$ mm. A corresponding Reynolds number is $\mathrm{Re}_\theta = 1400$ based on the momentum thickness.

In experiments, the near-wall streamwise velocity is measured by hot-film probe, and a two-component laser Doppler velocimeter (LDV) system is used to measure the streamwise and vertical velocity components. The boundary layer flow is visualized by hydrogen bubbles generated by hydrolysis of water, obtained by means of a platinum wire continuously energized as the cathode in an electric circuit and is recorded by video camera.

3.2.1.2 Experimental Results

Drag

Within the viscous sublayer ($y^+ < 5$), assuming a linear mean velocity profile, the wall shear stress is

$$\tau_w = \mu \frac{d\overline{U}}{dy} \approx \mu \overline{U}/y \tag{3.12}$$

Therefore, the drag on the wall can be obtained from the measured mean streamwise velocity for a position in the viscous sublayer.

The percentage of drag reduction rate is simply given by

$$Dr = (1 - \langle \tau_w \rangle / \langle \tau_{wn} \rangle) \times 100\% \tag{3.13}$$

where $\langle \tau_w \rangle$ denotes the average wall shear stress with spanwise oscillations and $\langle \tau_{wn} \rangle$ denotes the average wall shear stress without oscillation.

It has been indicated from the experiments that the drag reduction increases with increasing oscillation frequency or increasing oscillation amplitude.

Mean Streamwise Velocity

Figure 3.5 shows the mean streamwise velocity profiles normalized with u_τ, where black circles represent the fixed wall, and white circles represent the oscillating wall.

For the fixed wall, the profile in the viscous sublayer is described by a linear relationship

$$U^+ = y^+ \tag{3.14}$$

where the superscript "+" indicates quantities in wall units, that is, normalized with u_τ and ν, where the dimensional parameters u_τ and ν are the wall friction velocity for flows with a stationary wall and kinematic viscosity of the fluid, respectively.

And in logarithmic layer ($30 \le y^+ \le 200$), we have

$$U^+ = \frac{1}{\kappa} \ln y^+ + B \tag{3.15}$$

where, $\kappa = 0.41$ and $B = 5.0$. The dashed lines in the figure are plotted based on Eqs. (3.14) and (3.15), respectively.

The reduction in mean velocity occurs in the viscous sublayer due to the effects of the wall oscillation, which leads to the drag reduction.

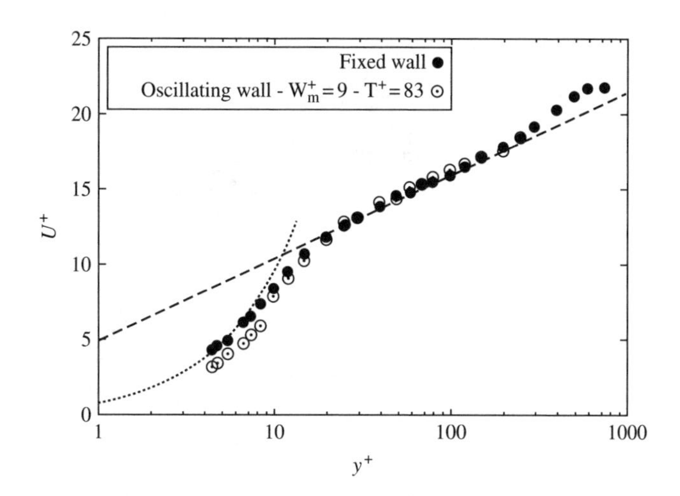

Figure 3.5 Mean velocity profiles of stationary wall and oscillating wall case.[17] *Source*: Ricco P 2004. Reproduced with permission of Elsevier Limited

Root Mean Square

Figure 3.6a and b shows root mean square (rms) streamwise and normal velocity profiles, respectively,where black circles present the fixed wall, and the other symbols represent the oscillating wall with different oscillatory parameters.

From Figure 3.6a, it has been observed that the rms streamwise velocity for the oscillating wall decreases up to the logarithmic-law region, whereas no influence is observed for $y^+ > 200$. The peak of the profile is reduced and shifts upward at a location of $y^+ \approx 25$, whereas the peak of the fixed-wall profile is at $y^+ \approx 15$. Figure 3.6b shows a decrease of the rms normal velocity and a shift of the profile peak similar to the case of the rms streamwise velocity profile.

Reynolds Stress

Reynolds stress profiles of stationary and oscillating wall case are shown in Figure 3.7, where black circles present the fixed wall and the white circles represent the oscillating wall. It can be seen that Reynolds stresses decrease more markedly than the mean velocity when the oscillations are imposed.

Probability Density Function (PDF) of Streamwise Velocity

The probability density functions (PDFs) of the streamwise velocity at $y^+ = 4, 9$, and 30 are presented in Figure 3.8a, b, c, respectively, where solid line denotes the oscillating wall, and dashed line denotes the stationary wall.

At the near-wall positions of $y^+ = 4$ and 9, there is a dramatic change in the PDF distributions for the oscillating wall relative to the stationary wall, but by $y^+ = 30$, there is no effect on PDF. The most significant change is a much larger peak in the low velocity range

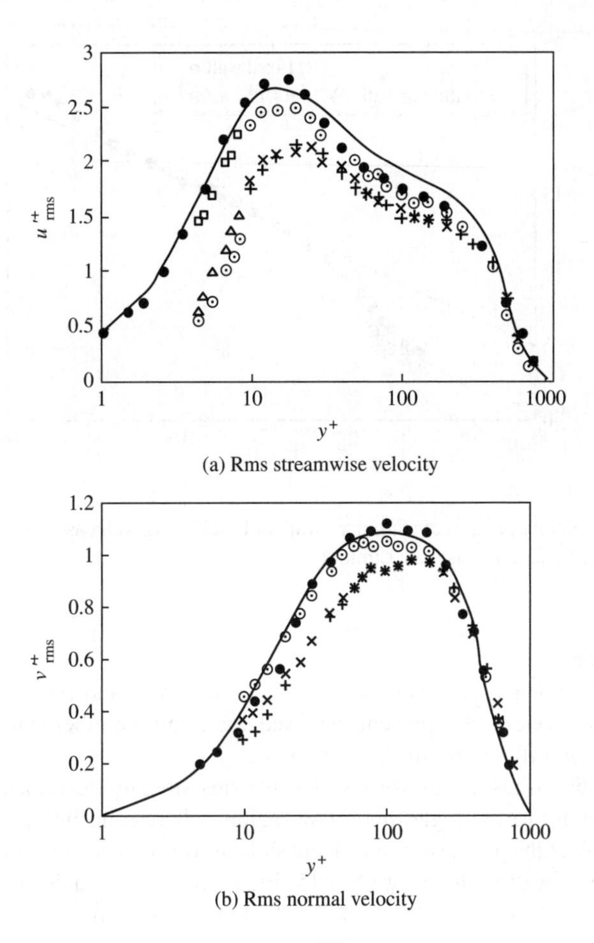

Figure 3.6 The rms velocity profiles scaled with u_τ.[15] *Source*: Ricco P 2004. Reproduced with permission of Elsevier Limited

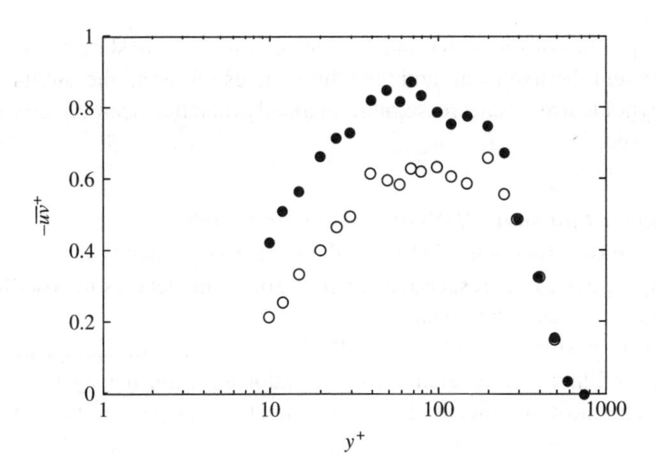

Figure 3.7 Reynolds stress profiles scaled with u_{τ_0}.[17] *Source*: Ricco P 2004. Reproduced with permission of Elsevier Limited

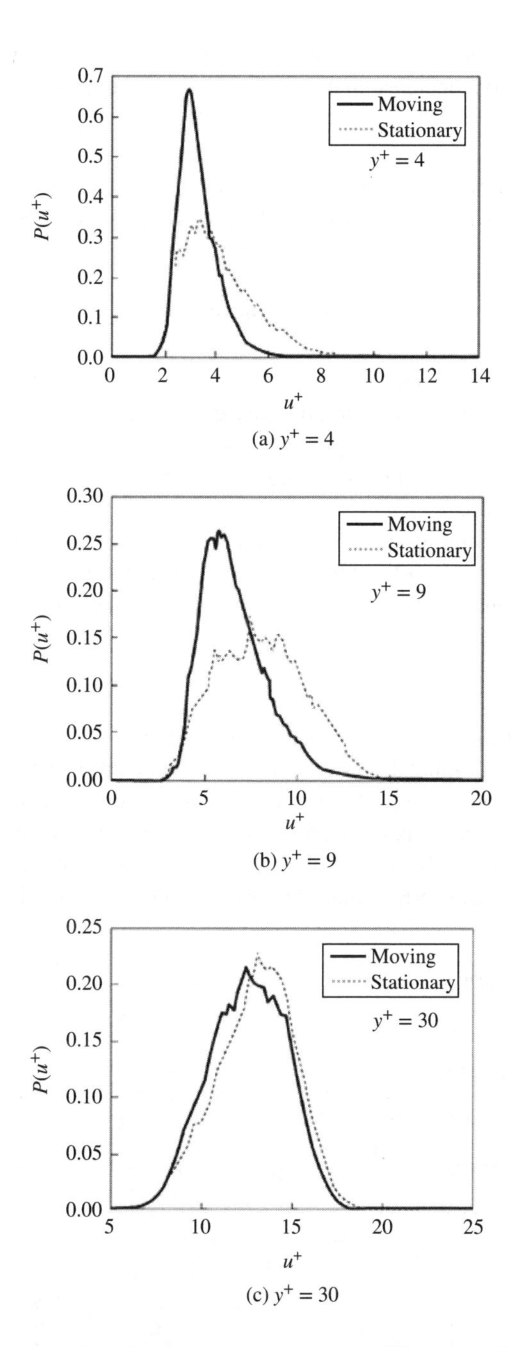

Figure 3.8 Probability density function for streamwise velocity.[16] *Source*: Trujillo S M 1997. Reproduced with permission of Cambridge University Press

for the oscillating wall case. Whereas in the range of higher velocities, there is a dramatic decrease, which leads to a decrease in mean velocity near the wall.

Burst

For near-wall turbulent flow, individual ejections occurring within a short time of each other can be designated as a single burst. The quadrant technique prescribes the occurrence of an ejection when $u' < 0$ and $v > 0$ (second quadrant), and

$$|u'v| > h|(\overline{u'v})_i| \tag{3.16}$$

where h is a threshold parameter, i indicates the uv quadrant. An analogous identification of sweeps is done based upon the detection of individual "injections," that is, events with $u' > 0$ and $v < 0$ (fourth quadrant) and the $|u'v|$ product exceeding the thresholds. Following the identification of all ejections occurring within a time record, the ejections are grouped into bursts based on a criterion.

The effects of the oscillating wall on the burst frequencies are present in Figure 3.9 for a distance from the wall of $y^+ = 30$ and for $x/\delta = 4.2$, where white circles represent the oscillating wall, and × represents the stationary wall. It shows that the oscillating wall does suppress bursts.

Flow Visualization

The alteration of the near-wall flow with wall oscillations can be described directly by flow visualization techniques. Figure 3.10 shows the streaky structures in the near-wall region visualized by hydrogen bubbles. Left image presents the stationary wall case, and right image presents the oscillating wall case. Flow direction is from right to left. When the wall oscillates, the streaky structure can still be identified, but it is more homogeneous, shorter in length, and convects downstream for a shorter time. The structures deviate from the streamwise direction.

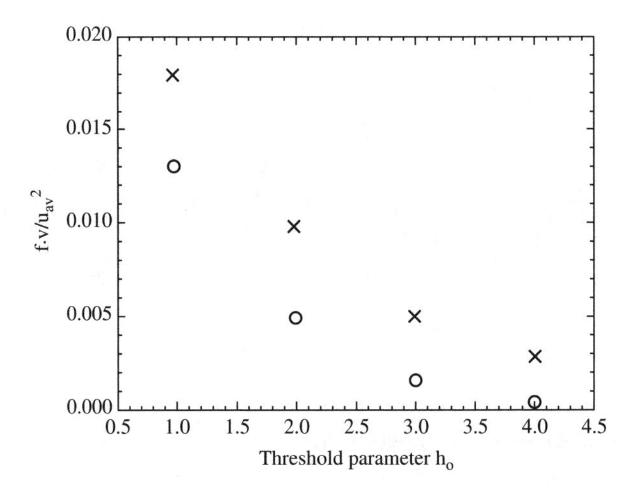

Figure 3.9 Effects of the oscillating wall on the burst frequencies scaled with stationary wall shear velocity.[16] *Source*: Trujillo S M 1997. Reproduced with permission of Cambridge University Press

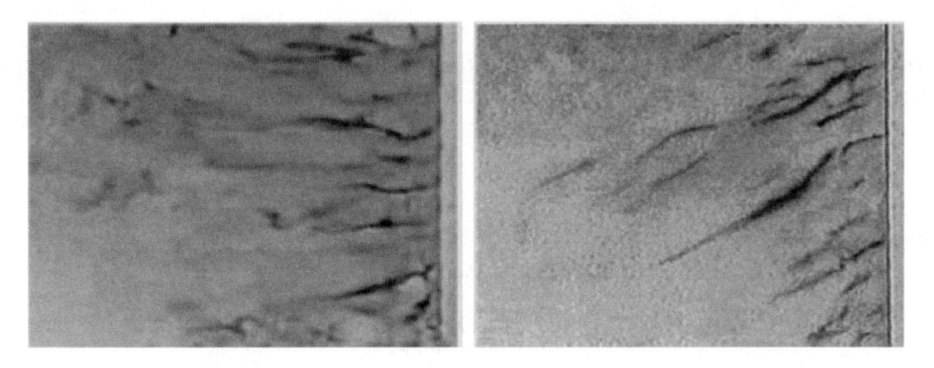

Figure 3.10 Streaky structures in the near-wall region.[17] *Source*: Ricco P 2004. Reproduced with permission of Elsevier Limited

3.2.2 Compressible Flow with Spanwise Wall Oscillation

3.2.2.1 Experimental Apparatus

The turbulent boundary layers in compressible flow can also be manipulated by oscillating a wall in the spanwise direction. The experiments are performed in wind tunnels as shown in Figure 3.11. The turbulent boundary layer flow is tripped at the inlet of the test section by a tripping wire. The leading edge of the oscillating plate is located downstream of the trip, set flush to the surrounding surface. The oscillation is produced by a crankshaft system. The drive system has a flywheel and a set of counter balances to maintain a constant angular velocity during the experiments with minimum vibrations. The pressure gradient along the length of the test section is nearly zero. When the freestream velocity is $U_c = 2.5\ \mathrm{m/s}$, the boundary layer thickness is $\delta = 60\ \mathrm{mm}$ at the trailing edge of the oscillating plate without wall oscillation. The corresponding Reynolds number is $\mathrm{Re}_\theta = 1190$ based on the momentum thickness.

In experiments, the near-wall velocity is measured by a hot-wire probe and a LDV system. The wall-surface temperature is measured by infrared thermography system. Flow

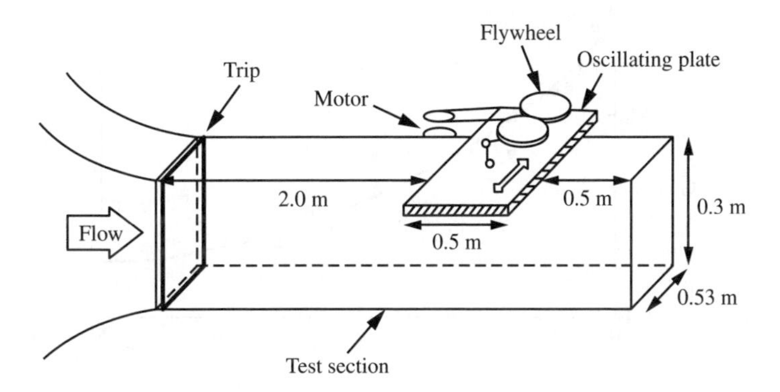

Figure 3.11 Schematic of wind tunnel facility.[18] *Source*: Choi K 1998. Reproduced with permission of Cambridge University Press

visualization of the near-wall turbulence is performed by a smoke-wire technique. A pulsed laser is used as a light source, which is fanned out with a cylindrical lens to produce a light sheet. The photographs are taken by video camera.

3.2.2.2 Experimental Results

Drag

The nondimensional wall speed can be expressed as

$$W^+ = \frac{D_m}{2}\omega/u_\tau$$

where $D_m/2$ is the oscillation amplitude, ω is the circular frequency, and u_τ is the friction velocity.

The drag reduction can be measured simply by C_f/C_{f0}, where C_f and C_{f0} are friction drag coefficients for the oscillating wall and the stationary wall, respectively. Figure 3.12 shows the variation of drag reduction by spanwise-wall oscillations with wall velocity W^+, where different symbols represent the data obtained from different experiments. It has been shown that the skin-friction drag reduction can be achieved by spanwise-wall oscillations, and nearly 45% drag reduction can be obtained at an optimum value of $W^+ = 15$.

Mean Streamwise Velocity

The mean-velocity profiles are shown in Figure 3.13 in a log-law plot. The curves drawn, covering the viscous layer, buffer layer, and log-law region, correspond to different frequencies of wall oscillation, but with the same oscillating amplitude of 70 mm. The profiles are seen to collapse onto a single curve in the region of the viscous sublayer but are otherwise shifted upward with an increase in oscillation frequency. Focusing only on the data in the viscous sublayer and the buffer layer as shown in Figure 3.14, it is clear that the mean velocity gradient in the near-wall region is significantly reduced with wall oscillation, which demonstrates that

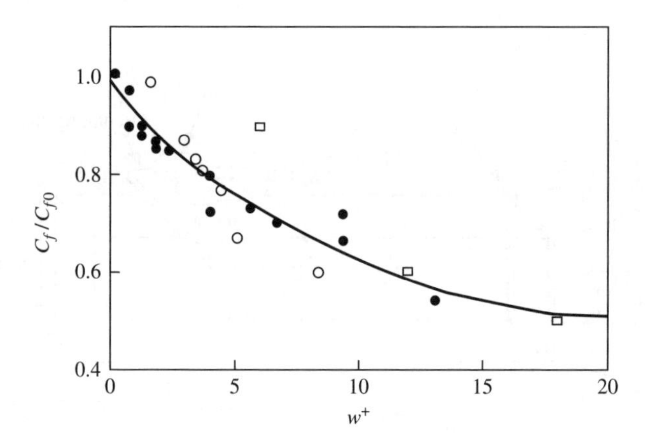

Figure 3.12 Variation of drag reduction by spanwise-wall oscillation with wall velocity.[19] *Source*: Choi K 2002. Reproduced with permission of Cambridge University Press

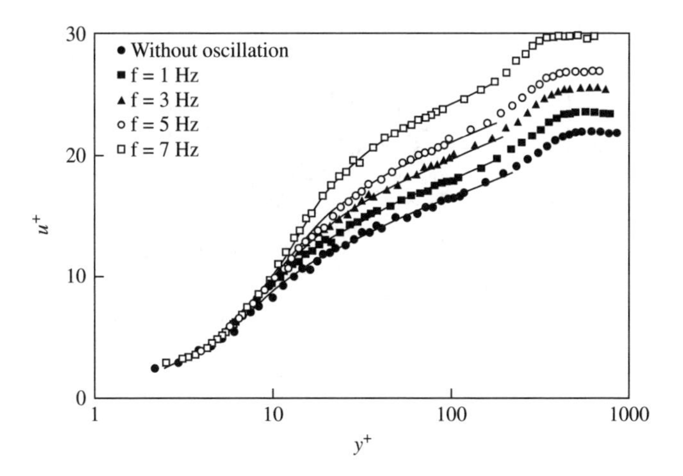

Figure 3.13 Mean-velocity profiles of boundary layer at different frequencies of wall oscillation.[18] *Source*: Choi K 1998. Reproduced with permission of Cambridge University Press

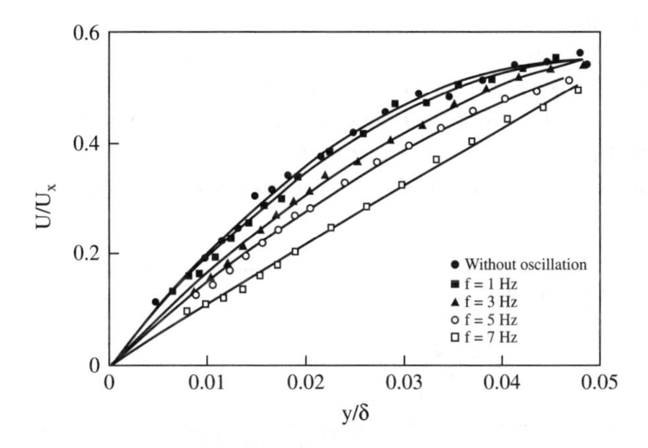

Figure 3.14 Mean velocity profiles in the near-wall region of boundary layer at different frequencies of wall oscillation.[18] *Source*: Choi K 1998. Reproduced with permission of Cambridge University Press

the wall-shear stress of the turbulent boundary layer is reduced. It is also shown that the extent of linear velocity region in the viscous sublayer is increased to $y^+ = 10$, whereas the viscous sublayer extends to only about $y^+ = 2.5$ for the stationary wall case.

Root Mean Square (rms)

The turbulent intensities, defined by rms velocity, of the boundary layer at different frequencies are shown in Figure 3.15. The reductions in the intensity values are evident within the inner region when wall is oscillated in the spanwise direction. The structure of the outer region of the boundary layer seems to be unaltered by the wall oscillation except that the boundary layer thickness seems to have been reduced.

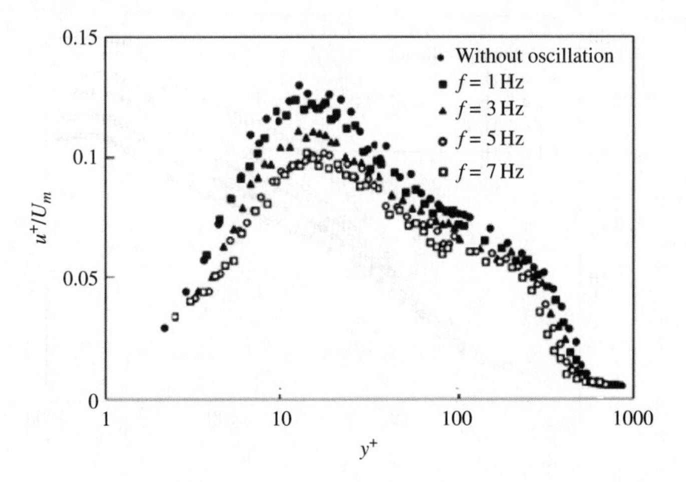

Figure 3.15 Turbulent intensity profiles of boundary at different frequencies of wall oscillation.[20] *Source*: Choi K 2001. Reproduced with permission of Cambridge University Press

Skewness and Kurtosis

The distributions of skewness and kurtosis of velocity fluctuations are presented in Figures 3.16 and 3.17. Both of the skewness and kurtosis seem to be increased within the near-wall region with an increase in turbulent drag reduction by the spanwise wall oscillation, which indicates that there is a good correlation between the increase in the higher moments of turbulence statistics in the near-wall region of the boundary layer and amount of drag

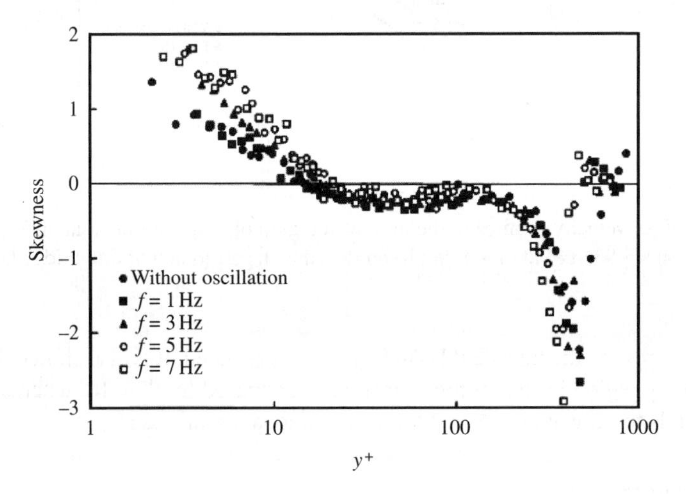

Figure 3.16 Skewness profile of boundary layer at different frequencies of wall oscillation.[20] *Source*: Choi K 2001. Reproduced with permission of Cambridge University Press

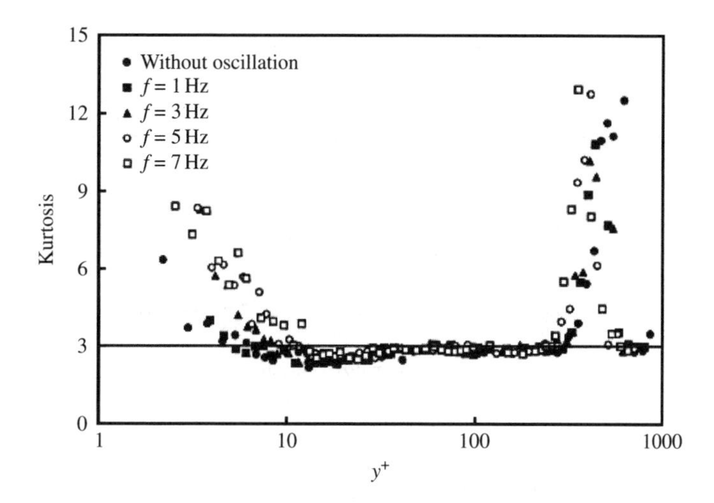

Figure 3.17 Kurtosis profile of boundary layer at different frequencies of wall oscillation.[20] *Source*: Choi K 2001. Reproduced with permission of Cambridge University Press

reduction. The increase in viscous sublayer thickness and the reduction in boundary layer thickness are also clearly seen in these figures.

Probability Density Function (PDF) of Streamwise Velocity

The probability density functions of the streamwise fluctuation velocity at $y^+ = 1.5, 4$, and 20 are presented in Figure 3.18a, b, c, respectively, where solid line denotes the oscillating wall, and dashed line denotes the stationary wall. In Figure 3.18c, presenting the PDF at $y^+ = 20$, about the value of the Stokes layer thickness, the difference between the solid and dashed lines is not as large as that in the viscous sublayer as shown in Figure 3.18b. The PDF of the velocity fluctuations at $y^+ = 4$, shown in Figure 3.18b, exhibits a long tail of positive probability, reflecting an increase in skewness and kurtosis within the viscous sublayer. It indicates that the positive velocity is predominant in this region. The elongation of the positive probability tail is not very clear in the PDF at $y^+ = 1.5$ as shown in Figure 3.18a. This is because the velocity signal at this near-wall location is strongly modulated by the oscillation of the Stokes layer. The location of maximum probability shifts to the positive side.

Conditionally Sampled Signature of Near-Wall Burst

The conditional averages of fluctuating streamwise velocity, obtained by using the variable interval time averaging (VITA) technique based on the detected bursting events at $y^+ = 1.5$, 4, and 20 are shown in Figure 3.19a, b, c, respectively, where solid line denotes the oscillating wall and dashed line denotes the stationary wall.

In Figure 3.19a, the change in the burst signature is remarkable in the near-wall region, where the duration of the burst is reduced to nearly one third of that without wall oscillation.

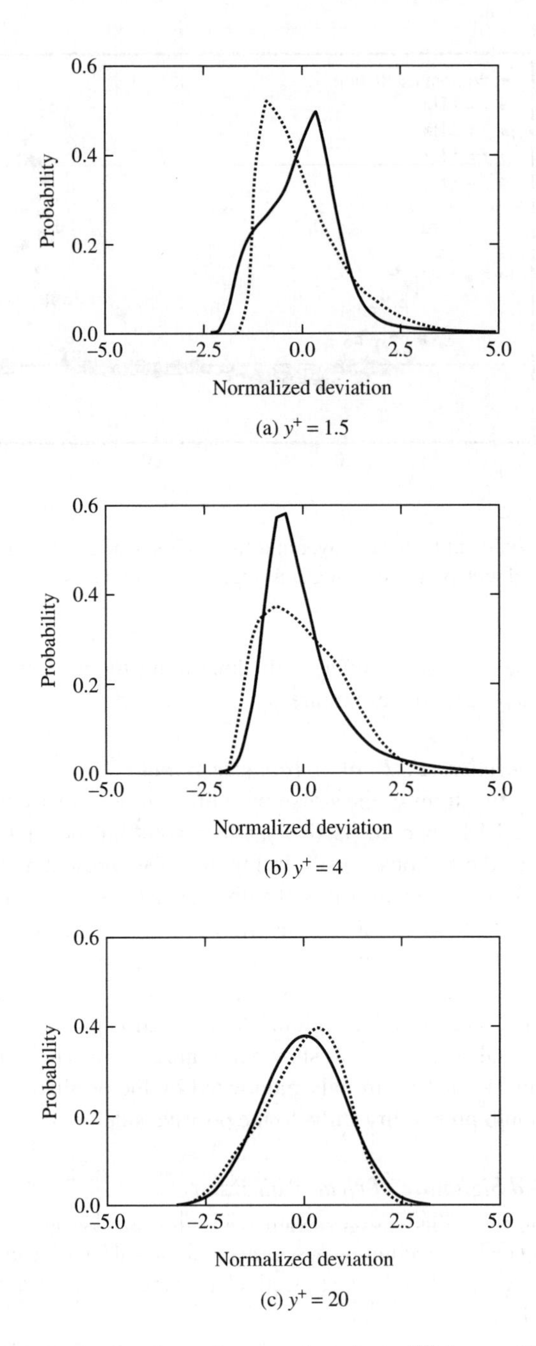

Figure 3.18 Probability density functions of streamwise velocity.[20] *Source*: Choi K 2001. Reproduced with permission of Cambridge University Press

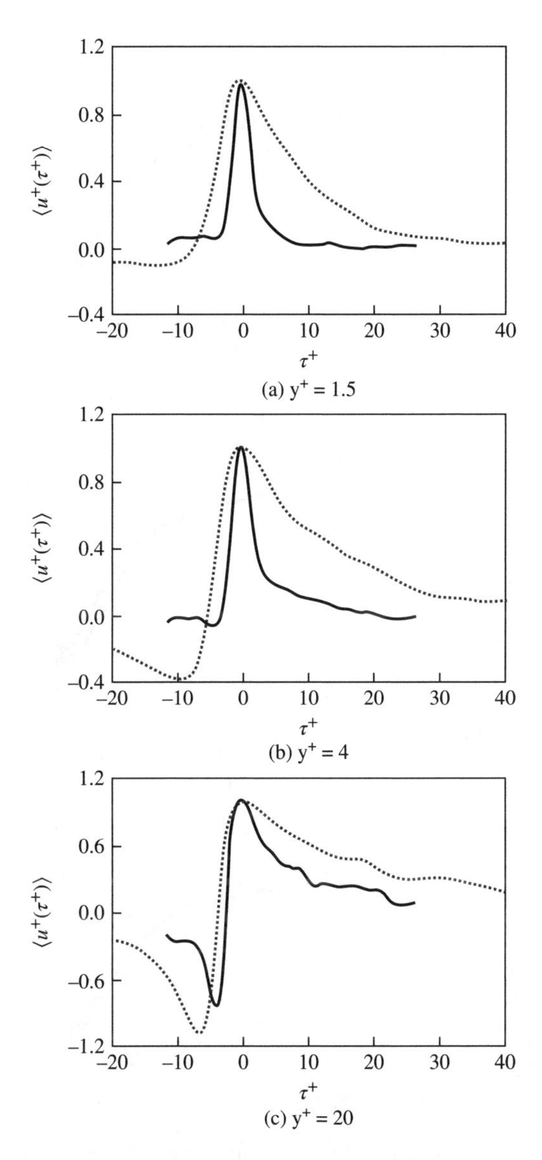

Figure 3.19 Conditionally sampled near-wall burst signatures.[20] *Source*: Choi K 2001. Reproduced with permission of Cambridge University Press

A similar reduction in burst duration is observed at $y^+ = 4$ as shown in Figure 3.19b. Even outside the viscous sublayer at $y^+ = 20$ shown in Figure 3.19c, the effect of wall oscillation on the burst signature is still significant, with the burst duration nearly a half of that without wall oscillation. It has been observed that the intensity of the near-wall bursts is greatly reduced by wall oscillation.

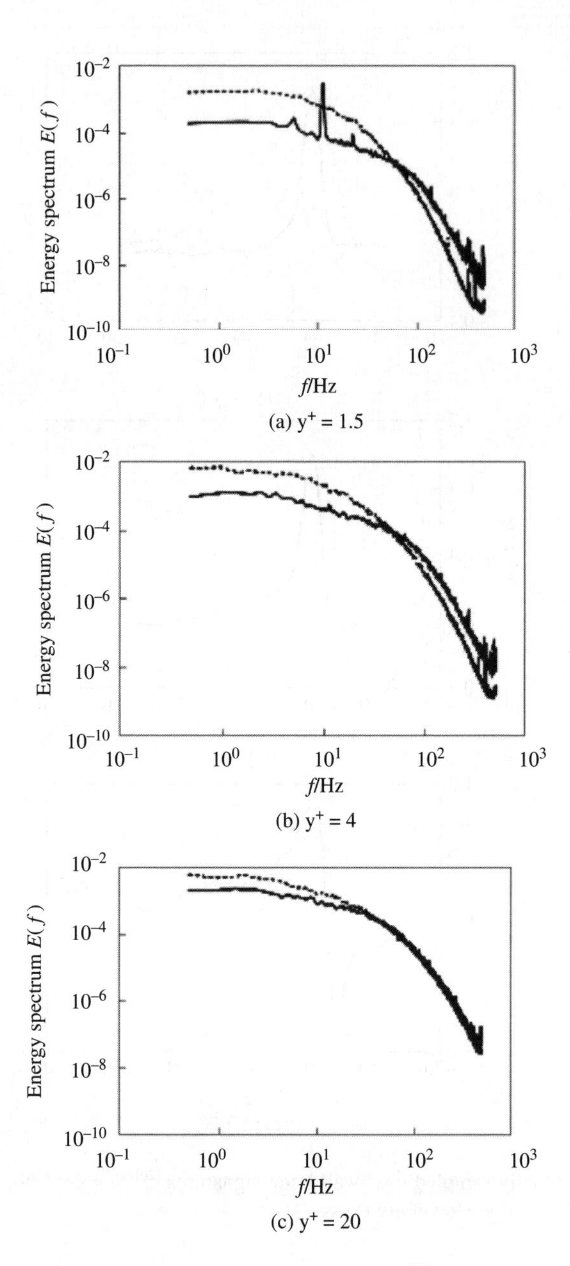

Figure 3.20 Energy spectra of streamwise velocity fluctuation.[20] *Source*: Choi K 2001. Reproduced with permission of Cambridge University Press

Energy Spectrum

The energy spectra of streamwise velocity fluctuations at $y^+ = 1.5,\ 4$, and 20 are shown in Figure 3.20a, b, c, respectively, where solid line denotes the oscillating wall and dashed line denotes the stationary wall. It has been clearly demonstrated that the turbulence energy is dramatically reduced at low frequencies (below 50 Hz) while the energy at higher frequencies is slightly increased. This seems to suggest that there is a transfer of energy from large-scale to small-scale turbulence eddies by the periodic Stokes layer developed over the oscillating wall. The spikes in the energy spectrum at $y^+ = 1.5$ are observed at the frequency of the wall oscillation from Figure 3.20a.

Flow Visualization

The photos of the near-wall structure of the turbulent boundary layer modified by the spanwise wall oscillation are shown in Figure 3.21. When an oscillating wall departs from its equilibrium with a positive displacement, the vortex sheet is tilted upward into the negative spanwise direction (upward) as shown in Figure 3.21a. However, as the wall displacement is negative, the vortex sheet is tilted downward into the positive spanwise direction (downward) as shown in Figure 3.21b.

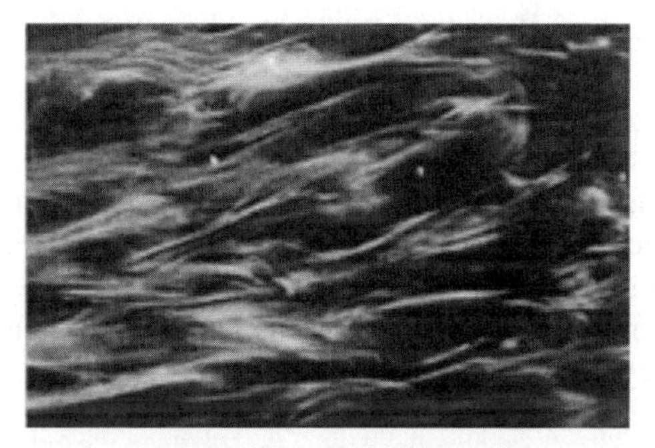

(a) Streamwise vortex tilted upward

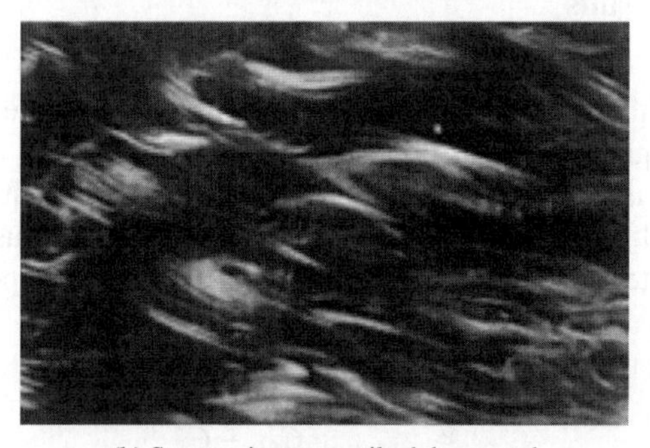

(b) Streamwise vortex tilted downward

Figure 3.21 Flow visualization of near-wall structure over oscillating wall.[19] *Source*: Choi K 2002. Reproduced with permission of Cambridge University Press

3.3 Numerical Simulation of Wall Turbulence with Spanwise Wall Oscillation

3.3.1 Wall Turblence with Spanwise Wall Oscillation

3.3.1.1 Governing Equations

The governing equations for an incompressible fully developed turbulent flow in a planar channel can be written as

$$\frac{\partial u_j}{\partial x_j} = 0 \tag{3.17}$$

$$\frac{\partial u_i}{\partial t} + \frac{\partial u_j u_i}{\partial x_j} - \frac{1}{\mathrm{Re}}\frac{\partial^2 u_i}{\partial {x_j}^2} + \frac{\partial p}{\partial {x_i}^2} = 0 \tag{3.18}$$

Here, all variables are normalized with respect to the channel half-width δ and the center line velocity U_c. p is the pressure, $\mathbf{u}$ is the velocity vector, ν is the kinematic viscosity, and Re is the Reynolds number. Subscript $i = 1, 2$, and 3 represent streamwise, wall-normal and spanwise directions, respectively. Repeated index j indicates a summation over all three values of the index.

The wall boundary condition for a laterally oscillating wall is

$$u_3|_W = W_m \sin(2\pi t/T) \tag{3.19}$$

The above equations are solved numerically based on standard Fourier–Chebyshev spectral method. Both a dealiazed Fourier method in the streamwise and spanwise directions and a Chebyshev-tau method in the wall-normal direction are used for the spatial derivatives. The time advancement is carried out by using a semi-implicit back-differentiation formula method with third-order accuracy.

3.3.1.2 Numerical Results

Drag

The time evolution of the friction drag on the wall subsequent to the start of oscillations at $t = 300$ is shown in Figure 3.22. It is observed that a drag reduction is obtained by the wall oscillation, whose size is strongly dependent upon the oscillation frequency. The largest drop is obtained at $T = 16$. It is also shown that these reductions can be sustained with a periodic oscillation in the long term after the flow has reached a statistically periodic steady state by the wall oscillation.

The percentage drag reduction $\%P_{\text{sav}}$ is defined as

$$\%P_{\text{sav}} = 100\frac{C_{f,0} - C_f}{C_{f,0}} \tag{3.20}$$

Figure 3.23 is a graphical representation of $\%P_{\text{sav}}$ in the $W_m^+ - T^+$ plane, Here, the superscript "+" indicates quantities in wall units. The percentage drag reduction, $\%P_{\text{sav}}$ is shown with a circle, its area being proportional to $\%P_{\text{sav}}$, the numerical value of which is

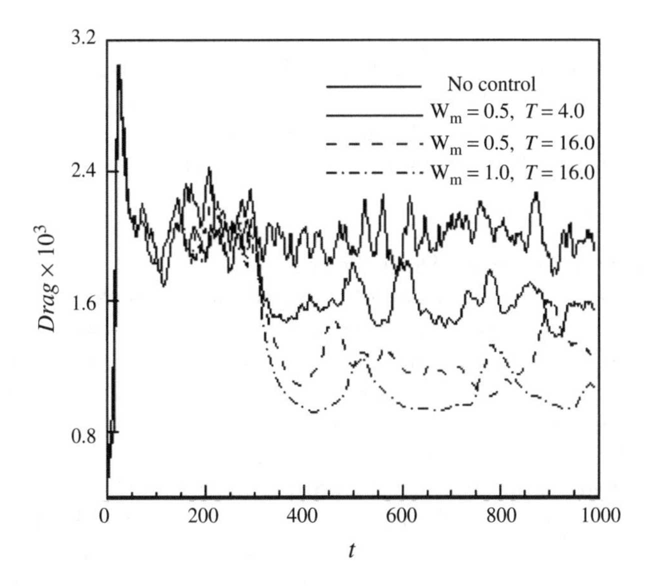

Figure 3.22 Time evolution of the friction drag on the oscillating wall.[14] *Source*: Huang L P 2010. Reproduced with permission of Elsevier

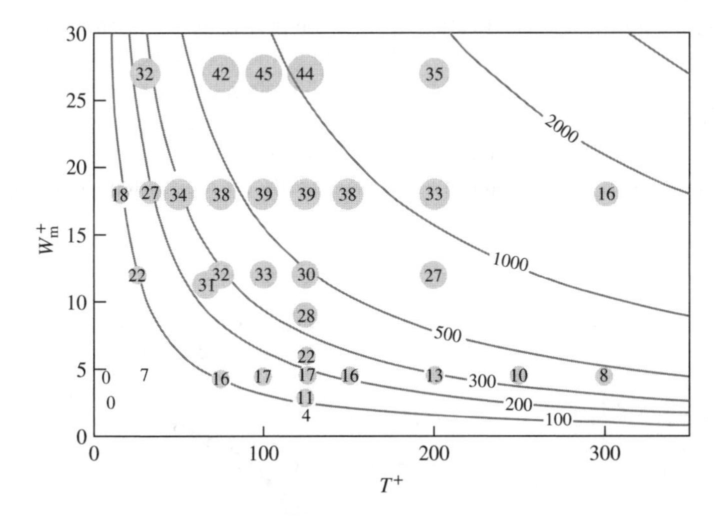

Figure 3.23 Three-dimensional plot of $\%P_{sav}$ in $W_m^+ - T^+$ plane.[21] *Source*: Quadrio M 2004. Reproduced with permission of Cambridge University Press

given inside. Hyperbolas in this plane are curves of constant maximum displacement. It has been observed that for a given value of W_m^+, the highest drag reductions can be attained in $100 \leq T^+ \leq 125$, and for a fixed T^+, drag reduction increases monotonically with W_m^+. Therefore, there exist optimum parameters, such as oscillation amplitude and period, leading to the maximum drag reduction.

Mean Velocity

Profiles of mean streamwise velocity are shown in Figure 3.24 in a log-law plot. The wall starts the oscillation of $T^+ = 100$ at $t^+ = 0$. It is seen that the profiles of mean streamwise velocity after $t^+ = 800$ when flow has reached a statistically periodic steady state, show a significant deviation from the regular law of the wall represented by a curve with $t^+ = 0$, which leads to the drag reduction.

Turbulent Intensity and Reynolds Shear Stress

Figure 3.25 shows the three components of turbulence intensities at $t^+ = 800$, 825, 850, and 875 in intervals of $T^+/4$, where the solid line represents the stationary wall and the white circles represent the experimental data. The turbulent intensities experience significant reductions by the wall oscillation, however, the percentage drops in v' and w' are significantly larger than the drop in u'. In addition to the drop in the magnitudes of turbulence intensities, the peaks of these quantities also move to the center of the channel.

Profiles of Reynolds shear stresses are shown in Figure 3.26 at $t^+ = 800$, 825, 850, and 875 in intervals of $T^+/4$, where the solid line represents the stationary wall and the white circles represent the experimental data. The streamwise component of Reynolds shear stress, $-\overline{uv}$ drops considerably; however, no significant changes are observed in the spanwise component of the Reynolds shear stress, $-\overline{vw}$.

Flow Visualization

The hydrogen bubble flow visualization can be simulated by calculating path of massless tracer particles introduced into the flow field along a line across the flow. Figure 3.27 shows instantaneous snapshots of the tracer particle distribution in the flow field. The flow direction is from the left to the right. Figure 3.27a, representing the stationary wall case, shows that the particles form well-defined streaks as they move downstream. Figure 3.27b, representing the oscillating wall case, shows that the tracer particles are shifted to the top as the wall

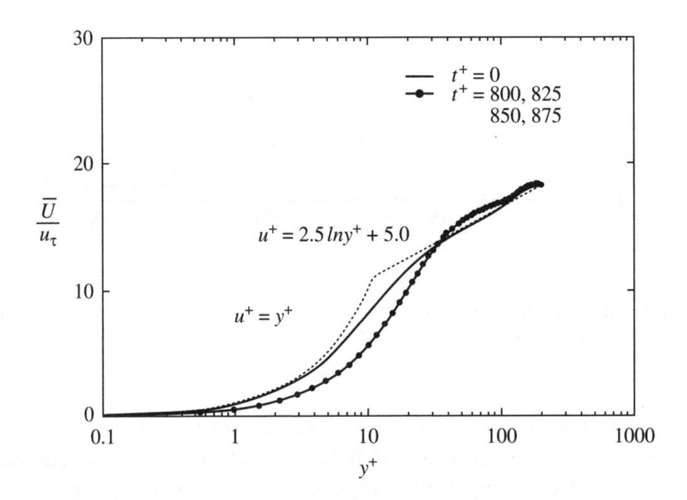

Figure 3.24 Mean streamwise velocity profiles.[22] *Source*: Jung W J 1992. Reproduced with permission of Cambridge University Press

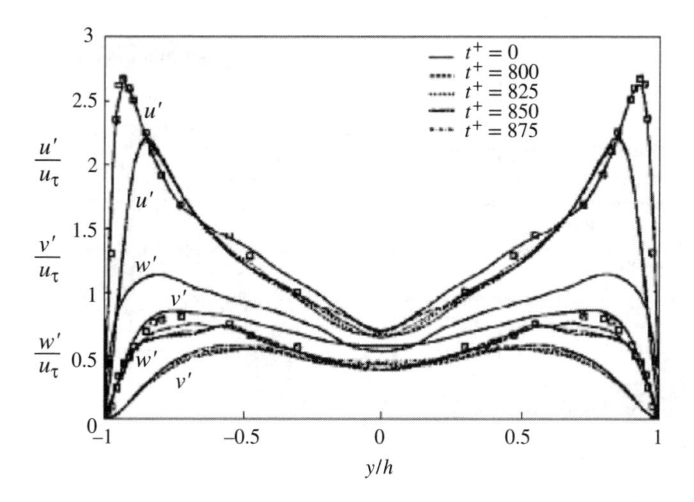

Figure 3.25 Profiles of three components of turbulence intensities.[22] *Source*: Jung W J 1992. Reproduced with permission of Cambridge University Press

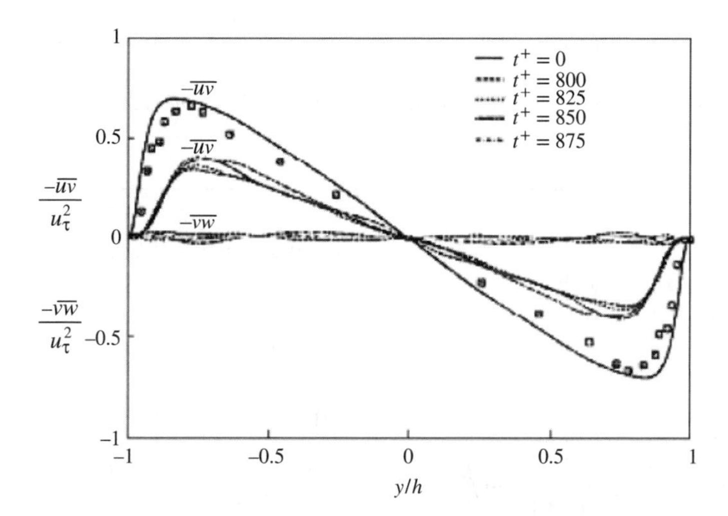

Figure 3.26 Profiles of Reynolds shear stress.[22] *Source*: Jung W J 1992. Reproduced with permission of Cambridge University Press

move in positive z-direction, and they are distributed much homogeneously than in stationary wall case.

3.3.2 *Control Mechanism of Spanwise Wall Oscillation*[14, 24, 25]

As mentioned in the Section 3.3.1.2, drag reduction is sustained and fluctuates periodically with elongated periodicities, while the flow is maintained upon a statistically periodic steady state under the manipulation of wall oscillations. Figure 3.28 shows a closer inspection of the

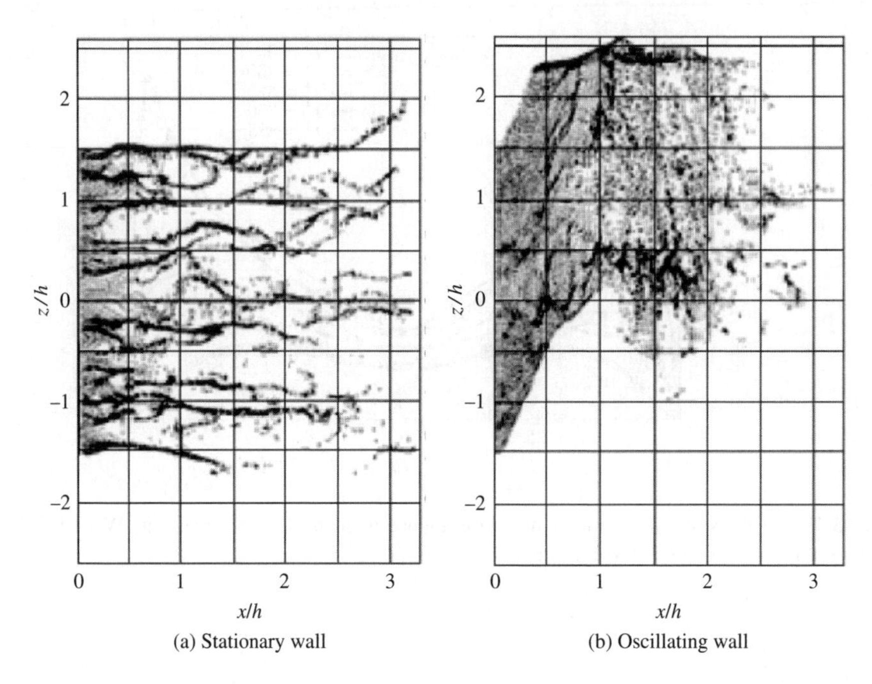

Figure 3.27 Distributions of tracer particles inserted at $y^+ = 6.5$.[23] *Source*: Bogard D G 2000. Reproduced with permission of Elsevier

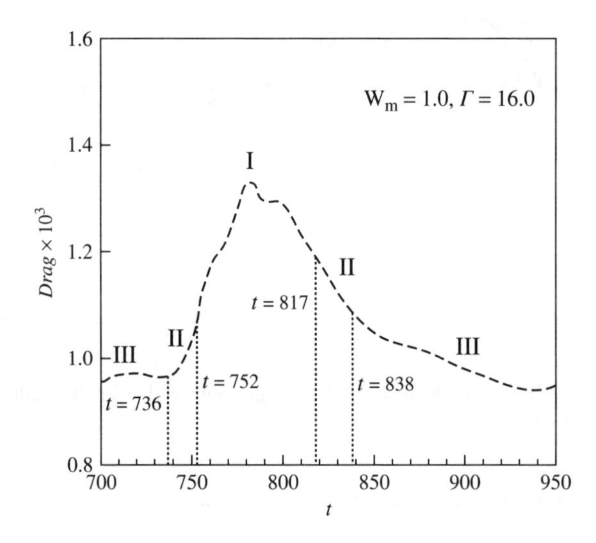

Figure 3.28 Friction drag history with $W_m = 1$ and $T = 16$ in $750 < t < 950$.[14] *Source*: Huang L P 2010. Reproduced with permission of Elsevier

friction drag history for the case of $W_m = 1$ and $T = 16$, in the time interval of $750 < t < 950$ in Figure 3.22, that is, near one period of the statistical periodic steady flow, during which the wall oscillates in many full cycles. This time interval can be divided into three regions denoted by I, II, and III as shown in Figure 3.28.

The flow configurations in region I at $t = 776$ and 784 are shown in Figure 3.29a and b, respectively. The distribution of streaky structures at $y^+ = 9$ is shown in the first row, where the gray area represents high-speed streak, denoted by $u' > 0$ and the black area represents low-speed streak, denoted by $u' < 0$. The vortex structures near the wall are shown in the second row, which are drawn by the imaginary part of complex eigenvalues of the velocity gradient tensor.

It is observed that the near-wall structures are tilted by the spanwise wall oscillation. As we know, the upstream part of a near-wall structure lies on the wall, while the downstream part is at a higher vertical location, so that the influence of the wall movement on the upstream part of a structure is greater than that on the downstream higher part. Therefore, the configuration of streaks and vortex structures mainly depends on the movement of structure upstream parts and dominated by the induced flow in the Stokes layer, showing a diagonal orientation with respect to the mean flow motion. When the wall displacement D is positive, that is, $\pi/2 < \alpha < 3\pi/2$ (see Figure 3.3), the near-wall structure is tilted to the negative spanwise direction. Otherwise, it is tilted to the positive spanwise direction, if D is negative. The tilting angle increases with the increasing D and the maximum angle occurs when the wall underneath is about to stop, that is $D = \pm D_m/2$.

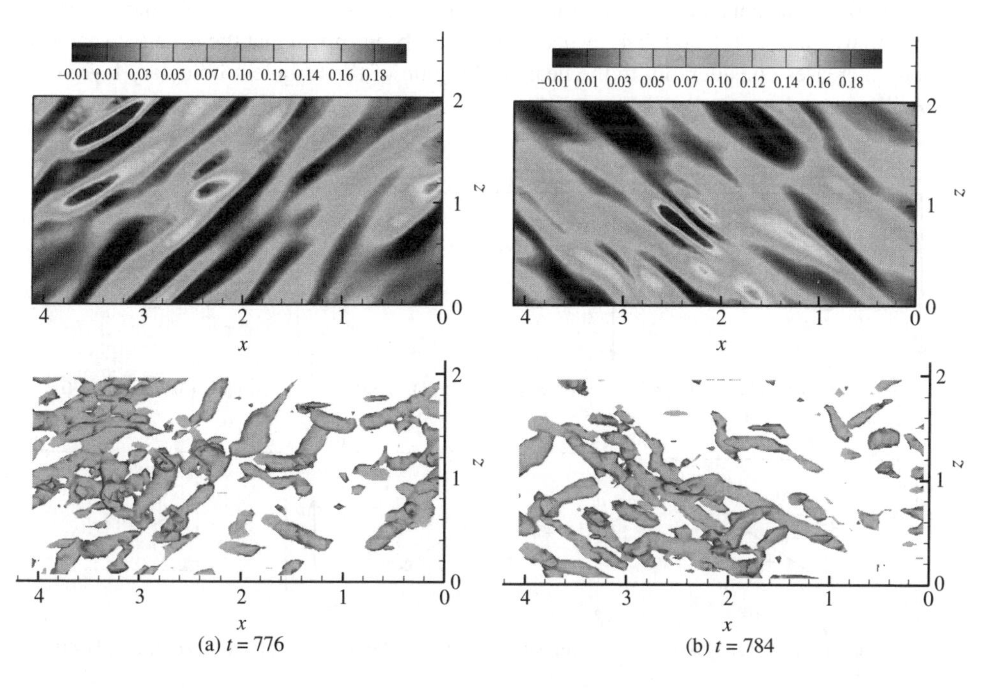

Figure 3.29 Streaky and vortex structures near the wall induced by wall oscillation.[14] *Source*: Huang L P 2010. Reproduced with permission of Elsevier

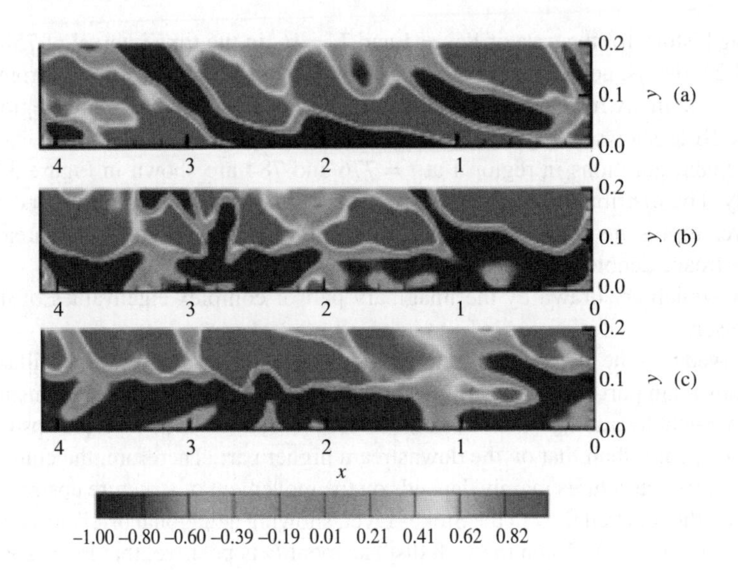

Figure 3.30 Snapshot distributions of spanwise vorticity near the wall.[14] *Source*: Huang L P 2010. Reproduced with permission of Elsevier

The shaded spanwise vorticity contours near the wall in the x–y plane are shown in Figure 3.30, where the gray areas refer to the positive vortex and black to negative vortex. Figure 3.30a represents the case of the stationary wall, and Figure 3.30b and c for the oscillating wall at $t = 776$ and 784, respectively. It has been shown that the negative spanwise vorticity increases considerably in the near-wall region of the turbulent boundary layer over the oscillating wall.

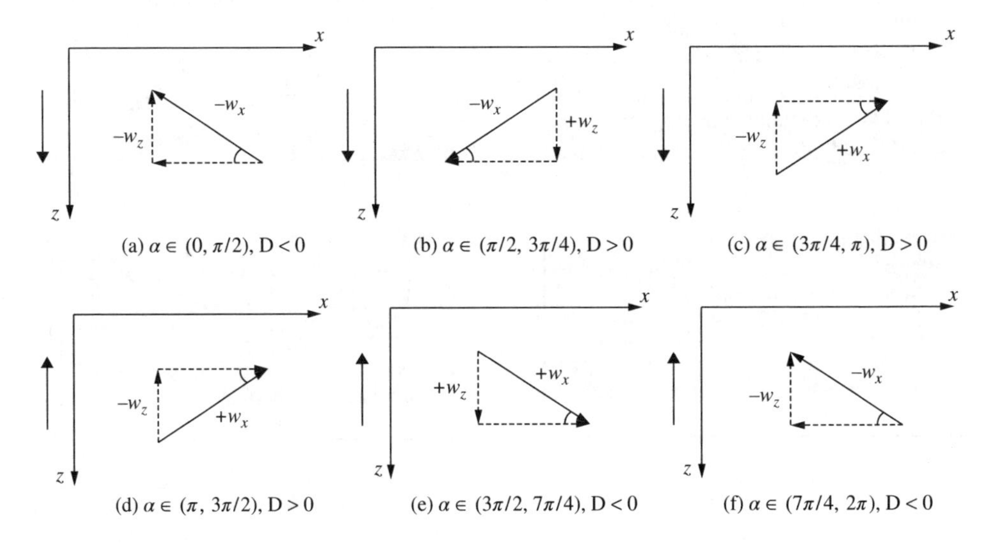

Figure 3.31 Schematic of tilting vortex structure in different phases.[24] *Source*: Huang L P 2010. Reproduced with permission of Elsevier

The creation of negative spanwise vorticity in the near-wall region during spanwise motion of the oscillating wall can be illustrated by the analysis of Stokes layer as shown in Figure 3.3, where the phase lag between profiles of ω_x and D is $\phi = \pi/4$. Based on the sign of ω_x, corresponding to its rotating direction, and the incline of vortex structure, depended on D, patterns of the tilting vortex structure in different phases are plotted in Figure 3.31 schematically. Arrow on the left side of figure indicates the wall moving direction. It is demonstrated that the negative spanwise vorticity generated is prevailing in a cycle of wall oscillation.

Profiles of averaged negative spanwise vorticity at $t = 776$ and 784 are shown in Figure 3.32. Solid line denotes the stationary wall and dashed line denotes the oscillating wall. The local maximum in the intensity of spanwise vorticity fluctuations is seen at around $y^+ \approx 15$.

The negative spanwise vorticity induced near the wall can affect the profile of mean streamwise velocity, reducing its gradient within the viscous sublayer and shifting the logarithmic velocity profile of the boundary upward, as shown in Figure 3.33, where solid line denotes the oscillating wall, and dashed line denotes the stationary wall. A concentrated spanwise vortex is superimposed at around $y^+ \approx 15$, leading to drag reductions.

As time proceeds from the region I to the region II, drag reduction is increased. The tilting turbulence structures are still present as shown in Figure 3.34, where Figure 3.34a indicates the distribution of streaky structures at $y^+ = 9$ and Figure 3.34b for vortex structures. But the number of vortex structures decreases significantly with increasing lateral spacing of the strakes.

Except the incline of vortex structure, the laterally oscillating wall can also result in a shift of vortices relative to streaks. Lessened streaks and its increased spacing shown in Figure 3.34 can be analyzed by examining the interrelation between streamwise vortices and near-wall low- and high-speed streaks.

The streamwise vortex is located in between the low- and high-speed streaks in the near-wall region of the turbulent boundary layer. For $\omega_x > 0$, when the wall moves into the negative spanwise direction (right), that is, from the high-speed side to the low-speed side, the fluid on the both sides of the streamwise vortex are driven to move to the right, so that the high-speed fluid intrudes beneath the vortex as shown in Figure 3.35a, where the gray areas represent high-speed streaks, the black areas represent low-speed streaks and the solid lines represent the vortex contours. However, when the wall moves into the left, that is from the low-speed side to the high-speed side, the near-wall high-speed fluid moves to the left, but the influence of the wall movement on the low-speed fluid is greatly reduced, since the vortex rotation counteracts the effect of the left moving wall. Therefore, the low-speed fluid hardly intrudes beneath the vortex as shown in Figure 3.35b. Since the variation of flow structure for $\omega_x < 0$ in a cycle of wall oscillation is almost the same as that for $\omega_x > 0$, with the opposite direction; it can be concluded that the laterally oscillating wall contributes to cause the high-speed fluid to intrude beneath the vortex, whether $\omega_x > 0$ or $\omega_x < 0$.

The shift of vortices with respect to streaks causes the vortices to eject high- rather than low-speed fluid away from the wall, and to sweep low- rather than high-speed fluid toward the wall, thereby homogenizing the u' fluctuations near the wall and disrupting the turbulent energy production in the turbulent boundary layer. The near-wall structures are weakened and lessened due to the spanwise motion of the oscillating wall, leading to drag reductions.

As time proceeds to the region III, drag reduction increases further, so that the maximum drag reduction is achieved when near-wall structures are almost eliminated as shown in

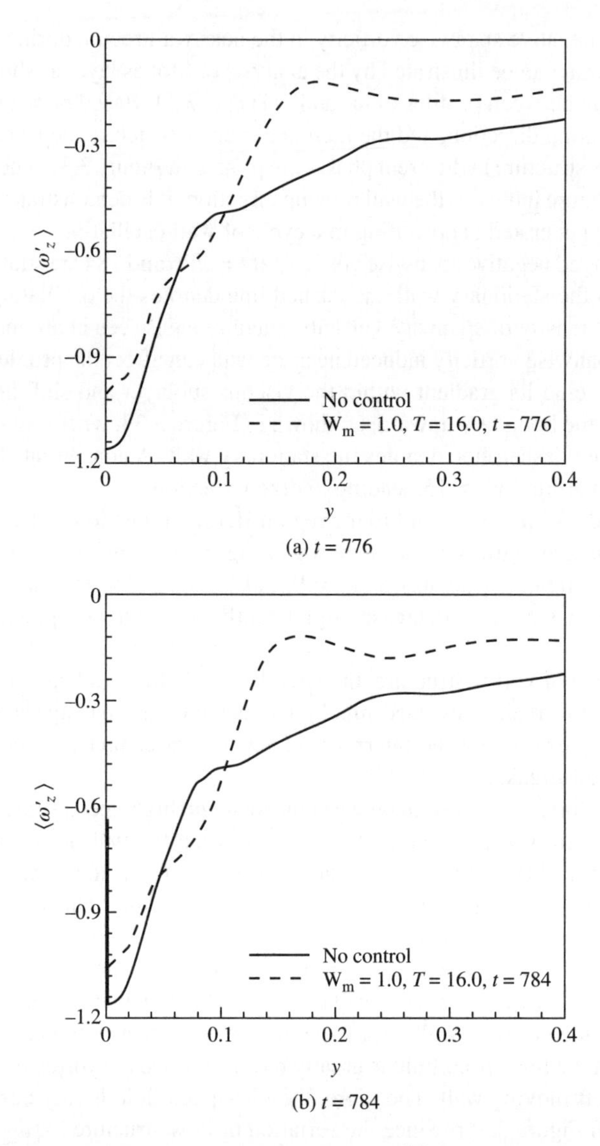

Figure 3.32 Profiles of averaged negative spanwise vorticity.[14] *Source*: Huang L P 2010. Reproduced with permission of Elsevier

Figure 3.36, where Figure 3.36a represents streaks in the x–z plane at $y^+ = 9$ and Figure 3.36b represents streamwise vortices near the wall.

Subsequently, more and more near-wall structures appear again, similar with that happening in region II in an inverse process. A statistically periodic fluctuation of drag deduction has completed as the minimum drag reduction is achieved with the flow configuration seeming to be almost the same as that in region I.

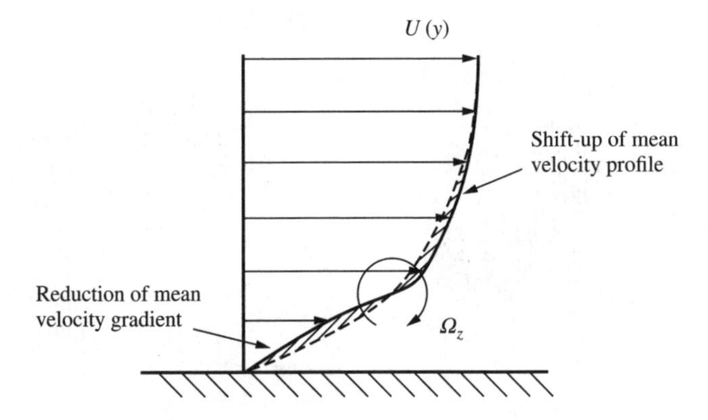

Figure 3.33 Negative spanwise vorticity effect on mean streamwise velocity profile.[20] *Source*: Quadrio M 2004. Reproduced with permission of Cambridge University Press

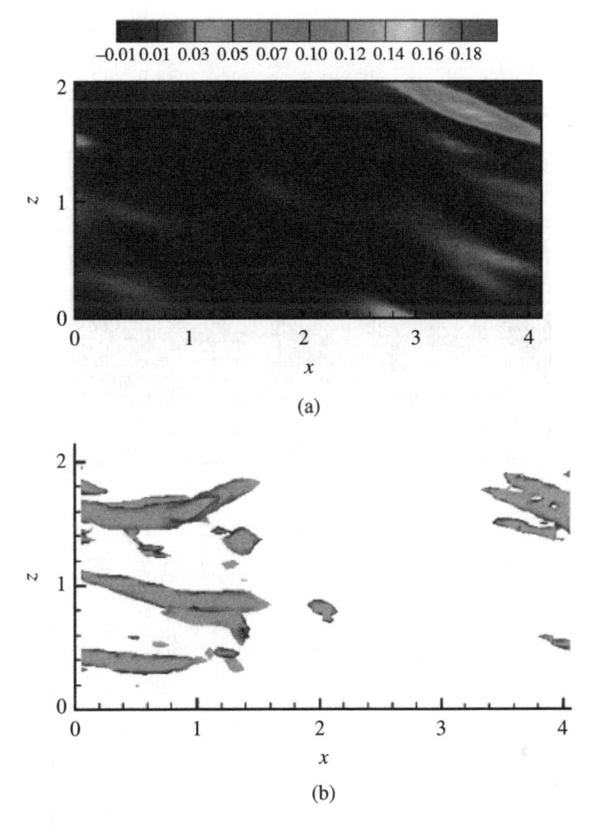

Figure 3.34 Streaky and vortex structures near the wall in region II.[14] *Source*: Huang L P 2010. Reproduced with permission of Elsevier

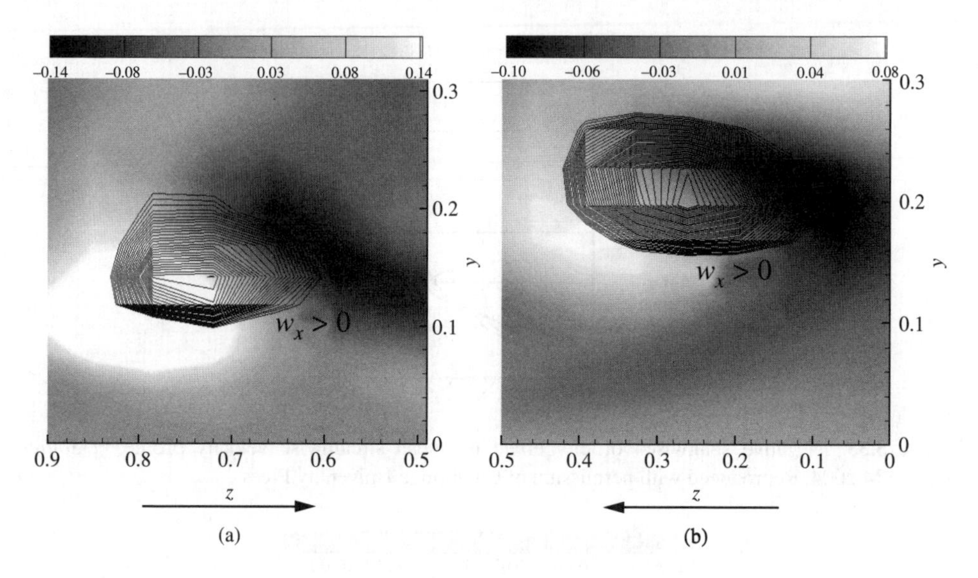

Figure 3.35 Response of vortex associated flow field with the oscillation of the wall.[14] *Source*: Huang L P 2010. Reproduced with permission of Elsevier

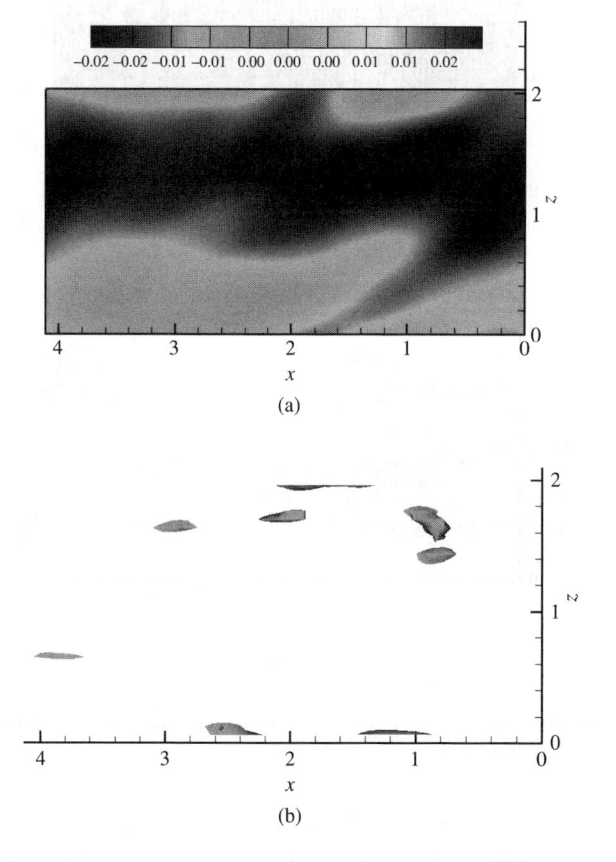

Figure 3.36 Streaky and vortex structures near the wall in region III.[14] *Source*: Huang L P 2010. Reproduced with permission of Elsevier

3.3.3 Wall Turbulence with Spanwise Traveling Wave on Wavy Wall[26, 27]

If the wall oscillates laterally in the form of spanwise-traveling waves, then the boundary condition for the governing equations (3.18) and (3.19) is given by

$$w|_w = W_m \cos(k_z z - \omega t) \tag{3.21}$$

$$\left.\frac{\partial v}{\partial y}\right|_W = -\left.\frac{\partial w}{\partial z}\right|_W$$

where $\omega = \frac{2\pi}{T}$ is the circular frequency, $k_z = \frac{2\pi}{\lambda_z}$ is the wave number in z direction, and λ_z is the wavelength in the z direction. A standard Fourier–Chebyshev spectral method is used for calculation, and the optimal control parameters used in the computation are $T^+ = 50$, $k_z^+ = 1$, and $W_m = 1$, where the superscript "+" indicates quantities in wall units.

Profiles of mean streamwise velocity are shown in Figure 3.37 in a log-law plot, where dash-dotted line denotes the wavy wall and dashed line denotes the stationary wall. The significant variation of velocity profile under the effect of wavy-wall is observed compared with the uncontrolled profile. The time history of drag is shown in Figure 3.38. Drag is reduced by the wavy wall.

The rms of the velocity and vorticity components on uncontrolled (right) and controlled (left) sides of the channel is shown in Figures 3.39 and 3.40, respectively, where solid, dashed, and dash-dotted lines represent components in streamwise, normal, and spanwise directions, respectively. It is shown from Figure 3.40 that the peak of rms streamwise vorticity disappears on the controlled side, which is identified as the sign of streamwise vorticities responsible for the streaks and drag enhancement.

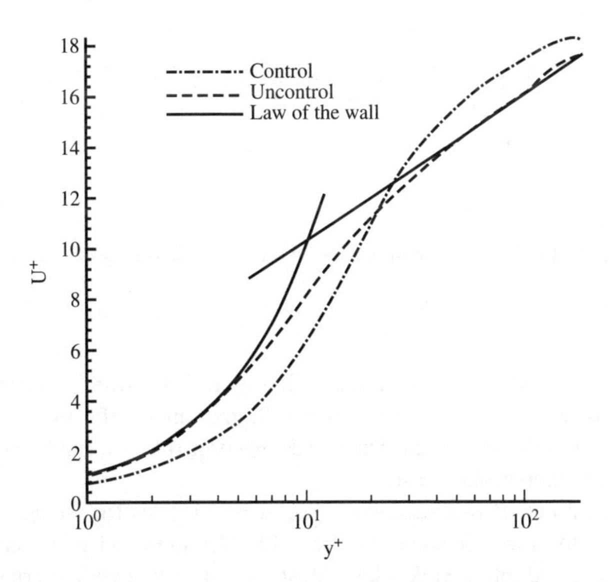

Figure 3.37 Mean streamwise velocity profiles.[27] *Source*: Zhao H 2004. Reproduced with permission of Elsevier

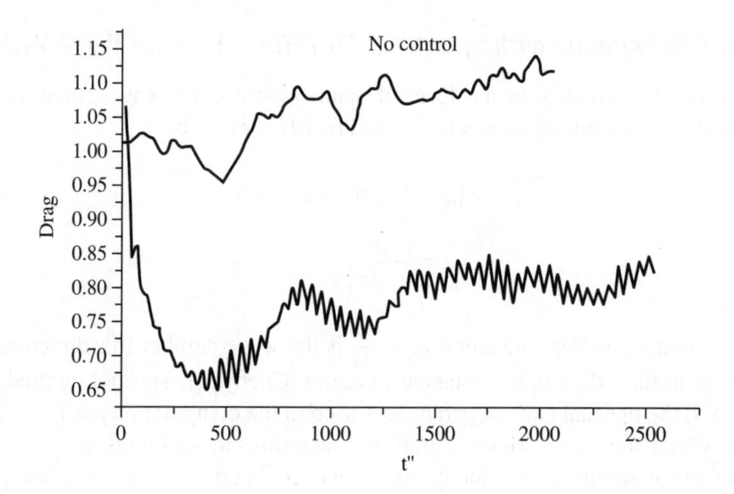

Figure 3.38 Time history of uncontrolled and controlled friction drag.[27] *Source*: Zhao H 2004. Reproduced with permission of Elsevier

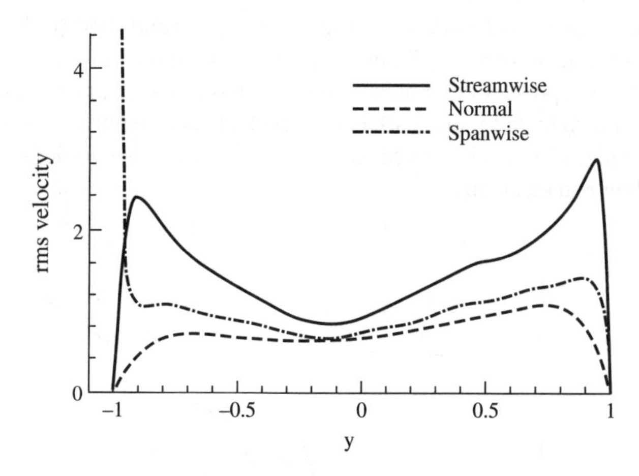

Figure 3.39 Profiles of rms velocity components.[27] *Source*: Zhao H 2004. Reproduced with permission of Elsevier

Streaks in the x–z plane at $y^+ = 5$ are shown in Figure 3.41, where Figure 3.41a represents stationary wall, Figure 3.41b represents wavy wall, gray areas refer to the high-speed streak and black areas refer to the low speed. The streaks become obscure and broaden for controlled flow compared with uncontrolled case.

The sectional contours of instantaneous streamwise vortices for uncontrolled (upper) and controlled (lower) flows are shown in Figure 3.42. The streamwise vortices are suppressed strongly and compressed into a Stokes layer near the wavy wall with alternative signs, which indicates that the Stokes layer generated by the wavy wall can absorb the random streamwise vorticity and alleviate the streaks.

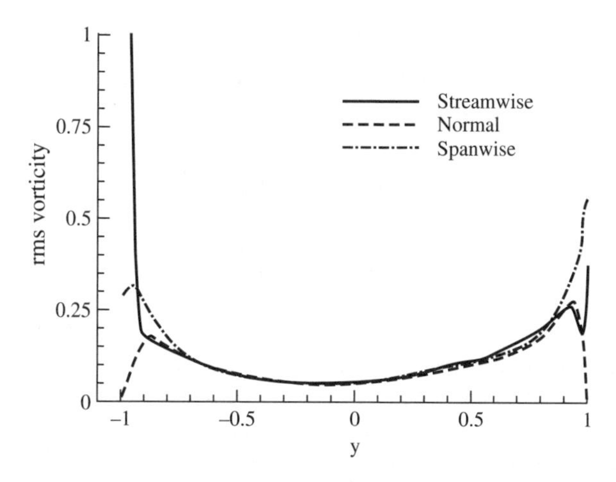

Figure 3.40 Profiles of rms vorticity components.[27] *Source*: Zhao H 2004. Reproduced with permission of Elsevier

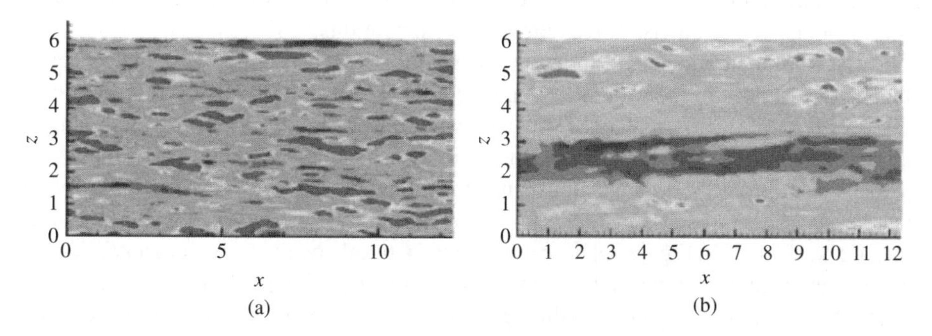

Figure 3.41 Streaks in the x–z plane at $y^{+} = 5$.[27] *Source*: Zhao H 2004. Reproduced with permission of Elsevier

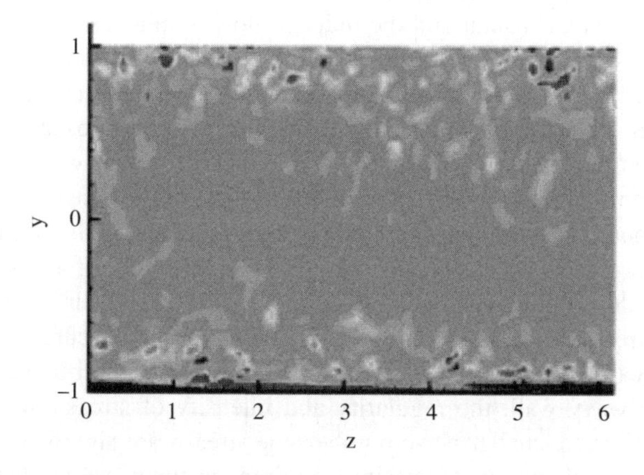

Figure 3.42 Sectional contours of instantaneous streamwise vortices in the y–z plane.[27] *Source*: Zhao H 2004. Reproduced with permission of Elsevier

3.3.4 Wall Turbulence with Streamwise Traveling Wave on Wavy Wall

If the wall oscillates laterally in the form of streamwise traveling waves, then the boundary condition for the governing equations (3.18) and (3.19), is given by

$$w|_w = W_m \cos(k_x x - \omega t) \tag{3.22}$$

$$\left.\frac{\partial v}{\partial y}\right|_W = -\left.\frac{\partial w}{\partial z}\right|_W$$

where $k_x = \frac{2\pi}{\lambda_x}$ is the wave number in the x direction and λ_x is the wavelength in the x direction. A standard Fourier–Chebyshev spectral method is used for calculation, with $U_m = 2/3$, where U_m is dimensionless bulk mean velocity, $\text{Re}_m = 2666$, $W_m = 1.0$, and $T = 8.0$.

The intrinsic coherent structures of near-wall turbulence are characterized by streamwise vortices and streaks, and the Stokes layer induced by the wavy wall can be described by spanwise velocities. When the induced flow in the Stokes layer is imposed to the near-wall turbulent flow, the intrinsic streaks and longitudinal vortex structures are modulated. Figure 3.43 shows snapshot distributions of spanwise velocity, streak, and vortex structure near the wall with different values of streamwise wave number for calculations.

The first column in Figure 3.43 shows the snapshots of the distribution of spanwise velocities in the x–z plane close to the wall ($y^+ = 9$), where the white and black areas represent, respectively, positive and negative values of spanwise velocities. In Figure 3.43a, for a stationary wall, a random and irregular instantaneous distribution of spanwise velocities is exhibited, and in Figure 3.43b the color is homogeneous since the signs of spanwise velocities are same under the action on laterally oscillating wall. For a controlled flow via the wavy wall, the aspects of white and black areas in Figure 3.43c–g are affected strongly by the wave number k_x. The shape of color area becomes increasingly regular as k_x increases for low-frequency waves, which indicates that the influence of the modulated turbulent flow on the induced flow becomes more and more weak. When $k_x = 3$, the configuration of spanwise velocity distributions consists of alternating white rectangle regions and black rectangle regions along the streamwise direction as shown in Figure 3.43e, which means that the flowing directions in the spanwise direction are homogeneous along $x =$ constant and the induced flow dominates the near-wall flow. However, as k_x increases further, the color areas become obscure, as shown in Figure 3.43f. For the high-frequency waves, the distribution of spanwise velocities becomes random again, and the white and black areas mingle and amalgamate with each other, as shown in Figure 3.43g, which indicates that the intrinsic turbulent flow dominates the near-wall flow.

The second column in Figure 3.43 shows the snapshots of the distribution of streaky structures in the x–z plane at $y^+ = 9$, where the white areas represent high-speed streaks and black areas represent low-speed streaks. As illustrated in Figure 3.43a for a stationary wall, the streaky structures do not always flow straight in the streamwise direction, but often meander in spanwise direction. In Figure 3.43b, the streaks become obscure and broaden, and the intensity is significantly weakened under the action on laterally oscillating wall. For the flow controlled via the wavy wall, the regularity and intensity of streaks are dependent on the wave number k_x. For the smaller wave number, the streaks are significantly sinusoidal with the increasing frequency and the decreasing amplitude as the wave number k_x increases, and its intensity decreases with the increase in k_x as shown in Figure 3.43c–g. Then for $k_x > 3$, the large-amplitude oscillation of streaks disappears gradually and the intensity of the streaks

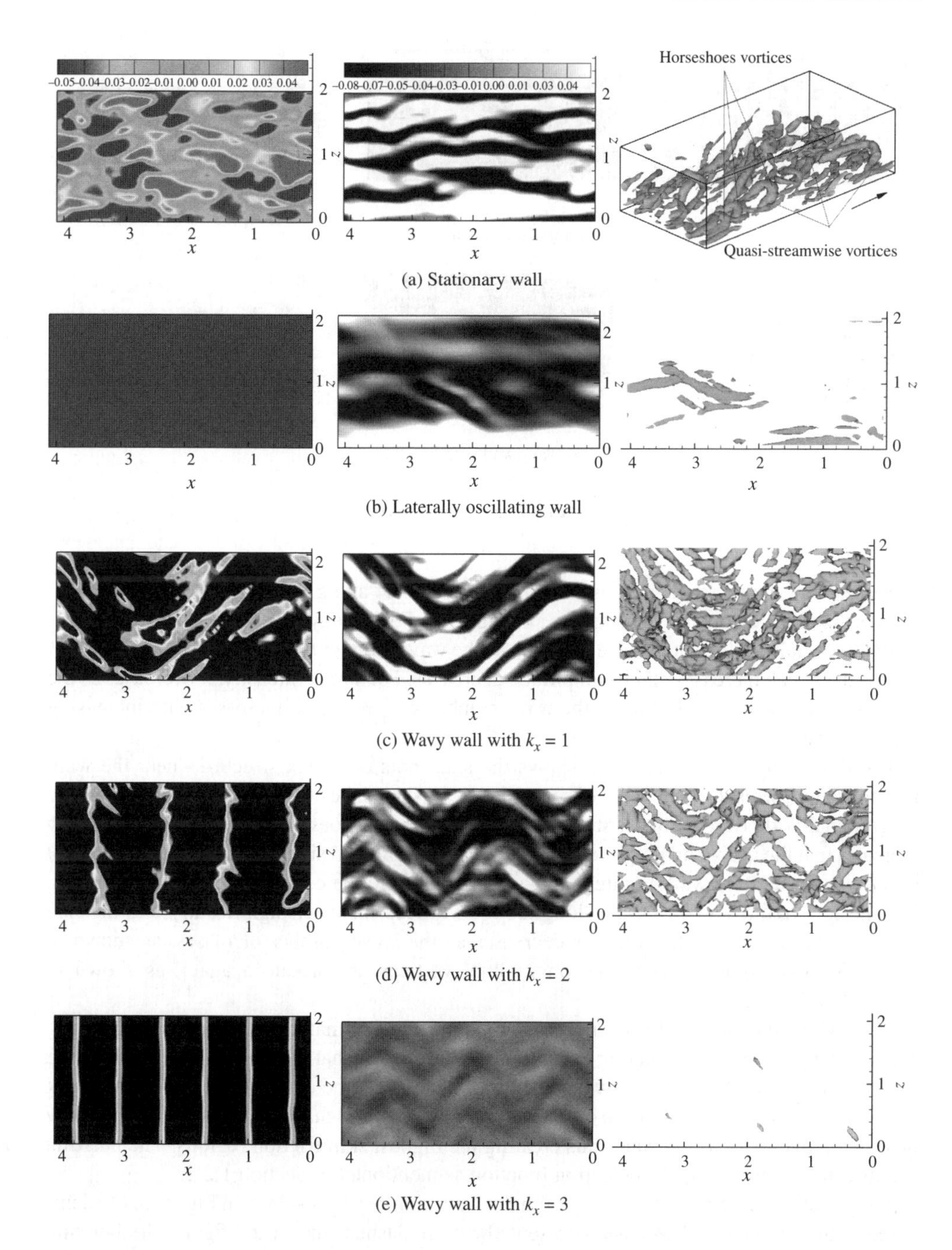

Figure 3.43 Snapshots of spanwise velocity, streak, and vortex structure near the wall.[28] *Source*: Huang L P 2011. Reproduced with permission of Elsevier

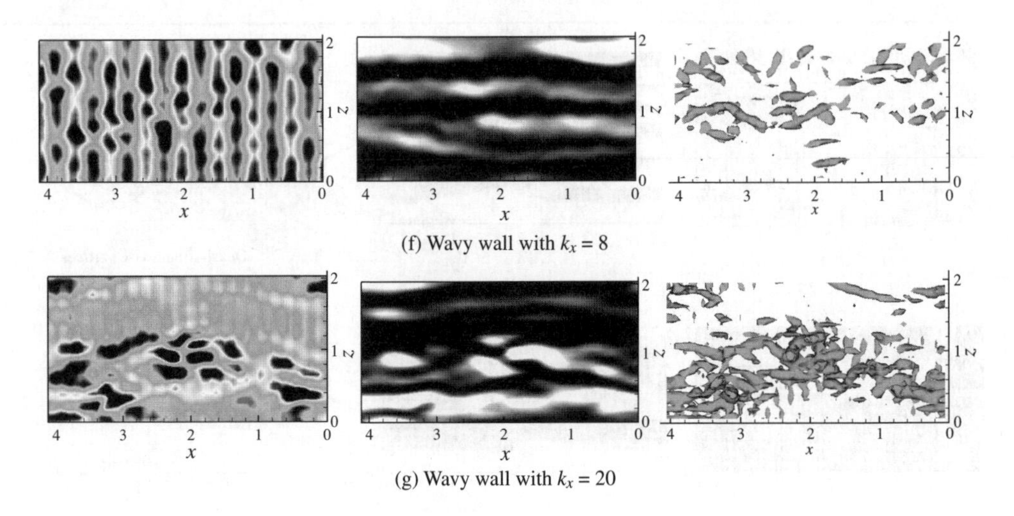

Figure 3.43 *(continued)*

increases with the increase in k_x as shown in Figure 3.43f and g. Moreover, the variations of streaks are revealed mainly in its aspect and intensity. For streaky aspect, the streaks are sinusoidal, especially in the small wave number. As the wave number k_x increases, the frequency of sinusoidal streak increases and its amplitude decreases. For streak intensity, it decreases as the increase in the wave number when the wave number $k_x \leq 3$, then it increases with increasing wave number.

The third column in Figure 3.43 shows the snapshots of vortex structures near the wall. Figure 3.43a, corresponding to the uncontrolled case, shows the characteristic structures of the longitudinal vortices (e.g., quasi-streamwise vortices, horseshoes vortices, etc.). Figure 3.43b indicates that the near-wall longitudinal vortices are suppressed significantly under the control of laterally oscillating wall. Figure 3.43c–g corresponds to the controlled case by wavy wall. For the low-frequency waves, the longitudinal vortices meander significantly and arrange in order; the number of the vortices decreases as the wave number increases, as shown in Figure 3.43c–e. When $k_x > 3$, the longitudinal vortices are produced again, as shown in Figure 3–45f and g.

The near-wall regions of turbulent boundary layers are dominated by a sequence of bursting events. The process of the bursting event can be described as that the parts of the streaks bring the low momentum fluid lift up away from the wall and become unstable, oscillate, and break down. This is followed by a sweep of the high momentum fluid that originates in the outer flow field and moves toward the wall, thus creating the highest skin friction regions. The bursting events can be detected by the detection function as mentioned in Section 1.2.2.

The variation of bursting frequency with the wave number k_x is shown in Figure 3.44. Compared with the uncontrolled case represented by the dashed line in the figure, the bursting frequency is significantly reduced by the control of the oscillating wall, represented by a black circle at $k_x = 0$. As the wavy wall is used to modify the near-wall flow, the bursting frequency decreases as the increase in k_x until $k_x = 3$. Then, it is maintained at an uncontrolled level.

The conditionally averaged fluctuating streamwise velocity $\langle u' \rangle$ is taken to describe the intensity of the bursting events. Then the variation of the bursting intensity with the wave

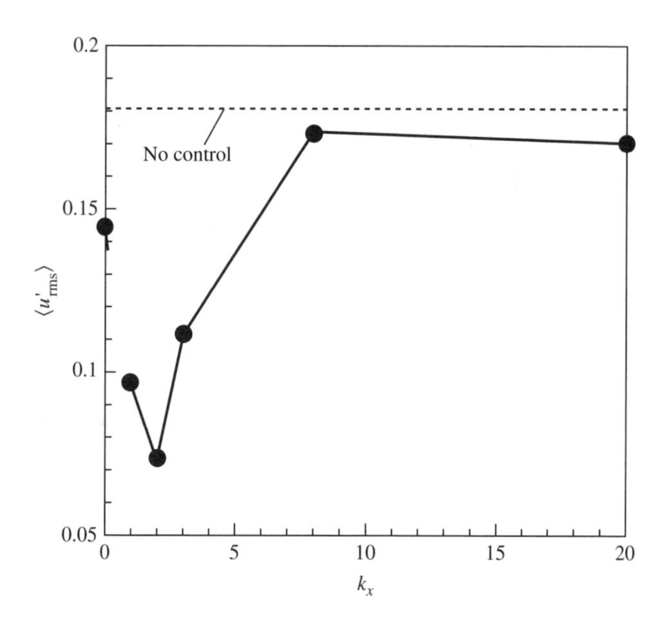

Figure 3.44 Variation of bursting frequency with wave number k_x.[28] *Source*: Huang L P 2011. Reproduced with permission of Elsevier

number k_x is shown in Figure 3.45. Compared with the uncontrolled case represented by the dashed line in the figure, the bursting intensity is reduced by the control of the spanwise oscillating wall ($k_x = 0$), or the wavy wall ($k_x > 0$). For the wavy wall, the intensity of turbulent bursts decreases when the value of wave number increases from 1 to 2. Then it increases with the increase in k_x.

The turbulent skin-friction drag is mainly contributed to the burst events. The increase in the frequency or intensity of the turbulent bursts can lead to an increase in the skin-friction drag on the wall. As depicted in Figures 3.44 and 3.45, the varying tendencies of the frequency and the intensity of the turbulent bursts with wave number k_x are unsynchronized with each other, which imply that there is an optimal parameter k_x corresponding to the largest amount of drag reduction. Figure 3.46 shows the drag reduction, defined by Eq. (3.13), versus k_x with $W_m = 1.0$ and $T = 8.0$. It is indicated that the largest amount of drag reduction occurs in $k_x = 3$.

3.4 Deformed Wall

A wall deformed to take a shape with normal deflection is specifically termed as deformed wall, or active skin here. The simplest method of generating such a deformed wall (a deformed thin plate) would be by subjecting the wall to a distributed load. For example, by subjecting the wall to equally spaced external moments, with the equal value and alternating sign, as shown in Figure 3.47a, a deformed wall with a static wave profile is generated. Then, a traveling surface deformation wave can be realized by shifting the points of application of the moments. Moreover, the deformed wall can also be generated by utilizing spaced vertical external forces,

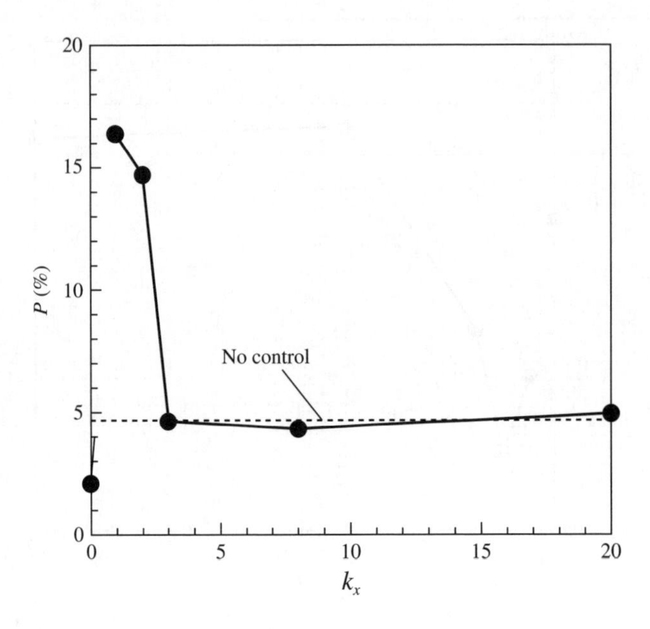

Figure 3.45 Variation of bursting intensity with wave number k_x.[28] *Source*: Huang L P 2011. Reproduced with permission of Elsevier

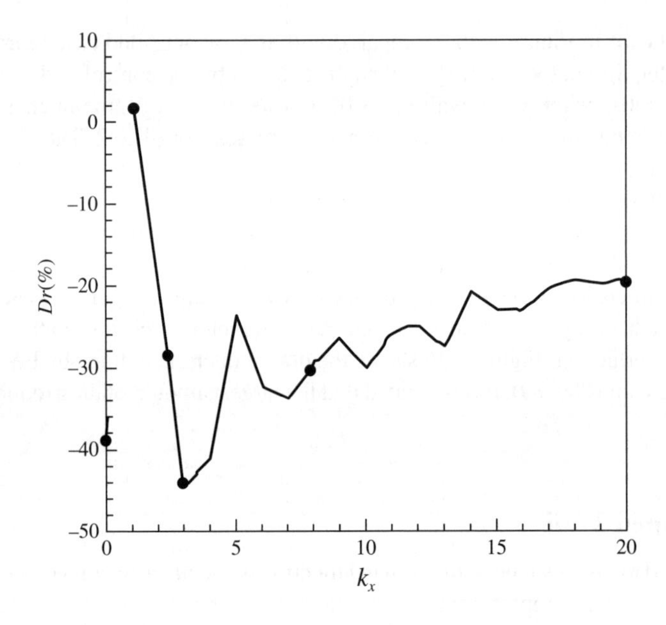

Figure 3.46 Variation of drag reduction with wave number k_x.[28] *Source*: Huang L P 2011. Reproduced with permission of Elsevier

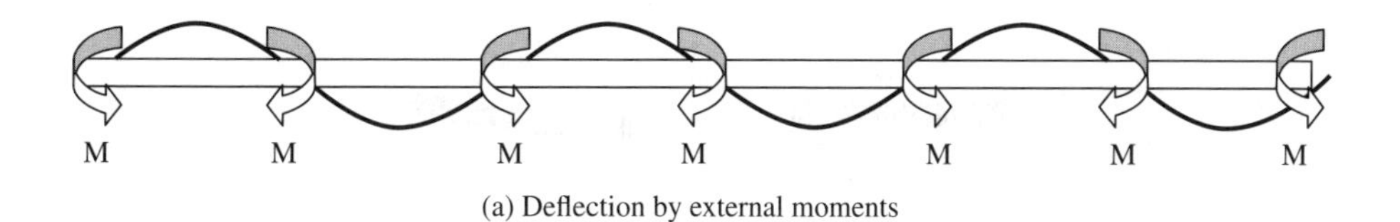

(a) Deflection by external moments

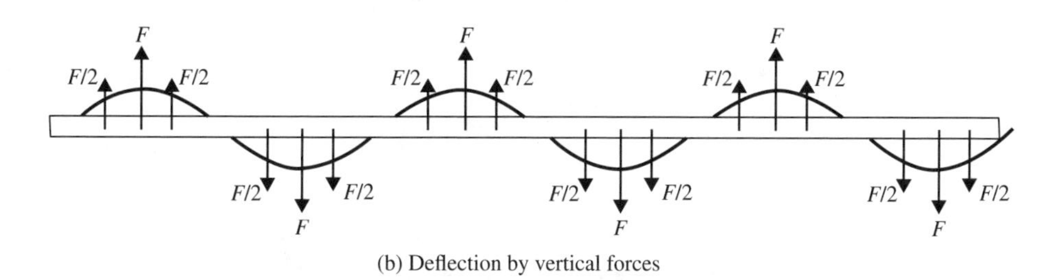

(b) Deflection by vertical forces

Figure 3.47 Deformed wall generated by distributed load.[29] *Source*: Rediniotis O K 2002. Reproduced with permission of Springer

periodically varying the amplitudes and the directions to match the plate deflection required as shown in Figure 3.47b.

In practice, the deformed wall with a required shape is generated by the use of actuators based on active material actuation or mechanical actuation. A traveling wave motion can then be achieved by changing the points of actuation in a serial fashion. There are many kinds of actuators used for active skin design that are discussed below.

3.4.1 Shape Memory Alloy

Figure 3.48 presents a cross section of an active skin in its nonactuated state, which consists of a top plate, a bottom plate, and legs. The top plate is exposed to the flow and the bottom plate is attached to the wall. The legs attached to the top plate can slide (left and right) with respect to the bottom plate. A shape memory alloy (SMA) wire runs through the legs, through small holes on their sidewalls. The direction of the SMA wire is in the spanwise direction, while the major dimension of the legs is along the streamwise direction. Within each leg, a circular flat disk is attached to the SMA wire, with its diameter significantly larger than the diameter of the holds on the sidewalls of the legs. Each SMA–disk joint is electrically connected to the electrical control circuit, and is powered independently. A current can be made to pass through a section of the SMA by applying a voltage difference between the discs at the two ends of

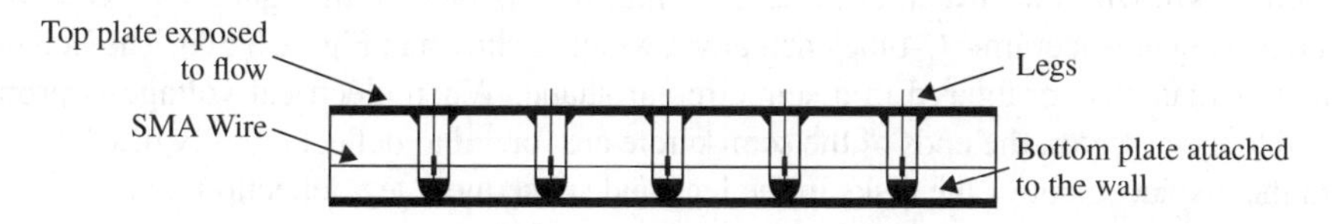

Figure 3.48 Cross section of SMA actuated active skin.[29] *Source*: Rediniotis O K 2002. Reproduced with permission of Springer

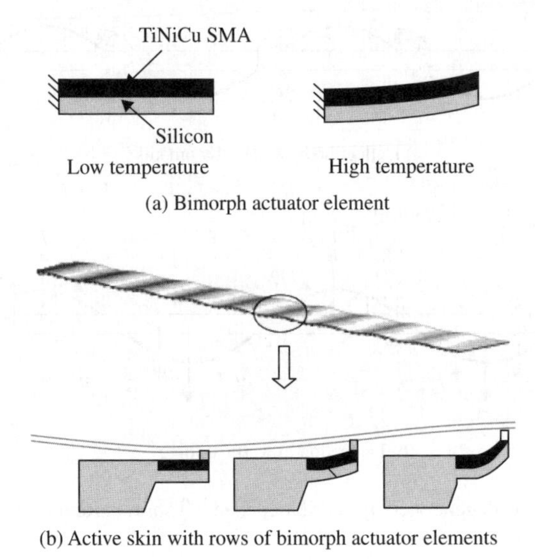

Figure 3.49 Active skin made by bimorph actuators.[30] *Source*: Mani R 2003. Reproduced with permission of Cambridge University Press

the SMA section. Then the actuated SMA section contracts and forces the disks at the ends of the SMA section to push against the respective legs. The moment thus generated, bends the section of the skin directly above the SMA section under actuation.

The other kind of active material actuators is made up of a silicon layer covered with a thin film TiNiCu SMA layer, called bimorph actuator, as shown in Figure 3.49a. Since silicon has a higher Young's modulus and a lower expansive coefficient than SMA, the silicon layer in effect acts like a biasing spring for the SMA layer. Hence, heating the bimorph element can result in the vertical deflection.

The active skin is designed in such a way that the vertical forces generated by deflection of the actuators are exerted on the polymer film which is exposed to the flow. The actuator distribution shown in Figure 3.49b is repeated at regular intervals, so that the resulted deflection is a good approximation of the required cylindrical bending. By phase shifting the thermal cycling of each of the individual bimorph actuators, it is possible to attain a traveling surface deformation wave.

3.4.2 *Piezoceramics*

The active skin with the structural design, similar to SMA shown in Figure 3.48, can also be actuated using piezoceramic C-block actuators, which is shown in Figure 3.50a. The individual C-block actuator is configured in a semicircular shape. When electrical voltage is properly applied to the actuator, the ends of the semi-circle are forced to deflect radially inward, resulting in the displacement of the disks in the legs and subsequent leg deflection.

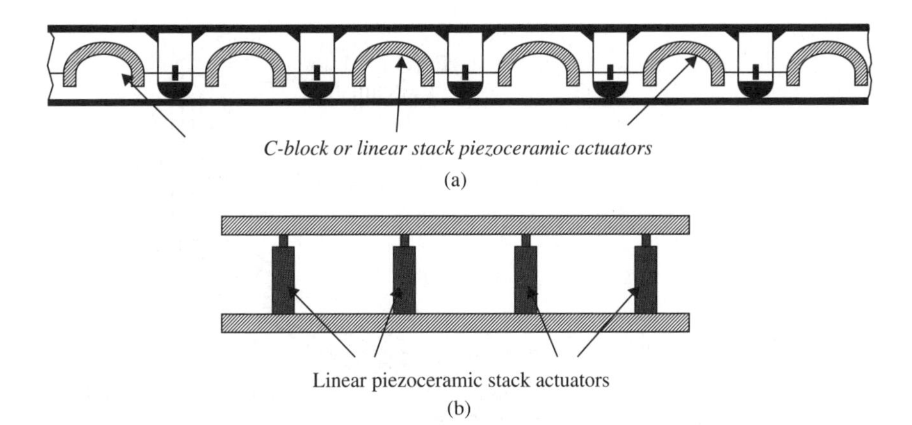

Figure 3.50 Active skin made by piezoceramic actuators.[29] *Source*: Rediniotis O K 2002. Reproduced with permission of Springer

On the force-based actuation principle, the legs shown in Figure 3.50a are replaced by linear piezoceramic stack actuator (PSA) as shown in Figure 3.50b. On actuation, the PSAs would exert a force on the skin in the vertical direction causing the skin to bend. By varying the force applied by each PSA with time in a periodic fashion, it would be possible to obtain a traveling wave.

3.4.3 Magnet

As shown in Figure 3.51, the magnetic actuator consists of an electromagnet, a miniature permanent magnet, and an elastic skin. When electrical voltage is properly applied to the electromagnet, which consists of a miniature copper coil and an iron core, the permanent magnet moves under the action of magnet force, resulting in the defection of the elastic skin attached.

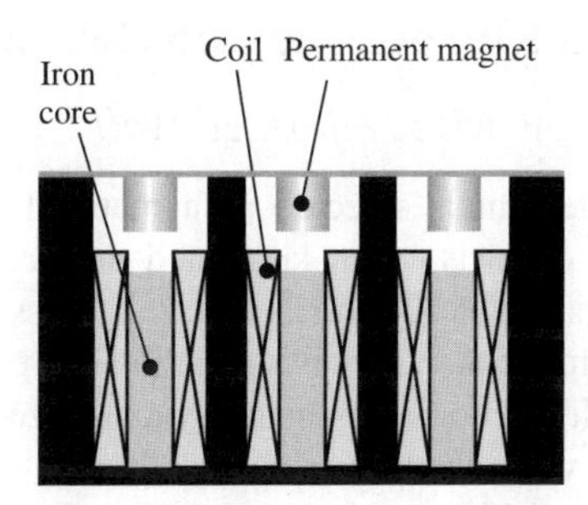

Figure 3.51 Active skin made by magnetic actuator.[31] *Source*: Suzuki Y 2005. Reproduced with permission of Elsevier

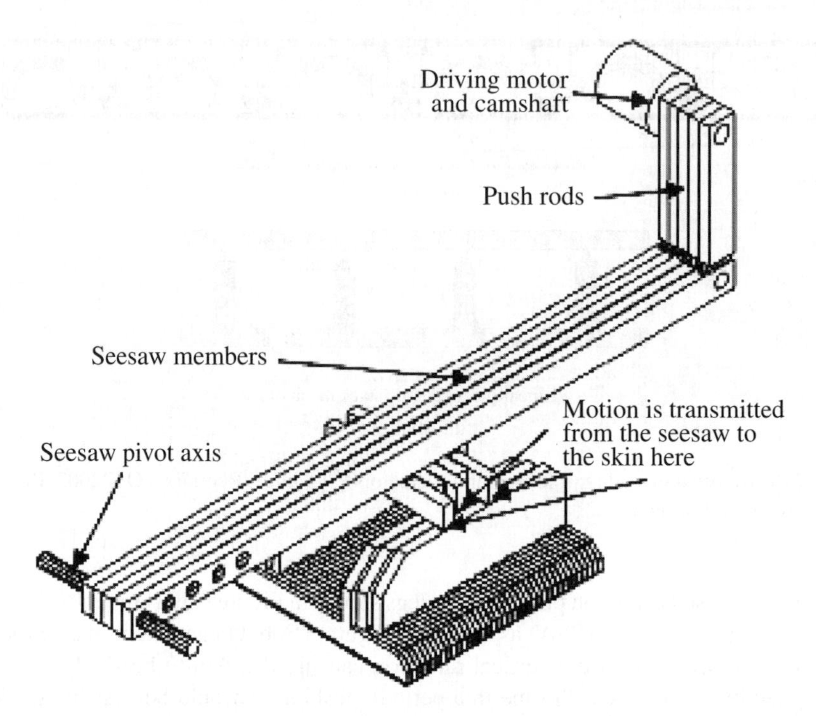

Figure 3.52 Mechanically actuated skin.[32] *Source*: Rediniotis O K 2002. Reproduced with permission of Springer

3.4.4 *Cam Mechanism*

The schematic of the mechanically actuated skin is shown in Figure 3.52. A cam mechanism is used. Each cam is attached to a seesaw member that isolates the linear (vertical) motion from the circular cam motion. Subsequently, the seesaw's motion is transmitted to the skin by the ribs resulting in the skin deformation.

3.5 Experiments of Wall Turbulence with Deformed Wall

3.5.1 *Incompressible Flow with Deformed Wall*

A laminated piezoceramic element is selected as a material of actuator, which stretches 17 μm at 160 V. An actuator consists of six laminated piezoceramic elements as shown in Figure 3.53a, with total elongation of 100 μm at 160 V. A cross section of an actuated active skin composed of six piezoceramic actuators is shown in Figure 3.53b, which can oscillate in a phase difference of 60° with each other by using synchronized signal generators, so that a traveling surface deformation wave can be attained.

The experiments are performed in a closed loop water channel as shown in Figure 3.54. When the freestream velocity is $U_c = 0.15\ \text{m/s}$, a corresponding Reynolds number is $\text{Re} = \frac{HU_\infty}{\nu} = 7500$, where H is height of the test section, and ν is the kinematic viscosity of the fluid. The friction velocity is $u_\tau = 0.01\,\text{m/s}$ at $y = 2\ \text{mm}$. The typical scale length λ is about 8 mm, based on the expression of $\lambda = 100\nu/u_\tau$. We chose actuators with a width less than $1/2\lambda$, that

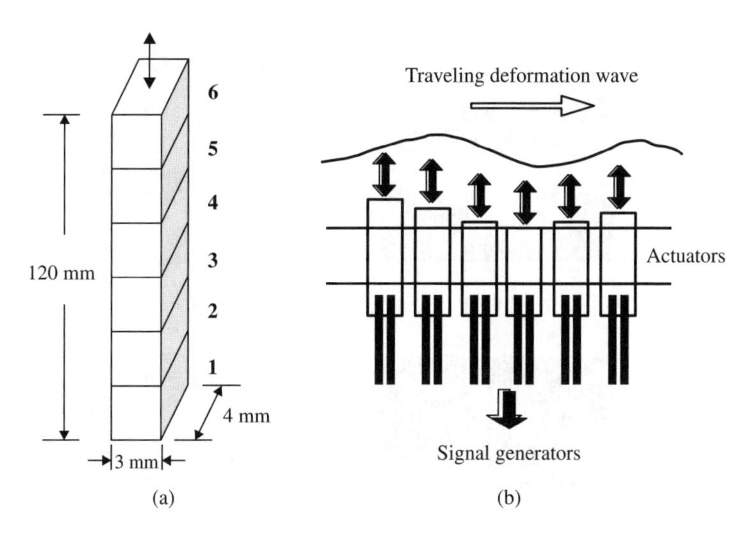

Figure 3.53 Active skin made by piezoceramic actuator.[33] *Source*: Segawa T 2002. Reproduced with permission of Cambridge University Press

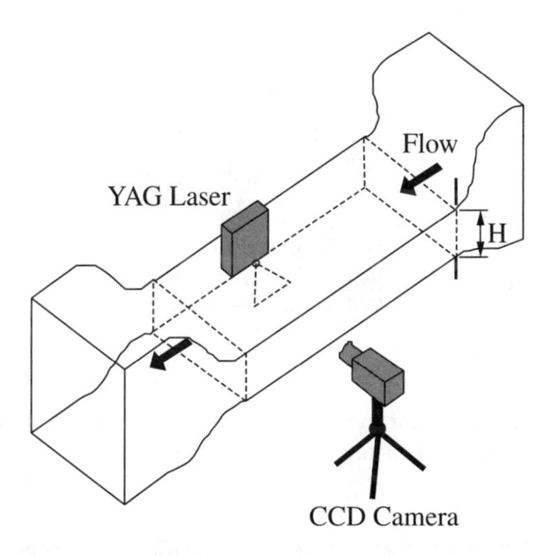

Figure 3.54 Schematic of water channel facility.[33] *Source*: Segawa T 2002. Reproduced with permission of Cambridge University Press

is 4 mm. The actuator array is mounted on a circular plate set in the top wall of the channel at $x = 3000$ mm downstream of the test section inlet.

To visualize the near-wall flow, a pair of yttrium aluminum garnet (YAG) laser sheets is set over the channel system. The images are photographed with a CCD camera and the velocity vectors are analyzed by particle image velocimetry (PIV) software.

Figure 3.55 shows the snapshot of velocity distributions in the x–z plane with $y = 5$ mm, that is $y^+ = 50$, measured by PIV, where Figure 3.59a–e represents the frequency of actuator

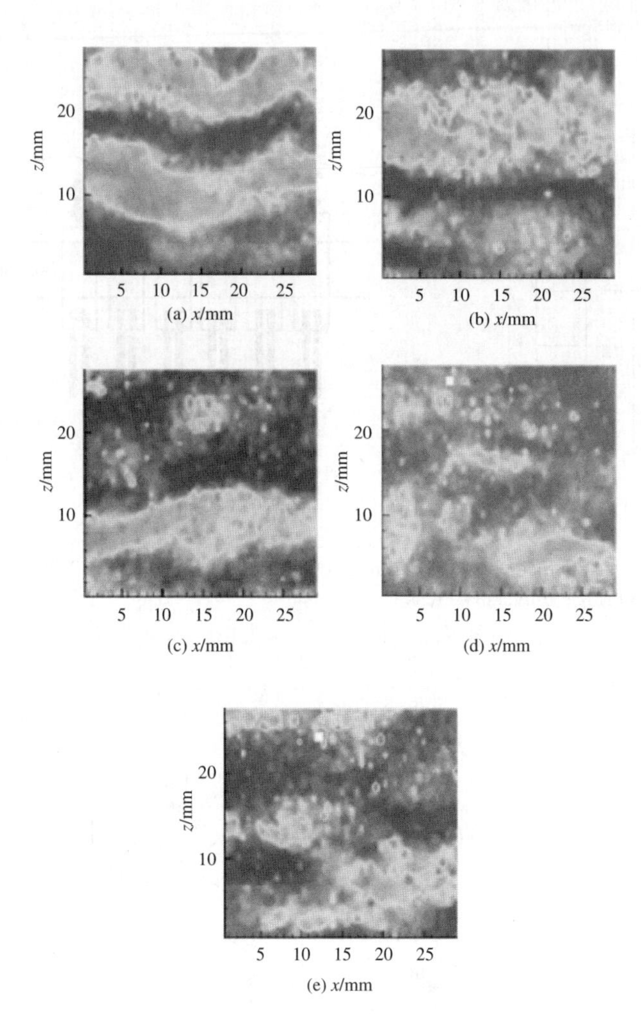

Figure 3.55 Measured velocity distribution in x–z plane under active control.[33] *Source*: Segawa T 2002. Reproduced with permission of Cambridge University Press

$f_a = 0$, 1.25, 6.25, 12.5, and 125 Hz, respectively. Dark areas represent high-speed streaks, and white areas represent low-speed streaks. Fine streaky structures with the lateral spacing about 10 mm are shown in Figure 3.55a for an uncontrolled case. However, the low-speed streaks become increasingly obscure with increase in the frequency of actuator for the controlled cases as shown in Figure 3.55b–d and are blurred as $f_a = 125$ Hz as shown in Figure 3.55e.

3.5.2 *Compressible Flow with Deformed Wall*

Coherent structures made a future of streamwise vortices and streaks play an important role in the production of turbulence and link closely with the high frictional drag. Therefore, the

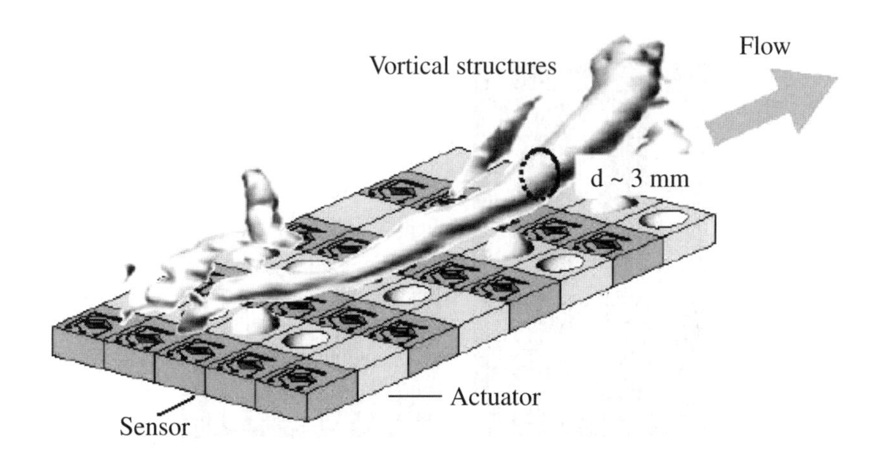

Figure 3.56 Schematic diagram of active feedback control system for wall turbulence.[31] *Source*: Suzuki Y 2005. Reproduced with permission of Elsevier

attempts to control turbulence flows have focused on the manipulation of the coherent structures. The best possible manipulation of the coherent structure with lower consumed energy can be performed by a feedback control system as shown in Figure 3.56, which consists of microsensors and actuators. The coherent structures should be detected by the sensors and manipulated by the actuators. Since the coherent structures have generally very small spatial-temporal scales, the actuators and the sensors contained in the control system should be with small size and high frequency response. Recent development of MEMS technology for sensing and actuation has made it possible to fabricate sensor and actuator of submillimeter scale required for practical implementation.[34, 35]

For the wall deformation magnetic actuator shown in Figure 3.57, a silicon rubber sheet of 0.1 mm in thickness is used as an elastic skin having dimensions of 2.4 mm and 14 mm, respectively, in the spanwise and streamwise directions. The resonance frequency is 800 Hz with maximum amplitude of about 50μm.

Magnified view and schematic of the cross section of the shear stress sensor are shown in Figure 3.58. A platinum hot film is deposited on a SiNx diaphragm ($400 \times 400\,\mu m^2$) of 1 μm in thickness, and a 200 μm-deep air cavity is formed underneath.

A control system consisting of four sensor rows and three actuator rows in between is shown in Figure 3.59. Each sensor row has 48 micro wall-shear stress sensors with 1 mm spacing, and each actuator row has 16 wall-deformation actuators with 3 mm spacing.

A digital signal processor system (MPC7410, MTT Inc.) with 224 analog input and 96 output channels is used as the controller of the system. The repetition frequency of the control loop is 5 kHz. An optimal control scheme based on genetic algorithm (GA) is employed.[36] Driving voltage of each wall deformation actuator E_a is determined by

$$E_a = \sum_{i=1,3} W_i \tau'_{wi} \tag{3.23}$$

where τ'_{wi} is measured with three sensors located upstream, and the control variables W_i are optimized in such a way that the mean wall shear stress measured with three sensors at the most

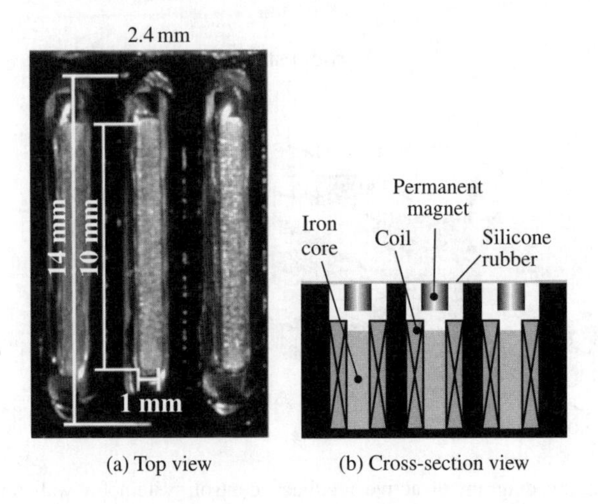

Figure 3.57 Wall deformation magnetic actuator.[31] *Source*: Suzuki Y 2005. Reproduced with permission of Elsevier

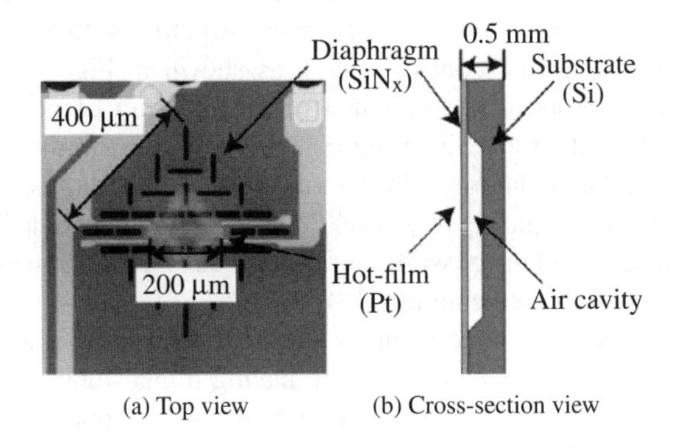

Figure 3.58 Hot-film shear stress sensor.[31] *Source*: Suzuki Y 2005. Reproduced with permission of Elsevier

downstream location is minimized. Note that the actuators move upward when E_a is positive, while downward when negative.

The experiments are performed in a turbulent air channel flow facility as shown in Figure 3.60. MEMS system is placed at the bottom wall of the test section. The bulk mean velocity $U_m = 3$ m/s and the Reynolds number Re_τ based on the wall friction velocity u_τ and the channel half-width, is $\mathrm{Re}_\tau = 300$. One viscous length and time unit correspond to 0.09 mm and 0.5 ms, respectively. Thus, the mean diameter of the near-wall streamwise vortices is estimated to be 2.7 mm and its characteristic time scale is 7.5 ms. The flow field is measured with a three-beam two-component LDV system.

Figure 3.61a–c shows profiles of the mean velocity, the rms values of velocity fluctuations and the Reynolds shear stress above the center of an actuator measured with LDV, respectively,

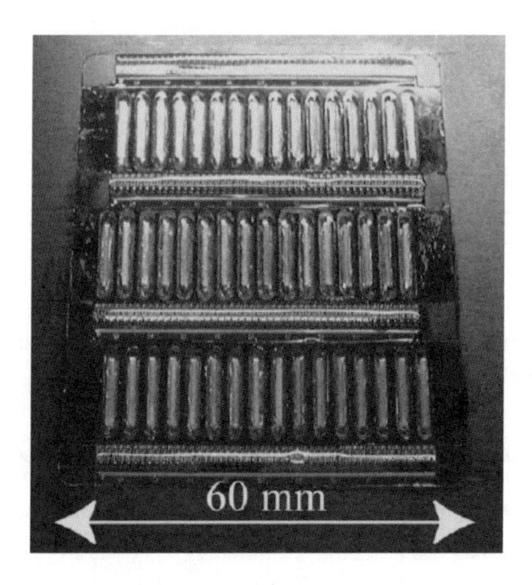

Figure 3.59 Feedback control system for wall turbulence.[31] *Source*: Suzuki Y 2005. Reproduced with permission of Elsevier

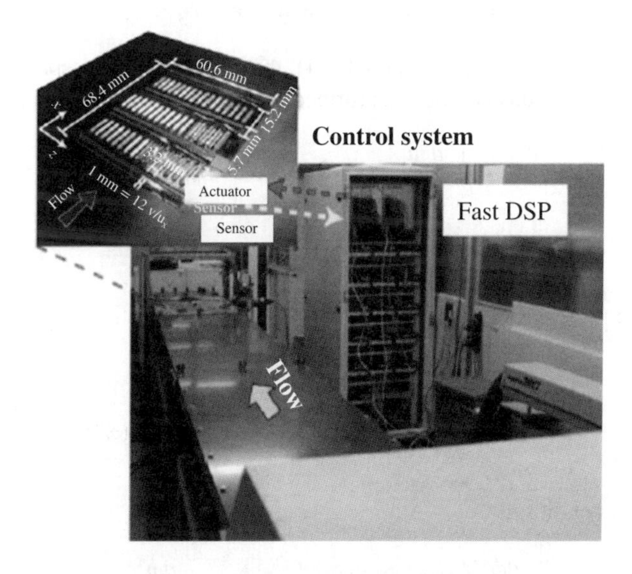

Figure 3.60 Turbulent air channel flow facility.[31] *Source*: Suzuki Y 2005. Reproduced with permission of Elsevier

where "+" denotes the stationary wall, empty circle denotes the GA-based control wall. When the present feedback control is applied, the mean velocity profile and rms values are unchanged. However, the Reynolds stress near the wall is decreased, which has a direct contribution to the drag reduction.

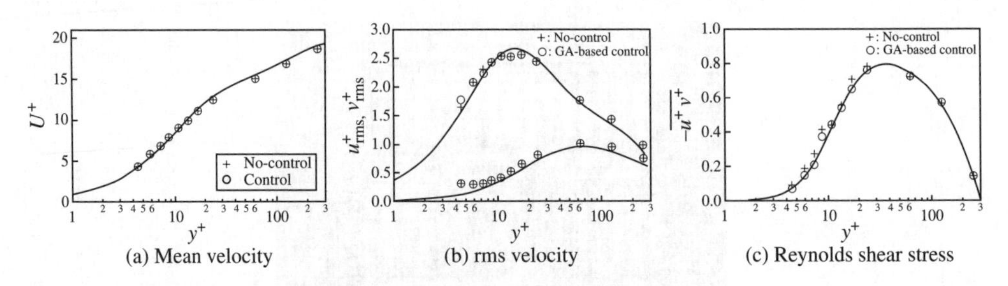

Figure 3.61 Profiles of mean velocity, rms velocity and Reynolds shear stress.[31] *Source*: Suzuki Y 2005. Reproduced with permission of Elsevier

3.6 Numerical Simulation of Wall Turbulence with Deformed Wall

3.6.1 Wall Turbulence with Streamwise-Traveling Surface Deformation Wave

To address the deformed walls, where the physical domain is nonrectangular in the simulation, a body-fitted coordinate set $(\tau, \ \boldsymbol{\xi})$ is introduced as follows:

$$\tau = t, \ \ \xi_1 = x_1, \ \ \xi_2 = \Phi(t, \mathbf{x}), \ \ \xi_3 = x_3$$

Using a time-split method (see Section 1.6.4), the N-S equation is split into Eqs. (1.224) and (1.225). These equations can be transformed further with the coordinate transformation, then

$$\frac{1}{R_e}\frac{\partial^2 \hat{u}_i}{\partial \xi_j^2} - \frac{2}{\Delta\tau}\hat{u}_i = -R - G(\hat{u}_i) \tag{3.24}$$

with the boundary condition

$$\hat{u}_i|_w = (1 + \Delta\tau\alpha_t^{n+1})u_i^{n+1}|_w + \Delta\tau\left(\left(1 - \delta_{2i}\right)\frac{\partial}{\partial \xi_i} + \alpha_i\right)\phi^n|_w \tag{3.25}$$

And

$$\frac{\partial^2 \phi^{n+1}}{\partial \xi_j^2} = \frac{1}{\Delta\tau}\left(\left(1 - \delta_{j2}\right)\frac{\partial}{\partial \xi_j} + \alpha_j\right)\hat{u}_j - G(\phi^{n+1}) \tag{3.26}$$

with the boundary conditions

$$\left(1 - \delta_{2i}\right)\left.\frac{\partial \phi^{n+1}}{\partial \xi_i}\right|_w + \alpha_i \phi^{n+1}|_w = \left(1 - \delta_{2i}\right)\left.\frac{\partial \phi^n}{\partial \xi_i}\right|_w + \alpha_i \phi^n|_w \tag{3.27}$$

$$u_i^{n+1}|_w = (1 - \Delta\tau\alpha_t^{n+1})\hat{u}_i|_w - \Delta\tau\left(\left(1 - \delta_{2i}\right)\frac{\partial}{\partial \xi_i} + \alpha_i\right)\phi^{n+1}|_w \tag{3.28}$$

where, $R = \frac{2}{\Delta\tau}u_i^n - 2\alpha_t^n u_i^n + (3H_i^n - H_i^{n-1}) + \nu\nabla^2 u_i^n$

$$H_i = -\left\{\left(1 - \delta_{2j}\right)\frac{\partial}{\partial \xi_j} + \alpha_j\right\}u_j u_i$$

$$\nabla^2 = \frac{\partial^2}{\partial \xi_j^2} + (1-\delta_{j2})2\gamma_j \frac{\partial^2}{\partial \xi_2 \partial \xi_j} - \delta_{j2}\frac{\partial^2}{\partial \xi_j^2} + \beta_j \ {}_j$$

$$G = 2(1-\delta_{j2})\gamma_j \frac{\partial^2}{\partial \xi_2 \partial \xi_j} - \delta_{j2}\frac{\partial^2}{\partial \xi_j^2} + \beta_j \ {}_j$$

$$\delta_{i2} = \begin{cases} 1 & \text{if } i = 2 \\ 0 & \text{if } i \neq 2 \end{cases}$$

$$\gamma_t = \frac{\partial \xi_2}{\partial t}, \ \gamma_i = \frac{\partial \xi_2}{\partial x_i}, \ \gamma_{ii} = \frac{\partial^2 \xi_2}{\partial x_i^2} \ \alpha_t = \gamma_t \frac{\partial}{\partial \xi_2}, \ \alpha_i = \gamma_i \frac{\partial}{\partial \xi_2}$$

$$\beta_i = \gamma_{i\ i}\frac{\partial}{\partial \xi_2} + \gamma_i^2 \frac{\partial^2}{\partial {\xi_2}^2}$$

where subscript i is the free index, and j is the dummy index, indicating a summation over all three values of the index.

Consider the lower wall undergoing an up-down oscillation in the form of a wave traveling in the streamwise direction as shown in Figure 3.62, the position of the wall is described by

$$y_w = a\sin(k_x x - \omega t) = a \sin k_x(x - ct) \tag{3.29}$$

where, y_w is the wall displacement, a is the magnitude of the displacement, c is the phase speed of the wave. Then, the boundary conditions on the lower wall are

$$u_w = 0 \tag{3.30}$$

$$v_w = -\omega a \cos k_x(x - ct) \tag{3.31}$$

$$w_w = 0 \tag{3.32}$$

In the body-fitted coordinates,

$$\xi_2 = \frac{y - y_w}{H - y_w} \tag{3.33}$$

$$\gamma_t = -v_w \left[\frac{1-\xi_2}{H - y_w}\right], \ \gamma_x = -\frac{\partial y_w}{\partial x}\frac{1-\xi_2}{H-y_w}, \ \gamma_y = \frac{1}{H-y_w}, \gamma_z = 0 \tag{3.34}$$

$$\gamma_{xx} = 2\left(\frac{\partial y_w}{\partial x}\right)^2 \frac{1-\xi_2}{(H-y_w)^2} + \frac{\partial^2 y_w}{\partial x^2}\frac{1-\xi_2}{H-y_w}, \ \gamma_{yy} = \gamma_{zz} = 0 \tag{3.35}$$

where H is the height of the upper boundary.

A free-slip boundary condition is applied at the upper boundary and a periodic boundary condition in horizontal direction.

A standard Fourier–Chebyshev spectral method is used for calculation, with Re = 10, 170, and wavy wall steepness $k_x a = 0.25$. The computational domain is 4λ (streamwise) $\times \frac{2\lambda}{\pi}$ (normal) $\times 2\lambda$ (spanwise),where λ is the wavelength.

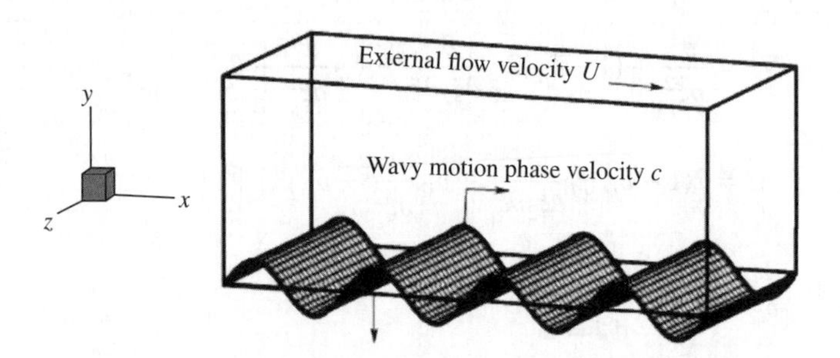

Figure 3.62 Deformed wall in form of a streamwise-traveling wave.[37] *Source*: Shen L 2003. Reproduced with permission of Cambridge University Press

3.6.1.1 Mean Flow

When $c/U = 0$, where U is the mean velocity at the upper boundary, the wavy wall is stationary. Figure 3.63 shows the streamline of the mean flow $(\langle u\rangle, \langle v\rangle)$, where $\langle\ \rangle$ denotes the mean value, the horizontal axis $x' = x - ct - n\lambda$, n is an integer and Figure 3.63a–c represents

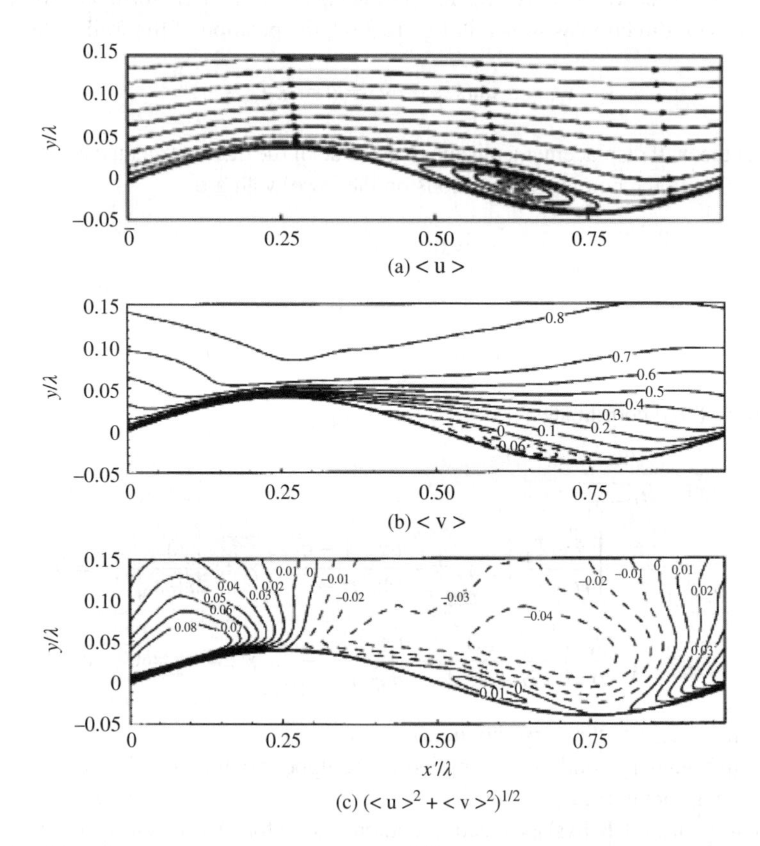

Figure 3.63 Mean flow over a stationary wavy wall.[37] *Source*: Shen L 2003. Reproduced with permission of Cambridge University Press

$\langle u \rangle$, $\langle v \rangle$, and $(\langle u \rangle^2 + \langle v \rangle^2)^{1/2}$, respectively. The streamlines in Figure 3.63a show a separation bubble located on the downhill side of the wavy wall. It should be noted that it only represents an averaged effect and the flow separation is highly intermittent actually. From Figure 3.63c, it is clear that the flow can be divided into four regimes: an outer flow, a separated region, an attached boundary on the uphill side of the wavy wall, and a free shear layer behind the crest.

When the wavy wall displacement travels in the streamwise direction, the wall boundary is no longer a streamline and there are streamlines that emanate from the wall, as shown in Figure 3.64, which plots the streamline pattern of the mean flow. Figure 3.64(1)–(3) represents $c/U = 0.4$, 1.2, and 2.0, respectively, and Figure 3.64a–c represents $\langle u \rangle$, $\langle v \rangle$, and $(\langle u \rangle^2 + \langle v \rangle^2)^{1/2}$, respectively. As indicated in contours of $\langle v \rangle$, the vertical velocity increases significantly, as c/U increases from 0.4 to 1.2 and then to 2.0. As a result, while the streamlines above the trough are concave at $c/U = 0$, they become flat at $c/U = 0.4$ and even convex at $c/U = 0.12$ and 2.0.

It is also shown from the contours of $\langle u \rangle$ and $(\langle u \rangle^2 + \langle v \rangle^2)^{1/2}$ that the wall traveling wave motion tends to suppress flow separation. No flow reversal region is found at $c/U = 0.4$. At $c/U = 0.12$, the flow becomes smoother and contour lines are parallel to the wave surface. At $c/U = 2.0$, the wavy wall forward of the crest pushes the fluid so strongly that there is a flow reversal region above the crest.

If the frame is moving with the wave, the wavy profiles become stationary in time. Then the velocity at the wall is

$$\begin{cases} \widetilde{u}_w = -c \\ \widetilde{v}_w = -ck_x a \cos\left(k_x x\right) \end{cases} \tag{3.36}$$

where hat "~" indicates the parameters in the moving frame. As a result,

$$(\widetilde{u}_w^2 + \widetilde{v}_w^2)^{1/2} = c(1 + (k_x a)^2 \cos^2(k_x x))^{1/2} \geq c \tag{3.37}$$

$$\frac{\widetilde{v}_w}{\widetilde{u}_w} = k_x a \cos(k_x x) = \frac{dy_w}{dx} \tag{3.38}$$

which means that fluid particles are gliding along the wall.

In the moving frame, the horizontal velocity component equals $-c$ at the wall and $U - c$ far away. Therefore, if $0 < c < U$, the mean velocity must change its sign as the wall is approached from the outer region. A trapped vortex located near the negative wave slope region is observed in Figure 3.65a and b for both $c/U = 0.4$ and 0.8. If $U < c$, such as $c/U = 1.2$, both $-c$ and $U - c$ are negative, all the streamlines are in the negative x-direction and there is no trapped vortex as shown in Figure 3.65c.

3.6.1.2 Turbulence Intensity

Figure 3.66 shows the turbulence intensities of each velocity component, that is, $\langle u'^2 \rangle$, $\langle v'^2 \rangle$, and $\langle w'^2 \rangle$, and turbulent energy $\frac{q^2}{2} = \langle (u'^2 + v'^2 + w'^2)/2 \rangle$ for a stationary wall. The maximum $\langle u'^2 \rangle$ exists above the trough and behind the crest, the location of which corresponds roughly to the free shear layer. The maximum $\langle v'^2 \rangle$ lags the streamwise one. However, the

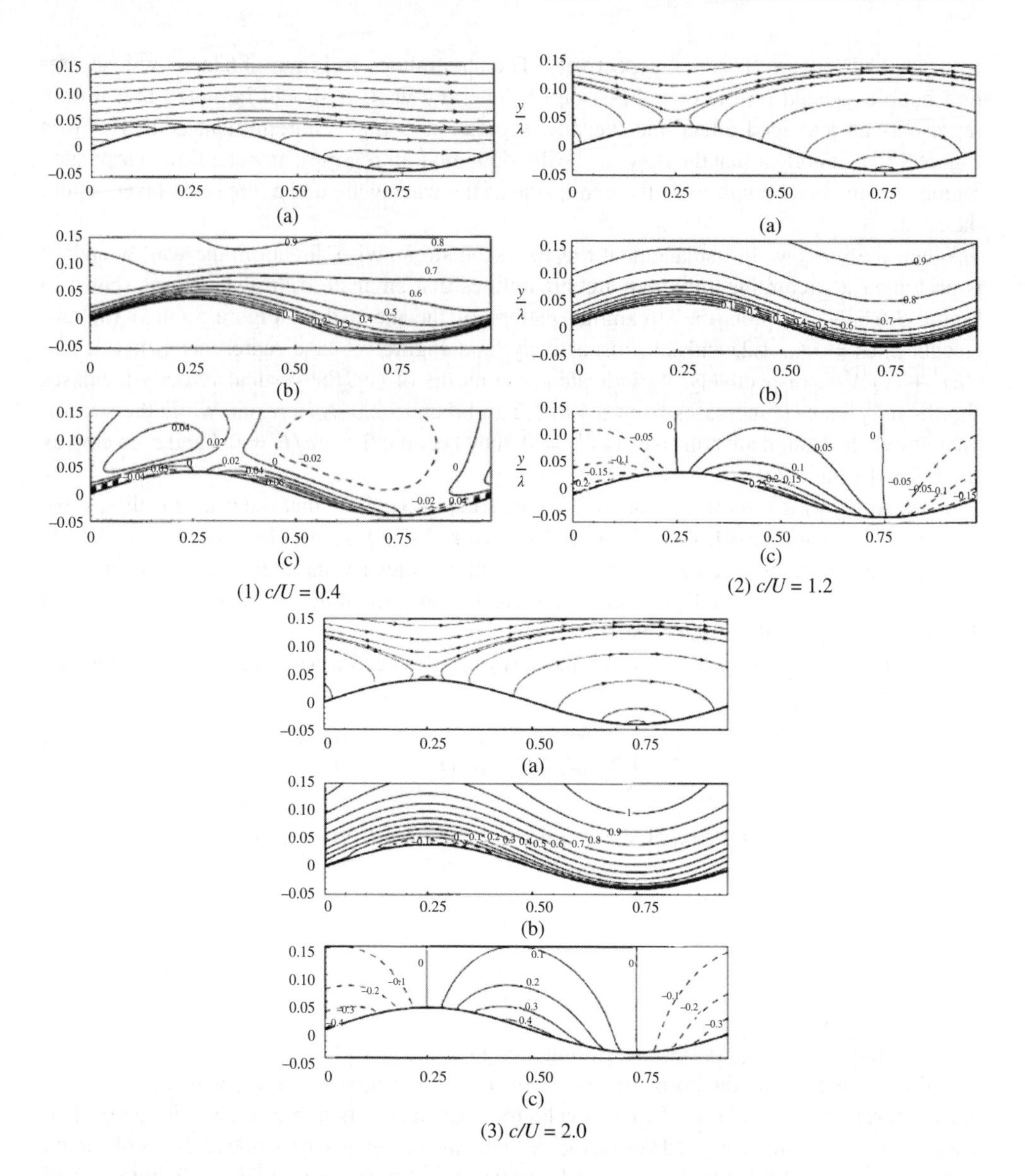

Figure 3.64 Mean flow over a wavy wall with different traveling velocity.[37] *Source*: Shen L 2003. Reproduced with permission of Cambridge University Press

maximum $\langle w'^2 \rangle$ locates near the uphill region where reattachment takes place. Among the three velocity components, the streamwise one contributes most to the turbulent kinetic energy.

The turbulence intensities for the traveling wavy walls with phase velocities $c/U = 0.4$ and 1.2 are shown in Figure 3.67(1) and (2), respectively. At $c/U = 0.4$, the locations of the maxima of $\langle u'^2 \rangle$ and $\langle v'^2 \rangle$ move upstream compared to the $c/U = 0$ case as shown in Figure 3.67(1)a and b, which is consistent with the upstream movement of the trapped vortex. There are two maxima of $\langle w'^2 \rangle$ appearing near the reattachment position and above the trough,

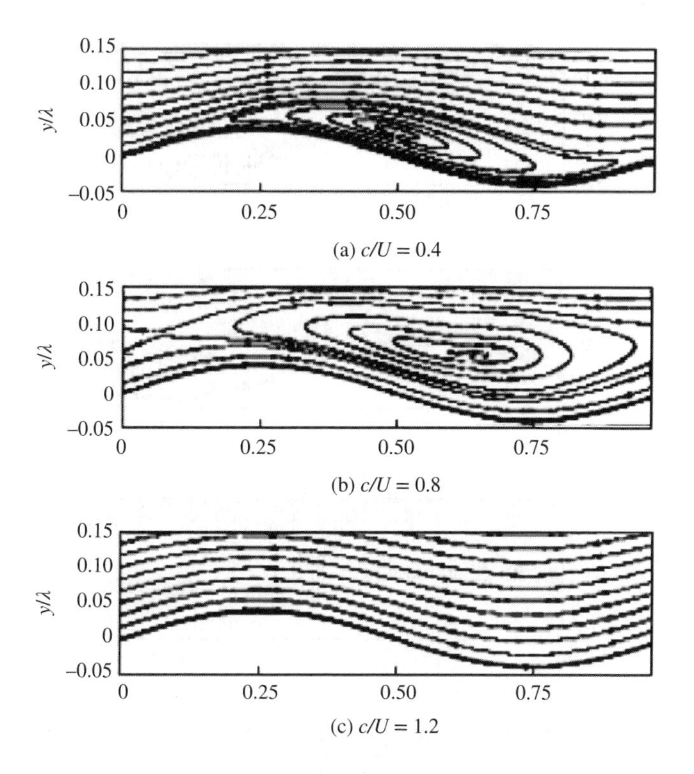

Figure 3.65 Contours of $\langle u \rangle$ in moving frame.[37] *Source*: Shen L 2003. Reproduced with permission of Cambridge University Press

respectively, as shown in Figure 3.67(1)c. The reason for the former can be attributed to the pressure–strain correlation as the fluid impacts on the backward face of the wavy wall, while the latter is related to energy transferred to spanwise direction from streamwise and vertical velocity fluctuations produced by the free shear layer. As shown in Figure 3.67(1) d, turbulence intensity is reduced compared to the $c/U = 0$ case, because of the weakening in the flow separation. As $c/U = 1.2$ shown in Figure 3.67(2), the turbulence intensity is substantially reduced in most of the region because of the elimination of flow separation.

The contours of Reynolds stress $\langle -u'v' \rangle$ are shown in Figure 3.68, where Figure 3.68a, b, and c represents $c/U = 0,\ 0.4$, and 1.2, respectively. For $c/U = 0$, the greatest Reynolds stress occurs near the location of the free shear layer. When $c/U = 0.4$, $\langle -u'v' \rangle$ is reduced and its maximum moves upstream. At $c/U = 1.2$, $\langle -u'v' \rangle$ becomes very small and its maximum is located upstream of the wave crest in contrast to the small c/U cases.

3.6.1.3 Vortex Structure

The instantaneous vortex structure is plotted in Figure 3.69, where Figure 3.69a and b represents $c/U = 0$ and 1.2, respectively. For $c/U = 0$, the dominant vortex structures are located above the trough and are generally less coherent than the $c/U > 0$ cases. For $c/U = 1.2$, the structures are substantially coherent and dominated by streamwise elements.

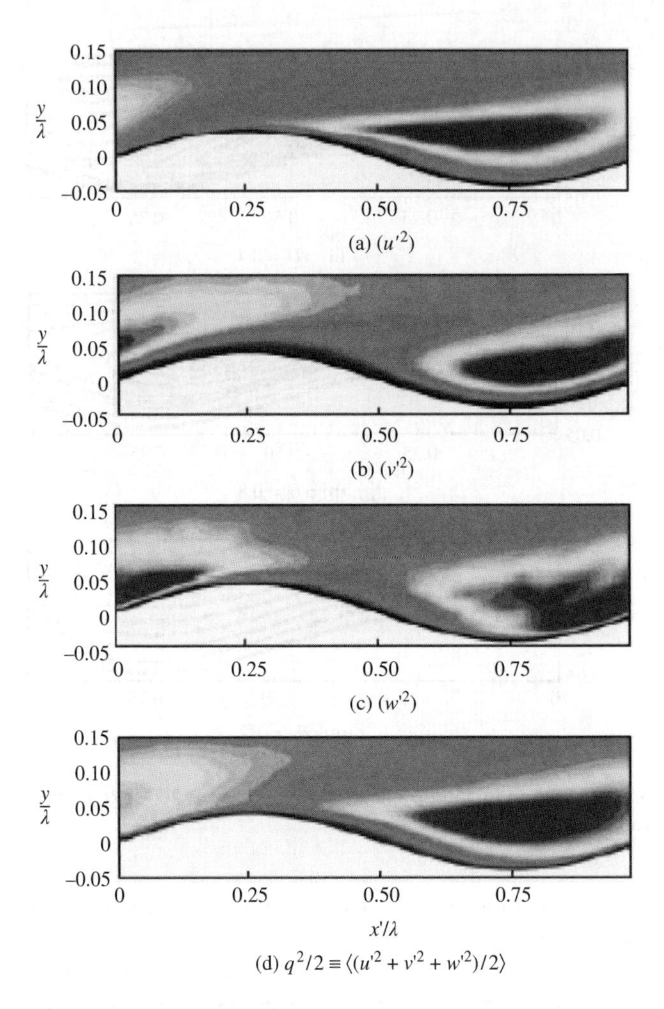

Figure 3.66 Contours of turbulence intensity for stationary wavy wall.[37] *Source*: Shen L 2003. Reproduced with permission of Cambridge University Press

3.6.1.4 Drag

The total drag force F_d on the wavy surface consisting of a friction drag F_f and a pressure drag F_p can be expressed as

$$F_d = F_f + F_p. \tag{3.39}$$

For an element of the wall surface shown in Figure 3.66, $ds = \left(1 + \left(\frac{dy_w}{dx}\right)^2\right)^{1/2}$, its tangential direction is $\mathbf{t} = (1,\ dy_w/dx)[1 + (dy_w/dt)^2]^{-1/2}$ and the wall-normal direction is $\mathbf{n} = (-dy_w/dx,\ 1)[1 + (dy_w/dx)^2]^{-1/2}$. The friction force and pressure force per unit area (projected

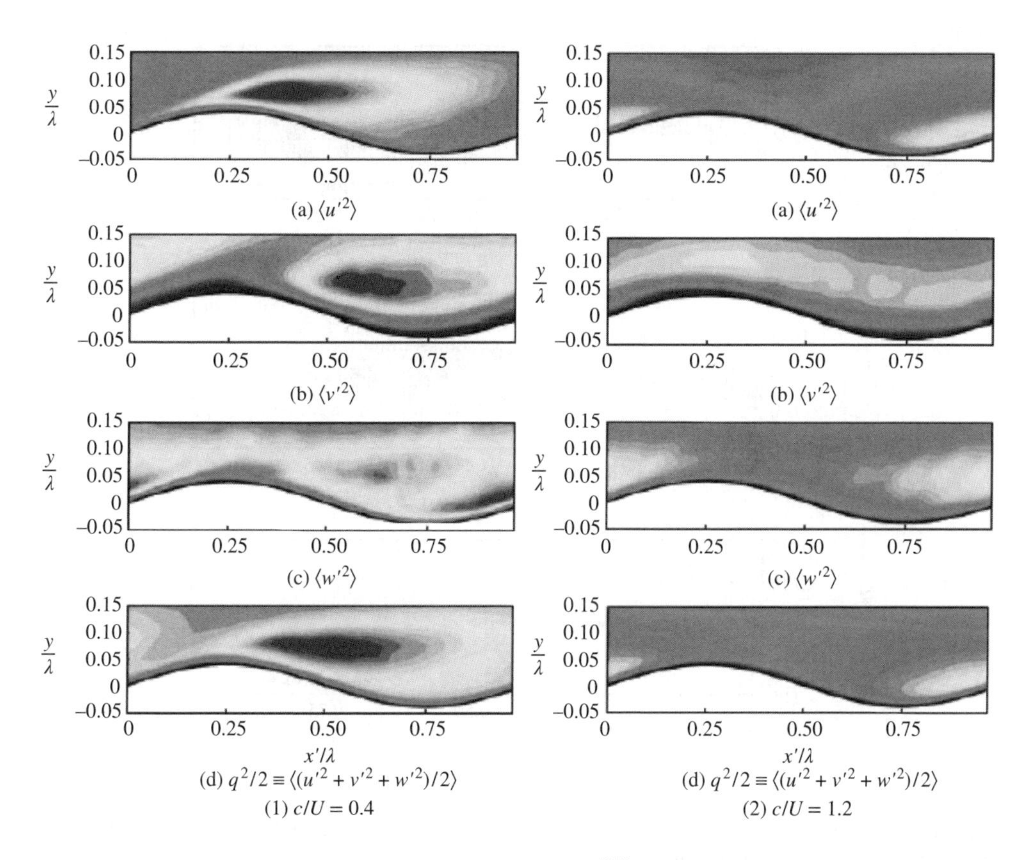

Figure 3.67 Turbulence intensities for traveling wavy wall.[37] *Source*: Shen L 2003. Reproduced with permission of Cambridge University Press

on the $(x,\ z)$-plane) are, respectively

$$f_x^f = \mu\left[-2\frac{\partial u}{\partial x}\frac{dy_w}{dx} + \left(\frac{\partial u}{\partial y} + \frac{\partial v}{\partial x}\right)\right]_w \tag{3.40}$$

$$f_x^p = p\frac{dy_w}{dx} \tag{3.41}$$

By integration, we obtain F_d, F_f, and F_p.

The variations of F_d, F_f, and F_p with the phase speed c/U are shown in Figure 3.70, where solid, dashed, and dotted lines represent F_d, F_p, and F_f, respectively. The friction drag F_f is always positive and its variation is relatively small. However, the pressure drag F_p decreases monotonically as c/U increases and changes sign from positive to negative, that is becomes a thrust, near $c/U \approx 1$. The total drag F_d decreases as c/U increases and becomes negative (thrust force) at large c/U (≈ 2).

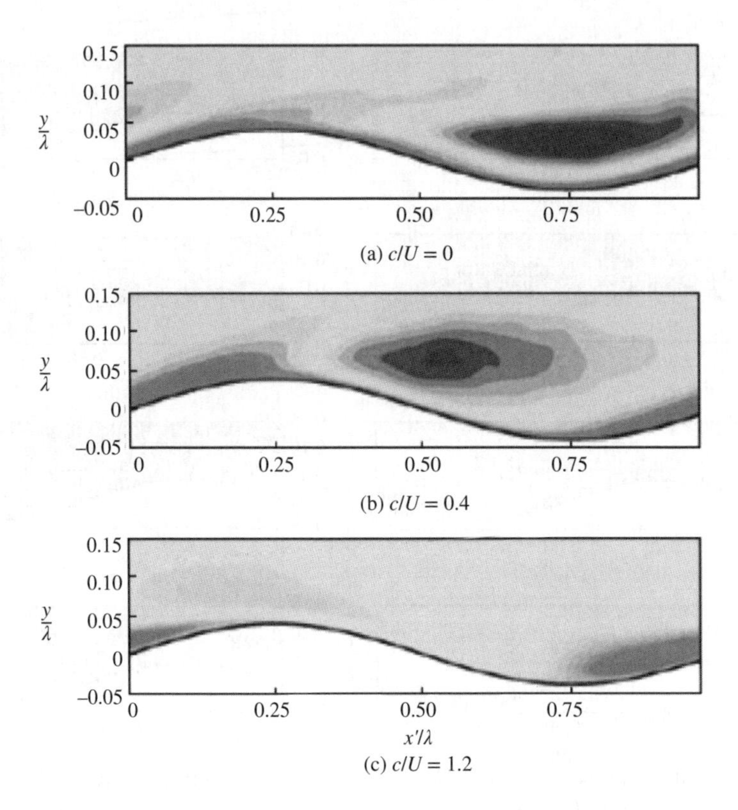

Figure 3.68 Contours of Reynolds stress $\langle -u'v' \rangle$ for wavy wall.[37] *Source*: Shen L 2003. Reproduced with permission of Cambridge University Press

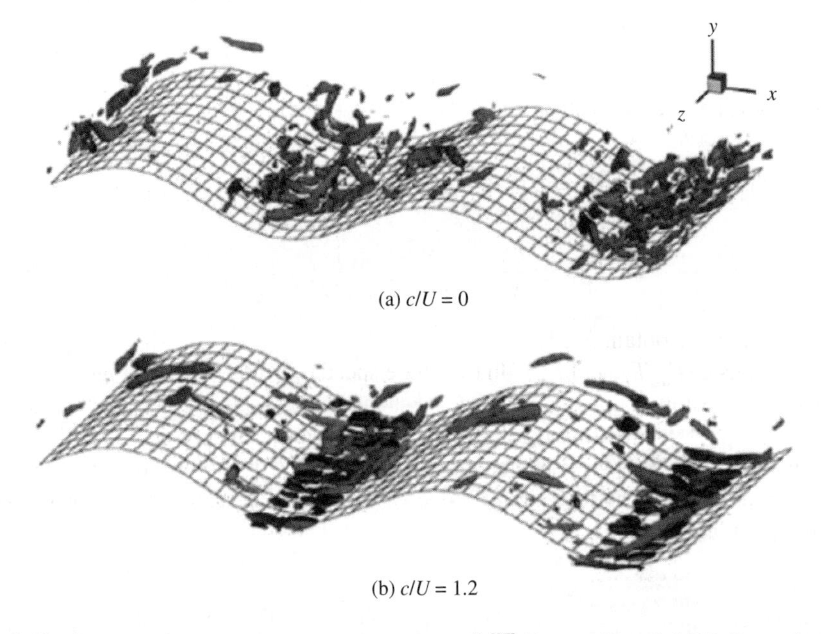

Figure 3.69 Instantaneous vortex structure for wavy wall.[37] *Source*: Shen L 2003. Reproduced with permission of Cambridge University Press

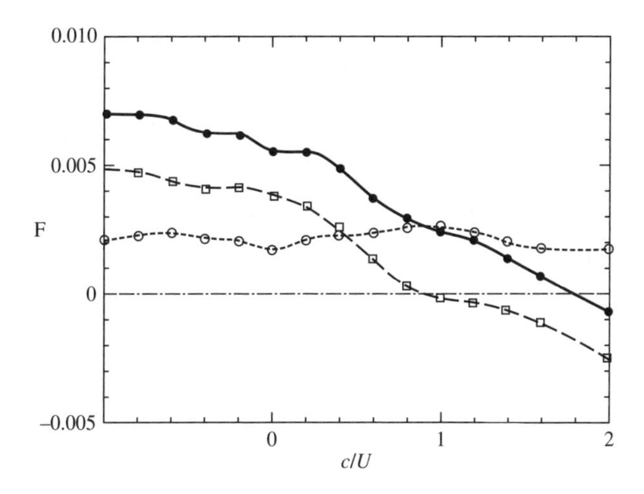

Figure 3.70 Variation of drag force with phase speed.[37] *Source*: Shen L 2003. Reproduced with permission of Cambridge University Press

3.6.2 Wall Turbulence with Sinusoidally Deformed Wall

Consider a plane channel, where the lower wall is deformed, while the upper wall is kept flat and stationary as shown in Figure 3.71. The deformed wall undergoes an up-down oscillation in the form of a standing wave, uniform in the streamwise direction, and sinusoidal in the spanwise direction, with the displacement given by

$$y_w = a \sin \frac{2\pi z}{s} \sin \frac{2\pi t}{T} \tag{3.42}$$

where a, s, and $T$are the amplitude, spanwise wavelength, and time period of the wall oscillation, respectively.

The wall deformation is described with a body-fitted coordinate system. A fractional-step method and a finite difference method are utilized for the temporal and spatial discretizations, respectively. In calculation, the Reynolds number is 4600 based on the bulk mean velocity U_m and the channel width 2δ.

3.6.2.1 Drag

The friction drag coefficient on the deformed wall,C_{f_l}, is defined as

$$C_{fl} = \left[\frac{1}{\Delta T} \int_{\Delta T} dt \cdot \int_{\Gamma_w} \tau_w d\Gamma / \Gamma_{w0} \right] / (\rho U_m^2 / 2) \tag{3.43}$$

where ΔT is the averaging time, Γ_w and Γ_{w0} denote the instantaneous wall surface area and its projection onto the x–z plane, respectively.

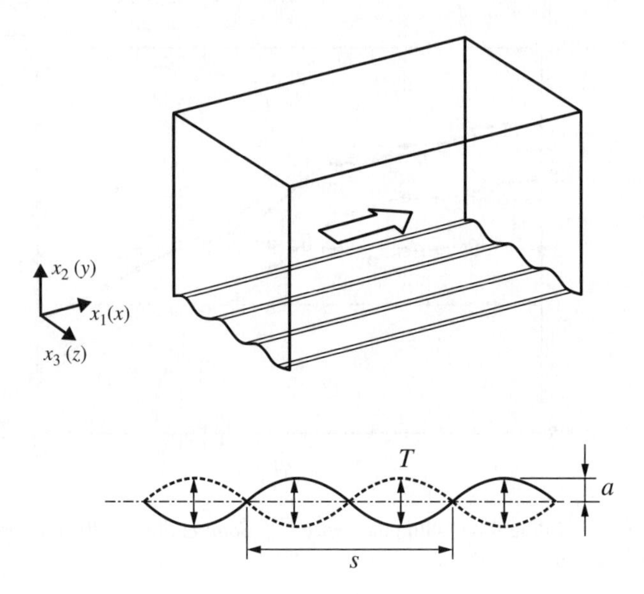

Figure 3.71 Deformed wall in form of a standing wave in spanwise direction.[38] *Source*: Mito Y 1998. Reproduced with permission of Elsevier Limited

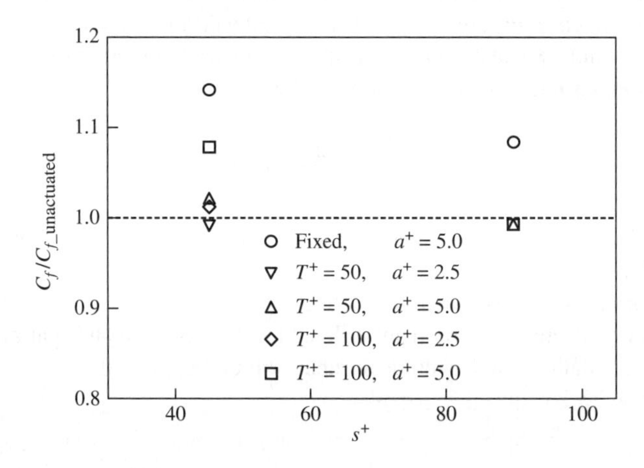

Figure 3.72 Skin friction coefficients for different parameters.[38] *Source*: Mito Y 1998. Reproduced with permission of Elsevier Limited

The ratios of the skin friction coefficients on the deformed walls and those in the unactuated case are shown in Figure 3.72, where $C_{f\_unactuated}$ denotes the skin friction coefficient on the unactuated wall, and the superscript "+" indicates quantities in wall units. It is clear that the skin friction can be reduced only by imposing a temporal effect on the wall deformation with the appropriate parameters, such as $(s^+, T^+, a^+) = (45, 50, 2.5)$, (90,50,5.0), and (90,100,5.0).

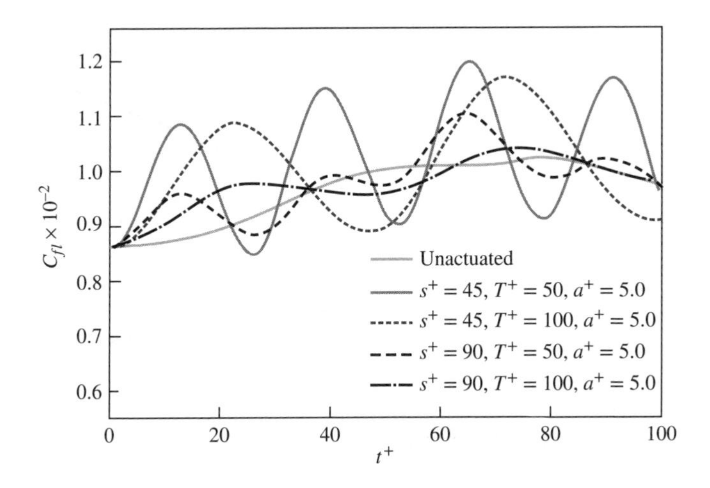

Figure 3.73 Evolutions of skin friction coefficient on deformed wall.[38] *Source*: Mito Y 1998. Reproduced with permission of Elsevier Limited

The initial evolutions of the skin friction coefficient on the deformed wall for different parameters are shown in Figure 3.73. Because of the symmetric effects of former and latter halves of the wall deformation period, each evolution has two fluctuations during each deformation period, in which the local minimum and maximum occur at the zero and maximum displacements, respectively. In addition to the small T, there exists a long-period fluctuation in the evolution of the skin friction coefficient, caused by structural alteration of turbulence. For the larger T, the skin friction coefficient sometimes shows larger variations than that exhibited in the unactuated case.

3.6.2.2 Mean Velocity

The time-averaged statistics have been obtained by interpolating data at each distance from the mean height of the deformed wall, thus the local zones covered by the wall deformation are excluded from the averaging. The mean velocity profiles on the deformed wall are shown in Figure 3.74. The mean flows are decelerated by wall deformation compared with the unactuated case.

3.6.2.3 Turbulence Intensity

Figure 3.75 shows the distributions of turbulence intensity on the deformed wall, where Figure 3.75a, b, c, and d represents rms velocity fluctuations, $u'_{\rm rms}$, $v'_{\rm rms}$ and $w'_{\rm rms}$, and Reynolds shear stress $-\overline{u'v'}$, respectively. Little difference is observed in u'_{rms}. However, $v'_{\rm rms}$ and $w'_{\rm rms}$ are much enhanced in the case of $(s^+, T^+, a^+) = (45, 100, 5.0)$, and reduced in the case of $(s^+, T^+, a^+) = (90, 100, 5.0)$, for which the skin friction has been similarly enhanced and reduced, respectively. In the near-wall region, $v'_{\rm rms}$ becomes larger than the unactuated case in accordance with the velocity component of wall deformation. Thus, $v'_{\rm rms}$

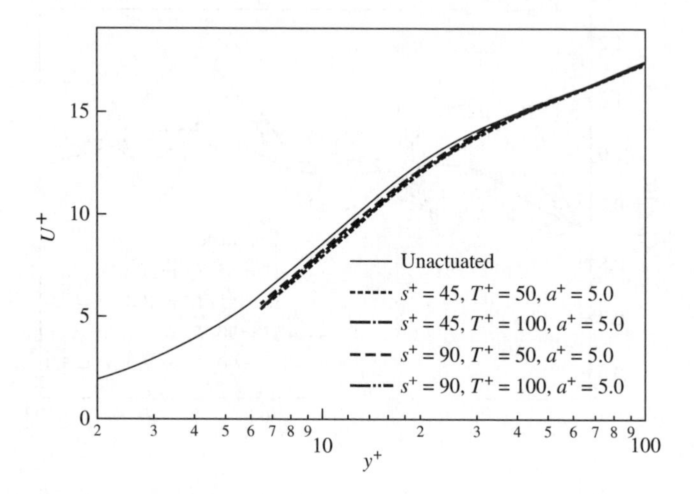

Figure 3.74 Mean velocity profiles on deformed wall.[38] *Source*: Mito Y 1998. Reproduced with permission of Elsevier Limited

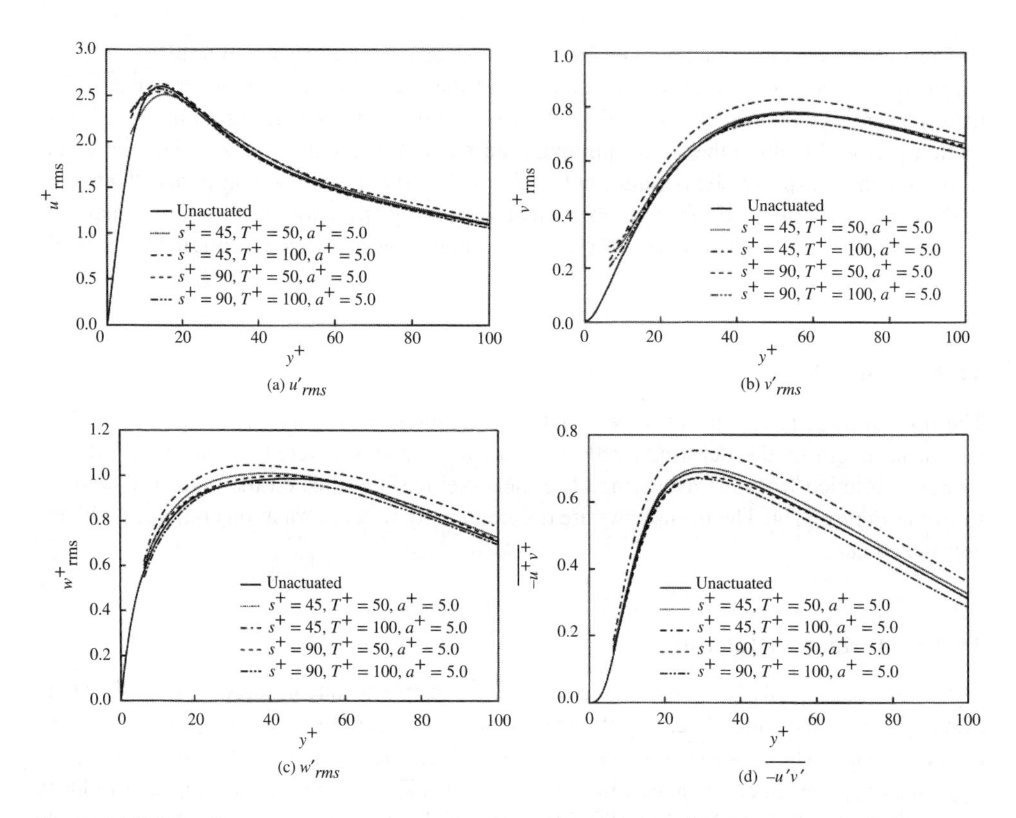

Figure 3.75 Distributions of turbulence intensity on the deformed wall.[38] *Source*: Mito Y 1998. Reproduced with permission of Elsevier Limited

in the case of $T^+ = 50$ is larger than in the case of $T^+ = 100$. Except for the near-wall region, $v'_{\rm rms}$ in the case of $T^+ = 50$ almost agrees with the unactuated case, whereas $v'_{\rm rms}$ could be enhanced and attenuated, respectively, in the cases of $(s^+, T^+, a^+) = (45, 100, 5.0)$ and $(s^+, T^+, a^+) = (90, 100, 5.0)$. The variation of $v'_{\rm rms}$ is reflected in the Reynolds shear stress distributions as shown in Figure 3.75d. In the case of the increased skin friction, the Reynolds shear stress is markedly increased, whereas it is decreased in the case of the reduced skin friction. The extent of the variation in the case of $T^+ = 50$ is smaller than the case of $T^+ = 100$.

3.6.2.4 Vortex Structure

The turbulent structures over the deformed walls with the maximum displacement at large skin friction instants are shown in Figure 3.76a–d. Figure 3.76e shows the turbulent structures in the unactuated case. The low-pressure regions, low-speed streaks and high-speed streaks are visualized by white, light gray and dark gray, respectively. The low-pressure regions mostly correspond to vortex core regions. The turbulent structures seem to be enhanced. For the smaller spanwise wavelength $s^+ = 45$, small-scale turbulent structures are frequently induced and intensified on the deformed walls.

The turbulent structures over the deformed walls with the maximum displacement at small skin friction instants are shown in Figure 3.77a–d. Figure 3.77e shows the turbulent structures in the unactuated case. Compared with Figure 3.76, the streaky structures are elongated and well-aligned with the streamwise uniform deformation of the wall, which implies that the intermittency of the turbulent structures is increased.

3.6.3 Wall Turbulence with Opposition Wall Deformation Control

3.6.3.1 Continuous Opposition Wall Deformation Control

The sinusoidally deformed wall hardly reduces drag in some cases as discussed in previous section. A more effective control in the drag reduction, called opposition wall deformation control, is proposed, which is a simple closed-loop control algorithm with control input continuously adjusted based on the instantaneous flow field. Its physical argument is schematically demonstrated in Figure 3.78. The wall is located deformed such that the velocity induced by the wall motion is equal and opposite to the wall-normal velocity component at $y^+ = 10$, to attenuate the motion of a streamwise vortex thereby reducing the transport of high momentum fluid toward the wall and reducing drag.

The governing equations are written as

$$
\begin{aligned}
&\frac{\partial u_i}{\partial t} = -\frac{\partial}{\partial x_k} u_i u_k + \frac{1}{\mathrm{Re}} \frac{\partial^2 u_i}{\partial x_k \partial x_k} - \frac{\partial p}{\partial x_i} - \frac{dP}{dx_i}\delta_{i1} \\
&\frac{\partial u_i}{\partial x_i} = 0
\end{aligned}
\tag{3.44}
$$

where $-dP/dx_1$ is the mean streamwise pressure gradient to impose a constant mass flux. (x_1, x_2, x_3), or (x, y, z), represent streamwise, wall-normal, and spanwise directions,

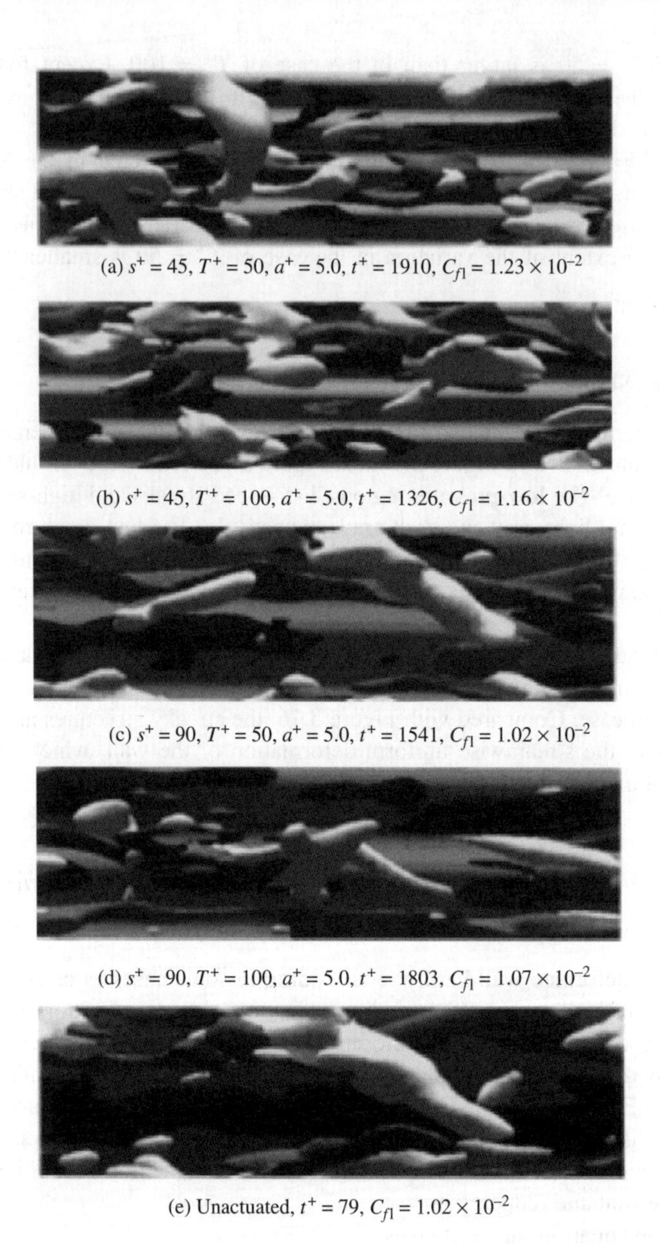

(a) $s^+ = 45$, $T^+ = 50$, $a^+ = 5.0$, $t^+ = 1910$, $C_{f1} = 1.23 \times 10^{-2}$
(b) $s^+ = 45$, $T^+ = 100$, $a^+ = 5.0$, $t^+ = 1326$, $C_{f1} = 1.16 \times 10^{-2}$
(c) $s^+ = 90$, $T^+ = 50$, $a^+ = 5.0$, $t^+ = 1541$, $C_{f1} = 1.02 \times 10^{-2}$
(d) $s^+ = 90$, $T^+ = 100$, $a^+ = 5.0$, $t^+ = 1803$, $C_{f1} = 1.07 \times 10^{-2}$
(e) Unactuated, $t^+ = 79$, $C_{f1} = 1.02 \times 10^{-2}$

Figure 3.76 Turbulent structures over deformed wall at large skin friction instants.[38] *Source*: Mito Y 1998. Reproduced with permission of Elsevier Limited

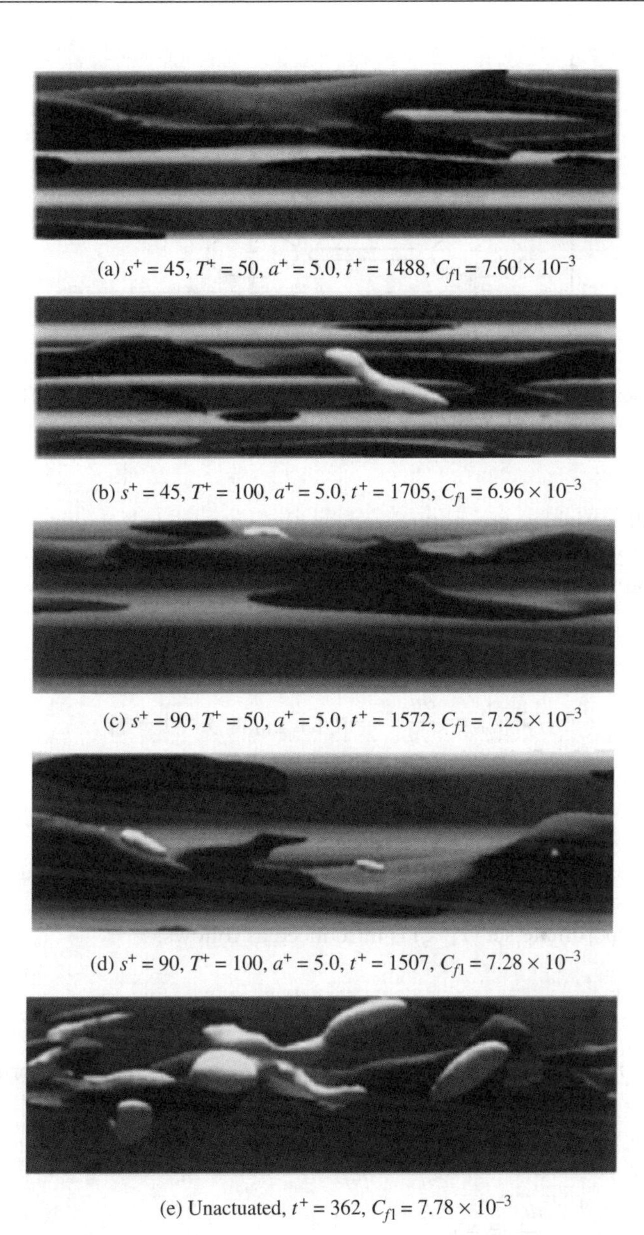

(a) $s^+ = 45$, $T^+ = 50$, $a^+ = 5.0$, $t^+ = 1488$, $C_{f1} = 7.60 \times 10^{-3}$

(b) $s^+ = 45$, $T^+ = 100$, $a^+ = 5.0$, $t^+ = 1705$, $C_{f1} = 6.96 \times 10^{-3}$

(c) $s^+ = 90$, $T^+ = 50$, $a^+ = 5.0$, $t^+ = 1572$, $C_{f1} = 7.25 \times 10^{-3}$

(d) $s^+ = 90$, $T^+ = 100$, $a^+ = 5.0$, $t^+ = 1507$, $C_{f1} = 7.28 \times 10^{-3}$

(e) Unactuated, $t^+ = 362$, $C_{f1} = 7.78 \times 10^{-3}$

Figure 3.77 Turbulent structures over deformed wall at small skin friction instants.[38] *Source*: Mito Y 1998. Reproduced with permission of Elsevier Limited

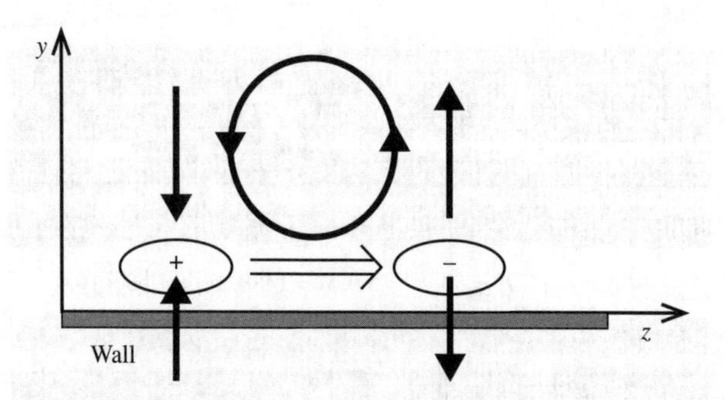

Figure 3.78 Schematic of opposition wall deformation control.[39] *Source*: Choi H 1994. Reproduced with permission of Cambridge University Press

respectively. All variables are normalized by the channel half-width δ and the laminar centerline velocity U_c.

For opposition wall deformation control, the boundary conditions are given as

$$u_1 = u_3 = 0 \tag{3.45}$$

$$u_2 = v_{wu} = \frac{\partial \eta_u}{\partial t} = -u_2|_{y^+{}_u=10} \text{ at } y = 1 + \eta_u \tag{3.46}$$

$$u_2 = v_{wd} = \frac{\partial \eta_d}{\partial t} = -u_2|_{y^+{}_d=10} \text{ at } y = -1 + \eta_d \tag{3.47}$$

where η_u and η_d are, respectively, the displacements of upper and lower walls with respect to the uncontrolled state and thus are only functions of variables (t, x_1, x_3).

A body-fitted coordinate set $(\tau,\ \xi)$ is introduced as follows:

$$t = \tau,\ x_1 = \xi_1,\ x_2 = \xi_2(1+\eta) + \eta_{0,}\ x_3 = \xi_3 \tag{3.48}$$

where $\eta = (\eta_u - \eta_d)/2$ and $\eta_0 = (\eta_u + \eta_d)/2$. Equation (3.44) is then transformed as

$$\begin{aligned} \frac{\partial u_i}{\partial \tau} &= -\frac{\partial u_i u_j}{\partial \xi_j} - \frac{\partial p}{\partial \xi_j} + \frac{1}{R_e}\frac{\partial^2 u_i}{\partial \xi_j^2} - \frac{dP}{d\xi_i}\delta_{i1} + S_i \\ \frac{\partial u_i}{\partial \xi_i} &= -S \end{aligned} \tag{3.49}$$

where

$$\begin{aligned} S_i = &-\gamma_t \frac{\partial u_i}{\partial \xi_2} - \phi_j \frac{\partial u_i u_j}{\partial \xi_2} - \phi_i \frac{\partial p}{\partial \xi_2} - \gamma_i \frac{\partial P}{\partial \xi_2}\delta_{i1} \\ &+ \frac{1}{\text{Re}}\left(2\left(1-\delta_{j2}\right)\gamma_j \frac{\partial^2 u_i}{\partial \xi_2 \partial \xi_j} - \delta_{j2}\frac{\partial^2 u_i}{\partial \xi_j^2} + \gamma_j\ {}_j\frac{\partial u_i}{\partial \xi_2} + \gamma_j^2 \frac{\partial^2 u_i}{\partial \xi_2^2}\right) \end{aligned} \tag{3.50}$$

$$S = \phi_j \frac{\partial u_i}{\partial \xi_2} \tag{3.51}$$

$$\phi_j = \gamma_j - \delta_{j2}$$

$$\gamma_t = -\frac{(1+\xi_2)v_{wu} + (1-\xi_2)v_{wd}}{2(1+\eta)}$$

$$\gamma_x = -\frac{\xi_2 \frac{\partial \eta}{\partial x} + \frac{\partial \eta_0}{\partial x}}{(1+\eta)}, \quad \gamma_y = \frac{1}{(1+\eta)}, \quad \gamma_z = -\frac{\xi_2 \frac{\partial \eta}{\partial z} + \frac{\partial \eta_0}{\partial z}}{2(1+\eta)}$$

$$\gamma_{xx} = -\frac{\xi_2 \frac{\partial^2 \eta}{\partial x^2} + \frac{\partial^2 \eta_0}{\partial x^2} + 2\gamma_x \frac{\partial \eta}{\partial x}}{(1+\eta)}, \quad \gamma_{yy} = 0, \quad \gamma_{zz} = -\frac{\xi_2 \frac{\partial^2 \eta}{\partial z^2} + \frac{\partial^2 \eta_0}{\partial z^2} + 2\gamma_z \frac{\partial \eta}{\partial z}}{(1+\eta)}$$

The mean pressure gradient is obtained by

$$-\frac{dP}{d\xi_1} = -\int b^* d\xi^3 / \int \frac{1}{1+\eta} d\xi^3 \tag{3.52}$$

where $d\xi^3 = d\xi_1 d\xi_2 d\xi_3$

$$b^* = u_1 \frac{\partial \eta}{\partial t} + \frac{1}{1+\eta}\left(-\frac{\partial (u_1 u_j)}{\partial \xi_j} - \frac{\partial P}{\partial \xi_1} + \frac{1}{\mathrm{Re}} \frac{\partial^2 u_1}{\partial \xi_j \partial \xi_j} + S_1\right)$$

The boundary conditions become

$$u_1 = u_3 = 0 \tag{3.53}$$

$$u_2 = v_{wd} = \frac{\partial \eta_d}{\partial t} = -u_2|_{y^+_d=10} \text{ at } \xi_2 = -1 \tag{3.54}$$

$$u_2 = v_{wu} = \frac{\partial \eta_u}{\partial t} = -u_2|_{y^+_u=10} \text{ at } \xi_2 = 1 \tag{3.55}$$

When a certain location of the wall reaches or approaches the maximum amplitude, the boundary conditions may not be physically valid due to the nonnegligible surface gradient, and the simulation finally broke down since rms wall deformation magnitude rapidly increases. Therefore the maximum amplitude of wall deformation is restricted to $|\eta_m^+| \leq 5$ ($m = u, d$).

The numerical method used is based on a semi-implicit, fractional step method: the diffusion and nonlinear terms are advanced in time with Crank–Nicolson method and a third-order Runge–Kutta method, respectively. All spatial derivatives are discretized with a second-order central-difference scheme. The computational domain is, respectively, 2.5 $\pi\delta$ and 0.75 $\pi\delta$ in the x- and z-directions.

Figure 3.79 shows the time history of the mean streamwise pressure gradient $-dP/dx_1$, where dashed line and solid line represent the cases of unactuated and deformed walls, respectively. The significant drag reduction is obtained by the opposition wall deformation control.

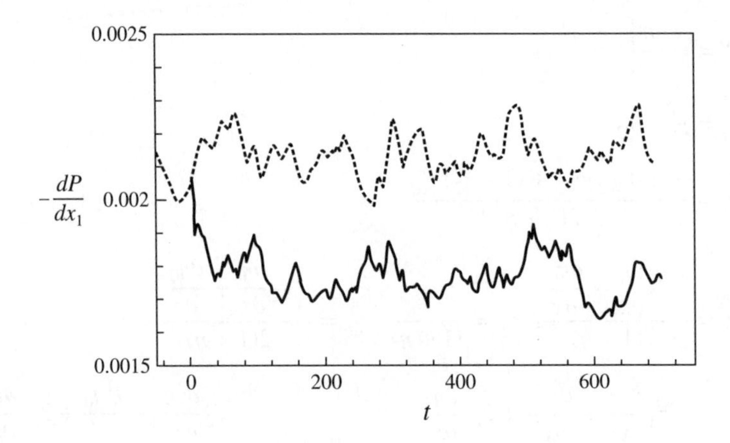

Figure 3.79 Time history of mean streamwise pressure gradient.[40] *Source*: Kang S 2000. Reproduced with permission of Springer

Contours of the instantaneous streamwise vorticity in a cross-flow plane for the cases of unactuated and deformed walls are shown in Figure 3.80a and b, respectively. It has been shown that the strength of the streamwise vorticity near the wall is significantly weakened by the opposition wall-deformation control.

Instantaneous wall shape of the deformed wall is shown in Figure 3.81, which is elongated in the streamwise direction and resembles riblet in appearance.

In order to suppress excessive wall deformation, the boundary conditions, expressed by Eqs. (3.54) and (3.55), can also be modified as follows:

$$u_{2W}(t_{n+1}) = -(u_{2s}(t_n) - \langle u_{2s}(t_n) \rangle) - 0.31\eta_m(t_n) \tag{3.56}$$

where t_n denotes time step n, u_{2w} denotes the wall-normal velocity at the wall, and u_{2s} is at $y/\delta = 0.1$ ($y^+ \approx 15$). $\langle\rangle$ denotes an ensemble average in the $x - z$ plane at each time step. The second term of right-hand side of Eq. (3.56) is devised to keep the total volume of the flow domain constant, and the third term is a damping term to restrict the amplitude of wall deformation.

Time histories of the drag coefficient normalized with its value for the uncontrolled plane channel flow, and the rms values of the wall displacement and velocity are shown in Figure 3.82, by solid line, dashed line and dotted-dashed line, respectively. The mean drag reduction rate is about 10%. The rms value of the wall velocity is $0.15u_\tau$ and almost same as that of the wall-normal velocity at $y^+ = 10$. The rms value of the wall displacement is about 1 wall unit, such that the deformed wall has minor effect on the turbulent flow field.

Figure 3.83 shows an instantaneous shape of wall deformation. It is evident that the deformation is much elongated in the streamwise direction.

Two-dimensional (x–z) spectrum of the wall velocity v_W is shown in Figure 3.84. Two marked peaks are observed at $(k_x, k_z) = (3, 3)$, and (5, 5). Thus, the typical length scales estimated from these peaks are $(m_x^+, m_z^+) = (200, 60)$ and $(120, 36)$, where the spanwise scale is in

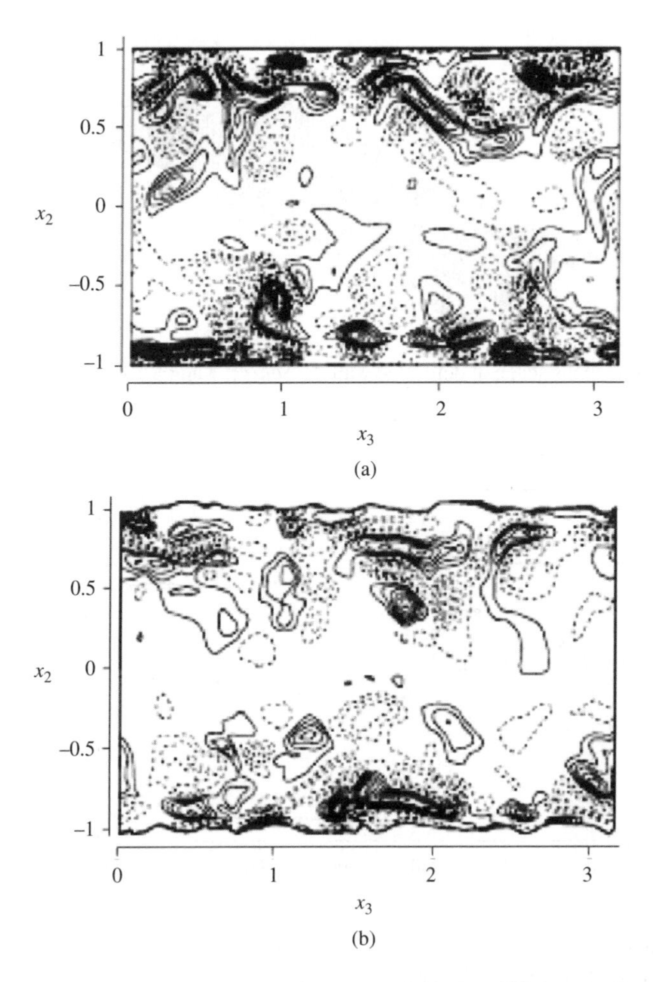

Figure 3.80 Contours of streamwise vorticity in a cross-flow plane.[40] *Source*: Kang S 2000. Reproduced with permission of Springer

between the mean diameter of the quasi-streamwise vortices and the spacing of the near-wall streaky structures.

3.6.3.2 MEMS Opposition Wall Deformation Control

For the opposition wall deformation control, the wall should be deformed in accordance with the information of instantaneous flow fields, and the wall deformation and information detection are actually performed by means of actuators and sensors with the scales comparable with those of the coherent structures. Recent development of MEMS technology enables us to fabricate prototypes of such micro devices.

Figure 3.81 Instantaneous shape of deformed wall.[40] *Source*: Kang S 2000. Reproduced with permission of Springer

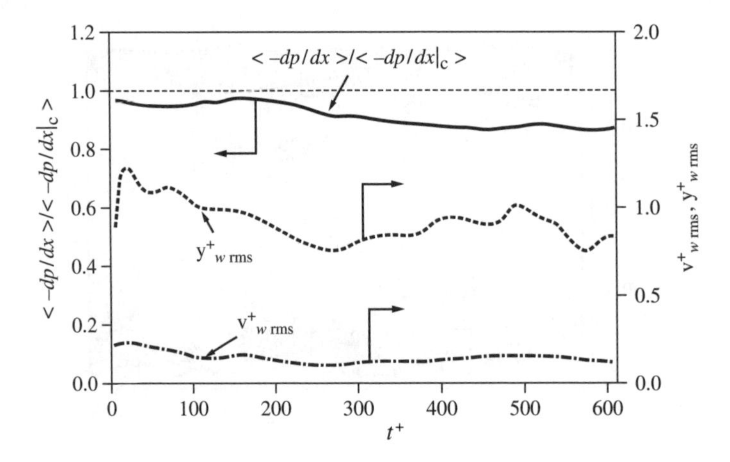

Figure 3.82 Time histories of normalized drag coefficient, rms wall displacement, and rms velocity.[41] *Source*: Endo T 2000. Reproduced with permission from Elsevier Limited

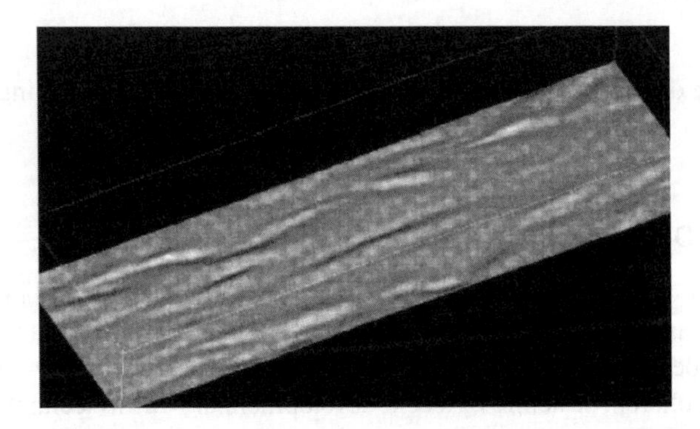

Figure 3.83 Instantaneous shape of wall deformation.[41] *Source*: Endo T 2000. Reproduced with permission from Elsevier Limited

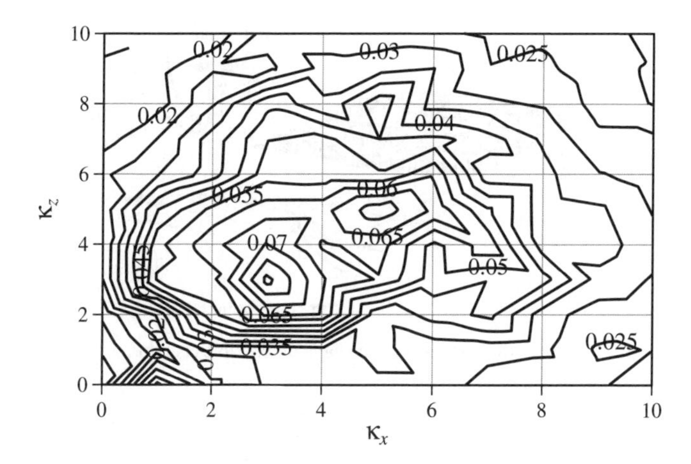

Figure 3.84 Two-dimensional spectrum of the wall velocity.[41] *Source*: Endo T 2000. Reproduced with permission from Elsevier Limited

In the previous section, actuators and sensors are assumed to be infinitely small, and placed at each corresponding computational grid point, which is unrealistic. Therefore, the finite dimensions of the control devices should be considered for an opposition wall deformation control with MEMS technology. A schematic of deformable actuator is shown in Figure 3.85a. The streamwise and spanwise dimensions of the actuator are chosen as 172 and 60 wall unit, respectively. Each actuator is assumed to be deformed only in the y-direction. The shape is determined with a sinusoid in the spanwise direction, in such way that the distance between the peak and trough is about the mean diameter of the quasi-streamwise vortices. A shear stress sensor is centered at 12.3 wall units upstream from the upstream end of the actuator. Hence, the streamwise distance between the sensor and the center of the actuator is 50 wall units.

An arrangement of arrayed shear stress sensors and deformable actuators is shown in Figure 3.85a. Thirty-six actuators (6×6 in the streamwise and spanwise directions) are distributed with a regular pitch on both walls of the channel.

Based on the spanwise gradients of the wall shear stresses, $\partial\tau_u/\partial z$ and $\partial\tau_w/\partial z$, measured by sensors, the velocity at the center of the peak/trough of the actuator is determined by

$$v_m^+(t_{n+1}) = \begin{cases} \alpha \tanh\left(\dfrac{\partial\tau_u^+(t_n)}{\partial z^+}\dfrac{1}{\beta}\right) - \gamma y_m^+(t_n) & \text{if } \dfrac{\partial\tau_w(t_n)}{\partial z} < 0 \\ -\gamma y_m^+(t_n) & \text{otherwise} \end{cases} \tag{3.57}$$

where, y_m is the wall displacement at the peak/trough, and α, β, and γ are control parameters, respectively. The wall velocity of each grid point on the actuator is given by

$$v_w^+(t_{n+1}) = v_m^+(t_{n+1}) f(x^+) \exp\left[-\frac{\left(z^+ - z_c^+\right)^2}{\sigma_z^{+2}}\right] \sin\left[\frac{2\pi\left(z^+ - z_c^+\right)}{m_z^+}\right] \tag{3.58}$$

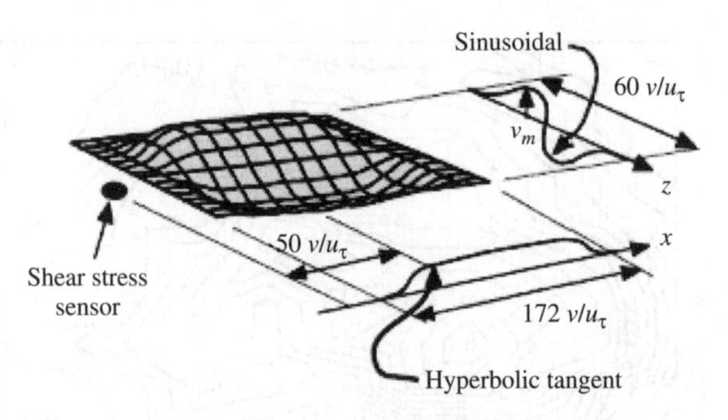

(a) Dimension of a single deformable actuator

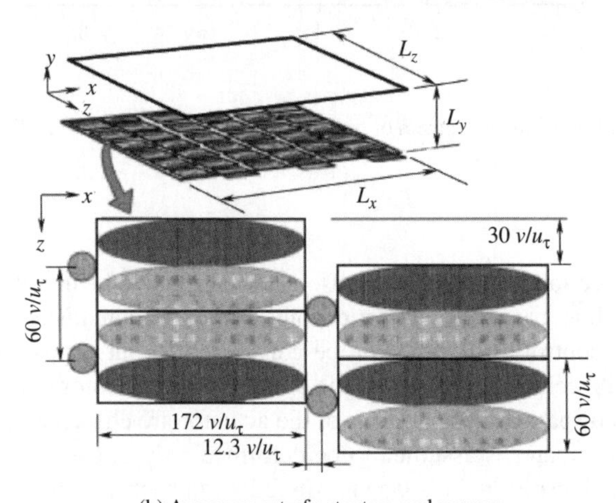

(b) Arrangement of actuators and sensors

Figure 3.85 Schematic of MEMS control device.[41] *Source*: Endo T 2000. Reproduced with permission from Elsevier Limited

where the function $f(x^+)$, introduced to keep the shape of actuator smooth in the streamwise direction, is given with a hyperbolic tangent as

$$f(x^+) = \begin{cases} \dfrac{1}{2}\left[1 + \tanh\left\{\dfrac{\left(x^+ - x_c^+\right) + 73.7}{\sigma_x^+}\right\}\right] & \text{if} - 86 \leq x^+ - x_c^+ \leq -61.5 \\ 1 & \text{if} - 61.5 \leq x^+ - x_c^+ \leq 61.5 \\ \dfrac{1}{2}\left[1 - \tanh\left\{\dfrac{\left(x^+ - x_c^+\right) - 73.7}{\sigma_x^+}\right\}\right] & \text{if } 61.5 \leq x^+ - x_c^+ \leq 86 \end{cases} \tag{3.59}$$

where, x_c^+ and z_c^+ denote the location of the center of actuator. The parameters are chosen as $\alpha = 2.3$, $\beta = 0.077$, $\gamma = 0.3$, $\sigma_x^+ = 6.14$, $\sigma_z^+ = 22.2$, and $m_z^+ = 60$, respectively.

Time histories of the normalized drag coefficient for continuous wall deformation and arrayed deformable actuators are shown in Figure 3.86, by thin solid line and thick

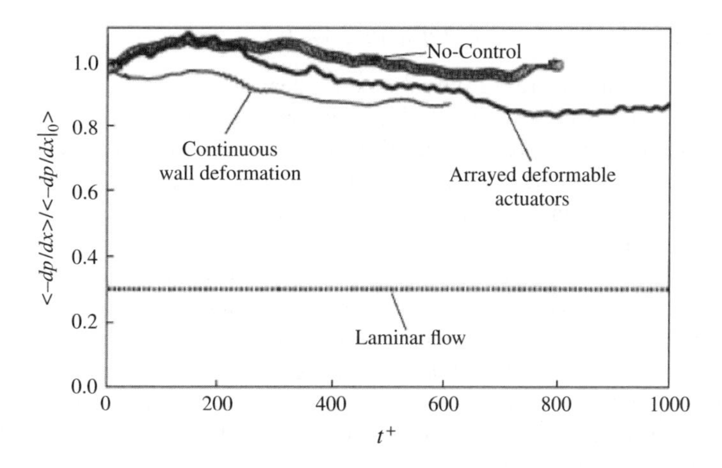

Figure 3.86 Time histories of normalized drag coefficient with MEMS control.[41] *Source*: Endo T 2000. Reproduced with permission from Elsevier Limited

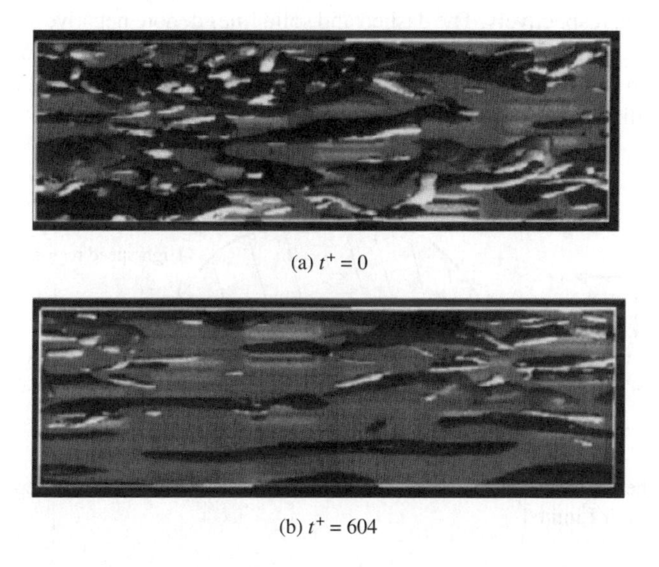

Figure 3.87 Near-wall structures under MEMS control.[41] *Source*: Endo T 2000. Reproduced with permission from Elsevier Limited

solid line, respectively. It is seen that for a MEMS control, no effect is exhibited until $t^+ = 200$, and then the effect appears to be efficient to decrease drag at $t^+ > 200$.

Figure 3.87a and b shows the top views of near-wall structures at $t^+ = 0$ and $t^+ = 604$, respectively, where white denotes the vortex structures, dark gray denotes $u'^+ = -3.5$ and light gray denotes $u'^+ = 3.5$. It is observed that under the MEMS control, the quasi-streamwise vortex becomes less populated and the meandering of low-speed streak is suppressed.

3.6.4 Control Mechanism of Deformed Wall

The dominant structural features of fully developed near-wall turbulence are streamwise vortices and streaks that link closely with the high frictional drag. In addition, streamwise vortex and streaky structure have a closed dynamical relationship with each other. The near-wall streaky structures often meander in spanwise direction, which are associated with streamwise vortex alternatively tilting in the x–z plane and play an important role in a quasi-cyclic process of turbulence regeneration.

A schematic of a modeled streaky structure is shown in Figure 3.88. When the velocity gradients in the two horizontal directions, that is, $\partial u'/\partial x$ and $\partial u'/\partial z$, are taken into account, the edges of the meandering streak can be grouped into four events as tabulated in Table 3.1, depending on the signs of $\partial u'/\partial x$ and $\partial u'/\partial z$, and detected by a threshold for velocity gradients. Note that E_1 and E_4 correspond to the downstream edge of the low-speed streak, while E_2 and E_3 to the upstream edge; moreover, E_1 and E_4, and also E_2 and E_3 are, respectively, of mirror symmetry in the spanwise direction.

A threshold of 0.32 is employed for both gradients to extract strong events, then the conditional average can be made for these events. Contours of conditionally averaged streamwise velocity, $\langle u'^+ \rangle$ of a plane channel flow at $y^+ = 15$ for events E1 and E2 are shown in Figure 3.89a and b, respectively. The dashed and solid lines denote negative and positive $\langle u'^+ \rangle$, respectively, and the detection point is at $x^+ = 0$ and $z^+ = 0$. The contours of negative $\langle u'^+ \rangle$, which correspond to the low-speed streaks, are elongated in the streamwise direction and tilted in the spanwise direction for both events.

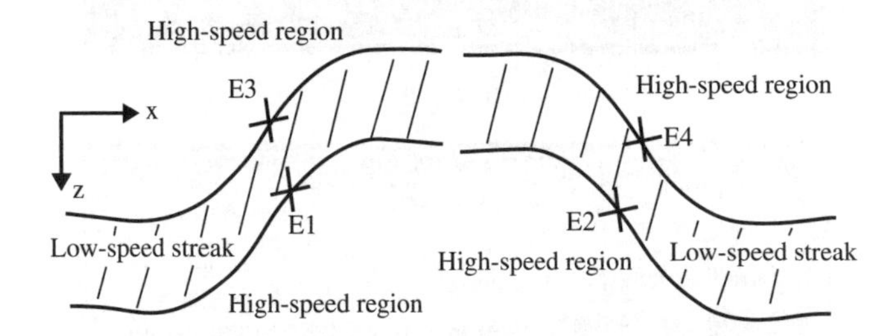

Figure 3.88 Schematic of modeled streaky structure.[41] *Source*: Endo T 2000. Reproduced with permission from Elsevier Limited

Table 3.1 Four events at the edge of the meandering streaks[41]

Event	$\partial u'/\partial x$	$\partial u'/\partial z$
E1	Positive	Positive
E2	Negative	Positive
E3	Negative	Negative
E4	Positive	Negative

Source: Endo T 2000. Reproduced with permission from Elsevier Limited.

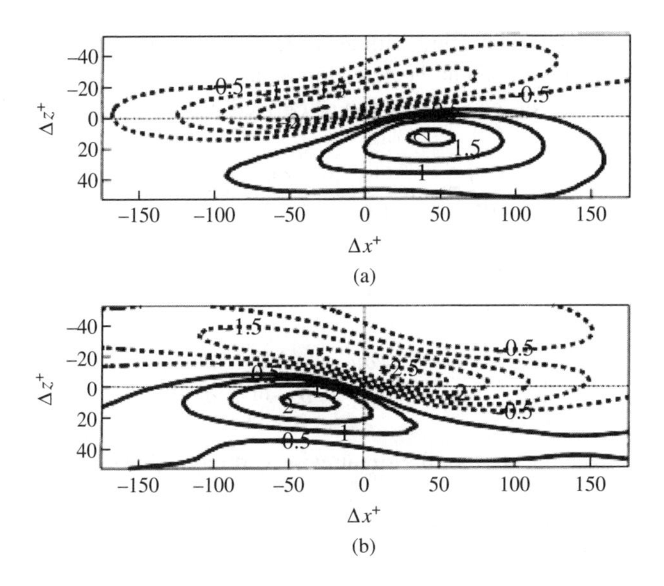

Figure 3.89 Contours of conditional-averaged streamwise velocity at $y^+ = 15$.[41] *Source*: Endo T 2000. Reproduced with permission from Elsevier Limited

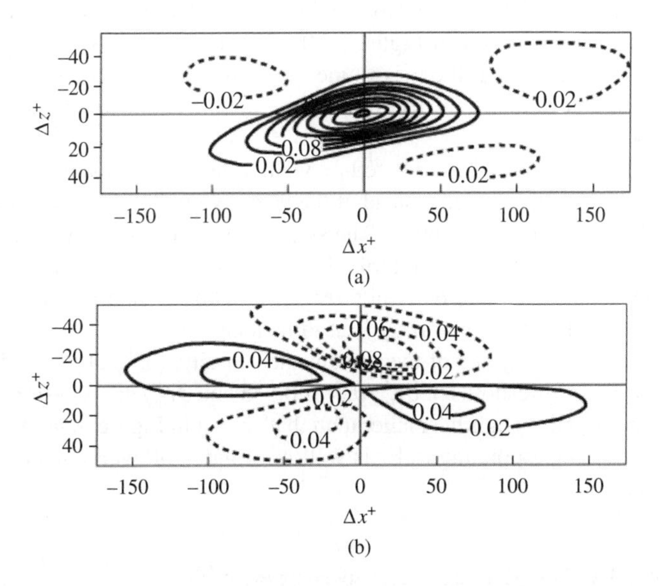

Figure 3.90 Contours of conditional-averaged streamwise vorticity at $y^+ = 15$.[41] *Source*: Endo T 2000. Reproduced with permission from Elsevier Limited

Contours of conditionally averaged streamwise vorticity $\langle \omega_x^+ \rangle$ at $y^+ = 15$ for events E1 and E2 are shown in Figures 3.90a and b, respectively. The dashed and solid lines denote negative and positive $\langle \omega_x^+ \rangle$, respectively and the detection point is at $x^+ = 0$ and $z^+ = 0$. It is indicated that the streamwise vorticity having strong positive $\langle \omega_x^+ \rangle$ often appears in event E_1. However, a negative peak in $\langle \omega_x^+ \rangle$ is associated with event E_4 (not shown). For events E_2 and E_3, the magnitude of $\langle \omega_x^+ \rangle$ is fairly small at the detection point.

The evolution equation of the vorticity can be derived from incompressible N-S equations

$$\frac{\partial \omega_i}{\partial t} = -u_j \frac{\partial \omega_i}{\partial x_j} + \omega_j \frac{\partial u_i}{\partial x_j} + \nu \frac{\partial^2 \omega_i}{\partial x_j \partial x_j} \tag{3.60}$$

where $\boldsymbol{\omega}$ is the vorticity vector.

In the near-wall region, we have

$$\omega_j \frac{\partial u}{\partial x_j} \approx \omega_x \frac{\partial u}{\partial x} - \frac{\partial w}{\partial x}\frac{\partial u}{\partial y} + \frac{\partial v}{\partial x}\frac{\partial u}{\partial z}$$

Thus, the normalized equation of the streamwise vorticity ω_x^+ is given as follows:

$$\frac{D\omega_x^+}{Dt^+} = \omega_x^+ \frac{\partial u^+}{\partial x^+} - \frac{\partial w^+}{\partial x^+}\frac{\partial u^+}{\partial y^+} + \frac{\partial v^+}{\partial x^+}\frac{\partial u^+}{\partial z^+} + \nabla^2 \omega_x^+ \tag{3.61}$$

where the first term of the right-hand side is the stretching term, while the second and third terms correspond to the tilting and twisting terms, respectively. Since both $\partial u^+/\partial x^+$ and ω_x^+ are positive at event E_1, ω_x^+ is further produced by the stretching term $\omega_x^+(\partial u^+/\partial x^+)$, which results in the peak of ω_x^+ shown in Figure 3.90a. However, $\partial u^+/\partial x^+$ is negative for event E_2, while ω_x^+ is positive, so that the magnitude of ω_x^+ should be decreased. Thus, the streak meandering should contribute to the evolution of ω_x^+ and hence the regeneration mechanism of quasi-streamwise vortices.

In order to identify the meandering based on wall variables, the spanwise shear stress gradients, $\partial \tau_u/\partial z$ and $\partial \tau_w/\partial z$, are used instead of the velocity gradients in the buffer layer, where $\tau_u^+ = \partial u'^+/\partial y^+|_w$ and $\tau_w^+ = \partial w'^+/\partial y^+|_w$. The signs of the shear stress gradients and the corresponding events are summarized in Table 3.2. A negative value of $\partial \tau_u/\partial z$ is observed at 50 wall units upstream of E_1 and E_4, while $\partial \tau_u/\partial z > 0$ for E_2 and E_3. Note that E_1 can be distinguished from E_4 by the sign of $\partial \tau_u/\partial z$ at same location.

Figure 3.91 shows the contours of the streamwise velocity, $\langle u'^+ \rangle$ at $y^+ = 15$, conditionally averaged for S_1 with the thresholds of $\partial \tau_u^+/\partial z^+ > 0.035$ and $\partial \tau_w^+/\partial z^+ < -0.005$. Although the tilting angle of the streak is slightly smaller than that shown in Figure 3.94, the meandering of the low-speed streak corresponding to E_1 is well-captured at 50 wall units downstream from the detection point.

Table 3.2 Four signals and corresponding events[41]

Signal	$\partial \tau_u/\partial z$	$\partial \tau_w/\partial z$	Event	ω'_x
S1	Positive	Negative	E1	Positive
S2	Positive	Positive	E2	
S3	Negative	Positive	E3	
S4	Negative	Negative	E4	Negative

Source: Endo T 2000. Reproduced with permission from Elsevier Limited.

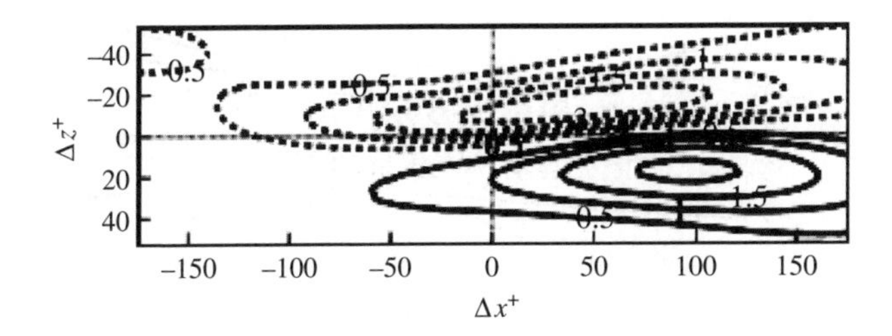

Figure 3.91 Contours of streamwise velocity at $y^+ = 15$, given conditions of $\partial\tau_u^+/\partial z^+ > 0.035$ and $\partial\tau_w^+/\partial z^+ < -0.005$.[41] *Source*: Endo T 2000. Reproduced with permission from Elsevier Limited

Contours of conditionally averaged streamwise vorticity $\langle\omega_x^+\rangle$ at $y^+ = 15$ are shown in Figure 3.92. A large positive peak of ω_x^+ at $y^+ = 15$ is observed in the same region, where the streak meanders. Similarly, a large negative peak of ω_x^+ is associated with S_4 ($\partial\tau_u^+/\partial z^+ < -0.035$ and $\partial\tau_w^+/\partial z^+ < -0.005$), which corresponds to E_4. Thus a streamwise vortex as well as its rotating direction can be detected by the combination of the signs of $\partial\tau_u^+/\partial z^+$ and $\partial\tau_w^+/\partial z^+$.

Contours of conditionally averaged wall-normal velocity $\langle v'^+\rangle$ at $y^+ = 15$ for S_1 are shown in Figure 3.93. Positive and negative peaks, corresponding to the ejection and sweep motions, respectively, are aligned side-by-side in spanwise direction. The spanwise distance of the positive and negative peaks is about 30 wall units, which is almost the same as the mean diameter of the streamwise vortices.

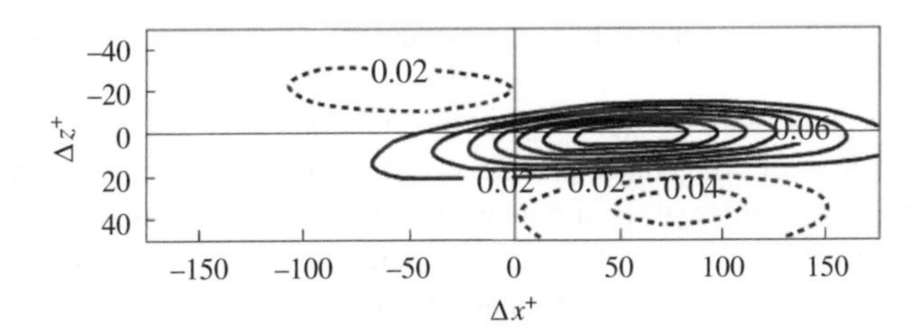

Figure 3.92 Contours of streamwise vorticity at $y^+ = 15$, given conditions of $\partial\tau_u^+/\partial z^+ > 0.035$ and $\partial\tau_w^+/\partial z^+ < -0.005$.[41] *Source*: Endo T 2000. Reproduced with permission from Elsevier Limited

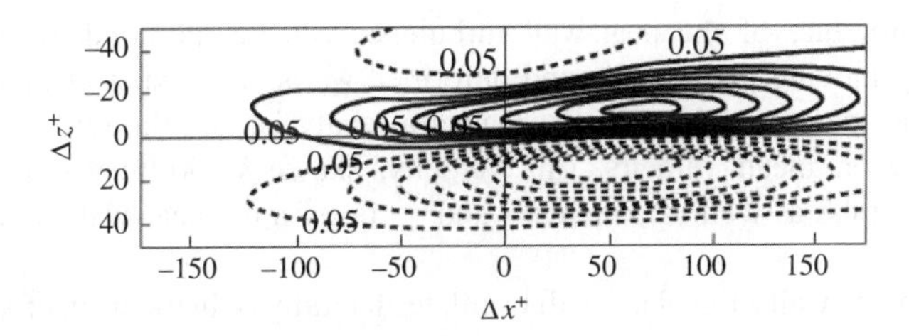

Figure 3.93 Contours of wall-normal velocity at $y^+ = 15$, given conditions of $\partial\tau_u^+/\partial z^+ > 0.035$ and $\partial\tau_w^+/\partial z^+ < -0.005$.[41] *Source*: Endo T 2000. Reproduced with permission from Elsevier Limited

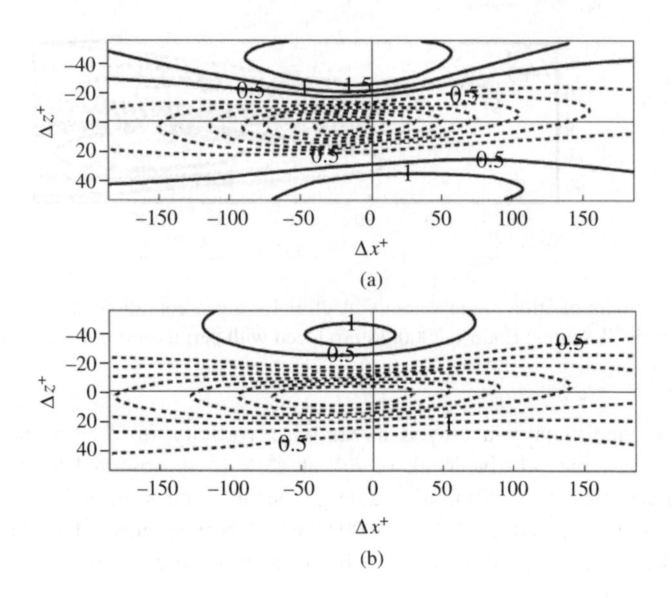

Figure 3.94 Conditional averaged streamwise velocity with ejection event.[41] *Source*: Endo T 2000. Reproduced with permission from Elsevier Limited

The quadrant analysis (see Section 1.2.2) is employed to detect the burst and to examine a conditionally averaged flow field. Conditionally averaged streamwise velocity at $y^+ = 15$ with the ejection event under the control with arrayed deformable actuators is shown in Figure 3.94b, while the velocity in uncontrolled case is shown in Figure 3.94a for comparison.

A low-speed streak is clearly observed near the ejection event, and it becomes longer in the streamwise direction under the control with arrayed deformable actuators. This fact indicates that the meandering of the streak is suppressed and the low-speed streak itself is much stabilized. Conditionally averaged displacement of actuators near the ejection event shows that the actuators are deformed as a groove beneath the low-speed streaks. The groove follows the low-speed streak as it is convected downstream with keeping a depth of about 1 wall unit.

3.7 Closed Remarks

The coherent structure of the near-wall turbulence can be modified by actively moving walls, including moving plane walls and deformed walls. The experimental and numerical data involved are dealt with statistical or spectral analysis, and the turbulence controls are discussed based on mean velocity, rms velocity, skewness, kurtosis, turbulent intensity, Reynolds shear stress as well as on streaks and vortices in the near-wall coherent structures, and others.

For moving plane walls, the plane wall oscillates laterally or in the form of a traveling wave, called oscillating wall and wavy wall, respectively. By the oscillating wall with the appropriate oscillation parameters, the near-wall vortex structures are tilted with alternately rotating and tiling direction, relative directly to the wall displacement. This structure tilt leads to the creation of negative spanwise vorticity in the near-wall region, resulting in drag reductions. Moreover, the oscillating wall can also result in a shift of vortices relative to streaks, which disrupts the turbulent energy production in the turbulent boundary layer. Thus, the near-wall structures weakened and lessened, so that drag is reduced. In the case of the wavy wall, that is, laterally oscillating wall in the form of streamwise-traveling waves, As k_x increases from $k_x = 0$, the streak intensity decreases first, then increases. Furthermore, the longitudinal vortices meander significantly and arrange in order for the low-frequency waves. There exists an optimal wave number $k_{x\ \mathrm{op}}$ corresponding to the largest amount of drag reduction. As $k_x < k_{x\ \mathrm{op}}$, number of the vortices decreases as k_x increases, and then the longitudinal vortices are produced again as $k_x > k_{x\mathrm{op}}$.

For deformed walls, the wall oscillates up-down or in the form of a traveling wave, called oscillating deformed wall and wavy deformed wall, respectively. The sinusoidally oscillating deformed wall hardly reduces drag, and an opposition wall deformation control, a simple closed-loop control algorithm, is generally more effective in the drag reduction. In this control, instantaneous wall shape of deformed wall is elongated in the streamwise direction and resembles riblet in appearance with keeping a depth of about 1 wall unit. Hence the low-speed streaks become longer in the streamwise direction, the meandering of the streak, which is responsible for sustaining the coherent structures, is suppressed and the low-speed streak itself is much stabilized. As a result, drag is reduced. In the case of the wavy deformed wall, that is, up-down oscillating wall in the form of streamwise-traveling waves, the wall traveling wave motion tends to suppress flow separation, occurring on the downhill side of the stationary wavy deformed wall. Moreover, the structures above the troughs in the wavy deformed wall are substantially coherent and dominated by streamwise elements. The total drag dominated by the pressure drag in this case, decreases monotonically as the traveling velocity c/U increases and even changes sign from positive to negative, that is becomes a thrust, at large c/U.

References

[1] Robinson S K. Coherent motions in the turbulent boundary layer. Annu. Rev. Fluid Mech., 1991, 23:601.

[2] Schoppa W and Hussain F. Coherent structures generation in near-wall turbulence. J. Fluid Mech., 2002, 453:57–108.

[3] Kim J. Control of turbulence boundary layer. Phys. Fluids, 2003, 15(5):1093–1105.

[4] Fukagata K, Iwamnoto K and Kasagi N. Contribution of Reynolds shear stress distribution to the skin friction in wall-bounded flows. Phys. Fluids, 2002, 14:L73–L76.

[5] Gad-el-Hak M, Pollard A and Bounet J P. Flow Control Fundamental and Practices, Springer, Berlin, 1998.

[6] Gad-el-Hak M. Flow Control, Passive, Active and Reactive Flow Management, Cambridge University Press, Cambridge, UK, 2000.

[7] Bushnell D M and McGinley C B. Turbulence control in wall flows. Annu. Rev. Fluid Mech., 1989, 21:1–20.

[8] Lumley J and Blossey P. Control of turbulence. Annu. Rev. Fluid Mech., 1998, 30:311–327.

[9] Lofdahl L and Gad-el-Hak M. MEMS applications in turbulence and flow control. Progress Aerospace Sci., 1999, 35:101–203.

[10] Choi K S. European drag-reduction research – recent developments and current status. Fluid Dyn. Res., 2000, 26:325–335.

[11] Kral L D. Active flow control technology, ASME Fluid Engineering Division Technical Brief, 2000.

[12] Schlichting H. Boundary-layer Theory, McGraw-Hill, New York, 1968.

[13] Choi J-II, Xu C X and Sung H J. Drag reduction by spanwise wall oscillation in wall-bounded turbulent flows. AIAA J., 2002, 40(5):842–850.

[14] Huang L P. Drag reduction in turbulent channel flows utilizing spanwise motions, Ph.D thesis, Nanjing University of Science and Technology, China, 2010.

[15] Ricco P and Wu S. On the effects of later wall oscillation on a turbulent boundary layer. Exp. Thermal Fluid Sci., 2004, 29:41–52.
[16] Trujillo S M, Bogard D G and Ball K S. Turbulent boundary layer drag reduction using an oscillating wall. AIAA paper, 1997.
[17] Ricco P. Modification of near wall turbulence due to spanwise wall oscillations. J. Turbulence, 2004, 4(1):024.
[18] Choi K, Roach P E, DeBisschop J and Clayton B. Turbulent boundary-layer by means of spanwise-wall oscillation. AIAA J. 1998, 36:1157.
[19] Choi K. Near-wall structure of turbulent boundary layer with spanwise-wall oscillation. Phys. Fluids, 2002, 14:2530–2542.
[20] Choi K and Clayton B R. The mechanism of turbulent drag reduction with wall oscillation. Int. J. Heat Fluid Flow, 2001, 22:1–9.
[21] Quadrio M and Ricco P. Critical assessment of turbulent drag reduction through spanwise wall oscillations. J. Fluid Mech., 2004, 521:251–271.
[22] Jung W J, Mangiavacchl N and Akhavan R. Suppression of turbulence in wall-bounded flows by high-frequency spanwise oscillations. Phys. Fluids, 1992, A 4(B): 1605–1607.
[23] Bogard D G, Ball K S and Wassen E. Drag reduction for turbulent boundary layer flows using an oscillating wall, Report TTCRL 00-2, 2000.
[24] Huang L P, Fan B C and Dong G. Mechanism of drag reduction due to spanwise wall oscillation in turbulent channel flow. J. Nanjing Univ. Sci. Technol., 2010, 34(3):361–366.
[25] Du Y, Karniadakis G E. Suppressing wall turbulence by means of a transverse travelling wave. Science, 2000, 288:1230–1234.
[26] Quadrio M, Ricco P and Viotti C. Streamwise-travelling waves of spanwise wall velocity for turbulent drag reduction. J. Fluid Mech., 2009, 627:161–178.
[27] Zhao H, Wu J Z and Luo J S. Turbulent drag reduction by travelling wave of flexible wall. Fluid Dyn. Res., 2004, 34:175–198.
[28] Huang L P, Fan B C and Dong G. Turbulent drag reduction via a streamwise traveling wave induced by spanwise wall oscillation. Chinese J. Theor. Appl. Mech., 2011, 43(2):277–281.
[29] Rediniotis O K, Lagoudas D C, Mani R and Karniadakis G E. Active skin for turbulent drag reduction, smart structures and materials 2002: Smart electronics, MEMS, and nanotechnology, Varadan, V.K. Editor, Proceedings of SPIE, Vol. 4700, 2002.
[30] Mani R, Lagoudas D C and Rediniotis O K. MEMS based active skin for turbulent drag reduction, smart structures and material 2003: Smart structures and integrated systems, Baz A.M. Editor, Proceedings of SPIE, Vol. 5058, 2003.
[31] Suzuki Y, Yoshyino T, Yamagami T and Kasagi N. Drag reduction in a turbulent channel flow by using a GA-based feedback control system, Proceedings of 6th Symposium on Smart Control of Turbulence, Tokyo, 2005.
[32] Rediniotis O K, Lagoudas D C, Mani R, Traub L, Allen R and Karniadakis G E. Computational and experimental studies of an active skin for turbulent drag reduction, AIAA, 2002, 2002–2830.
[33] Segawa T, Li P, Kawagucdhi and Yoshida H. Visualization of wall turbulence under artificial disturbance by piezo actuator array. J. Turbul., 2002, 3(15):1–14.
[34] Ho C M and Tai Y C. Micro-electro-mechanical-systems (MEMS) and fluid flows. Annu. Fluid Mech., 1998, 30:579–612.
[35] Yamagami T, Suzuki Y and Kasagi. Development of feedback control system of wall turbulence using MEMS devices, Proceedings of 6th Symposium on Smart Control of Turbulence, Tokyo, 2005.
[36] Goldberg D E. Genetic algorithms with sharing for multi-model function optimization, Proceedings of the 2nd International Conference on Genetic Algorithms, 1987, 41–49.
[37] Shen L, Zhang X, Yue K P and Traiantafyllou M S. Turbulent flow over a flexible wall undergoing a streamwise traveling wave motion. J. Fluid Mech., 2003, 484:197–221.
[38] Mito Y and Kasagi N. DNS study of turbulence modification with streamwise-uniform sinusoidal wall-oscillation. Int. J. Heat Fluid Flow, 1998, 19:470–481.
[39] Choi H, Moin P and Kim J. Active turbulence control for drag reduction in wall-bounded flows. J. Fluid Mech., 1994, 262:75–110.
[40] Kang S and Choi H. Active wall motions for skin-friction drag reduction. Phys. Fluids, 2000,12(12) 3301–3304.
[41] Endo T, Kasagi N and Suzukin Y. Feedback control of wall turbulence with wall deformation. Int. J. Heat Fluid Flow, 2000, 21:568–575.

4

Control of Turbulence by Lorentz Force

As mentioned in Chapters 2 and 3, turbulent boundary layers typically contain streaks and streamwise vortices, which play a dominant role in the near-wall turbulent transport phenomena. Meanwhile, the skin-friction drag of a turbulent wall-bounded flow is directly attributed to the existence of near-wall vertical structures and the associated ejection/sweep events.

The evolution equation for vorticity obtained from the incompressible Navier–Stokes equations is

$$\frac{D\omega_i}{Dt} = \omega_j \frac{\partial u_i}{\partial x_j} + \nu \frac{\partial^2 \omega_i}{\partial x_j \partial x_j} \tag{4.1}$$

where ω_i and u_i represent vorticity and velocity, respectively, with $i = 1,\ 2,\ 3$ or interchangeably $x,\ y,\ z$ denoting the streamwise, wall-normal, and spanwise directions, respectively. The first term on the right-hand side represents the tilting and stretching effect of vortical structures. The greatest contributions to this term in the near-wall region for fully developed channel flow are from $-\frac{\partial w}{\partial x}\frac{\partial u}{\partial y}$ and $-\frac{\partial v}{\partial z}\frac{\partial u}{\partial y}$ in the ω_x-equation and ω_y-equation, respectively. Neglecting the contribution of advection terms in either production or dissipation yields

$$\frac{\partial \omega_x}{\partial t} \simeq -\frac{\partial w}{\partial x}\frac{\partial u}{\partial y} + \nu\left(\frac{\partial^2 \omega_x}{\partial x^2} + \frac{\partial^2 \omega_x}{\partial y^2} + \frac{\partial^2 \omega_x}{\partial z^2}\right) \tag{4.2}$$

$$\frac{\partial \omega_y}{\partial t} \simeq -\frac{\partial v}{\partial z}\frac{\partial u}{\partial y} + \nu\left(\frac{\partial^2 \omega_y}{\partial x^2} + \frac{\partial^2 \omega_y}{\partial y^2} + \frac{\partial^2 \omega_y}{\partial z^2}\right) \tag{4.3}$$

where u and v are streamwise and wall-normal components of velocity, respectively.

Multiplying Eqs. (4.2) and (4.3) by ω_x and ω_y, respectively, and adding the equations together, followed by averaging over an x–z plane gives

$$\frac{\partial}{\partial t}\left(\frac{\overline{\omega}_x^2 + \overline{\omega}_y^2}{2}\right) \simeq P_{\omega_x} + P_{\omega_y} - \varepsilon_\omega \tag{4.4}$$

Principles of Turbulence Control, First Edition. Baochun Fan and Gang Dong.

$$P_{\omega_x} = -\overline{\omega_x \frac{\partial w}{\partial x}} \frac{dU}{dy} \tag{4.5}$$

$$P_{\omega_y} = -\overline{\omega_y \frac{\partial v}{\partial z}} \frac{dU}{dy} \tag{4.6}$$

$$\varepsilon_\omega = \nu \left(\overline{\left(\frac{\partial \omega_x}{\partial x}\right)^2} + \overline{\left(\frac{\partial \omega_x}{\partial y}\right)^2} + \overline{\left(\frac{\partial \omega_x}{\partial z}\right)^2} + \overline{\left(\frac{\partial \omega_y}{\partial x}\right)^2} + \overline{\left(\frac{\partial \omega_y}{\partial y}\right)^2} + \overline{\left(\frac{\partial \omega_y}{\partial z}\right)^2} \right) \tag{4.7}$$

where the overbar denotes averaging over an x–z plane. $\overline{\omega_i^2}$ is enstrophy in the i direction, defined as a squared component of vorticity. This equation is similar to the turbulent kinetic energy equation. The role of each term on the right-hand side is similar to that in the turbulent kinetic energy equation, that is, the first two terms represent the production of enstrophy, whereas the last term represents the dissipation of enstrophy. P_{ω_x} and P_{ω_y} act as the sources for ω_x^2 and ω_y^2, respectively. Both P_{ω_x} and P_{ω_y} are proportional to $\omega_x \omega_y (dU/dy)$, since $-\partial w/\partial x$ and $-\partial v/\partial z$ contribute to ω_y and ω_x, respectively. This implies that near-wall structures with the same sign of ω_y and ω_x can grow and survive, whereas structures with the opposite sign of ω_y and ω_x decay, when $\frac{dU}{dy}$ is positive.

Distributions of P_{ω_x}, P_{ω_y}, and ε_ω near the wall in a fully developed channel flow at $\mathrm{Re}_\tau = 110$ are shown in Figure 4.1. It can be seen that P_{ω_x} and P_{ω_y} are positive everywhere except very close to the wall, indicating that when $\frac{dU}{dy}$ is positive, ω_x and $-\partial w/\partial x$ (or ω_y and $-\partial v/\partial z$) are positively correlated almost everywhere. Meanwhile, P_{ω_y} contributes twice as much to the total production as P_{ω_y} does. Total production is greater than dissipation in the range $6 < y^+ < 17$, and a large amount of production occur in the viscous sublayer due to the large value of $\frac{dU}{dy}$ near the wall.

The tilt of vortex in the viscous sublayer represented by $-\partial w/\partial x$ (or $-\partial v/\partial z$) in the equation extracts streamwise (or wall-normal) enstrophy from the mean flow. Since the

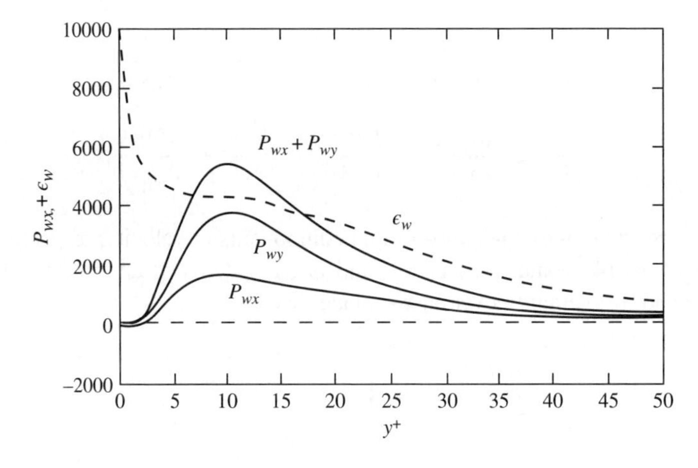

Figure 4.1 Vorticity production and dissipation distributions near the wall.[1] *Source*: Lee C 2002. Reproduced with permission of Cambridge University Press

streamwise vortices are usually slanted outward from the wall, ω_y and ω_x are have the same sign. The same signed ω_y and ω_x are continuously produced together and one cannot maintain itself without the other, which is essential for sustaining near-wall turbulence structures. In addition, streamwise vortices typically grow in size and get stronger as the mean shear stretches them. Full-grown streamwise vortices create opposite signed small streamwise vortices in the region between the parent vortices and the wall through viscous interaction with the wall. Newly generated vortices will continuously grow and generate another one. Some of these new vortices may grow into streamwise vortices in the buffer layer, and some disappear quickly due to viscous dissipation. Overall, the viscous sublayer can be viewed as a birthplace of vorticity. Therefore, preventing the production of vorticity in the sublayer alone might suppress the self-maintaining mechanism of turbulent boundary layers. The manipulation of flow only within the viscous sublayer can control the wall-bounded turbulence and result in a drag reduction effectively.

The near-wall turbulence can be modified by actively moving walls known in Chapter 3, which is directly related to the Stokes layer confined in the viscous sublayer. The control mechanism can be viewed as a modulation of the near-wall turbulence via the sublayer flow induced by the moving walls. Obviously, the flow utilized to modulate the near-wall turbulence can also be created directly by a body force, when it decays dramatically away from the wall. Therefore, this body force should be capable of modifying the near-wall flow, which corresponds to the control strategy based on altering the governing equations.

If the considered fluid has a nonzero electrical conductivity, the application of suitable magnetic fields and/or external electric currents allows one to organize a body force (called Lorentz force) distribution inside the fluid. For low conducting fluids, like seawater or electrolytes, a separate external electric field has to be applied to produce a sufficient electrical current density in the field, necessary for the desired flow control. Electrical currents fed via electrodes will only slightly penetrate into the fluid, so that the Lorentz force is noticeable in some vicinity of the electrodes and decays dramatically away from the surface. Electromagnetic (EM) actuators made of electrodes and magnets are used usually to produce the desired Lorentz force distribution.[2–7] The near-wall force distribution is dependent on the EM actuator arrangement. Therefore, in order to achieve a specific goal, one has to choose the appropriate configuration of magnets and electrodes.

In this chapter, the turbulence control with Lorentz force will be discussed. Section 4.1 introduces the fundamental equations to describe the electric and magnetic fields, the induced Lorentz force field and electro-magnetohydrodynamic flow field. Then, the experiments of a near-wall turbulence flow controlled by the spanwise Lorentz force, performed in a water channel, are presented in Section 4.2. Finally, in Section 4.3, numerical simulations of turbulence control with oscillating or wavy Lorentz forces are provided and the detailed mechanisms are also discussed. In addition, the turbulence controlled by wall-normal Lorentz force is also provided briefly in Section 4.4.

4.1 Lorentz Force[8–12]

A charged particle moving in a magnetic field will be affected by an electromagnetic force, called the Lorentz force. When its influence on the magnetic field is negligible, the Lorentz force $\mathbf{f}$ is given by Ampere's law

$$\mathbf{f} = q\mathbf{u} \times \mathbf{B}$$

where $\mathbf{u}$ is the particle velocity vector and q is the electric charge quantity of the particle, and $\mathbf{B}$ is the magnetic flux density.

For a fluid with a nonzero electrical conductivity, such as an electrolyte and a plasma, a Lorentz force distribution inside the fluid can be generated by the application of magnetic fields and/or electric fields and is expressed as

$$\mathbf{F} = \mathbf{J} \times \mathbf{B} \tag{4.8}$$

where $\mathbf{F}$ is the Lorentz force, $\mathbf{J}$ is the current density, given by the generalized Ohm's law

$$\mathbf{J} = \sigma(\mathbf{E} + \mathbf{u} \times \mathbf{B}) \tag{4.9}$$

where σ is the electrical conductivity and $\mathbf{E}$ is the electric field vector. The second term on the right-hand side of the equation denotes the induced current, which together with the external magnetic field generate the Lorentz force.
Thus,

$$\mathbf{F} = \sigma(\mathbf{E} \times \mathbf{B}) + \sigma(\mathbf{u} \times \mathbf{B}) \times \mathbf{B} \tag{4.10}$$

It is obvious from the above equation that the Lorentz force can be divided into two parts generated by a magnetic field alone and by a combination of the magnetic field and electric field, respectively.

For high-conducting fluids ($\sigma \sim 10^6\,\mathrm{S\,m^{-1}}$), such as plasma, liquid metals or semiconductor melts, and an external magnetic field along can have a strong influence on the flow, which is called the magnetohydrodynamic (MHD) effect. Unlike MHD, for electrically low conducting fluids like seawater or electrolytes, the Lorentz force generated from the induced current is too low to produce any noticeable effect on the flow. Therefore, an external electric field should be applied to produce a sufficient electrical current density in the fluid, necessary for the desired fluid flow control, called electro-magnetohydrodynamic (EMHD) effect. These electrical currents will only slightly penetrate into the fluid, so the Lorentz force will be noticeable only in some vicinity of the electrodes. In addition, the induced currents are much less than those due to the applied electrical field, such that

$$\mathbf{J} = \sigma\mathbf{E} \tag{4.11}$$

Thus,

$$\mathbf{F} = \sigma(\mathbf{E} \times \mathbf{B}) \tag{4.12}$$

For the incompressible EMHD flows, the momentum equations including the Lorentz force can be expressed as

$$\frac{\partial \mathbf{u}}{\partial t} + \mathbf{u} \cdot \nabla \mathbf{u} = -\frac{1}{\rho}\nabla p + \nu \nabla^2 \mathbf{u} + \frac{1}{\rho}(\mathbf{J} \times \mathbf{B}) \tag{4.13}$$

In addition, for the magnetic and electric fields, we have

Faraday' law

$$\nabla \times \mathbf{E} = -\frac{\partial \mathbf{B}}{\partial t} \tag{4.14}$$

Ampere's law

$$\nabla \times \mathbf{B} = \mu \mathbf{J} \tag{4.15}$$

where μ is the magnetic permeability and Kirchhoff's law

$$\nabla \cdot \mathbf{J} = 0 \quad \text{or} \quad \nabla \cdot \mathbf{B} = 0 \tag{4.16}$$

From Eqs. (4.14), (4.15), and (4.9), we obtain the Maxwell equation represented by

$$\frac{\partial \mathbf{B}}{\partial t} = \nabla \times (\mathbf{u} \times \mathbf{B}) + \frac{1}{\mu\sigma}\nabla^2 \mathbf{B} \tag{4.17}$$

One fundamental quantity of EMHD is the magnetic Reynolds number Re_m as the ratio between the two terms on the right-hand side of the above equation, that is, the ratio between the characteristic time for magnetic diffusion and the transit time of fluid particles

$$\mathrm{Re}_m = \mu\sigma[U][L]$$

When Re_m is small ($\leq 10^{-5}$), the magnetic field is not influenced by the velocity field, and we have

$$\frac{\partial \mathbf{B}}{\partial t} = \frac{1}{\mu\sigma}\nabla^2 \mathbf{B}$$

If permanent magnets are chosen to produce the magnetic field further, that is, $\frac{\partial \mathbf{B}}{\partial t} = 0$, then

$$\nabla^2 \mathbf{B} = 0 \tag{4.18}$$

and from Eq. (4.14),

$$\mathbf{E} = -\nabla \Phi \tag{4.19}$$

which implies that the electric field can be represented by a potential function Φ.

Combined with Eq. (4.9), this yields

$$\nabla^2 \Phi = \nabla \cdot (\mathbf{u} \times \mathbf{B}) \tag{4.20}$$

For weakly conductive fluids, the influence of the induced current is negligible, and then

$$\nabla^2 \Phi = 0 \tag{4.21}$$

and

$$\mathbf{J} = \sigma \mathbf{E} \tag{4.22}$$

With the suitable boundary conditions, Eqs. (4.18) and (4.20) (or (4.22)) can be solved. Then substituting these solutions into Eq. (4.11), the Lorentz force distribution inside the fluid is determined. In addition, the Lorentz force distribution will be independent from the flow field, when Eq. (4.22) is applied.

4.2 Experiments of Wall Turbulence with Spanwise Lorentz Force

4.2.1 Control with Uniform Spanwise Oscillating Lorentz Force[13–19]

4.2.1.1 Experimental Apparatus

Among the three direction forces, the spanwise Lorentz force seems to be the most effective one. Hence, in this section the EM actuators are designed, so that only the spanwise force is generated.

Planar EM actuators used usually are made of a series of long rectangular magnets and electrodes in an interleaved way with a suitable thick insulator in between, as shown in Figure 4.2, where N and S indicate north and south magnet poles, and " + " and " − " indicate the positive and grounded electrodes, respectively. The electric field lines indicated by the dash line and magnetic field lines indicated by the solid line always cross each other, such that the direction of the Lorentz force can be determined by right-hand rule. The spanwise Lorentz force will be produced when the linear arrays of magnets are oriented normal to the flow direction as shown in Figure 4.2. By switching the polarity of direct current to the electrodes, the direction of the spanwise Lorentz force is altered periodically, which introduces an oscillatory flow motion in the near-wall region. It should be noted that an ideally uniform force distribution cannot be obtained by such a simple configuration of magnets and electrodes, and in the low-conducting fluid, the force will rapidly decay in the normal direction.

In order to characterize the induced flow in a generalized Stokes layer, the actuator is placed in a still salt water, such that the spanwise velocity profiles (averaged in the spanwise direction), induced by the uniform oscillatory spanwise Lorentz force using the actuator, are measured by particle image velocimetry (PIV) as shown in Figures 4.3 and 4.4. Figure 4.3 shows a plot of the spanwise velocity profiles, where several profiles are plotted at different phases during the oscillating cycle. Figure 4.4 shows the variation of spanwise velocity at a fixed height from the wall over a complete cycle. It is observed that the velocity reaches a maximum at approximately 1 mm above the surface and dies out at a distance of about 4 mm.

The experiments of the near-wall turbulence flow controlled by the spanwise oscillating Lorentz force are performed in water channels. The complete control surface in the experiments is assembled by covering with an array of EM actuator tiles as shown in Figure 4.5, where the total control area is made up by 16 small EM actuator tiles, each tile is made of 5 mm wide, 10 mm high, and 50 mm long rectangular bar of Nd-Feb permanent magnet with 1.2 T, interleaved with copper electrodes of the same size with a 1.63-mm-thick insulator in between, giving the effective size of magnets and electrodes of 6.63 mm. The assembled

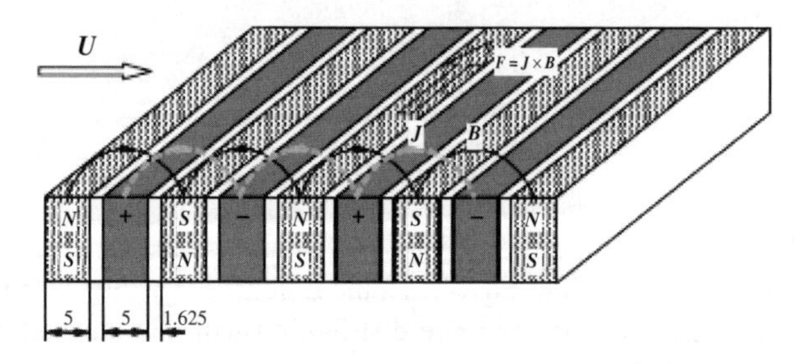

Figure 4.2 Schematic of an EM actuator.[13] *Source*: Pang J 2004. Reproduced with permission from Cambridge University Press

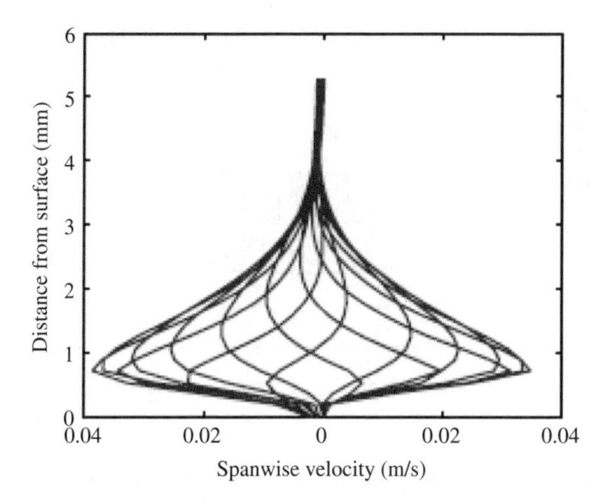

Figure 4.3 Spanwise velocity profiles induced by uniform oscillatory spanwise Lorentz force.[14] *Source*: Park J 2004. Reproduced with permission from Springer

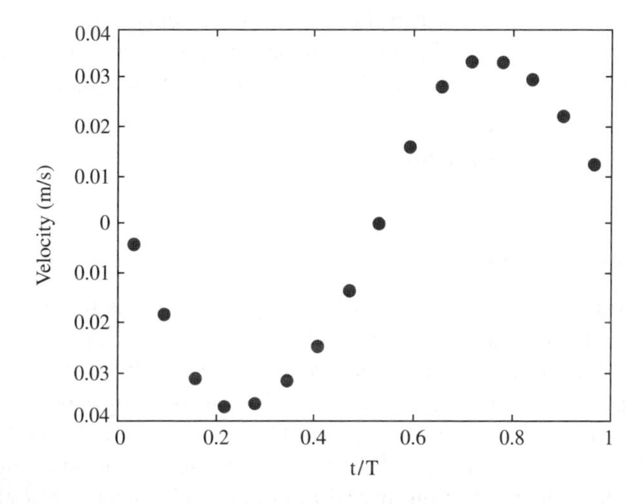

Figure 4.4 Variation of spanwise velocity at a fixed height from the wall over a complete cycle.[14] *Source*: Park J 2004. Reproduced with permission from Springer

control actuators are placed in the replaceable part of a test plate, and kept flush to the wall surface. The 4-m-long test plate with 20:1 ellipse leading edge is placed in the channel. The turbulent boundary layer is tripped just after the leading edge by an artificial surface toughness, and $CuSO_4$ is introduced to fresh water through a long injection slot set on the plate, rather than conducting experiments in a sea water channel, to increase the conductivity of the fluid over the working area and to reduce the corrosion problem.

Velocity measurements are made with a hot wire/film anemometer system and confined at the 5 mm downstream of the EM actuator arrays.

The estimated mean magnetic strength at the surface of the actuator is $B_0 = 0.78$ T. By adjusting the voltage applied to the EM actuators between $V_0 = 3.2$ and 12.9 V, the current

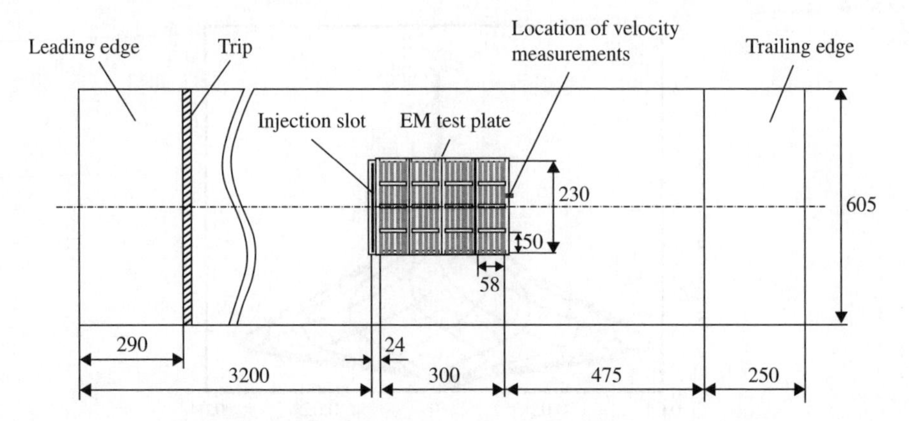

Figure 4.5 Schematic of test plate and positions of electromagnetic (EM) actuator.[13] *Source*: Pang J 2004. Reproduced with permission from Cambridge University Press

density $J_0 = \frac{\pi}{4a}\sigma V_0$ is set between 94 and 376 A m^{-2}, where a is the width of magnets and electrodes, and σ is the conductivity at the hot-film measurement position. Then Stuart number St is given from 105 to 422; here the Stuart number represents the strength of the magnetic force relative to the inertial force of the fluid as defined by $St = J_0 B_0 \delta/(\rho u_\tau^2)$, where δ and $u_\tau = \sqrt{\tau_w/\rho} = \sqrt{\nu \frac{du_x}{dy}}$ are the boundary layer thickness and the friction velocity of undisturbed boundary layer, respectively.

4.2.1.2 Experimental Results

The nondimensional period is defined as $T^+ = Tu_\tau^2/\nu$, where T is the period of oscillation and ν is the kinematic viscosity. Under different combinations of the oscillation period ($T^+ = 49, 73, 97, 146,$ and 195) and the Stuart number ($St = 105, 158, 210, 315,$ and 422), 5×5 matrix of data points will be generated from all corresponding experimental conditions. From these data points, a surface plot of skin-friction reductions by uniform oscillatory spanwise Lorentz forces as a function of St and T^+ is created as shown in Figure 4.6, where there is a band of high drag reduction (over 40%) in the region near $St \approx 200$ and $T^+ \approx 100$ (shown by bright color).

In analyzing spanwise-wall oscillation data, the spanwise wall velocity has been introduced (see Section 3.2.2). Here, we chose the "equivalent spanwise-wall velocity" defined by

$$W_{\text{eq}}^+ = \frac{St \cdot T^+}{2\pi \text{Re}_\tau}$$

where $\text{Re}_\tau = u_\tau \delta/\nu$ is the Reynolds number based on the boundary layer thickness δ and the friction velocity u_τ. The introduction of W_{eq}^+, which is proportional to the envelope profile of the Stokes layer, makes it possible to link wall oscillation and Lorentz oscillation together and make comparisons under the same spanwise wall velocity scale.

Measured drag reduction against W_{eq}^+ is plotted in Figure 4.7, where the hollow square "□" denotes spanwise-wall oscillation data, and the other symbols denote Lorentz oscillation data.

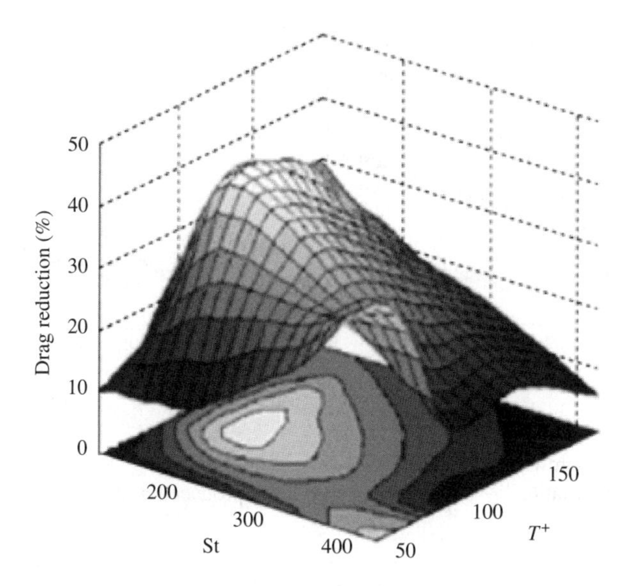

Figure 4.6 Contour of drag reduction as a function of St and T^+.[15] *Source*: Pang J 2004. Reproduced with permission from Cambridge University Press

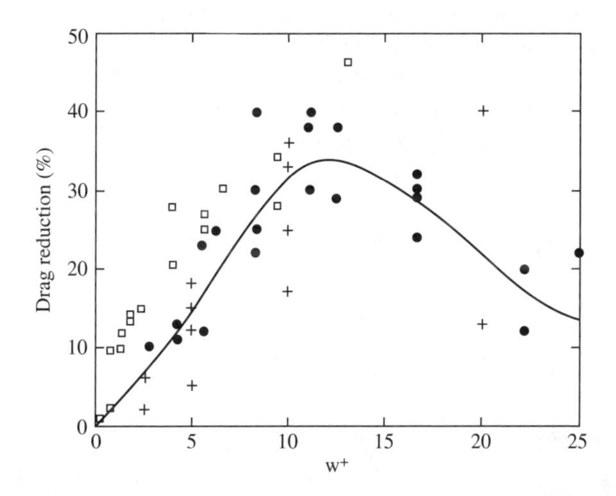

Figure 4.7 Drag reduction as a function of W^+_{eq}.[15] *Source*: Pang J 2004. Reproduced with permission from Cambridge University Press

The excellent agreement among these results has been shown, which seems to suggest that wall oscillation and Lorentz force oscillation may work in a very similar mechanism for drag reduction. In addition, the maximum drag reduction by spanwise Lorentz force oscillation can be obtained when $W^+_{eq} = 10–15$.

Figure 4.8 shows the mean streamwise velocity profiles of turbulent boundary layers, normalized with u_τ, where hollow square "□" denotes the case without the Lorentz force, and black circles represent the case with the Lorentz force oscillation at $S_t = 210$ and $T^+ = 97$, a

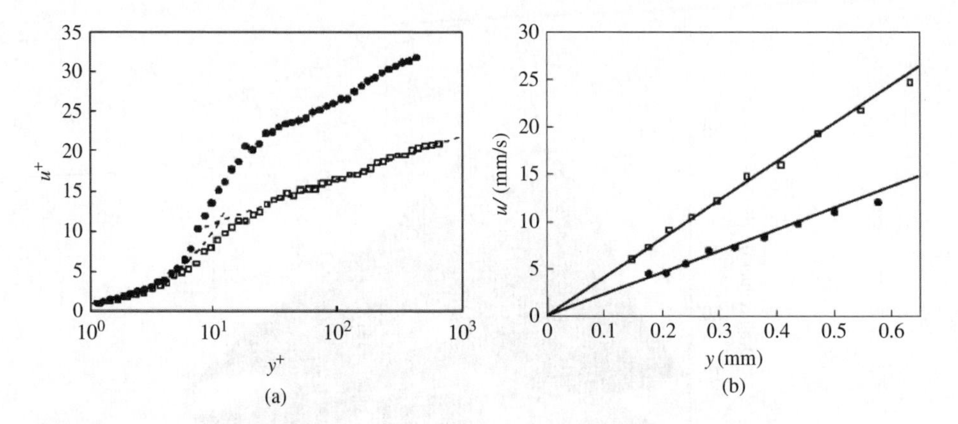

Figure 4.8 Mean streamwise velocity profiles of turbulent boundary layers with and without spanwise Lorentz force oscillation.[13] *Source*: Pang J 2004. Reproduced with permission from Cambridge University Press

condition with high drag reduction. The left frame shows logarithmic velocity profiles and the right frame shows the comparison of the near-wall velocity slopes. It can be observed that the logarithmic velocity profile is clearly shifted upward, suggesting that the viscous sublayer is thickened, and the mean velocity gradient is reduced in the condition with the Lorentz force oscillation. The drag reduction is about 35.0%.

Figure 4.9 shows the turbulence intensity profiles, that is, the root-mean-square (rms) streamwise velocity profiles. It can be seen that the turbulence intensity under the Lorentz force oscillation decreases not only in the near-wall region but also in the logarithmic region of the boundary layer.

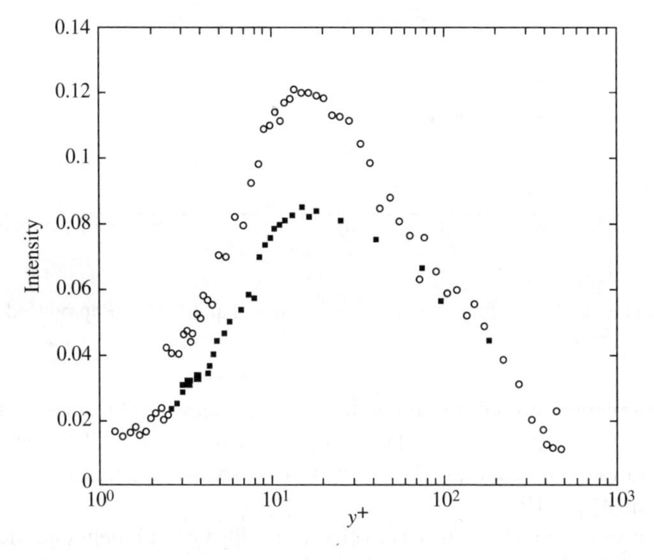

Figure 4.9 Turbulence intensity profiles of turbulence boundary layers with and without the spanwise Lorentz force oscillation.[20] *Source*: Xu P 2007. Reproduced with permission from Cambridge University Press

The distribution of skewness and that of the kurtosis of velocity fluctuations are presented in Figures 4.10 and 4.11. Both of the skewness and kurtosis seem to be increased within the near-wall region by the spanwise Lorentz force oscillation. These trends agree with the results on spanwise-wall oscillation shown in Figures 3.18 and 3.19.

Using the VITA technique, the distributions of conditionally averaged fluctuating streamwise velocity of the near-wall burst events at $y^+ = 2, 6, 12$, and 20 are shown in Figure 4.12a, b, c, and d, respectively, where the solid line denotes the case with control, and the dashed line represents the case without control. Here, the threshold level is $k = 1$ and the window size is $T^+ = 10$. At $y^+ = 2$ and 6 (Figure 4.12a and b), the changes in the burst signature are

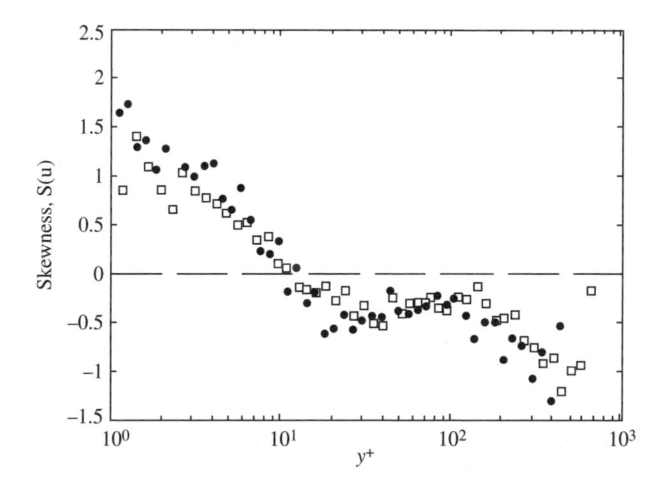

Figure 4.10 Skewness profile of streamwise velocity fluctuation in turbulent boundary layer with and without the Loretz force oscillation.[14] *Source*: Park J 2004. Reproduced with permission from Springer

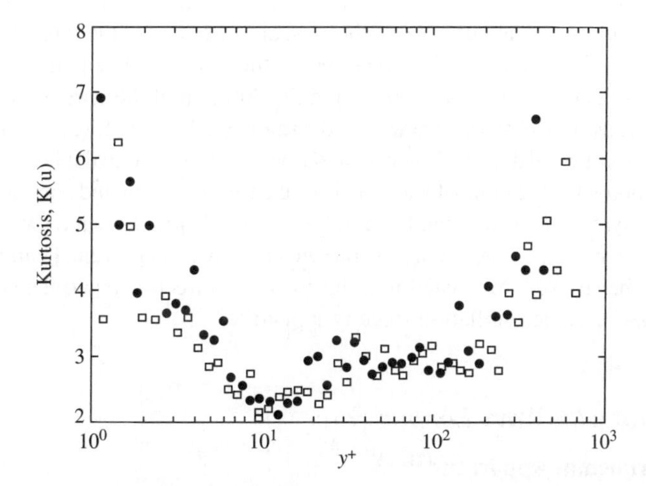

Figure 4.11 Kurtosis profile of streamwise velocity fluctuation in a turbulent boundary layer with and without the Loretz force oscillation.[14] *Source*: Park J 2004. Reproduced with permission from Springer

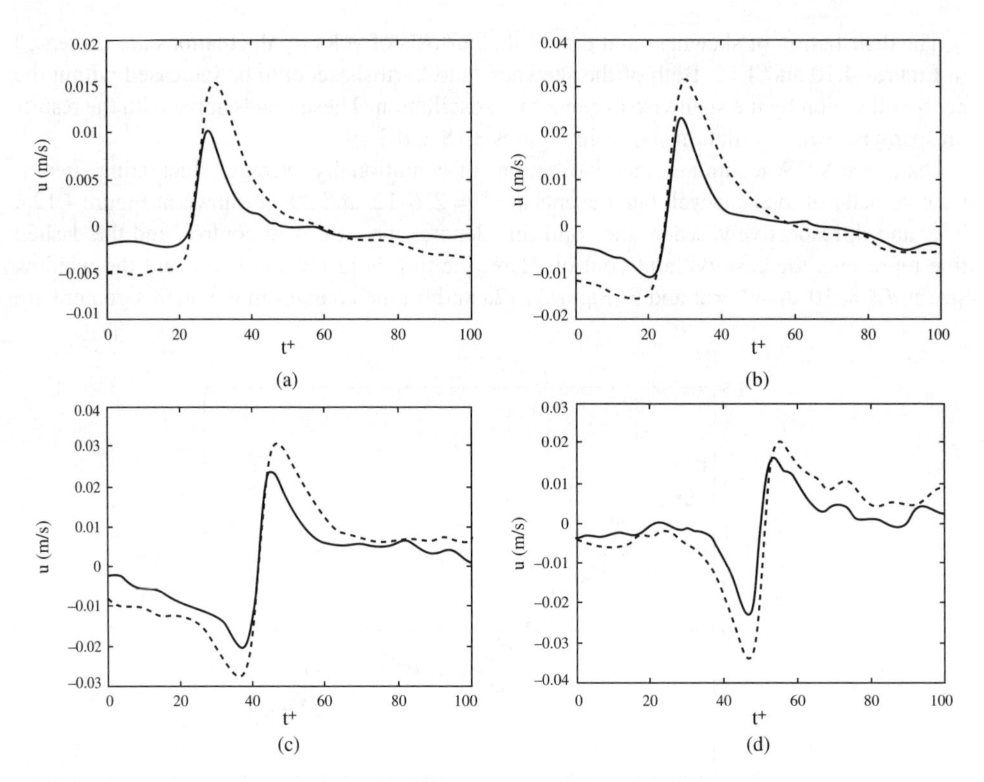

Figure 4.12 Conditionally sampled near-wall burst signatures with and without the Lorentz force oscillation.[14] *Source*: Park J 2004. Reproduced with permission from Springer

obvious in the near-wall region of the boundary layer, and the burst duration has been reduced to about half of that without control and the intensity is also reduced by 30–40%. At $y^+ = 12$ (Figure 4.12c), where the effect of the oscillation Lorentz force on the burst signature is still strong, the burst duration is about 70% of the noncontrol case and the intensity is reduced by about 20%. At $y^+ = 20$ (Figure 4.12d), the effect of the Lorentz force to the near-wall burst is less, but we can still see about 15% reduction in the duration of the burst.

To visualize the near-wall structures with and without the Lorentz force oscillation, $KMnO_4$, instead of $CuSO_4$ as a good dye for flow visualization, is introduced to the flow field through the injection slot. Photos for the control and noncontrol cases are taken and shown in Figure 4.13a and b, respectively. Some local breakdown of the near-wall structure and sweep events represented by brighter sports, where the dye is blown away by sweep events, can be observed. In Figure 4.13(b), the streaks are twisted into the spanwise directions, and the spacing between streaks is increased, as the oscillation force is applied.

4.2.2 Control with Wavy Lorentz Force

4.2.2.1 Experimental Apparatus[21, 22]

A schematic of a unit of a traveling wave EM actuator, called the actuator tile, is shown in Figure 4.14. The 16 short discrete electrode stripes are sparsely arranged to be four-lined arrays

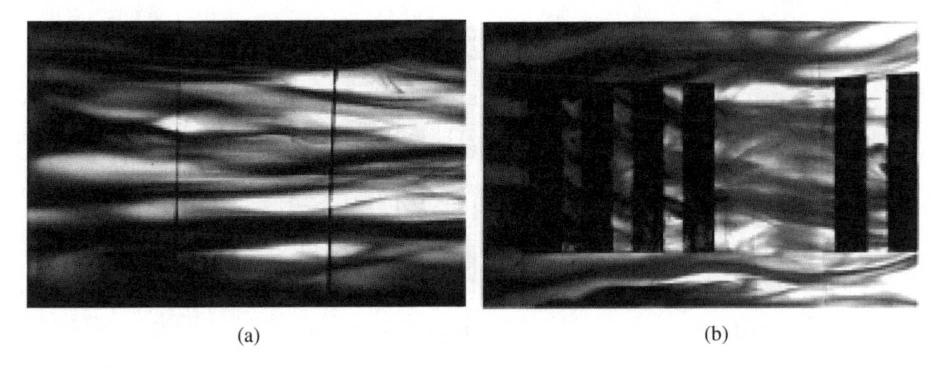

Figure 4.13 Streaky structures in a near-wall region with and without the Lorentz force oscillation.[14] *Source*: Park J 2004. Reproduced with permission from Springer

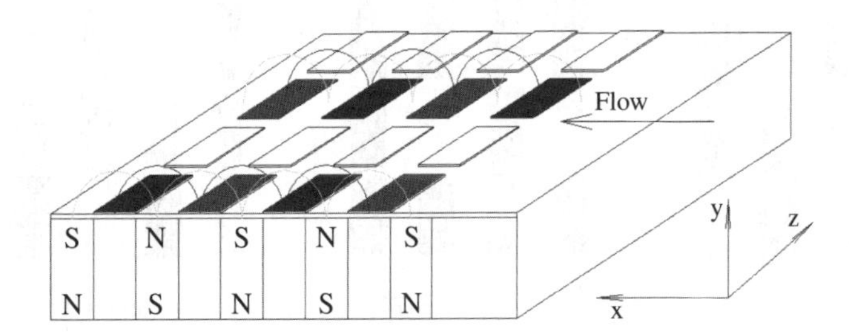

Figure 4.14 Configuration of the EM actuator tile for traveling wave excitation.[21] *Source*: Xu P 2009. Reproduced with permission from Cambridge University Press

aligned with the magnet bars. Each array consists of four short electrodes, instead of the long electrodes, as shown in Figure 4.2, and each electrode can be activated individually following a preset phase change via a PLV controller, as shown in Figure 4.14, where the gray color indicates a positively activated electrode, the black one a negatively activated electrode, and the white a not-activated electrode.

The required two-dimensional traveling wave can be generated via the traveling EM actuators by the multiphase excitation. For example, when the polarity of the electrodes alternates periodically with the phase changing strategy as shown in Figure 4.15, where black color denotes positively activated electrodes, light gray denotes negatively activated electrodes, and dark gray not activated electrodes, then the Lorentz forces are generated in every two alternate rows of electrodes in the spanwise direction and the induced local forcing directions will alter periodically denoted by solid arrows. When flow is from right to left, the expected traces of the local flow are denoted by the hollow arrows.

The total control area is made up by eight actuator tiles in an array pattern as shown in Figure 4.16, where the instantaneous polarity of the electrodes is described by different colors, that is, the dark gray denotes the positively activated electrodes, black denotes the negatively activated electrodes, and gray denotes the not-activated electrodes. The light gray denotes the magnetic tiles.

These assemble control actuator tiles are placed in the replaced part of a 4-m-long, 600-mm-wide, and 20-mm-thick test plate with a 20:1 ellipse leading edge shown in

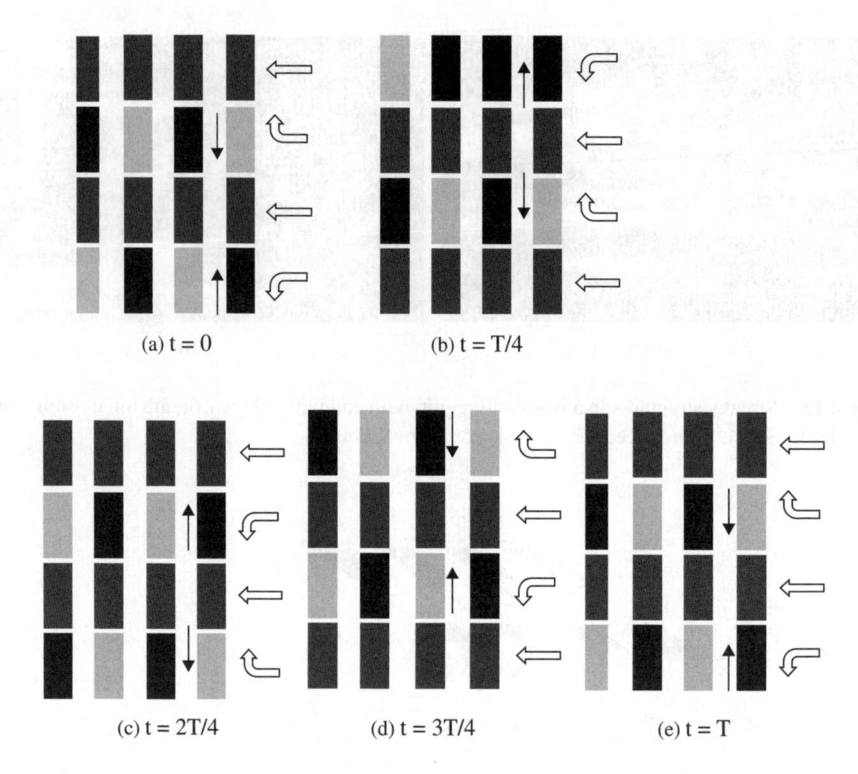

Figure 4.15 Four phases of a traveling wave EM actuator.[21] *Source*: Xu P 2009. Reproduced with permission from Cambridge University Press

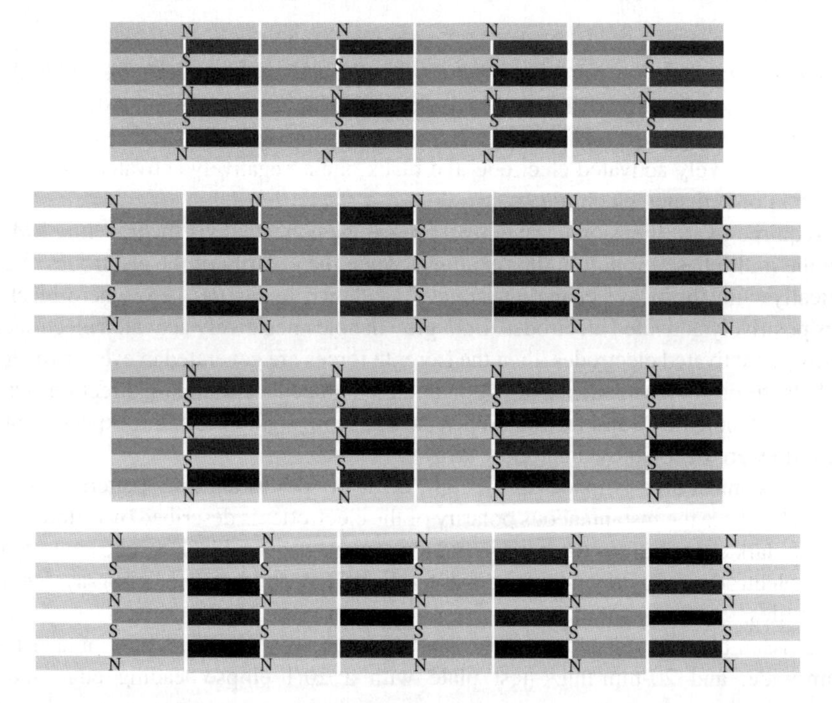

Figure 4.16 Assemble control actuators consisting of eight EM tiles.[21] *Source*: Xu P 2009. Reproduced with permission from Cambridge University Press

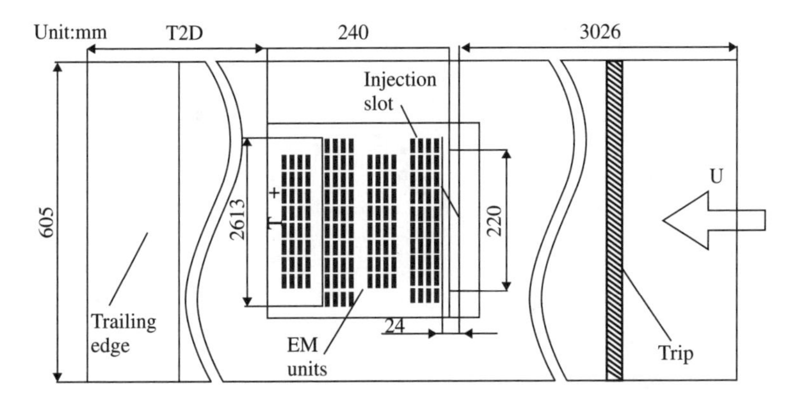

Figure 4.17 Schematic of a test plate and positions of assemble control actuators.[21] *Source*: Xu P 2009. Reproduced with permission from Cambridge University Press

Figure 4.17, which is placed in channel. The turbulent boundary layer is tripped by a trip device set 150 mm downstream of the leading edge. The $CuSO_4$ solution is injected through a slot set on the plate, while the $KMnO_4$ solution, instead of the $CuSO_4$ solution, is employed for the flow visualization.

The experiments are carried in an open water channel. A hot-film anemometer is used as the basic data-collecting device.

4.2.2.2 Experimental Results

Figure 4.18 shows the mean velocity profiles in the near-wall region, where black squares represent the case without control and hollow squares denote the case with the Lorentz forcing traveling wave control. It is observed that the velocity profile is shifted upward, the sublayer is thickened, and the mean velocity gradient is reduced, so that the drag is decreased.

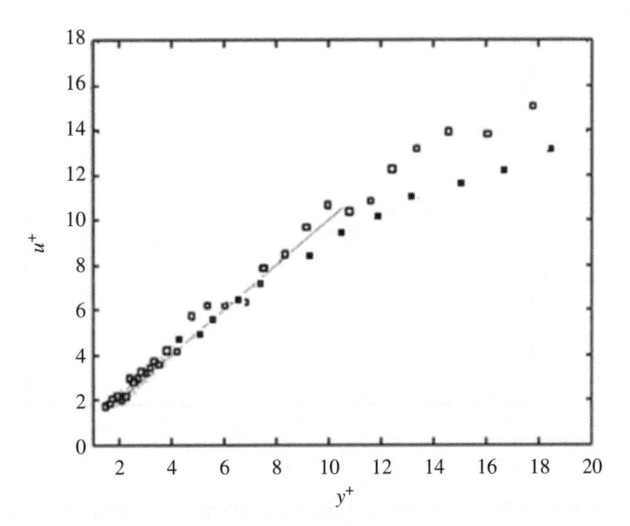

Figure 4.18 Mean velocity profiles in a near-wall region for cases with and without control.[21] *Source*: Xu P 2009. Reproduced with permission from Cambridge University Press

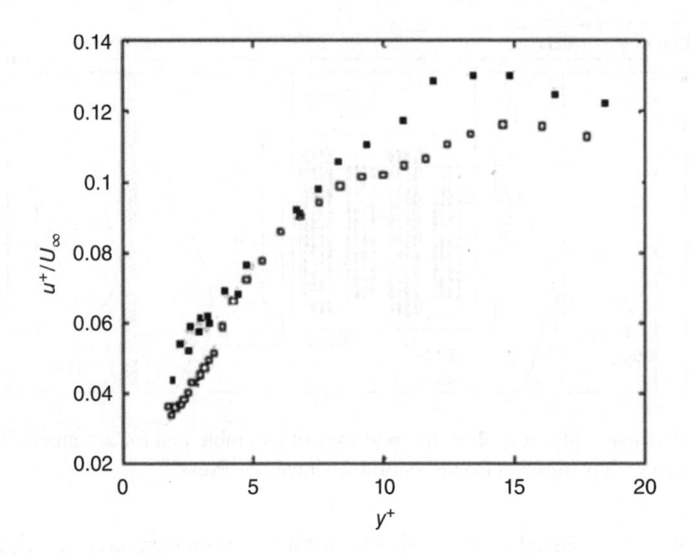

Figure 4.19 RMS streamwise velocity profiles of turbulence boundary layers with and without spanwise Lorentz forcing traveling wave.[21] *Source*: Xu P 2009. Reproduced with permission from Cambridge University Press

Figure 4.19 shows the rms streamwise velocity profiles in the near-wall region, where the black and hollow squares represent the cases without and with the Lorentz forcing traveling wave control, respectively. The results suggest that the largest streamwise velocity fluctuation amplitudes occur at $y^+ \approx 13$ in the no-control case, and the turbulent intensity is reduced by the Lorentz forcing traveling wave control.

Figure 4.20 shows the skewness of the velocity fluctuations, where the black and hollow squares represent the cases without and with the Lorentz forcing traveling wave control, respectively. For control case, the skewness is increased in the near-wall

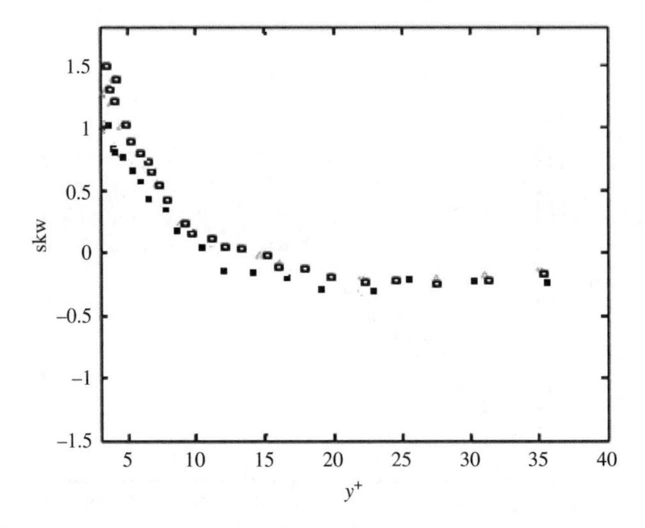

Figure 4.20 Skewness profile of streamwise velocity fluctuation in the turbulent boundary layer with and without the spanwise Lorentz forcing traveling wave.[21] *Source*: Xu P 2009. Reproduced with permission from Cambridge University Press

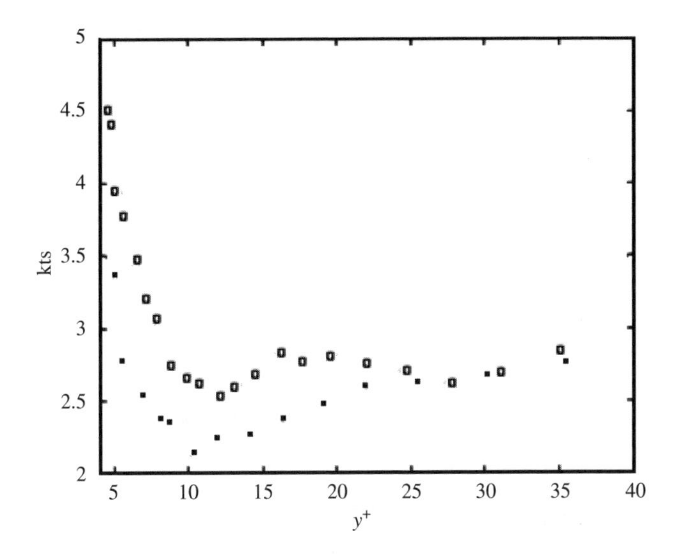

Figure 4.21 Kurtosis profile of streamwise velocity fluctuation in the turbulent boundary layer with and without the spanwise Lorentz forcing travelling wave.[21] *Source*: Xu P 2009. Reproduced with permission from Cambridge University Press

region especially when $y^+ < 14$. It means that the u' with positive values occur more frequently in this region.

The distribution of kurtosis of velocity fluctuations is presented in Figure 4.21. For the control case, the kurtosis values are increased, which means that the probability distribution of the velocity fluctuations is much sharper. The increasing skewness and kurtosis of velocity fluctuations suggest that the turbulent activities have been modified by the spanwise traveling wave actuation.

Using the VITA technique, the conditionally averaged fluctuating streamwise velocity of the near-wall burst events at $y^+ = 5, 12$, and 20 are shown in Figure 4.22a, b, and c, respectively, where the solid line denotes the case with control and the dashed line represents the case without control. At $y^+ = 5$, the changes in the burst signature are clear in the near-wall region of the boundary layer, the burst duration has been reduced to about half of that without control, and the intensity is also reduced by 40%. At $y^+ = 12$, where the effects of the traveling Lorentz force on the burst signature are still strong, the burst duration is reduced by about 28%, and the intensity is reduced by 38%. At $y^+ = 20$, the effect of the Lorentz force is less, but we can still see about 18% reduction in the duration and 30% reduction in the intensity.

Figure 4.23 shows a series of flow pictures visualized by $KMnO_4$ dye, illustrating the time evolution of a flow field in a control period, where Figure 4.23a–e correspond to five phases of the EM units, as shown in Figure 4.15, respectively. Flow is from right to left, and the traces of the local flow are denoted by the hollow arrows on the right-hand side of the pictures. It is clearly seen that the intake flow clusters to the first row of the EM units at the beginning (shown in Figure 4.23a), then to the second row, and, in turn, to the third row and the fourth row following the Lorentz force oscillation as shown in Figure 4.23b, c, and d, respectively. Finally, the new cluster of dye traces is formed at the first row again as shown in Figure 4.23e, and then a cycle is completed. It is also observed that the trace of the flow is significantly sinusoidal in the streamwise direction.

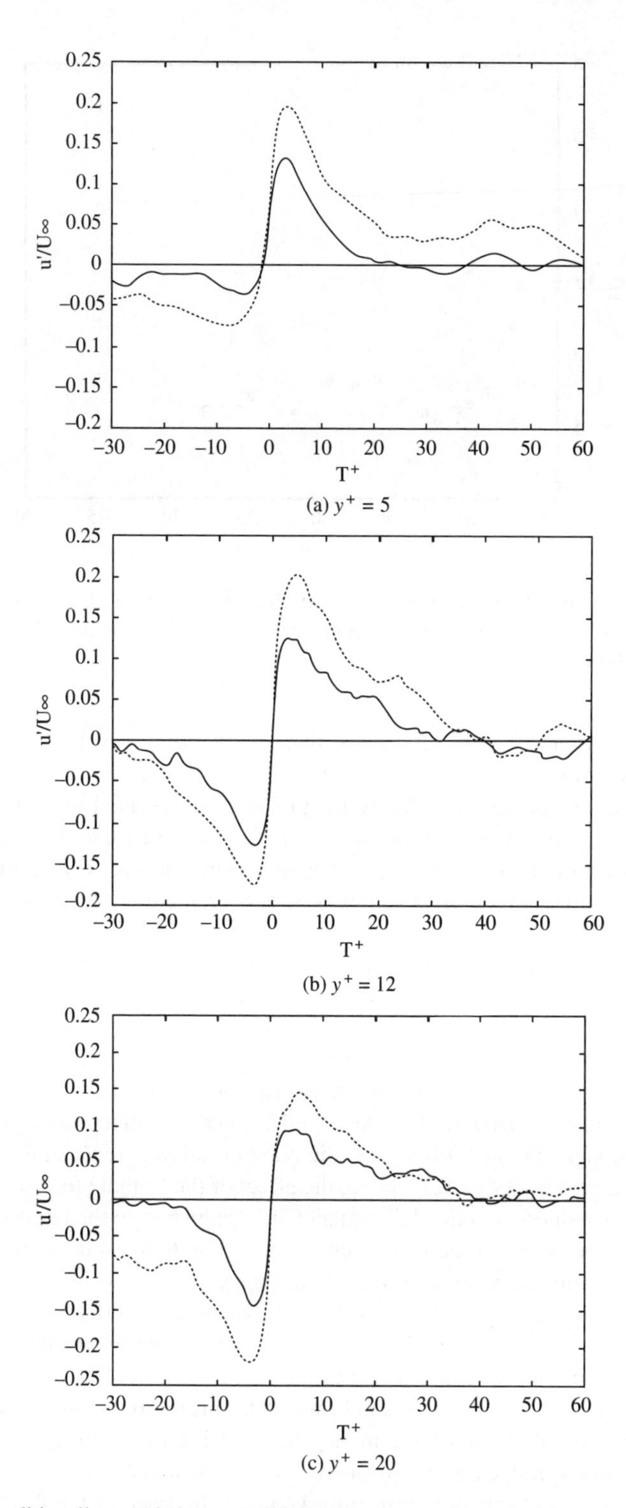

Figure 4.22 Conditionally sampled near-wall burst signatures with and without the spanwise Lorentz forcing traveling wave.[21] *Source*: Xu P 2009. Reproduced with permission from Cambridge University Press

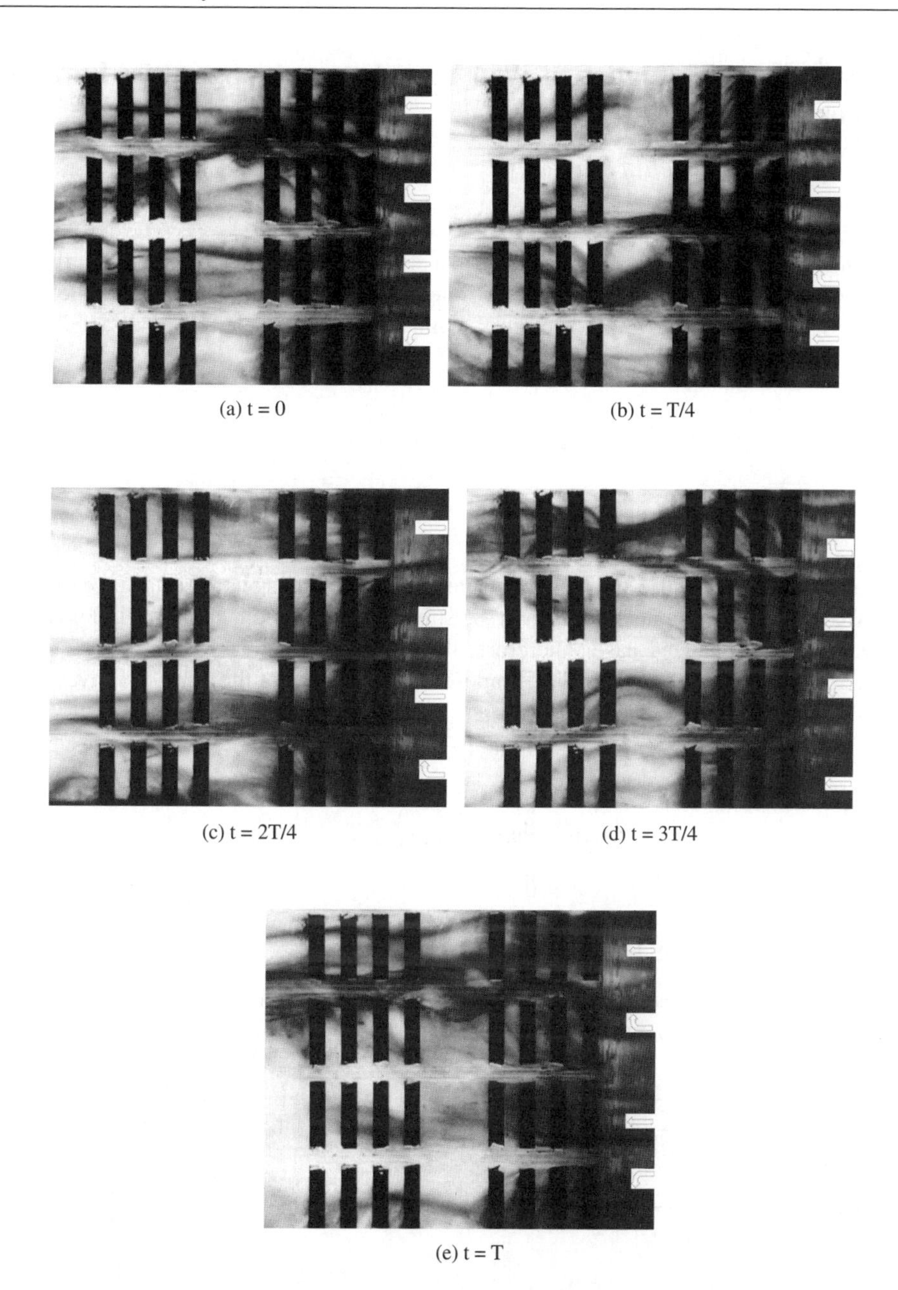

(a) t = 0
(b) t = T/4
(c) t = 2T/4
(d) t = 3T/4
(e) t = T

Figure 4.23 Visualization of a near-wall flow field controlled by the spanwise Lorentz forcing traveling wave.[21] *Source*: Xu P 2009. Reproduced with permission from Cambridge University Press

4.3 Numerical Simulation of Wall Turbulence with Spanwise Lorentz Force

4.3.1 Spanwise Lorentz Force

4.3.1.1 Idealized Spanwise Lorentz Force

The spanwise Lorentz force seems to be the most effective to control the near-wall turbulence. It can be created generally by placing electrodes and magnets side by side, parallel to one another, as shown in Figure 4.2. The corresponding magnetic and electric fields are described by the Laplace equations written as

$$\nabla^2 \mathbf{B} = 0 \tag{4.18}$$

$$\nabla^2 \Phi = 0 \tag{4.21}$$

and

$$\mathbf{J} = \sigma \mathbf{E} = -\sigma \nabla \Phi \tag{4.22}$$

Assuming that the electric and magnetic potential distributions on the surface are sinusoidal in the spanwise direction, no variations in the streamwise direction, as shown in Figure 4.24, referred to as an idealized case. Therefore, the Neumann boundary conditions for Φ and Dirichlet boundary conditions for $\mathbf{B}$ are imposed at the surface of the array, we have

$$\mathbf{J}_y|_{y=0} = \mathbf{J}_0 \sin\left(\frac{\pi}{2a}x\right) = -\sigma \left.\frac{\partial \Phi}{\partial y}\right|_{y=0} \tag{4.23}$$

$$\mathbf{B}_y|_{y=0} = \mathbf{B}_0 \cos\left(\frac{\pi}{2a}x\right) \tag{4.24}$$

$$\mathbf{J}_y|_{y=2h} = 0 \tag{4.25}$$

$$\mathbf{B}_y|_{y=2h} = 0 \tag{4.26}$$

where a is the width of magnets and electrodes, and h is the channel half-width.

The current density and magnetic flux density distributions become

$$\mathbf{J}_x(x,y) = -\frac{J_0}{\tanh\left(\frac{\pi}{a}h\right)} \cos\left(\frac{\pi}{2a}x\right)\left[\cosh\left(\frac{\pi}{2a}y\right) - \tanh\left(\frac{\pi}{a}h\right)\sinh\left(\frac{\pi}{2a}y\right)\right] \tag{4.27}$$

$$\mathbf{J}_y(x,y) = -\frac{J_0}{\tanh\left(\frac{\pi}{a}h\right)} \sin\left(\frac{\pi}{2a}x\right)\left[\sinh\left(\frac{\pi}{2a}y\right) - \tanh\left(\frac{\pi}{a}h\right)\cosh\left(\frac{\pi}{2a}y\right)\right] \tag{4.28}$$

$$\mathbf{B}_x(x,y) = -B_0 \sin\left(\frac{\pi}{2a}x\right)\left[\sinh\left(\frac{\pi}{2a}y\right) - \frac{\cosh\left(\frac{\pi}{2a}y\right)}{\tanh\left(\frac{\pi}{a}h\right)}\right] \tag{4.29}$$

$$\mathbf{B}_y(x,y) = B_0 \cos\left(\frac{\pi}{2a}x\right)\left[\cosh\left(\frac{\pi}{2a}y\right) - \frac{\sinh\left(\frac{\pi}{2a}y\right)}{\tanh\left(\frac{\pi}{a}h\right)}\right] \tag{4.30}$$

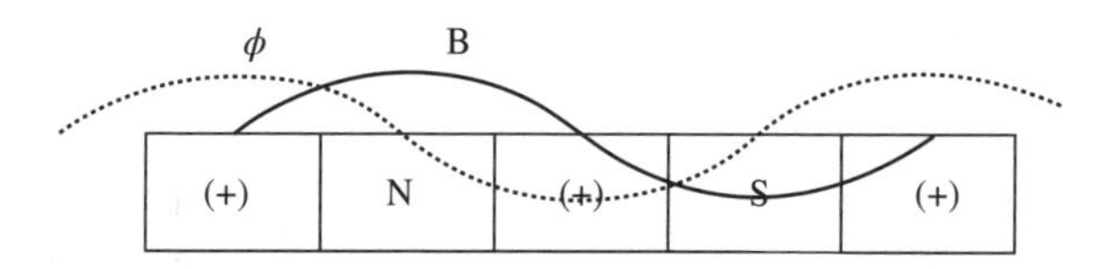

Figure 4.24 Sinusoidal electric/magnetic field distribution for the idealized case.[23] *Source*: Berger T W 2000. Reproduced with permission of Cambridge University Press

Taking the vector product of **J** and **B**, by substituting Eqs. (4.27)–(4.30) into Eq. (4.8), the resulting force distribution acts only in the spanwise direction and is only a function of the wall-normal distance y,

$$\mathbf{F}_z = J_0 B_0 \left[\sinh\left(\frac{\pi}{2a}y\right) - \frac{\cosh\left(\frac{\pi}{2a}y\right)}{\tanh\left(\frac{\pi}{a}h\right)} \right] \left[\cosh\left(\frac{\pi}{2a}y\right) - \frac{\sinh\left(\frac{\pi}{2a}y\right)}{\tanh\left(\frac{\pi}{a}h\right)} \right] \tag{4.31}$$

When $h/a \to \infty$,

$$\mathbf{F}_i = \delta_{i3} J_0 B_0 e^{-\frac{\pi}{a}y} \tag{4.32}$$

which indicates that the created force field is constant both in the streamwise and spanwise directions and decays exponentially in the wall-normal direction. The rate of decay is determined by the width of the electrodes and magnets.

At the wall, the nondimensional spanwise force is written in terms of the applied voltage, V_0, such that

$$F_z^+ = \frac{\pi}{4}\frac{\delta}{a}\frac{\sigma V_0 B_0}{\rho u_\tau^2} e^{-\frac{\pi}{a^+}y^+} = St e^{-\frac{\pi}{a^+}y^+} \tag{4.33}$$

where $V_0 = \frac{4a}{\pi\sigma} J_0$, and the superscript " + " indicate variables normalized by wall-unit scales from the undisturbed flow.

4.3.1.2 Realistic Spanwise Lorentz Force

As described in the previous section, the idealized spanwise Lorentz force is obtained from artificial boundary conditions. However, this boundary condition is not physical, since it creates a normal component on each nonconducting surface. In order to calculate a realistic force distribution, the exact boundary conditions for the electrode and magnet surfaces for the EM actuator, as shown in Figure 4.2, should be utilized.

Assuming that the electric and magnetic fields are constant in the streamwise direction, when the electrode and magnet are long enough, then Eq. (4.15) can be simplified as

$$\nabla \times \mathbf{B} = \left(\frac{\partial \mathbf{B}_y}{\partial x} - \frac{\partial \mathbf{B}_x}{\partial y} \right) \mathbf{k} = \mu \mathbf{I}_z = 0$$

where **k** is the unit normal vector. Therefore, the magnetic field can be represented by a potential function Ψ, defined as

$$\mathbf{B} = -\nabla \Psi$$

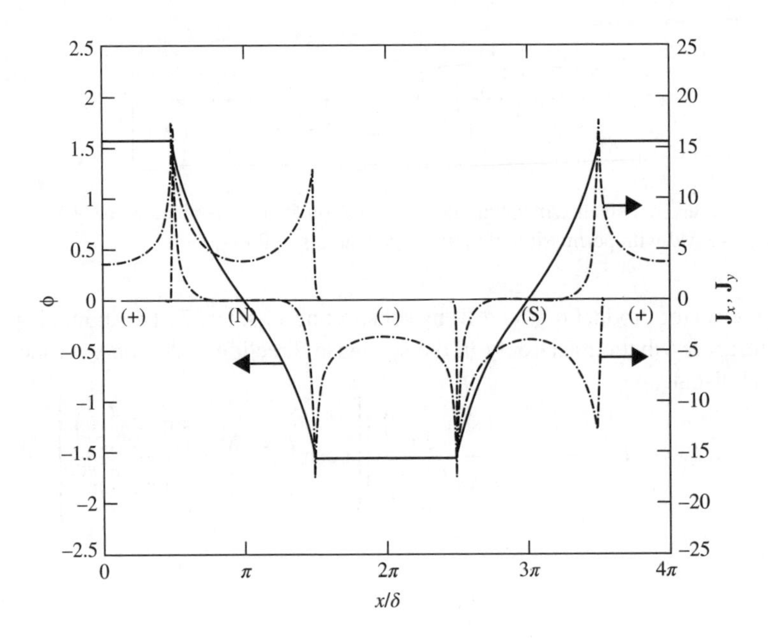

Figure 4.25 Electric field potential and current density distributions on the boundary surface.[23] *Source*: Berger T W 2000. Reproduced with permission of Cambridge University Press

Since $\nabla \cdot \mathbf{B} = 0$, (Eq. (4.16))

$$\nabla^2 \Psi = 0$$

Here, both electric and magnetic fields are described by the Laplace equation of the potential function. It is assumed that the electrode can conduct only the electric current and likewise for the magnet with regard to magnetic flux. Then, for the Laplace equation, Neumann boundary conditions are imposed on the nonconducting surfaces and Dirichlet boundary conditions on the conducting surfaces. For example, a step-function approach is used on conducting surfaces and a tangent distribution is used between conductors, as shown in Figure 4.25 by solid lines.

The electric field potential Φ and current density $\mathbf{J}$ for the electric field are shown in Figure 4.25, where the solid line represents Φ, the dashed line and dash-dotted line represent $\mathbf{J}_x$ and $\mathbf{J}_y$, respectively. It is observed that only $\mathbf{J}_y$ exists on the electrode surface and only $\mathbf{J}_x$ exists on the magnet surface. Similar results are derived for the magnetic flux density field, as shown in Figure 4.26, where the solid line and the dashed line represent $\mathbf{B}_x$ and $\mathbf{B}_y$, respectively.

The spanwise Lorentz force distribution defined by $\mathbf{J}$ and $\mathbf{B}$ is shown in Figure 4.27. It is observed that the force decays in the wall-normal direction, with a rate determined by the width of the electrodes and magnets, and oscillates periodically in the spanwise direction, the maximum force occurs at the interspace between the electrode and magnet.

4.3.2 Generalized Stokes Layer Induced by Oscillating Lorentz Force

4.3.2.1 Flow Induced by Temporally Oscillating Lorentz Force

By alternating the polarity of the electrodes in time, Lorentz force oscillates periodically, introducing an oscillatory flow in the near-wall region. The spanwise Lorentz force oscillating in

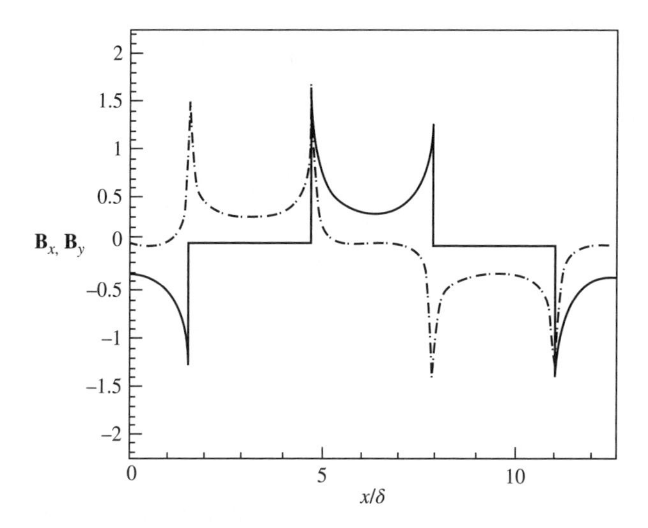

Figure 4.26 Magnetic flux density distribution on the boundary surface.[23] *Source*: Berger T W 2000. Reproduced with permission of Cambridge University Press

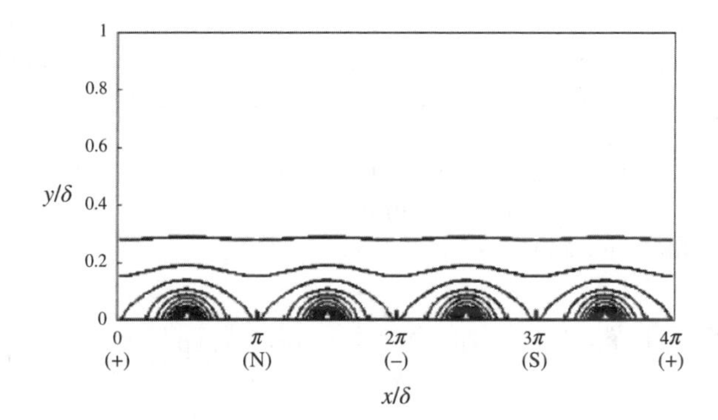

Figure 4.27 Realistic spanwise Lorentz force distribution.[23] *Source*: Berger T W 2000. Reproduced with permission of Cambridge University Press

time and decaying exponentially in the wall-normal direction can be expressed as

$$f_z^+ = St\, e^{\left(-\frac{\pi y^+}{a^+}\right)} \sin\left(\frac{2\pi t^+}{T^+}\right) \tag{4.34}$$

Considering the force described by Eq. (4.34) imposing on a laminar flow, the governing equation for spanwise velocity then becomes

$$\frac{\partial W^+}{\partial t^+} = \frac{\partial^2 W^+}{\partial y^{+2}} + \frac{St}{\mathrm{Re}_\tau} e^{\left(-\frac{y^+}{\Delta^+}\right)} \sin\left(\frac{2\pi t}{T^+}\right) \tag{4.35}$$

with the boundary conditions

$$W^+(y^+ = 0, t^+) = 0 \tag{4.36}$$

$$W^+(y^+ \to \infty, t^+) = 0 \tag{4.37}$$

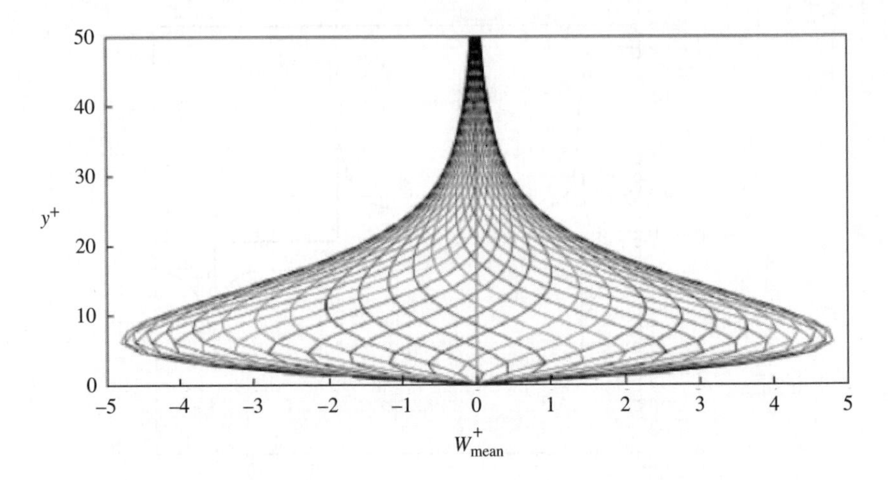

Figure 4.28 Spanwise velocity profiles induced by the oscillatory spanwise Lorentz force.[23] *Source*: Berger T W 2000. Reproduced with permission of Cambridge University Press

where Δ^+ is the penetration depth representing a thickness within the influence of the Stokes layer, which is dependent on the width of magnets and electrodes. This problem is similar to Stokes' oscillating plate problem, whereas here the wall is fixed in time with the fluid being oscillated by the Lorentz force.

The spanwise velocity profiles induced by the oscillatory spanwise Lorentz force based on the solutions of Eq. (4.35) are shown in Figure 4.28.

The envelope of the solution for W^+ is

$$W^+_{\text{env}} = \frac{St\widetilde{\Delta}^2}{\text{Re}_\tau}\left\{\exp\left(-2\widehat{y}^+\right) + \exp(-2\widetilde{y}^+) - 2\cos(\widehat{y}^+)\exp\left[-\left(\widetilde{y}^+ + \widehat{y}^+\right)\right]\right\}^{1/2} \tag{4.38}$$

where $\widehat{y}^+ = y^+\sqrt{\frac{\pi}{T^+}}$, $\widetilde{y}^+ = \frac{y^+}{\Delta^+}$, $\widetilde{\Delta}^2 = \frac{\Delta^{+2}}{\sqrt{1+\frac{4\Delta^{+4}\pi^2}{T^{+2}}}}$.

The maximum spanwise velocity near the wall is proportional to the magnitude of the forcing and inversely proportional to the Reynolds number, expressed as

$$W^+_{\text{env}}\Big|_{\max} \propto \frac{St\widetilde{\Delta}^2}{\text{Re}_\tau} \tag{4.39}$$

When $\left(\frac{2\Delta^{+2}\pi}{T^+}\right)^2 \gg 1$, the above equation becomes

$$W^+_{\text{env}}\Big|_{\max} \propto \frac{St \cdot T^+}{\text{Re}_\tau} \tag{4.40}$$

4.3.2.2 Flow Induced by Spatially Oscillating Lorentz Force

By altering the arrangement of the electrodes, the resulting force distribution can be made to oscillate spatially, expressed as

$$f_z^+ = St \exp\left(-\frac{y^+}{\Delta^+}\right) \sin\left(\frac{2\pi x^+}{\lambda_x^+}\right) \tag{4.41}$$

where λ_x^+ is the wavelength in the streamwise direction.

When the force described by Eq. (4.41) is imposed on a laminar flow, the governing equation for the mean spanwise velocity then becomes

$$U^+\frac{\partial W^+}{\partial x^+} = \frac{\partial^2 W^+}{\partial x^{+2}} + \frac{\partial^2 W^+}{\partial y^{+2}} + \frac{St}{\mathrm{Re}_\tau} \exp\left(-\frac{y^+}{\Delta^+}\right) \sin(k_x^+ x^+) \tag{4.42}$$

with the boundary conditions

$$W^+(x^+, y^+ = 0) = 0$$

$$W^+(x^+, y^+ \to \infty) = 0$$

where $k_x^+ = 2\pi/\lambda_x$ is the streamwise wave number and the boundary conditions are periodic in the streamwise direction.

When the form of the solution is chosen as $f(y^+)\cos(k_x^+x^+) + g(y^+)\sin(k_x^+x^+)$ and $U^+/k_x^+ \gg 1$, the solution for W^+ is found to be

$$W^+(x^+, y^+) = \frac{St}{\mathrm{Re}_\tau k_x^+ U^+}\left[\exp\left(-y^+\sqrt{\frac{k_x^+U^+}{2}}\right) \times \cos\left(y^+\sqrt{\frac{k_x^+U^+}{2}} - k_x^+x^+\right) - \exp\left(-\frac{y^+}{\Delta^+}\right)\cos(k_x^+x^+)\right] \tag{4.43}$$

The maximum velocity near the wall is proportional to the magnitude of the forcing and inversely on the mean velocity, the Reynolds number, and the streamwise wave number

$$W_{\max}^+ \propto \frac{St}{\mathrm{Re}_\tau k_x^+ U^+} \tag{4.44}$$

Induced spanwise velocity profiles based on Eq. (4.43) are shown in Figure 4.29.

4.3.3 *Control with Spanwise Oscillating Lorentz Force*

The flow is governed by the incompressible Navier–Stokes equations with an externally imposed body force term written as

$$\frac{\partial \mathbf{u}}{\partial t} + \mathbf{u}\cdot\nabla\mathbf{u} = -\nabla p + \frac{1}{\mathrm{Re}}\nabla^2\mathbf{u} + \mathbf{f} \tag{4.45}$$

$$\nabla\cdot\mathbf{u} = 0 \tag{4.46}$$

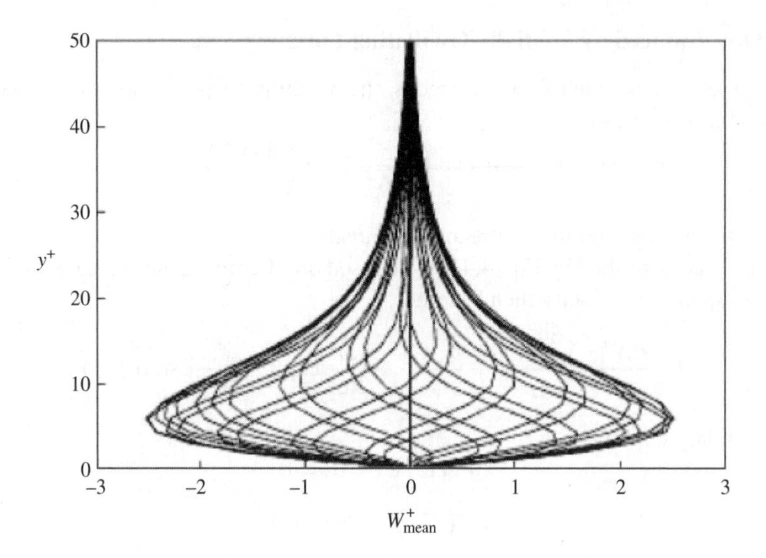

Figure 4.29 Spanwise velocity profiles induced by the spatially oscillating Lorentz force.[23] *Source*: Berger T W 2000. Reproduced with permission of Cambridge University Press

Here, all variables are normalized with the channel half-width h and the center line velocity, U_c. $\mathbf{f}$ is the Lorentz force per unit mass, given by

$$\mathbf{f} = (0, 0, f_z)$$

At the wall, the usual no-slip and no-penetration conditions are applied, and the periodic conditions are imposed for the homogeneous directions.

Equations (4.45) and (4.46) are solved numerically by standard the Fourier–Chebyshev spectral method discussed in Section 1.6. The time advancement is carried out by using a semi-implicit back-differentiation formula method with third-order accuracy. In order to ensure that the computed solutions satisfy both the incompressibility constraint and the momentum equation, a Chebyshev- τ influence-matrix method, including a τ -correction step, is employed for the linear term and the pressure term. Aliasing errors in the streamwise and spanwise directions are removed by spectral truncation method referred to as $3/2$-rule.

The spanwise Lorentz force can be generated by EM actuators made of a series of long magnets and electrodes side by side in an interleaved way. The temporally oscillating Lorentz force and the spatially oscillating Lorentz force are obtained by alternating the polarity of the electrodes in time and altering the arrangement of the electrodes, respectively, and approximately expressed individually as

$$f_z = St\,e^{\left(-\frac{y}{\Delta}\right)} \sin\left(\frac{2\pi t}{T}\right) \tag{4.47}$$

and

$$f_z = St\,e^{\left(-\frac{y}{\Delta}\right)} \sin\left(\frac{2\pi x}{\lambda_x}\right) \tag{4.48}$$

or

$$f_z = St\, e^{\left(-\frac{y}{\Delta}\right)} \sin\left(k_x \frac{2\pi x}{L_x}\right)$$

where k_x, defined as $k_x = L_x/\lambda_x$, is the streamwise wave number, L_x is the streamwise length of the computational domain.

4.3.3.1 Temporally Oscillating Lorentz Force[24–27]

After a force expressed by Eq. (4.47), referred to as temporally oscillating Lorentz force, has been applied to a turbulent channel flow, with $\mathrm{Re}_\tau = 100$ and $\Delta^+ = 10$, the drag histories are shown in Figure 4.30, where the thick solid line denotes the no control case presented for comparison purposes and the dotted and dashed line for $T^+ = 100$ and $S_t = 20\pi$. It is clearly seen that a drag reduction is achieved and sustained with periodical fluctuation by an oscillating Lorentz force.

When the control is applied only to the upper wall, the mean velocity profile is shown by the dashed line in Figure 4.31, where the solid line represents the no-control. The slope of the mean profile near the control wall is reduced.

The profiles of rms velocity fluctuation are shown in Figure 4.32, where the solid line denotes streamwise velocity fluctuation, u_{rms}/u_τ; the dashed line for v_{rms}/u_τ, and the dotted line for w_{rms}/u_τ. A significant decrease in all three velocity components is observed, which implies that a Lorentz force to act upon the near-wall structures can profoundly reduce the kinetic energy of the turbulence.

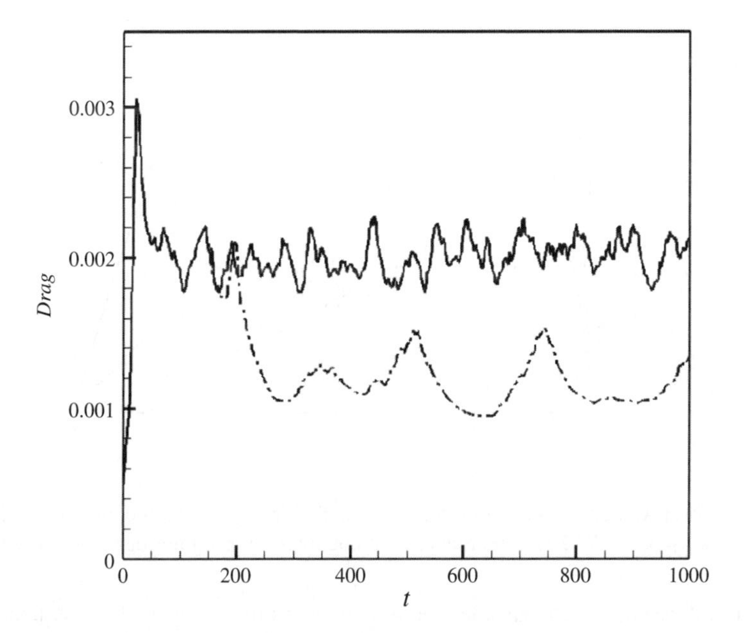

Figure 4.30 Drag histories for controlled flow by the temporally oscillating Lorentz force.[30] *Source*: Du Y 2002. Reproduced with permission of Cambridge University Press

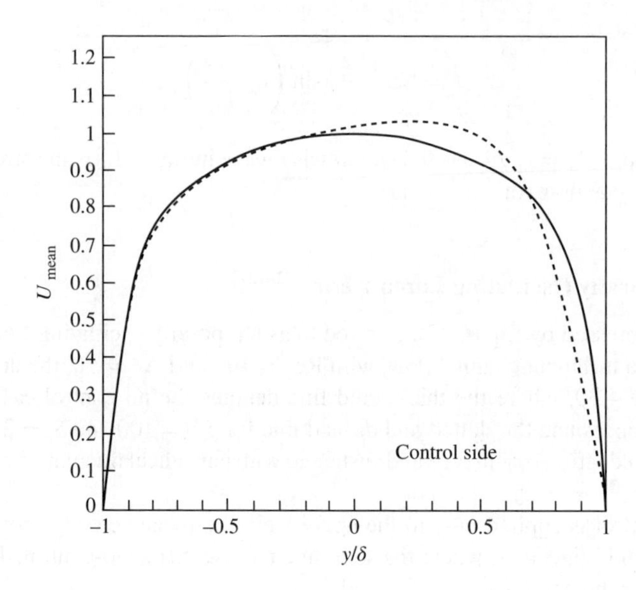

Figure 4.31 Mean velocity profile for controlled flow by temporally oscillating Lorentz force.[23] *Source*: Berger T W 2000. Reproduced with permission of Cambridge University Press

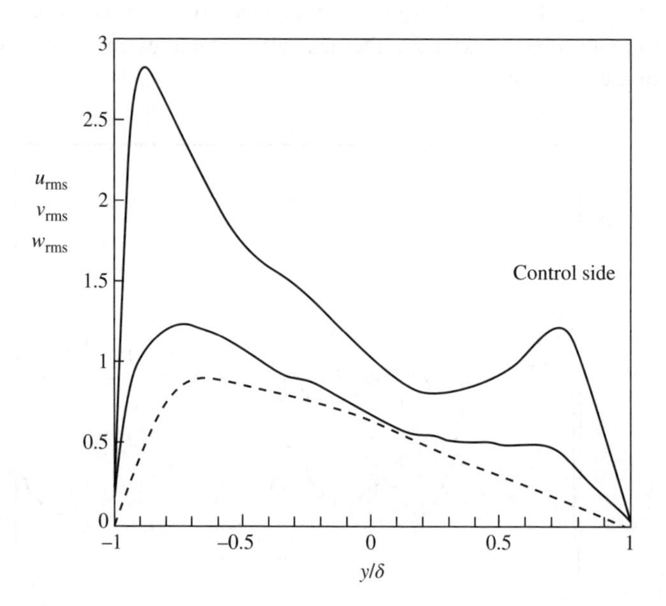

Figure 4.32 RMS velocity fluctuations for controlled flow by the temporally oscillating Lorentz force.[23] *Source*: Berger T W 2000. Reproduced with permission of Cambridge University Press

The near-wall regions of the turbulent boundary layer are dominated by a sequence of bursting events associated with the most of the turbulent production and the coherent structure of wall turbulence. The process of the streaks bring the low-momentum fluid lift up away from the wall and become unstable, oscillate, and break down. This is followed by a sweep of the

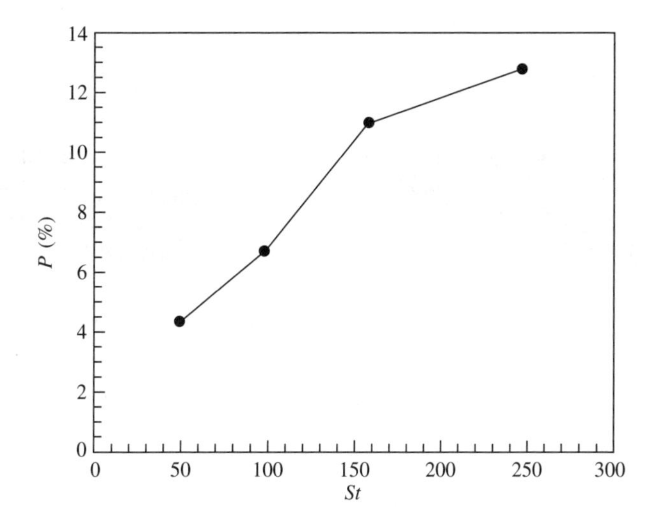

Figure 4.33 Variations of the averaged bursting frequency with St at $y^+ = 13.5$.[26] *Source*: Mei D J 2010. Reproduced with permission of Cambridge University Press

high-momentum fluid that originates in the out flow field and moves toward the wall, thus creating the most high skin-friction regions.

The variation of the bursting frequency P with the Stuart number St at $y^+ = 13.5$ and $T^+ = 113$ is shown in Figure 4.33, where $P = C/N$, where N is the total number of samples and C corresponds to the number of points in space at which the burst is detected. It reveals that the bursting frequency increases with the increase in St.

Once a reference position form each burst has been determined by the detection function, the conditional averaging techniques can be employed to study the burst phenomenon. The distributions of the conditionally averaged fluctuating streamwise velocity $\langle u'(\xi, z)\rangle$ in the (ξ, z) plane at $y^+ = 13.5$ are shown in Figure 4.34. It has been seen that the strength of the abrupt change of $\langle u'(\xi, z)\rangle$ near the location of detection decreases as St increases. The decrease in the bursting intensity with the St increasing can also be observed from Figure 4.35, which shows the variation of the bursting intensity with St at $T^+ = 113$.

The turbulent skin-friction drag is mainly contributed to the burst events. The increase in the frequency or intensity of the turbulent bursts can lead to an increase in the skin-friction drag on the wall. As depicted in Figures 4.33 and 4.35, the varying tendencies of the frequency and the intensity of the turbulent bursts with the Stuart number St are unsynchronized with each other, which implies that there is an optimal parameter St corresponding to the largest amount of drag reduction. Figure 4.36 shows the drag reduction versus St with $T^+ = 113$. It is indicated that the largest amount of the drag reduction occurs in $St = 99$.

For further illustrations, the distributions of the skin-friction drag, described by the streamwise velocity gradient on the wall, $\frac{du}{dy}$, are shown in Figure 4.37, where Figure 4.37a, b, and c, respectively, represent the cases with St = 80, St = 99, and St = 127, corresponding to the points denoted by a, b, and c individually in Figure 4.35. In Figure 4.37, the fluctuating frequency and intensity of the distribution surface correspond to the bursting frequency and intensity. It can be seen that the lowest frequency and the strongest intensity occur at point

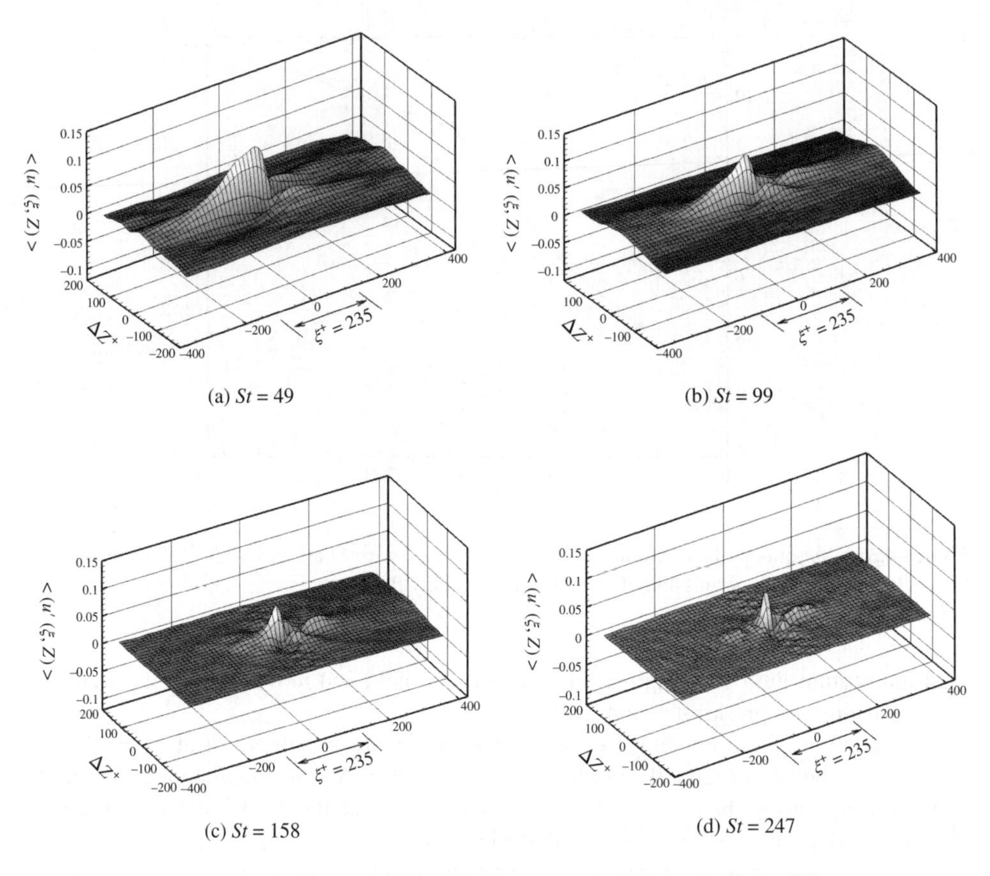

Figure 4.34 Conditional average of fluctuating streamwise velocities at $y^+ = 13.5$.[26] *Source*: Mei D J 2010. Reproduced with permission of Cambridge University Press

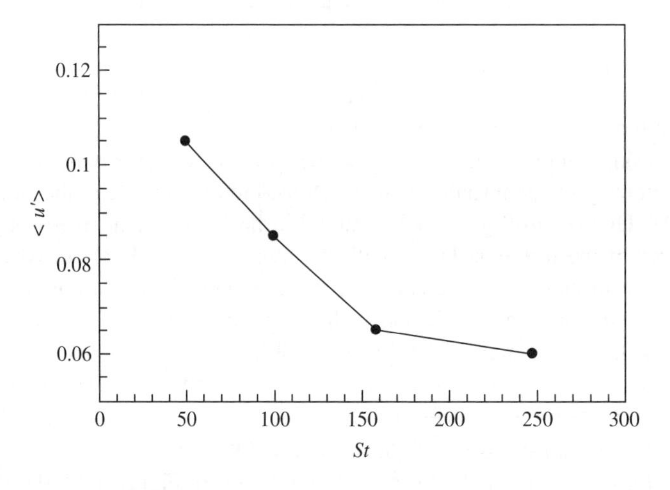

Figure 4.35 Variations of bursting intensity with *St* at $T^+ = 113$.[26] *Source*: Mei D J 2010. Reproduced with permission of Cambridge University Press

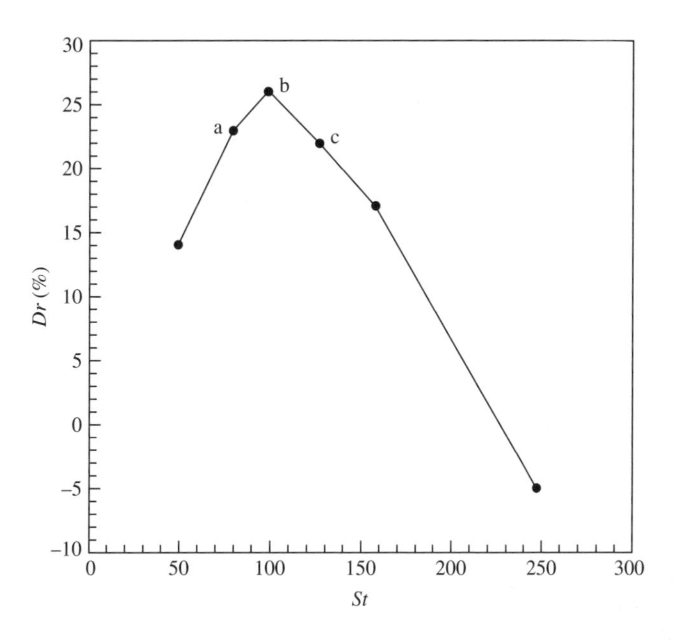

Figure 4.36 Variations of drag reduction with St at $T^+ = 113$.[26] *Source*: Mei D J 2010. Reproduced with permission of Cambridge University Press

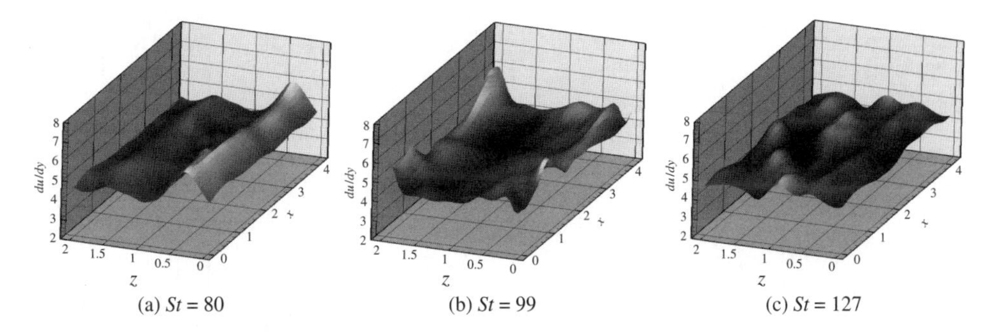

Figure 4.37 Distributions of skin-friction drag on a wall.[26] *Source*: Mei D J 2010. Reproduced with permission of Cambridge University Press

a, while the highest frequency and weakenest intensity at point c. However, the largest amount of the drag reduction occurs at point b.

The effects of T^+ on the bursting events with $St = 173$ are shown in Figures 4.38 and 4.39, respectively. It reveals that the bursting frequency increases with the increase in T^+, while the bursting intensity decreases with the increasing T^+. This contrary tendencies lead to the existence of an optimal T^+ corresponding to the largest amount of drag reduction as shown in Figure 4.40.

The drag reduction versus the "equivalent spanwise-wall velocity" defined by $W^+_{eq} = St \cdot T^+/(2\pi \mathrm{Re}_\tau)$ is shown in Figure 4.41, where the black circle denotes $T^+ = 113$, and black triangle denotes $St = 173$. The drag reduction increases with the increasing W^+ for lower W^+, and then decreases with the increasing W^+, after $W^+ = 10$ where a maximum drag reduction the largest amount of the drag reduction occurs in $St = 99$.

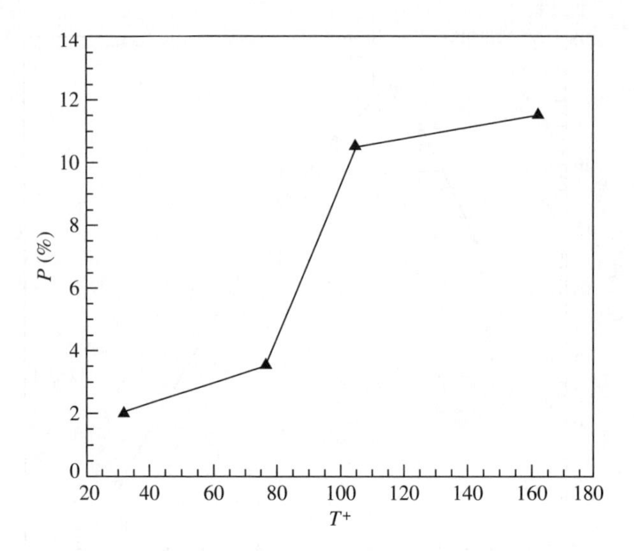

Figure 4.38 Variations of the averaged bursting frequency with T^+ at $y^+ = 13.5$.[26] *Source*: Mei D J 2010. Reproduced with permission of Cambridge University Press

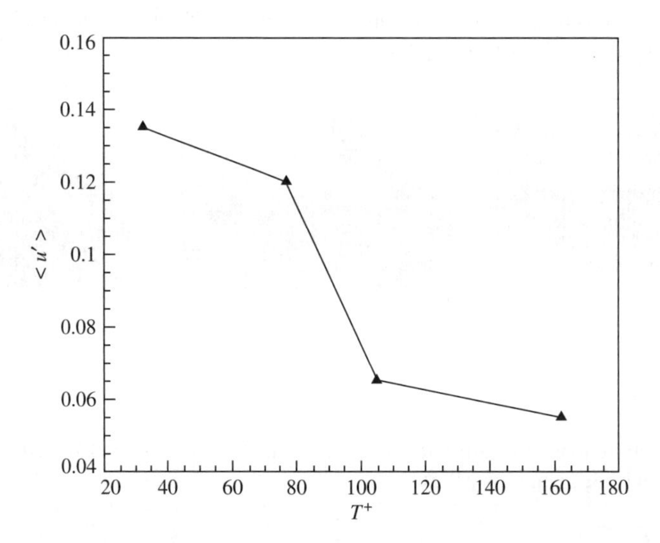

Figure 4.39 Variations of the bursting intensity with T^+ at $y^+ = 13.5$.[26] *Source*: Mei D J 2010. Reproduced with permission of Cambridge University Press

Drag reduction can be sustained with periodical fluctuation, when the flow is finally maintained effectively under the Lorentz force control. In one period of drag oscillation, whereas the Lorentz force oscillates in many full cycles, the near-wall structures are tilted upward and downward alternatively and then the number of vortex structures decreases with increasing its lateral spacing; finally, more and more tilted structures appear again, which process is similar to that occurring in the oscillating wall control discussed in Section 3.2.2.

For tilted near-wall structures, the streamwise vortex is positive and is tilted to the negative spanwise direction as f_z is negative. On the other hand, the negative vortex is tilted to the

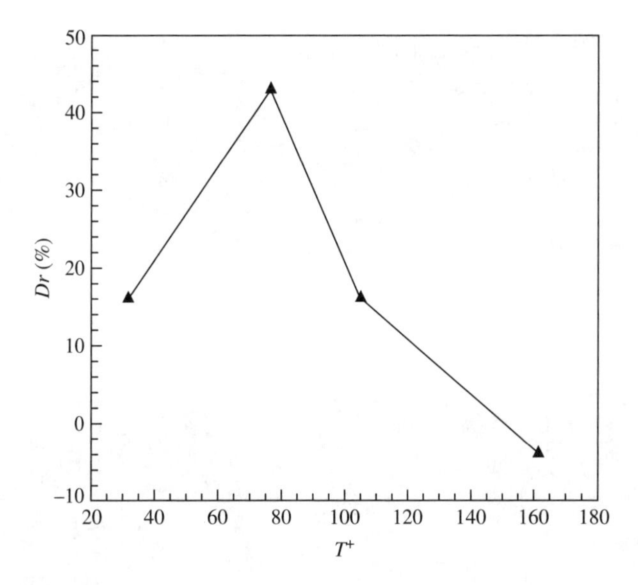

Figure 4.40 Variations of drag reduction with T^+.[26] *Source*: Mei D J 2010. Reproduced with permission of Cambridge University Press

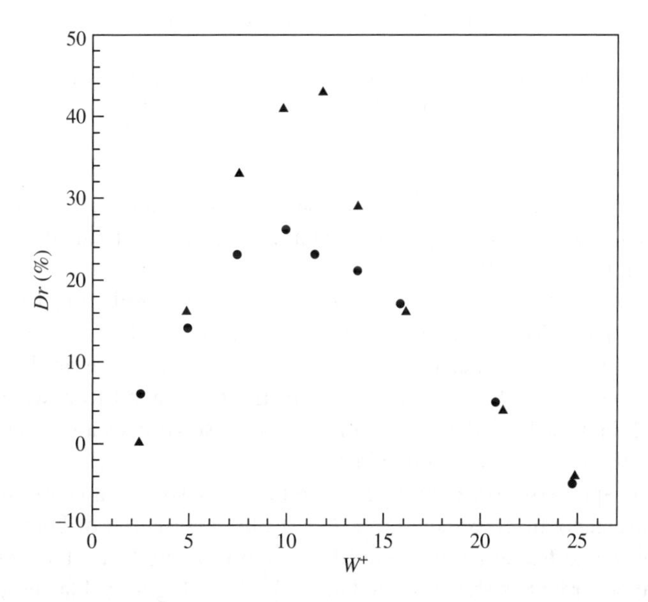

Figure 4.41 Variations of drag reduction with W^+.[26] *Source*: Mei D J 2010. Reproduced with permission of Cambridge University Press

positive spanwise direction, when f_z is positive. The tilting angle increases with the increasing f_z and the maximum angle occurs when $|f_z| = St$ on the wall.

The maximum spanwise velocity induced by the oscillatory Lorentz force occurs in the sublayer, at approximately 1 mm above the surface, where the influence of the Lorentz force on a streak is greatest. But in large part of a streak, the influence of the Lorentz force on the

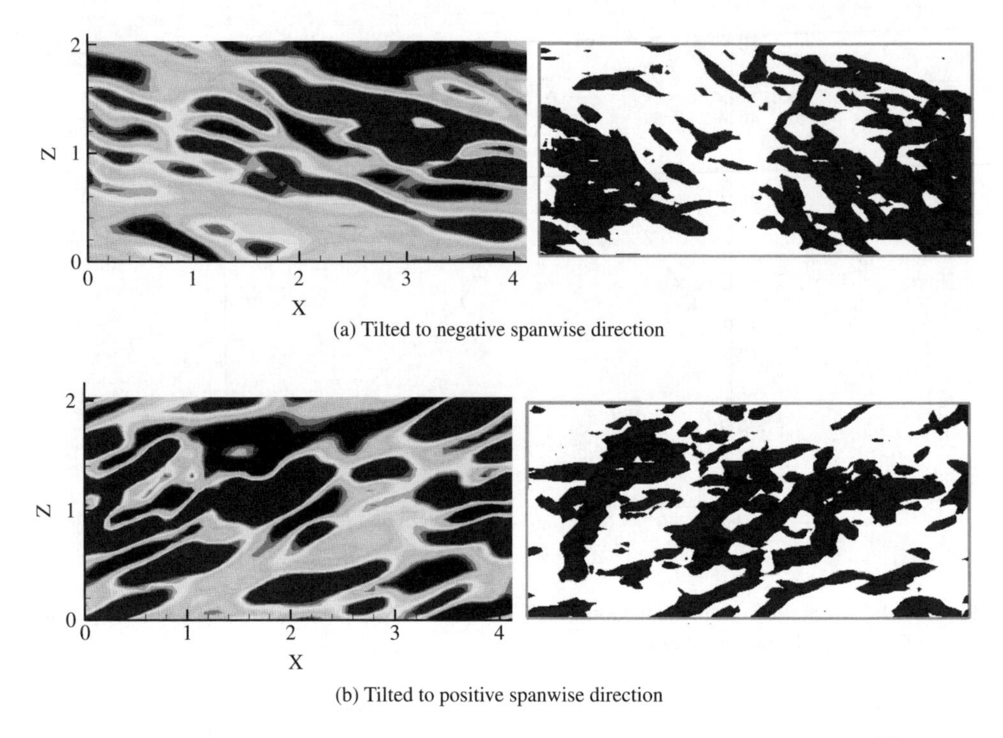

Figure 4.42 Streaky and vortex structures near the wall induced by the oscillation force.[27] *Source*: Mei D J 2011. Reproduced with permission of Cambridge University Press

upstream part of the streak is greater than that on the downstream higher part. Therefore, the configuration of streaks shows a diagonal orientation with respect to the mean flow direction, with a bended tail near the wall.

The instantaneous tilted near-wall structures induced by oscillating Lorentz forces are shown in Figures 4.42. The distribution of streaky structures is shown in the first row, where the gray areas represent high-speed streaks and the black areas represent low-speed streaks. The vortex structures near the wall are shown in the the second row, where few hairpins vortices can be found. Instead, the same signed streamwise vortices are observed tilting with almost the same tilt direction at the same instant.

The oscillating spanwise force can be regarded as a constant source of vorticity of alternate signs for the generalized Stokes layer. Due to the alternation of both the rotating and tilt directions of the vortex structure, induced by the oscillating force, the negative spanwise vorticity will be generated as shown in Figure 4.43, where Figure 4.43a and b correspond to Figure 4.42a and b, respectively. Figure 4.43 is a plot of shaded spanwise vorticity contours near the wall in the x–y plane, where the gray areas refer to the positive vortex and black areas refer to the negative vortex. It has been shown that the negative spanwise vorticity increases considerably in the near-wall region compared with no control case, as shown in Figure 3.33a, affecting the profile of mean streamwise velocity and then leading to the drag reductions.

The laterally oscillating force can also result in a shift of vortices relative to streaks, so that the near-wall structures are weakened and lessened, in many cases eliminates completely, as

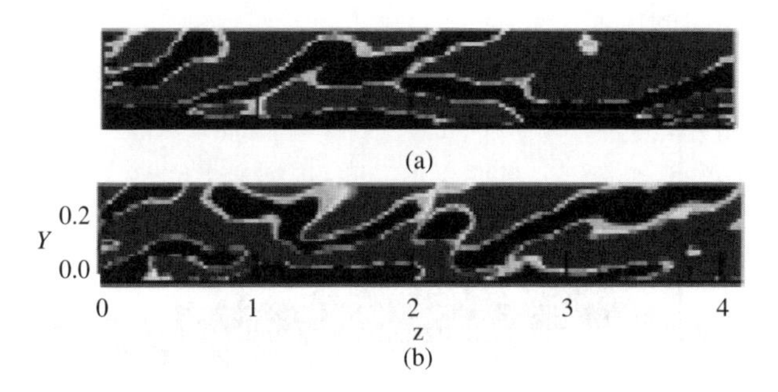

Figure 4.43 Distributions of spanwise vorticity near the wall.[27] *Source*: Mei D J 2011. Reproduced with permission of Cambridge University Press

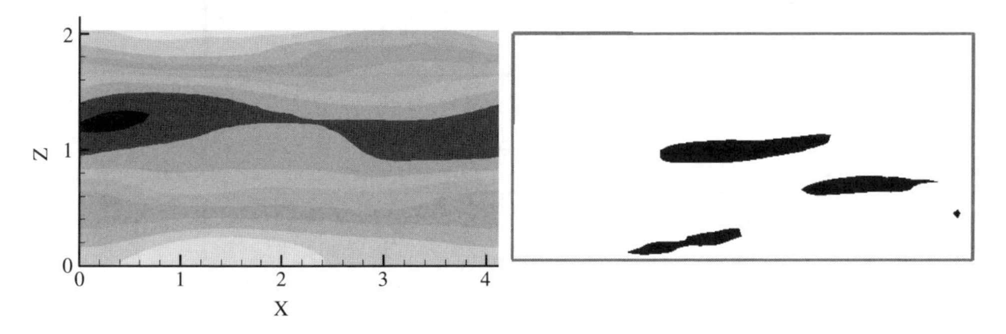

Figure 4.44 Streaky and vortex structures near the wall induced by the oscillation force.[27] *Source*: Mei D J 2011. Reproduced with permission of Cambridge University Press

shown in Figure 4.44, which leads to maximum drag reductions in effectively reducing the drag force.

Based on Eq. (3.4), the skin-friction drag in a turbulent channel flow is directly dependent on the Reynolds shear stress, produced by ejection and sweep activities. Hence, the quadrant analysis of the Reynolds shear stress can be used to further discuss the drag reduction under the turbulence control.

The scatter plot of the fluctuating velocities u' and v' at all discrete grid points on the $y^+ = 11.3$ plane for an uncontrolled flow is shown in Figure 4.45a. The ejection and sweep activities correspond to the second (Q2) and fourth (Q4) quadrants, respectively. The probabilities of the Reynolds shear stress are 15.2%, 35.8%, 14.3%, and 34.7%, in the four quadrants, respectively, for the uncontrolled case. It can be seen that Q2 and Q4 activities contribute to the most of the Reynolds stress compared with Q1 and Q3 activities.

The scatter plot of u' and v' at $y^+ = 11.3$ for an oscillating Lorentz force control with $W^+ = 11.8$ is shown in Figure 4.45b. The discrete Reynolds stress is clearly redistributed in the four quadrants. The appearance probabilities of the four quadrants are 19.1%, 31.9%, 26.3%, and 22.7%, respectively. The Q3 events are significantly increased, while the Q2 and Q4 events are reduced, which leads to the drag reduction in terms of Eq. (3.4).

To more clearly illustrate the oscillating Lorentz force effect, a comparison is made of $u' - v'$ distributions from the controlled and uncontrolled flow. We define the subtraction of

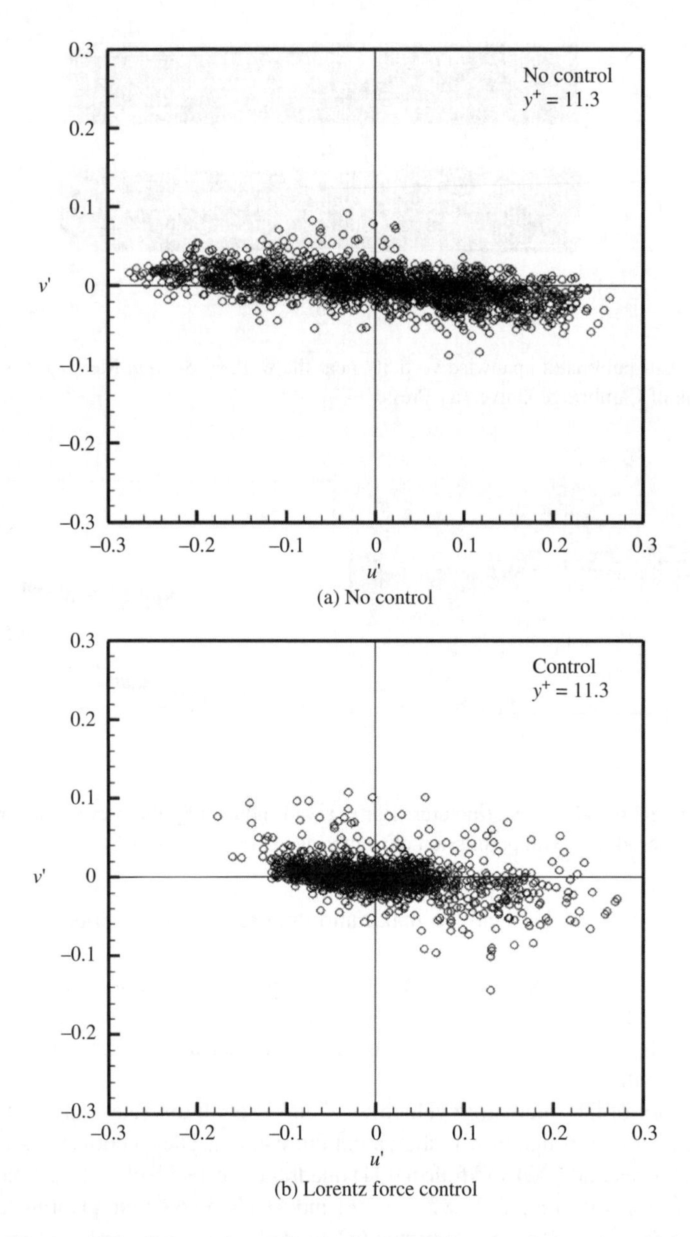

Figure 4.45 Scatter plot of fluctuating velocities u' and v' at $y^+ = 11.3$

the scatter plot as the set difference between Figure 4.45a and b, as shown in Figure 4.46, that is, the white and black circles correspond, respectively, to data points in Figure 4.45a that are not in Figure 4.45b and data points in Figure 4.45b are not in Figure 4.45a. It is clear from Figure 4.46 that the action of the oscillating Lorentz force is to move the data points from the white circle locations to the black circle locations. Compared with the white circles, large numbers of black circles are concentrated in the vicinity of the origin of coordinates, with the

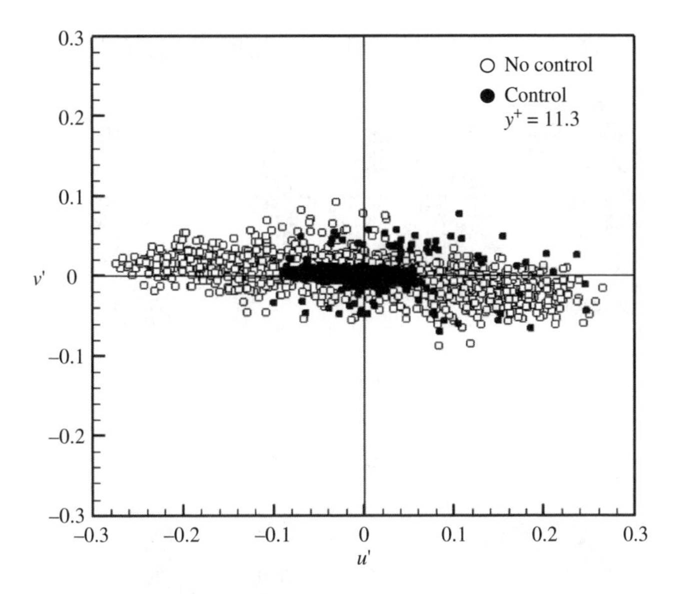

Figure 4.46 Subtraction of scatter plot of u' and v' at $y^+ = 11.3$ as a set difference between controlled and uncontrolled flows

smaller fluctuating velocities u' and v'. This indicates that the ejection and sweep activities induced by the streamwise vortices have been weakened due to control, thereby resulting in the drag reduction.

The instantaneous distributions of Reynolds stress at the y–z plane and corresponding skin-friction drag at the x–z plane for the uncontrolled and controlled cases are shown in Figure 4.47a and b, respectively, to illustrate the contributions of Reynolds stress to the skin-friction drag. Obviously, the high drag areas on the wall are connected directly to Q4 of Reynolds stress. For the controlled case, Q4 events adjacent to the wall are significantly reduced, while Q1 events are increased.

4.3.3.2 Spatially Oscillating Lorentz Force

The spatially oscillating Lorentz force, f_z, defined by Eq. (4.48), is independent of time. The distribution of f_z on the wall is shown in Figure 4.48a. When this force is introduced into a laminar flow, a generalized Stokes layer is then created. The distribution of the induced spanwise velocity and its shaded contour map in the x–y plane are shown in Figure 4.48b and c, respectively. In Figure 4.48c, the white areas refer to the positive velocity, and the black areas to the negative velocity. Due to the main flow, the white and black areas are overlaid at $\lambda_x/2$, λ_x, etc. to form the inclined shear layers as shown in Figure 4.48c. Note that there is no vortex in the induced Stokes layer.

When the induced flow is imposed to the near-wall turbulent flow, the intrinsic streaks and longitudinal vortex structures are modulated. Figure 4.49 shows snapshot distributions of the spanwise velocity, streak, and vortex structure near the wall with different values of the streamwise wave number for calculations.

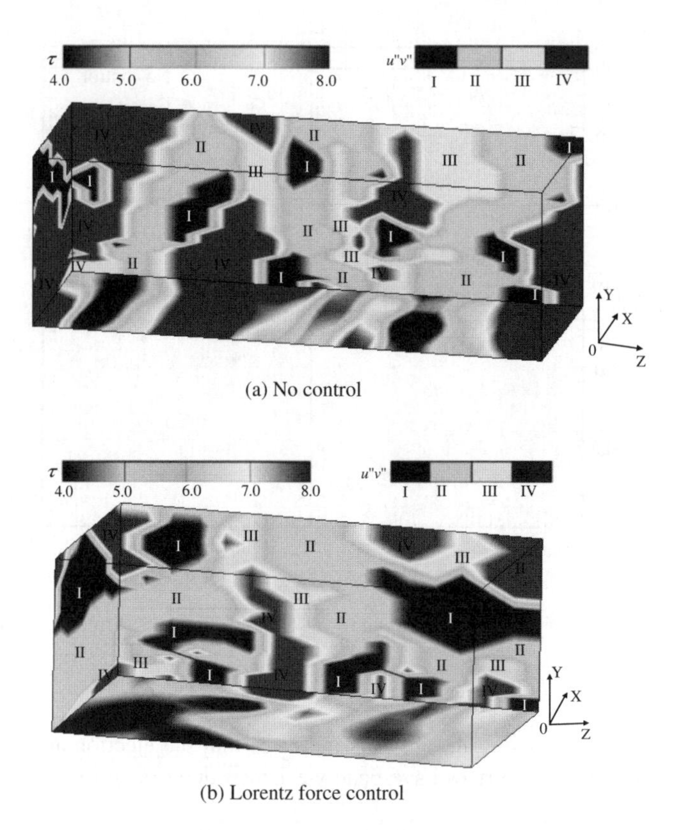

(a) No control

(b) Lorentz force control

Figure 4.47 Instantaneous distributions of Reynolds stresses at the *y–z* plane and the wall shear stress at *x–z* plane

The first column in Figure 4.49 shows the snapshots of the distribution of spanwise velocities in the *x–z* plane close to the wall ($y^+ = 5.4$), where the gray and black areas represent, respectively, positive and negative values of spanwise velocities. The shape of color area becomes more and more regular as k_x increases for low-frequency oscillation, which indicates that the influence of the induced flow on the turbulence becomes more and more strong. When $k_x = 3$, the configuration of spanwise velocity distributions consists of alternating gray rectangle regions and black rectangle regions along the streamwise direction as shown in Figure 4.49c, which means that the induced flow dominates the near-wall flow. However, as k_x increases further, the instability of the modulated flow field is induced by the high-frequency oscillation, so that the color areas become obscure, which indicates that the intrinsic turbulent flow dominates the near-wall flow.

The second column in Figure 4.49 shows the snapshots of the distribution of streaky structures in the *x–z* plane at $y^+ = 5.4$, where the white areas represent high-speed streak and black areas represent low-speed streak. For the smaller wave number, the streaks are sinusoidal with the decreasing frequency and amplitude as the wave number k_x increases as shown in Figure 4.49a–c. Then for $k_x > 3$, the large-amplitude oscillation of streaks disappears gradually and the intensity of the streaks increases with the increase in k_x as shown in Figure 4.49d and e.

The third column in Figure 4.49 shows the snapshots of vortex structures near the wall. For the low-frequency waves, the longitudinal vortices meander significantly and arrange in order;

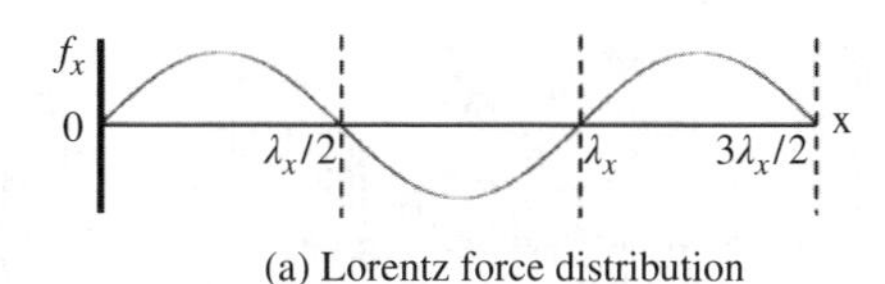

(a) Lorentz force distribution

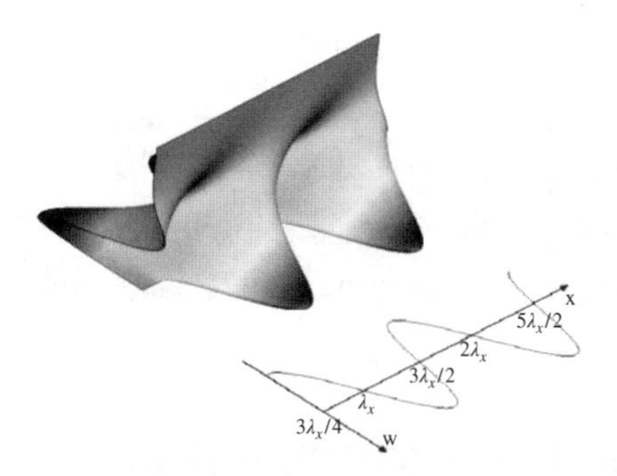

(b) Distribution of induced spanwise velocity

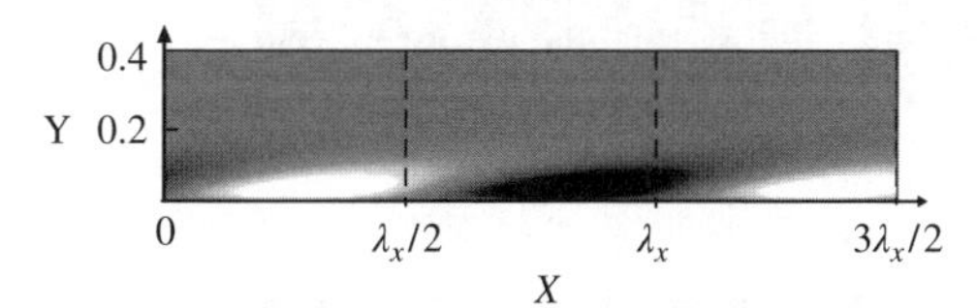

(c) Shaded contours of induced spanwise velocity

Figure 4.48 Distributions of the Lorentz force and induced spanwise velocity.[28] *Source*: Wu W T 2014. Reproduced with permission of Elsevier

the number of the vortices decreases as the wave number increases, as shown in Figure 4.49a–c. When $k_x > 3$, the longitudinal vortices are produced again, as shown in Figure 4.49d–e.

The variations of the bursting frequency and intensity with the wave number k_x at $y^+ = 14$ and $St = 1.0$ are shown in Figure 4.50a and b, respectively. It reveals that the minimum bursting frequency and intensity occure at $k_x = 3$, so that the largest amount of the drag reduction is achieved at $k_x = 3$ for $St = 1.0$.

Similarly, the variations of the bursting frequency and intensity with the Stuart number St at $y^+ = 14$ and $k_x = 3$ are shown in Figure 4.51a and b, respectively. The largest amount of the drag reduction is achieved at St $= 1.0$ for $k_x = 3$.

Therefore, the spatially oscillating Lorentz force with $St = 1.0$, $k_x = 3$ can result in a significant drag reduction. The drag history for this force, when imposed to a turbulent channel flow with $\mathrm{Re}_\tau = 150$, is shown by a dashed line in Figure 4.52, where the thick solid line denotes the no control case presented for comparison purposes. It is obvious that a drag reduction is achieved and sustained by the control.

This drag reduction is stained with periodical fluctuation, while the flow is finally maintained upon a statistically steady state via the oscillating Lorentz force. The drag reduction achieves the maximum value when the vortex shape is most similar to each and all vortices are organized well. The typical controlled near-wall structures are shown in Figure 4.53, where Figure 4.53a shows the distribution of f_z on the wall in the x direction and Figure 4.53b shows

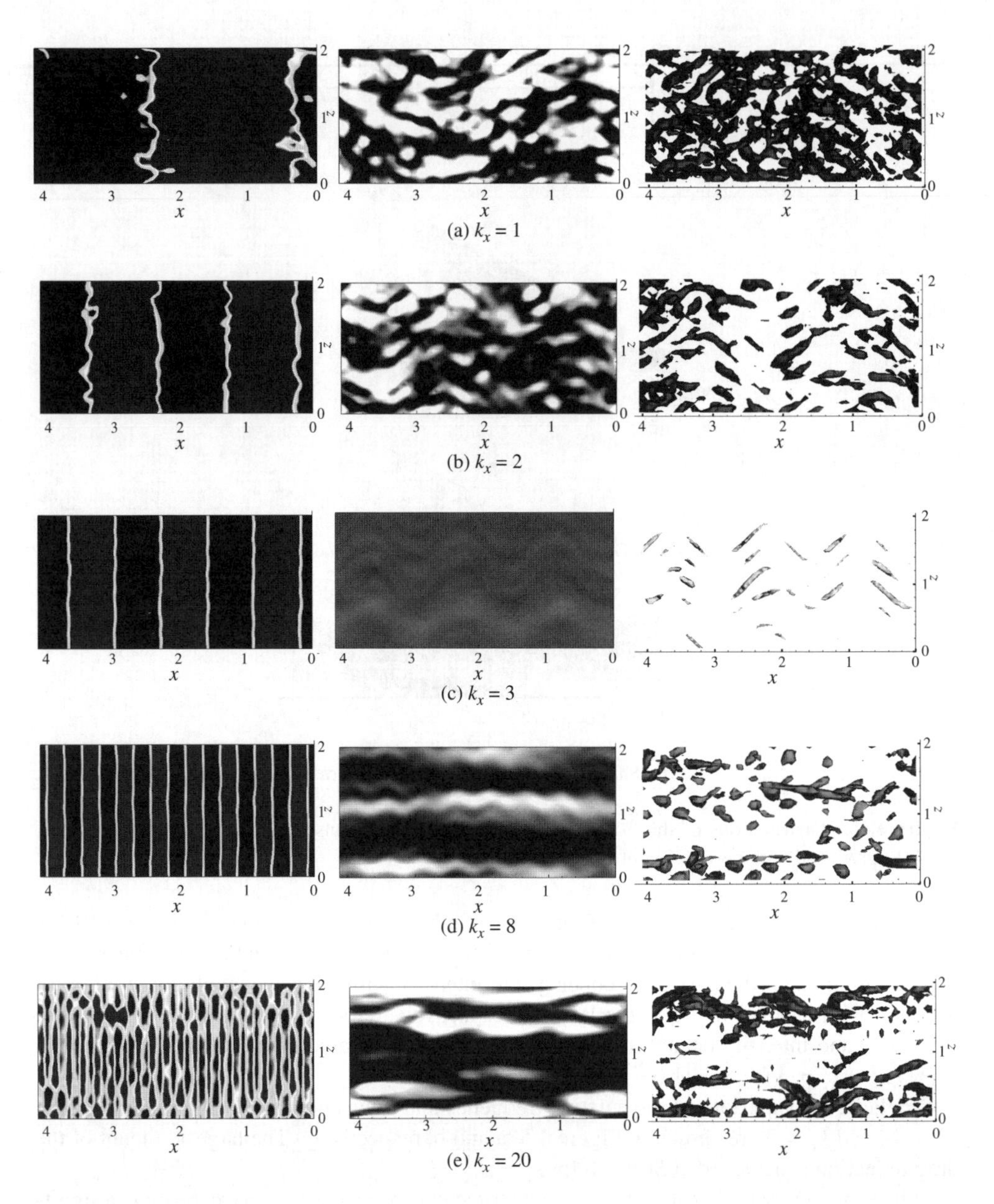

Figure 4.49 Snapshots of spanwise velocity, streak, and vortex structure near the wall.[29] *Source*: Guo C F 2013. Reproduced with permission of Springer

the distribution of the streamwise vortex, where the dark and light areas represent, respectively, the positive value and negative value of vorticities. In Figure 4.53c, the black and gray shading areas represent low- and high-speed streakes, respectively, and solid contours represent the streamwise vortex.

As mentioned previously, the spanwise Lorentz forces lead to the tilt of the near-wall structures. The negative f_z generates positive streamwise vortexes tilted to the negative spanwise

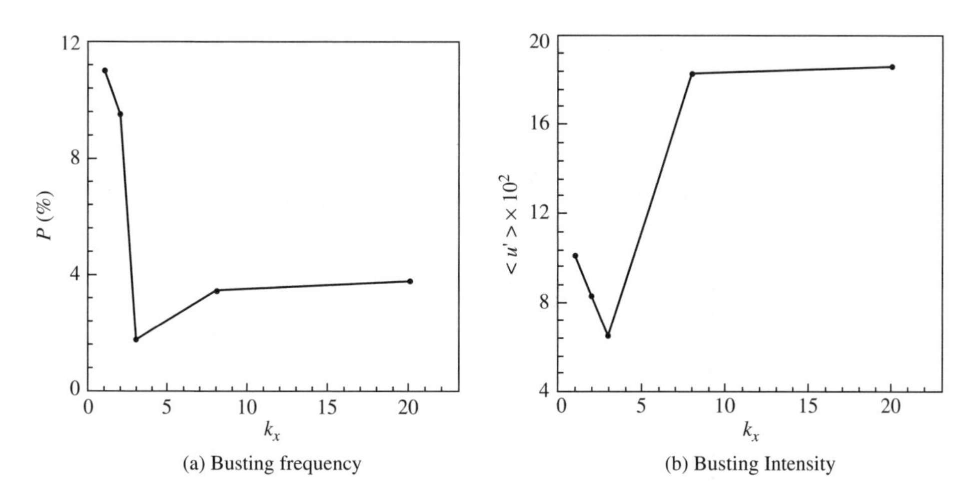

Figure 4.50 Variations of averaged bursting frequency and intensity with k_x at $y^+ = 14$.[29] *Source*: Guo C F 2013. Reproduced with permission of Springer

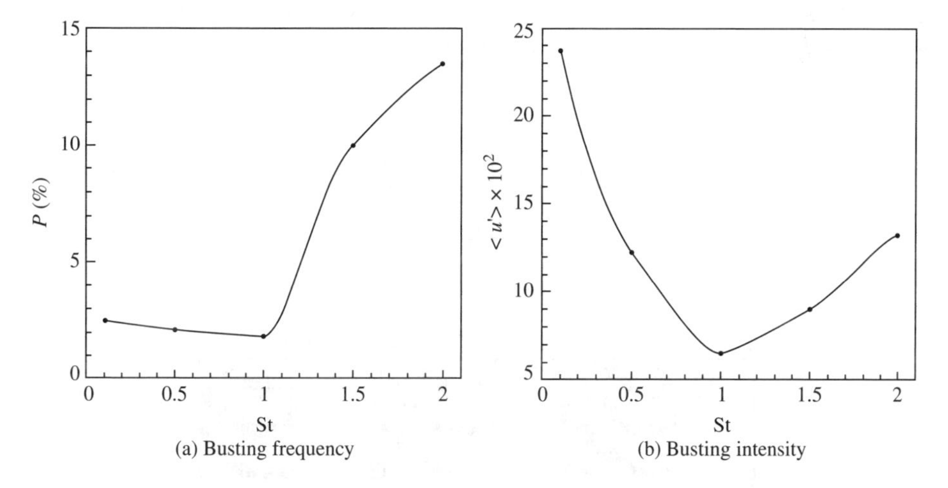

Figure 4.51 Variations of averaged bursting frequency and intensity with *St* at $y^+ = 14$.[29] *Source*: Guo C F 2013. Reproduced with permission of Springer

direction, while the positive f_z generates negative vortexes tilted to the negative spanwise direction. Hence, it is clearly seen from Figure 4.53 that the controlled boundary layer is dominated by stabilized, well-organized quasi-streamwise vortices with similar shapes, whereas a few hairpin vortices can be found. Meanwhile, the streamwise vortices with the same signs are arrayed, along the spanwise direction, to be a column, and the columns are arranged in an interleaved way with alternating sign. In addition, the positive vortices cross the low-speed streakes form its right-hand side when facing the downstream, whereas the negative one from its left-hand side. Each vortex has an averaged streamwise length of 0.5 λ_x, and the intervals of 0.5 λ_x and 0.3λ_x in the streamwise and spanwise directions respectively, and overlays with the other with the opposite sign at the shear layers located at $\lambda_x/2$, λ_x, etc.

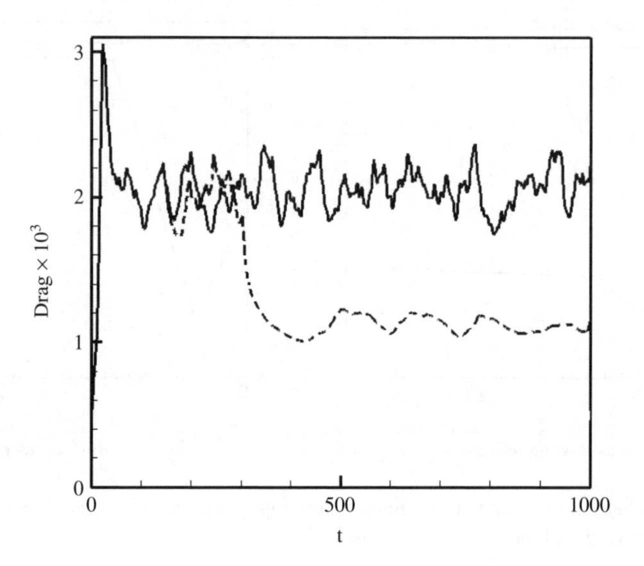

Figure 4.52 Drag histories for controlled flow by the spatially oscillating Lorentz force.[29] *Source*: Guo C F 2013. Reproduced with permission of Springer

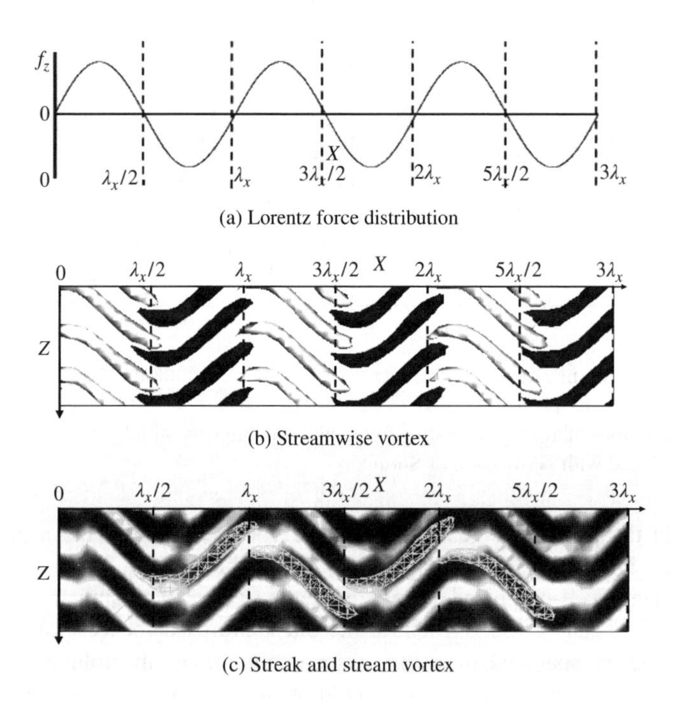

Figure 4.53 Typical near-wall structures controlled by the spatially oscillating Lorentz force.[28] *Source*: Wu W T 2014. Reproduced with permission of Elsevier

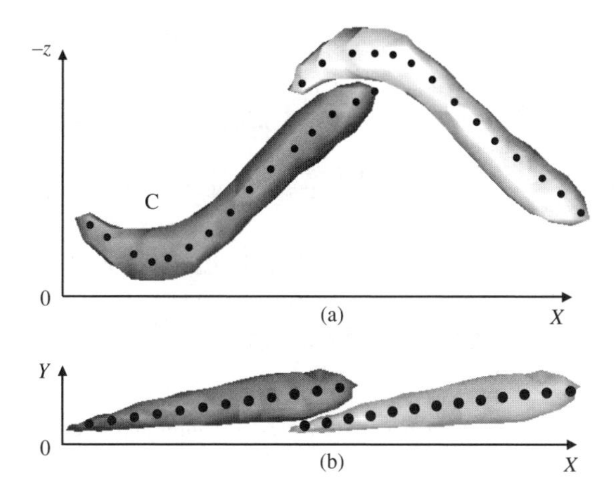

Figure 4.54 Averaged streamwise vortex structures in (a) top view and (b) side view.[28] *Source*: Wu W T 2014. Reproduced with permission of Elsevier

The averaged streamwise vortex structures are shown in Figure 4.54a and b as top view and side view, respectively. The dot represents the location of the vortex axis, defined as a line connecting local maxima of the imaginary part of complex eigenvalues of the velocity gradient tensor. The black area and the gray area represent the positive value and negative value of vorticities, respectively. To describe the orientation of vortices, inclination and tilting angles are introduced, defined by the angle between the x-axis and the projection of the vortex axis on the x–y plane and the x–z plane, respectively. It is observed that the positive vortex is tilted downward into the positive spanwise direction first and then at point C, where the spanwise velocity is zero (see also Figure 4.56), tilted upward into the negative direction with the tilting angle $\theta_1 \approx -32.5°$, vice versa. In addition, the vortex is lifted with the inclination angle $\theta_2 \approx 5.2°$.

As shown in Figure 4.54, all vortices bend to the ejection sides due to the changing sign of spanwise velocity, which would focus its ejections and strengthen the local low-speed streakes. Hence, the bursting events are triggered, when the vortices cross these bent low-speed streakes from the upstream high-speed streakes.

The averaged streamwise vortex structures and its shaded contour maps of streamwise voticity in the induced flow field in the y–z planes are shown in Figure 4.55, where SP and SN denote the positive and negtive ω_x, respectively, and solid contours in the cross sections represent the streamwise vortex in the controlled boundary layer. At the section D–D, the streamwise vortex is stong enough (see also Figure 4.56) to roll up the near-wall vorticity sheets with opposite vorticity, which are induced by the Lorentz force and attached to a no-slip wall. Then a new vortex with the opposite vorticity will be generated on the ejection side of a parent vortex as shown in section D_1–D_1.

The distributions of spanwise velocity and streamwise voticity along the vortex axes projected on the x-axis in the induced flow field are shown in Figure 4.56 by a dash-dotted line and dashed line, respectively, where the solid line denotes the distribution of the Lorentz force on the wall. It is clear that the pattern of well-organized near-wall structures in the controlled flow field is dominated mainly by the induced flow.

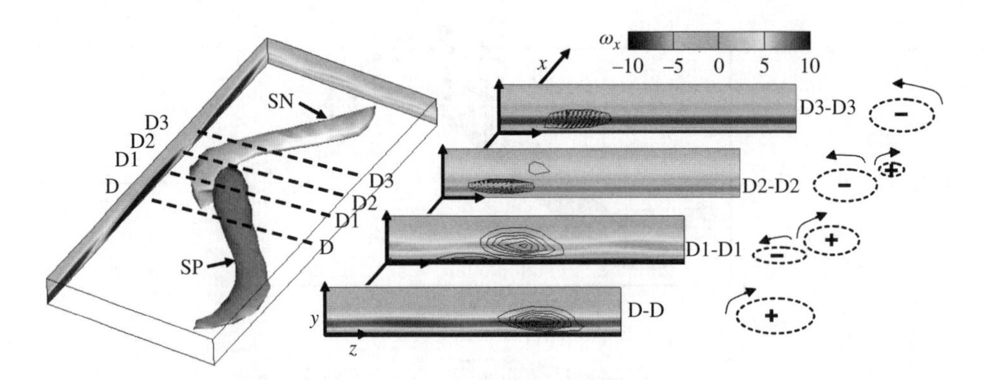

Figure 4.55 Averaged streamwise vortex structures and its cross sections

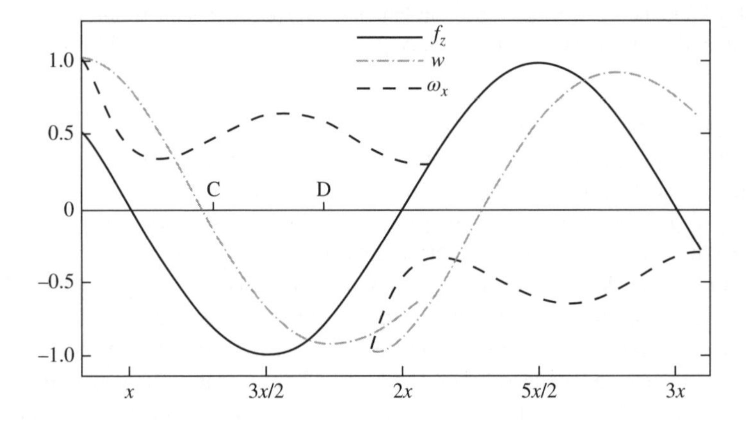

Figure 4.56 Distributions of spanwise velocity and streamwise voticity along the vortex axes projected on the x-axis, in the induced flow field

The subtraction of the scatter plot between the uncontrolled and controlled cases is shown in Figure 4.57, where the white and black circles correspond, respectively, to data points only occurring in the uncontrolled and controlled cases. The action of the oscillating Lorentz force is to move the data points from the white circle locations to the black circle locations. Large numbers of black circles concentrating in the vicinity of the origin of coordinates indicate that the ejection and sweep activities induced by the streamwise vortices have been weakened, which results in the drag reduction.

The instantaneous distributions of the Reynolds stress at the y–z plane and the corresponding skin-friction drag at the x–z plane for the controlled case are shown in Figure 4.58. The appearance probabilities of the four quadrants are 24.6%, 25.5%, 24.5%, and 25.4%, respectively. Obviously, Q4 events adjacent to the wall are significantly reduced due to the control case, compared with the uncontrolled case, as shown in Figure 4.47a.

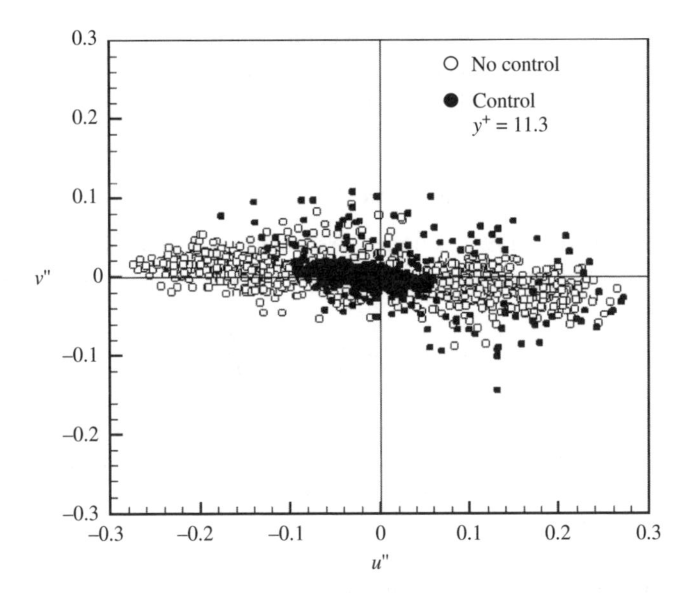

Figure 4.57 Subtraction of the scatter plot of u' and v' at $y^+ = 11.3$ between the controlled and uncontrolled flows

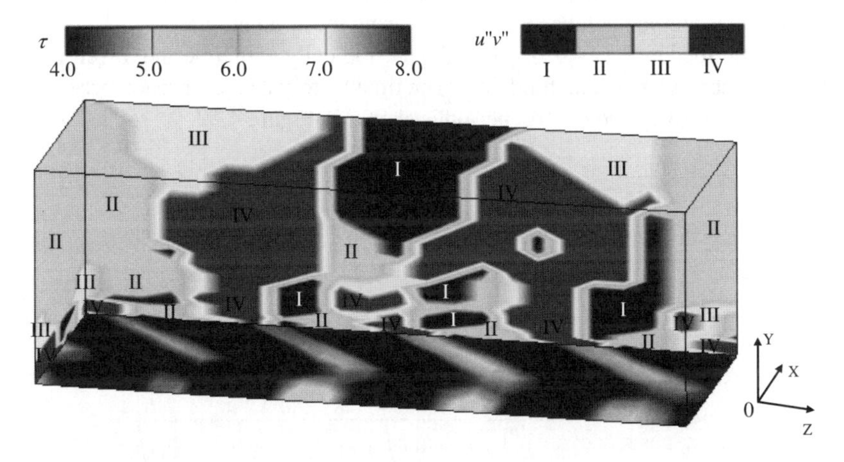

Figure 4.58 Instantaneous distributions of Reynolds stresses at the y–z plane and wall shear stress at the x–z plane

4.3.4 *Control with Wavy Lorentz Force*

One appealing approach for electrically conducting fluids is to use the Lorentz force to induce motion in the fluid which might then lead to turbulence suppression. It is possible to carry out many different types of Lorentz force control by simply alternating the polarity of the electrodes and rearranging the electrodes and magnets. The laterally oscillating Lorentz force traveling in the spanwise (or streamwise) direction, called the wavy Lorentz force, can be generated via the traveling EM actuators, and is the most effective one to manipulate the near-wall turbulence among the different types of Lorentz force.

4.3.4.1 Longitudinal Wavy Lorentz Force[30–34]

The force expressed by Eq. (449) is referred to as the longitudinal wavy Lorentz force that transmits energy resembling a traveling wave along the spanwise direction in electrically conducting fluids:

$$f_z = St\, e^{-\frac{y}{\Delta}} \sin\left(\frac{2\pi}{\lambda_z} z - \frac{2\pi}{T} t\right) \tag{4.49}$$

where λ_z is the wavelength along the spanwise direction; k_z, defined as $k_z = L_z/\lambda_z$, is the spanwise wave number; here L_z is the spanwise length of the computational domain. When $\lambda_z \to \infty$, that is, $k_z = 0$, Eq. (4.49) corresponds to the temporally oscillating spanwise Lorentz force expressed by Eq. (4.47).

According to the discussion in Section 4.3.3, the most effective control can be achieved as $St = 1.0, T = 8.0$, when using the temporally oscillating Lorentz force to control the near-wall turbulence. Hence, for the control with the longitudinal wavy Lorentz force, the oscillating parameters, St and T, are set by $St = 1.0, T = 8.0$, and the effects of the spanwise wave number k_z on the near-wall turbulence are then discussed numerically here. The flow rate is kept constant and the Reynolds number $\mathrm{Re}_\tau = 180$ during the simulations.

The force described by Eq. (4.49), oscillating periodically with time and travelling along the z direction, is independent of the fluid flow. The time history of the Lorentz force at $z = \lambda_z/2$, (i.e., $\pi/3$), on the wall in two control period is shown in Figure 4.59.

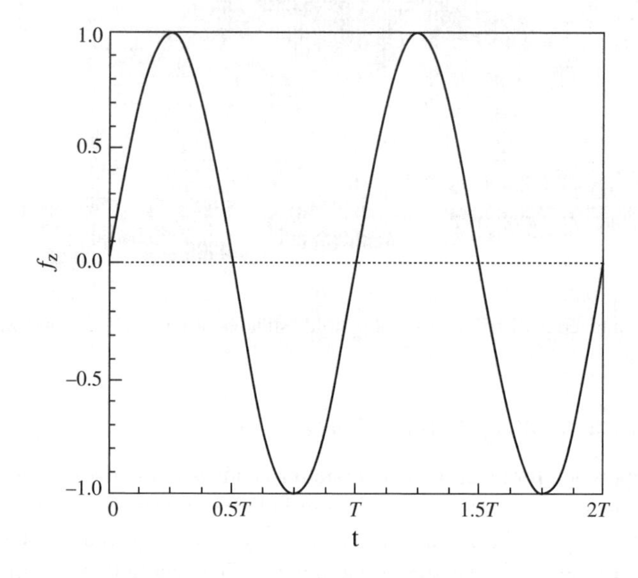

Figure 4.59 Time history of the imposed spanwise Lorentz force on the wall at $z = \lambda_z/2$.[32] *Source*: Huang L P 2010. Reproduced with permission of Springer

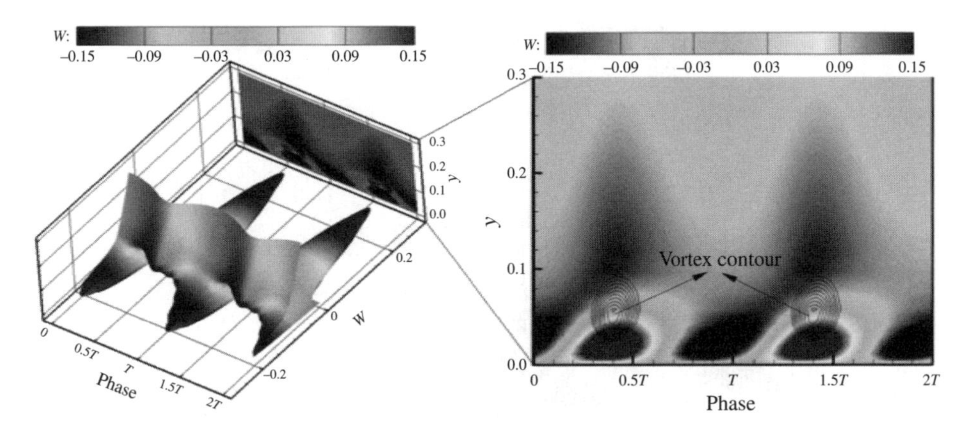

Figure 4.60 Time evolution of spanwise velocity and corresponding vortex structures in the Stokes layer induced by the longitudinal wavy Lorentz force at $z = \lambda_z/2$.[32] *Source*: Huang L P 2010. Reproduced with permission of Springer

As this force is imposed on a laminar flow with Re = 4000, a generalized Stokes layer is then created. When the induced flow field has been developed fully, the time evolution of spanwise velocity and corresponding longitudinal vortex structures induced in two control periods at $z = \lambda_z/2$ is shown in Figure 4.60. For $t = 0$–$0.5T$, flow particles at $z = \lambda_z/2$ are accelerated along the positive spanwise direction by the positive Lorentz force. Meanwhile, the fluid on the wall always keeps at rest. Just after the flow field is acted by the negative Lorentz force for a half of the oscillating period, the spanwise velocity is negative at $t = 0$ and the Lorentz force decays exponentially away from the wall, so that the sign of the spanwise velocity of the fluid along the vertical direction changes from positive to negative over time, resulting in the generation of a negatively signed vortex structure, as shown in Figure 4.60. When $t = 0.5T$, flow particles at $z = \lambda_z/2$ are accelerated along the negative spanwise direction by the negative Lorentz force, so that the opposite-signed spanwise velocity is reduced gradually, and finally resulting in the disappearance of the generated vortex structure at $t \approx 0.75T$. With the spanwise travelling Lorentz force, this generated vortex structure evolves in a periodic fashion, as shown in Figure 4.60. It is generated again when $t > T$, and then disappears at $t \approx 1.75T$.

The instantaneous longitudinal vortex structures and streaks induced are shown in Figure 4.61, where the gray areas represent high-speed streaks, the black areas represent low-speed streaks, and the iso-surfaces of longitudinal vortex structures are hatched by the solid circles. The negatively signed longitudinal vortices, stretching through the entire computational domain, travel from the left to the right with a constant velocity λ_z/T, and draw the low-speed fluid from its leading edge (on the right) to wrap it around, forming a wide region of low-speed fluid. The spanwise spacing of the longitudinal vortices as well as the streaks is equal to the wavelength of the Lorentz force.

The scatter plot of u' and v' at $y^+ = 5.4$ in the Stokes layer induced by a longitudinal wavy Lorentz force is shown in Figure 4.62. All data of the discrete Reynolds shear stress ($u'v'$) constitute a closed line, which indicates that the ejection and sweep activities do exist due to the induced longitudinal vortex. However, the y-weighted integral of $\overline{u'v'}$ is zero based on calculations, namely, the ejection and sweep activities have no contribution on the drag in the induced laminar flow.

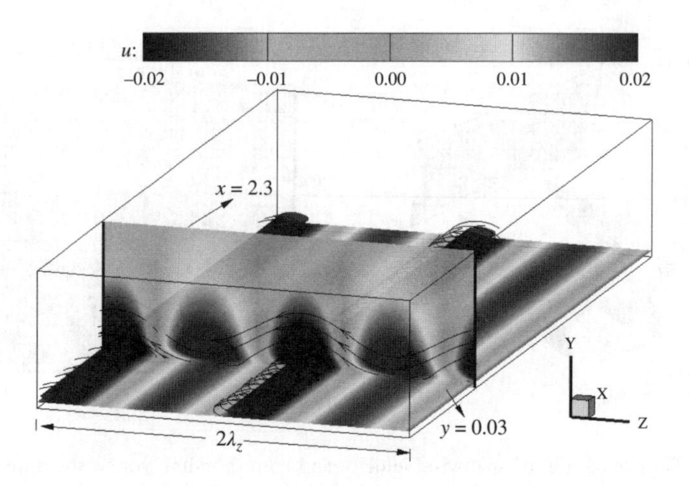

Figure 4.61 Instantaneous streaks and longitudinal vortex structures with streamlines in the Stokes layer induced by the longitudinal wavy Lorentz force.[32] *Source*: Huang L P 2010. Reproduced with permission of Springer

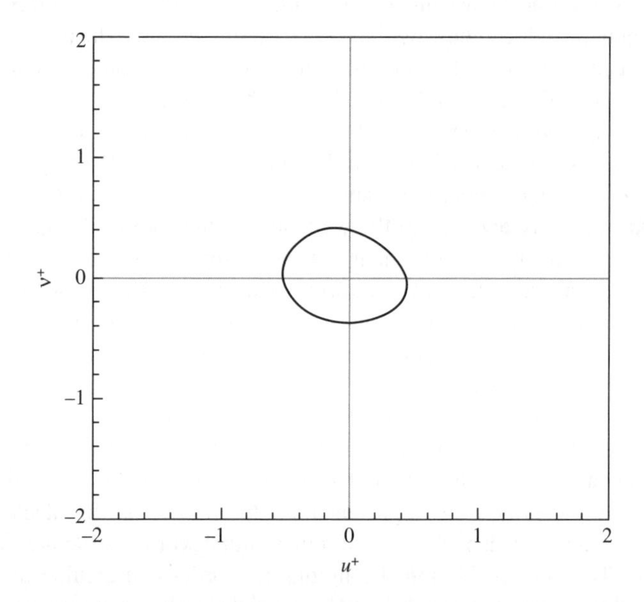

Figure 4.62 Distribution of u' and v' in the induced laminar Stokes layer.[32] *Source*: Huang L P 2010. Reproduced with permission of Springer

The induced flow utilized to modulate the turbulent flow near the wall is also modified by the intrinsic near-wall turbulent flow in the control process due to the two-way coupling. Figure 4.63 shows the snapshot distributions of the spanwise velocity, streak, and vortex structure near the wall for the controlled flows, where the spanwise velocity is closely related to the induced flow and the streaky and vortical structures are closely related to the intrinsic flow.

The first column in Figure 4.63 shows the snapshots of the distribution of spanwise velocities in the x–z plane close to the wall ($y^+ = 9$), where the gray and black areas represent,

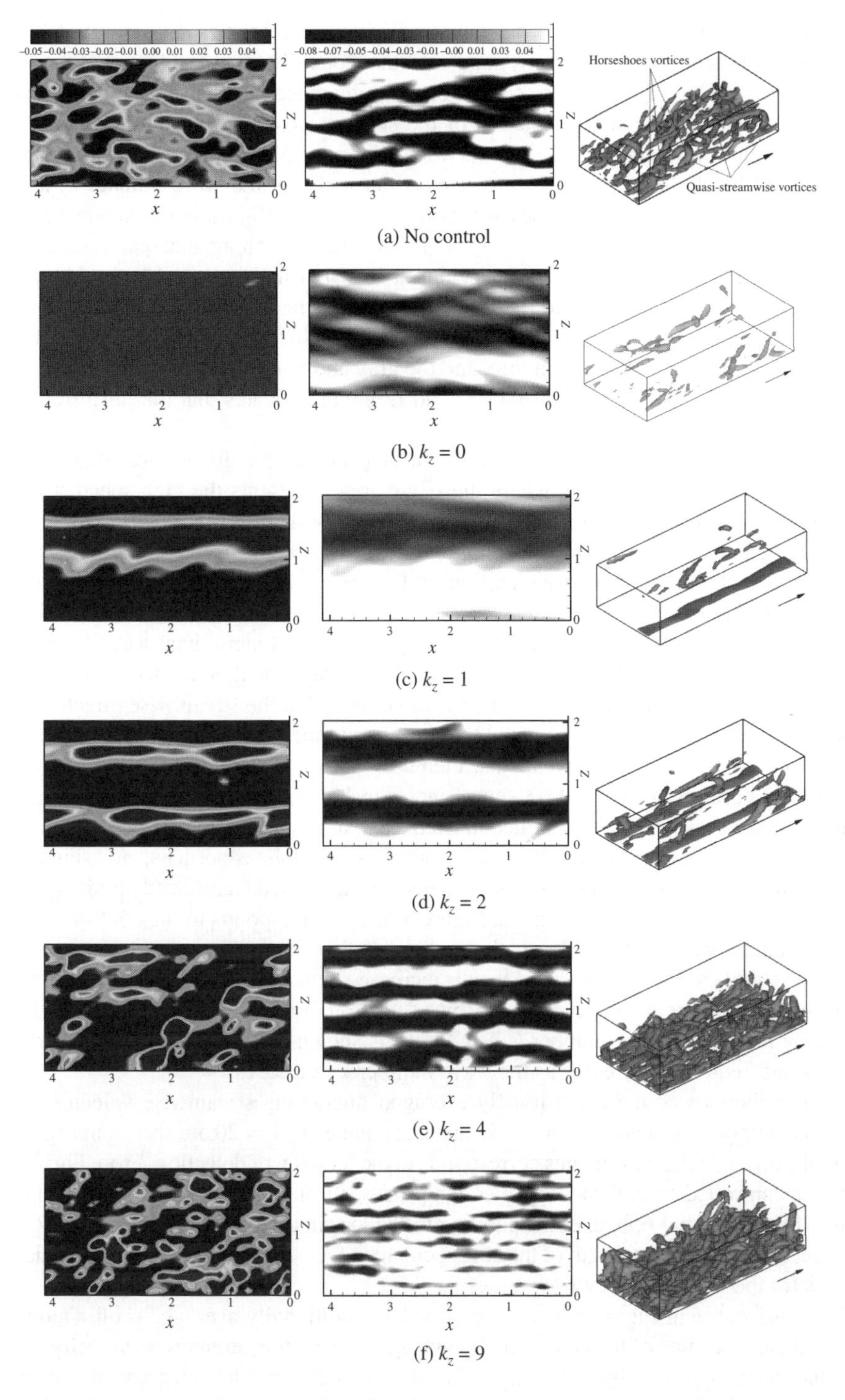

Figure 4.63 Snapshot distributions of the spanwise velocity, streak and vortex structure near the wall for the controlled flows by the longitudinal wavy Lorentz force.[33] *Source*: Huang L P 2012. Reproduced with permission of Springer

respectively, the positive and negative values of spanwise velocities. In Figure 4.63a, for an uncontrolled flow, a random and irregular instantaneous distribution of spanwise velocities is exhibited, and in Figure 4.63b, the color is homogeneous since the signs of spanwise velocities are the same under the action of the spanwise oscillating Lorentz force ($k_z = 0$). For a controlled flow via the longitudinal wavy Lorentz force (i.e., $k_z > 0$), the aspects of the gray and black areas in figures are affected strongly by the wave number k_z as shown in Figure 4.63c–f. For low-frequency waves ($k_z < 4$), the configuration of spanwise velocity distributions consists of alternating gray and black regions along the spanwise directions. The shape of the colored area is more regular at $k_z = 1$, which means that the induced flow dominates the near-wall flow, and then the colored areas become more and more obscure as k_z increases, as shown in Figure 4.63c and d. For the high-frequency waves, the distribution of spanwise velocities becomes random, and the gray and black areas mingle and amalgamate with each other, as shown in Figure 4.63e and f, which indicates that the intrinsic turbulent flow dominates the near-wall flow.

The second column in Figure 4.63 shows the snapshots of the distribution of streaky structures in the x–z plane at $y^+ = 9$, where the white area represents the high-speed streaks and the black area represents the low-speed streaks. As illustrated in Figure 4.63a for an uncontrolled flow, the streaky structures do not always flow straight in the streamwise direction, but often meander in the spanwise direction. In Figure 4.63b, the streaks become obscure and broaden and the intensity is significantly weakened under the control of the spanwise oscillating Lorentz force ($k_z = 0$). For the flow controlled by the longitudinal wavy Lorentz force (i.e., $k_z > 0$), the regularity and intensity of streaks are dependent on the wave number k_z. For low-frequency waves, the streaky structures are elongated in the streamwise direction and the number of the streaks seems to be equal to the wave number as shown in Figure 4.63c–e. The intensity of streaks at $k_z = 1$ is weak, and then it increases with the increase in k_z.

The third column in Figure 4.63 shows the snapshots of vortex structures near the wall. Figure 4.63a, corresponding to the uncontrolled case, shows the characteristic structures of the longitudinal vortices (e.g., quasi-streamwise vortices, horseshoes vortices, etc.). Figure 4.63b indicates that the near-wall longitudinal vortices are suppressed significantly under the control by spanwise oscillating Lorentz forces. Figure 4.63c–f corresponds to the controlled cases of the longitudinal wavy Lorentz forces. It can be seen that the number of longitudinal vortices decreases significantly at $k_z = 1$, and then it increases with the increase in k_z.

The bursting events are detected usually by a detection function. The variation of the bursting frequency P with the wave number k_z at $y^+ = 17$ is shown in Figure 4.64. It is observed that the bursting frequency increases as the wave number k_z increases.

The distributions of the conditionally averaged fluctuating streamwise velocity $\langle u'(\xi, z)\rangle$ and its corresponding shaded contours in the (ξ, z) plane at $y^+ = 20$ are shown in Figure 4.65, where the origin of the coordinates corresponds to the location of detection. From Figure 4.65a, for the uncontrolled case, the sweep–ejection process through the detection point can be seen clearly. From Figure 4.65b–e, corresponding to the longitudinal wavy Lorentz force controls, it has been seen that the strength of the abrupt change of $\langle u'(\xi, z)\rangle$ near the location of detection reduces for the low-frequency wavy cases.

The white curve in Figure 4.65a represents the conditionally averaged profile along the ξ axis, and the rms value of the conditionally averaged fluctuating streamwise velocity $\langle u'(\xi, z)\rangle$ over the curve, denoted by $\langle u'(\xi)\rangle_{\text{r.m.s.}}$, is taken to describe the intensity of the bursting events. Then the variation of the intensity with the wave length number k_z at $y^+ = 17$ is shown in Figure 4.66. Compared with the uncontrolled case represented by the dashed line in

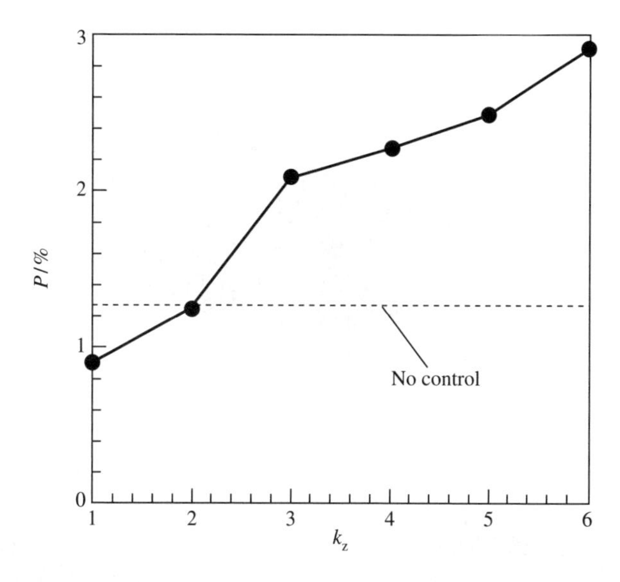

Figure 4.64 Variations of averaged bursting frequency with k_z at $y^+ = 17$.[33] *Source*: Huang L P 2012. Reproduced with permission of Springer

Figure 4.66, the bursting intensity is reduced greatly by the control of the longitudinal wavy Lorentz force.

The variation of the drag reduction D_r, defined by Eq. (3.13), with the wave number k_z is shown in Figure 4.67. It is indicated that the drag reduction can be achieved by the longitudinal wavy Lorentz force control with $k_z < 4$, and the largest amount of the drag reduction occurred in $k_z = 1$ is 33.6%.

As the longitudinal wavy Lorentz force with $St = 1.0$, $T = 8.0$, and $k_z = 1$ is utilized to control the near-wall turbulent flow, the intrinsic near-wall turbulent flow will be modified by an additional induced flow. Evolutions of the streaks and the vortex structures in the near-wall region in the modulating process are illustrated in Figure 4.68, where t is the time from the start of forcing. The first column shows the streaky structures, where the black area represents the low-speed streak and the gray area represents the high-speed streak. The second column shows the vortex structures in $0 < y^+ < 100$. It can be seen that a set of induced streaks and longitudinal vortices, stretching through the entire computational domain and traveling from the left to the right, have been introduced into the turbulent boundary layer and become more and more evident, whereas the intrinsic coherent structures are reduced over time. The induced flow finally dominates the near-wall flow and consequently relaminarizes the turbulent flow.

Essentially, there exist two kinds of streamwise vorteices in the near-wall-modulated turbulent flow field, that is, a set of induced streamwise vortices and the intrinsic random streamwise vortices. When the set of induced vortices stretching through the entire flow field sweep over the near-wall flow field in the spanwise direction, they will collide and interact with intrinsic longitudinal vortices continually and finally absorb those random streamwise vortices, whatever rotation direction, leading to consequently relaminarizing the near-wall turbulent flow.

The subtraction of the scatter plot between the uncontrolled and controlled cases at $y^+ = 11.3$ is shown in Figure 4.69 where the white and black circles correspond, respectively,

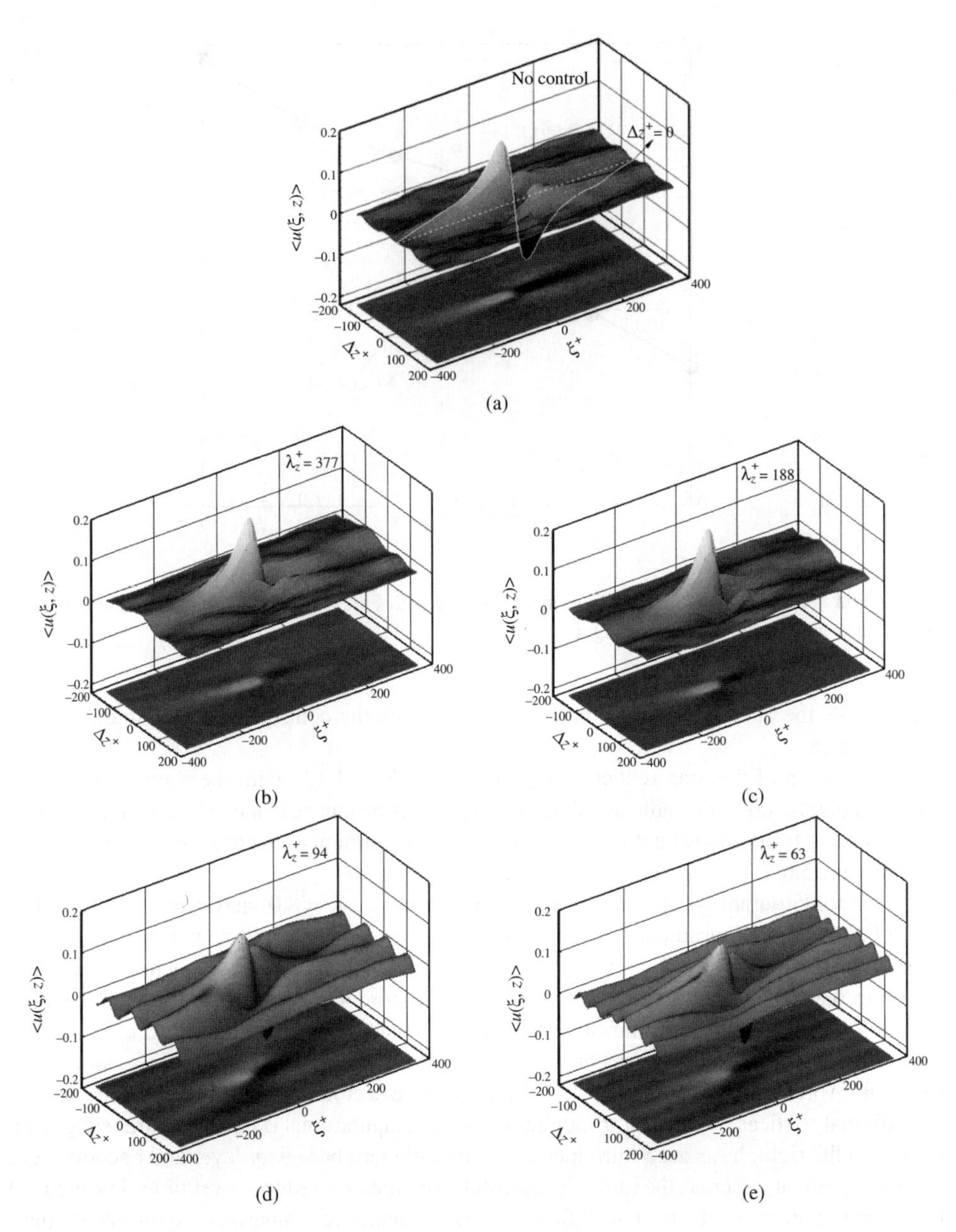

Figure 4.65 Conditional average of fluctuating streamwise velocities with a contour map in the (ξ, z) plane.[33] *Source*: Huang L P 2012. Reproduced with permission of Springer

to data points only occurring in the uncontrolled or controlled cases. The action of the longitudinal wavy Lorentz force is to move the data points from the white circle locations to the black circle locations. Large numbers of black circles distribute in a circular area, since increasing in v' and decreasing in u' by the control. Therefore, the ejection and sweep activities

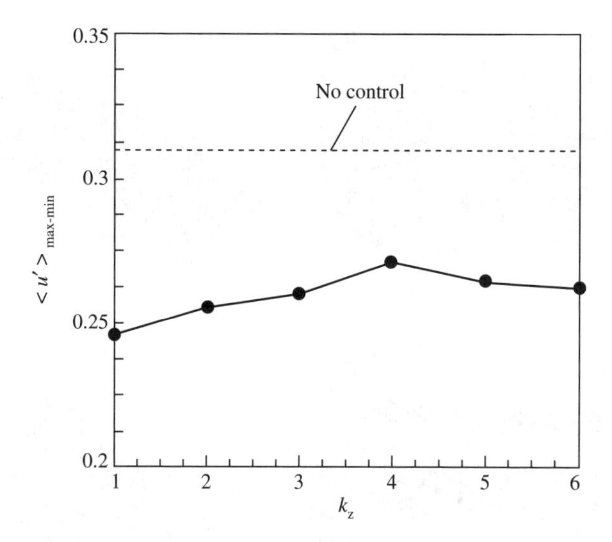

Figure 4.66 Variation of intensity of bursts with k_z at $y^+ = 17$.[33] *Source*: Huang L P 2012. Reproduced with permission of Springer

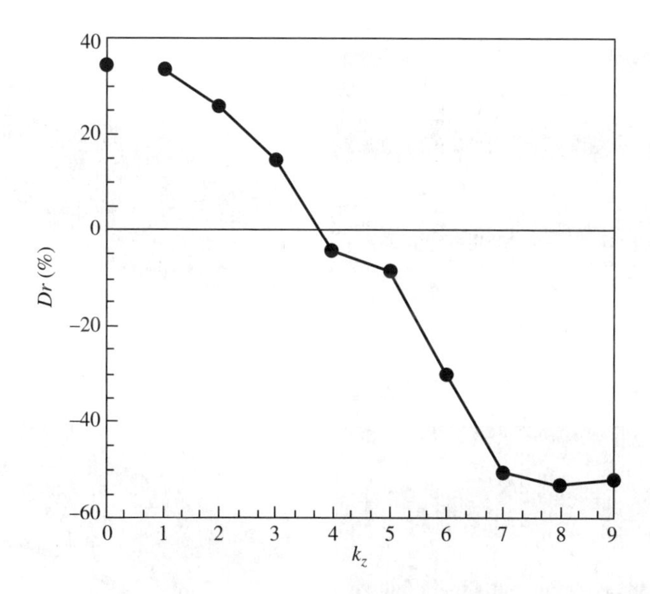

Figure 4.67 Variation of drag reduction with the wave number k_z.[33] *Source*: Huang L P 2012. Reproduced with permission of Springer

are strengthened, whereas the streamwise momentum transform in the wall-normal direction induced by these activities are weakened significantly, which results in the drag reduction.

The instantaneous distributions of the Reynolds stress at the y–z plane and the corresponding skin-friction drag at the x–z plane for the controlled case are shown in Figure 4.70. The appearance probabilities of the four quadrants are 24.0%, 23.7%, 22.9%, and 29.4%, respectively.

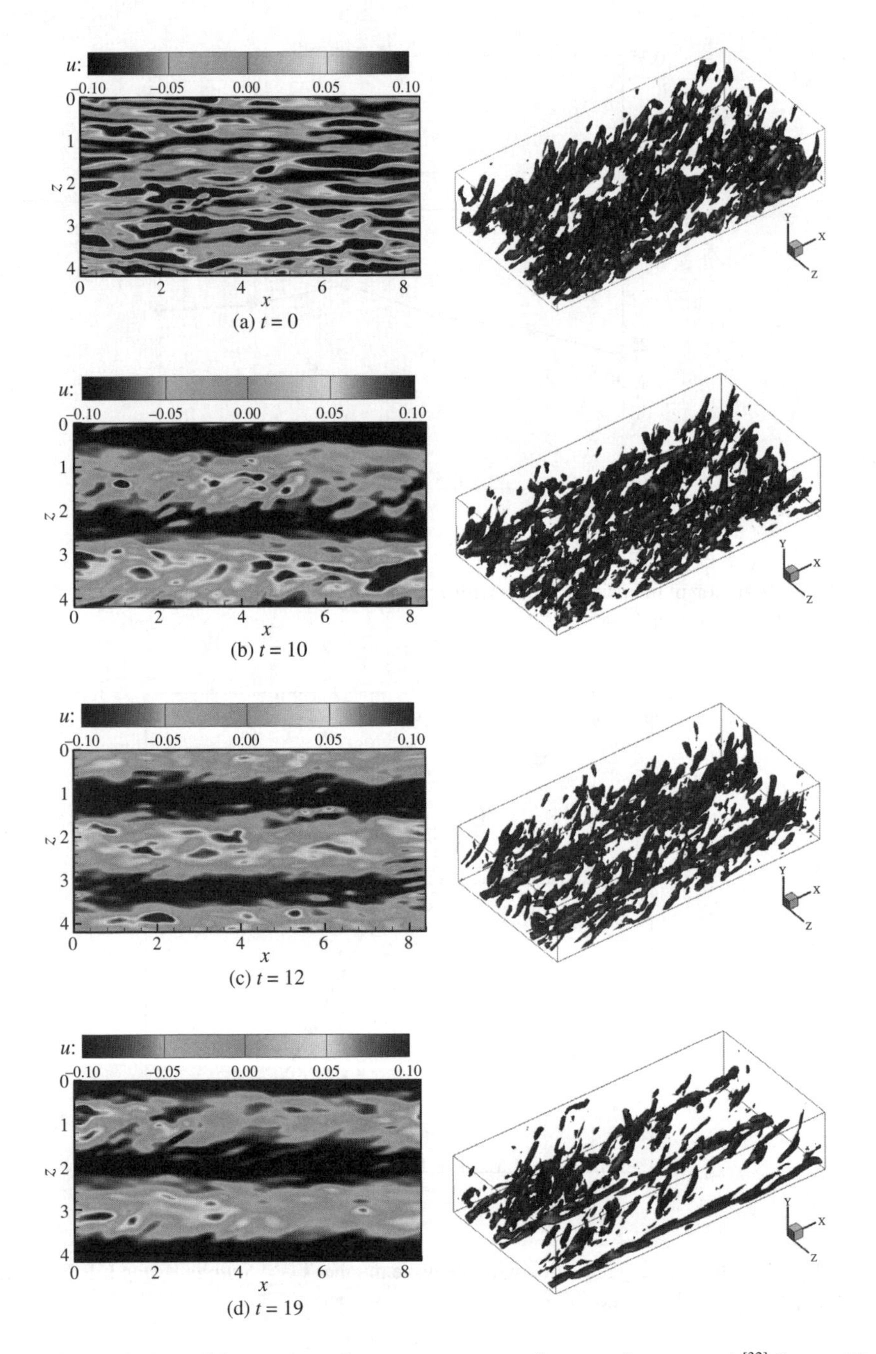

Figure 4.68 Evolutions of the streaks and vortex structures under wavy force control.[32] *Source*: Huang L P 2010. Reproduced with permission of Springer

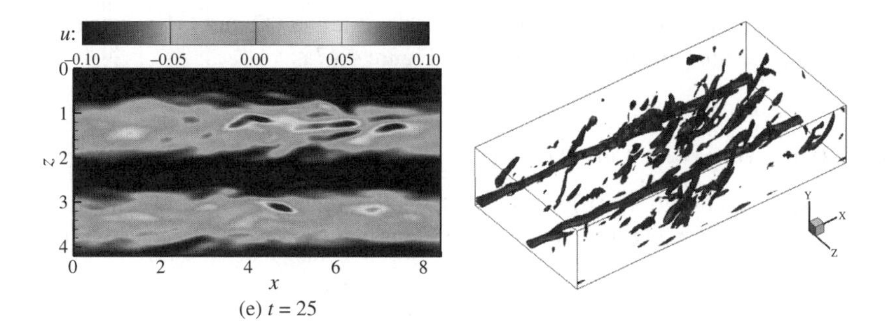

Figure 4.68 (*continued*)

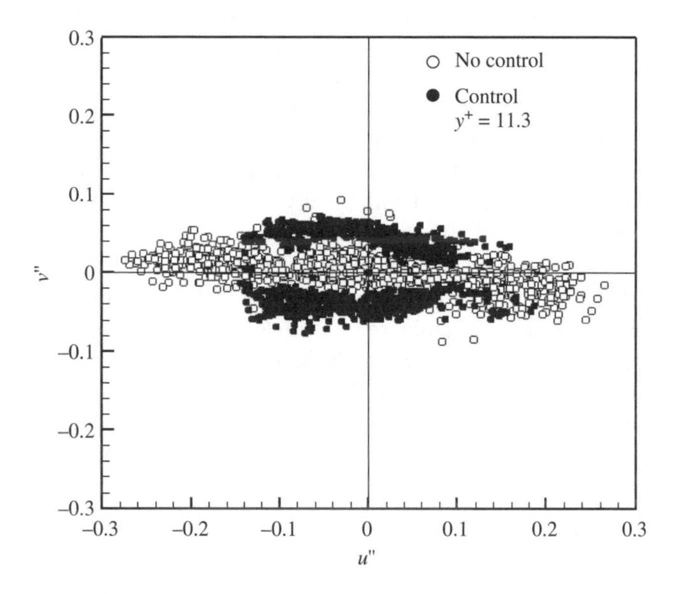

Figure 4.69 Subtraction of the scatter plot of u' and v' at $y^+ = 11.3$ between the controlled and uncontrolled flows

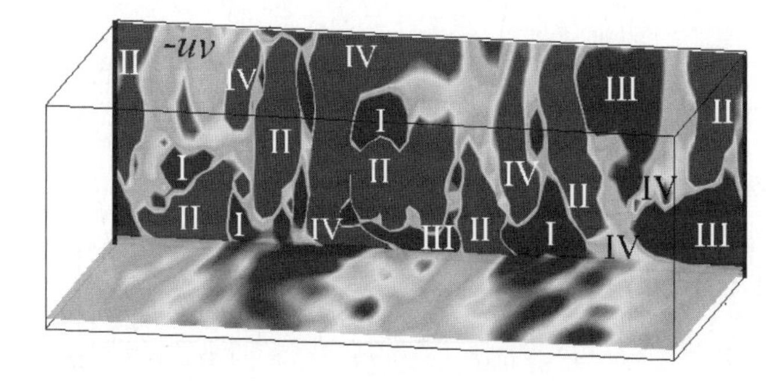

Figure 4.70 Instantaneous distributions of the Reynolds stress and wall drag.[32] *Source*: Huang L P 2010. Reproduced with permission from Cambridge University Press

The Q1 and Q3 events are significantly increased, while the Q2 and Q4 events are reduced, compared with the uncontrolled case shown in Figure 4.47a.

4.3.4.2 Transverse Wavy Lorentz Force

The force expressed by Eq. (4.50) is referred to as the transverse wavy Lorentz force, which transmits energy resembling a traveling wave along the streamwise direction in electrically conducting fluids:

$$f_z = St\, e^{-\frac{y}{\Delta}} \sin\left(k_x x - \frac{2\pi}{T} t\right) \tag{4.50}$$

where λ_x is the wavelength along the streamwise direction, k_x, defined as $k_x = L_x/\lambda_x$, is the streamwise wave number; here L_x is the streamwise length of the computational domain. When $\lambda_x \to \infty$, that is, $k_x = 0$, Eq. (4.50) corresponds to the temporally oscillating spanwise Lorentz force expressed by Eq. (4.47). On the other hand, when $T \to \infty$, Eq. (4.50) corresponds to the spatially oscillating spanwise Lorentz force expressed by Eq. (4.48).

For the control with the transverse wavy Lorentz force, the oscillating parameters, St and T, are also set by $St = 1.0$, $T = 8.0$; the effects of the streamwise wave number, k_x, on the near-wall turbulence are then discussed numerically here.

The force described by Eq. (4.50), oscillating periodically with time and traveling along the x direction, is independent of the fluid flow. The time history of the Lorentz force at $x = 2\pi/3$ ($L_x = 4\pi/3$) on the wall in two control period is shown in Figure 4.71.

When this force is imposed on a laminar flow with $\mathrm{Re}_\tau = 180$, a generalized Stokes layer is then created. The time evolution of profiles of the induced spanwise velocity at $x = 2\pi/3$, after the induced flow field has been developed fully, is shown in Figure 4.72. When $t = 0$–$0.5T$, flow particles at $x = 2\pi/3$ are accelerated along the positive spanwise direction by the positive Lorentz force. Meanwhile, the fluid on the wall always keeps motionless. When $t = 0.5T$, flow

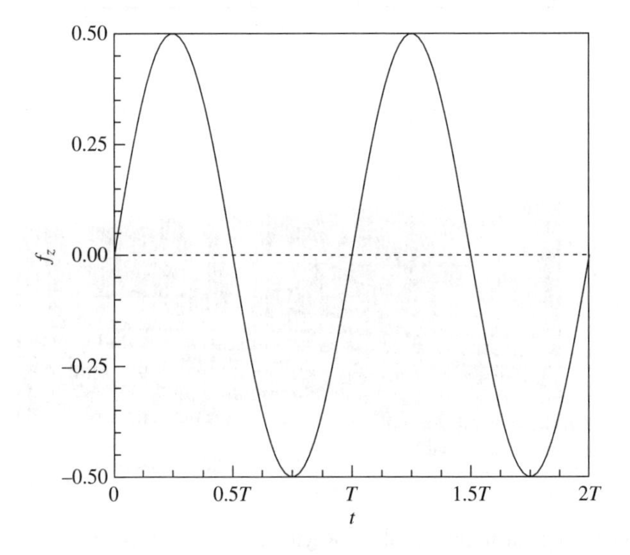

Figure 4.71 Time history of the imposed spanwise Lorentz force on the wall at $x = 2\pi/3$.[32] *Source*: Huang L P 2010. Reproduced with permission from Cambridge University Press

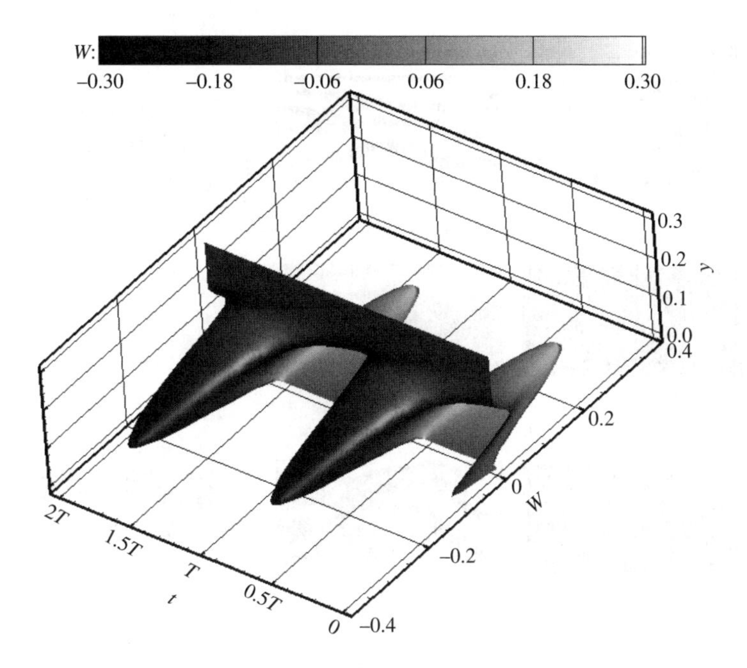

Figure 4.72 Time evolution of the induced spanwise velocity at $x = 2\pi/3$.[32] *Source*: Huang L P 2010. Reproduced with permission from Cambridge University Press

particles are accelerated along the negative spanwise direction by the negative Lorentz force, which leads to the opposite-signed spanwise velocity reduced gradually.

Snapshot distributions of the spanwise velocity, streak and vortex structure for the flows modulated by the induced flow are shown in Figure 4.73.

The first column in Figure 4.73 shows the snapshots of the distribution of the spanwise velocities at $y^+ = 9$, where the white area and black area represent, respectively, the positive value and negative value of spanwise velocities. Figure 4.73a and b corresponding to the no control and control with the spanwise oscillating Lorentz force, respectively, have been discussed previously. For a controlled flow via the transverse wavy Lorentz forces (i.e., $k_x > 0$), the aspects of the white and black areas in figures are affected strongly by the wave number k_x as shown in Figure 4.73c–g. The shape of the colored area becomes more and more regular as k_x increases for low-frequency waves (i.e., $k_x = 1$–4 corresponding to $\lambda_x^+ \approx 188$–754), which indicates that the influence of the modulated turbulent flow on the induced flow becomes more and more weak. When $k_x = 4$ ($\lambda_x^+ = 188$), the configuration of spanwise velocity distributions consists of alternating white rectangle regions and black rectangle regions along the streamwise directions as shown in Figure 4.73e, which means that the flowing directions in the spanwise direction are homogeneous along $x =$ const, and the induced flow dominates the near-wall flow. However, as k_x increases further, the colored areas become obscure, as shown in Figure 4.73f. For the high-frequency waves, the distribution of spanwise velocities becomes random again, and the white and black areas mingle and amalgamate with each other, as shown in Figure 4.73g, which indicates that the intrinsic turbulent flow dominates the near-wall flow.

The second column in Figure 4.73 shows the snapshots of the distribution of streaky structures in the x–z plane at $y^+ = 9$, where the white areas represent the high-speed streak and the black areas represent the low-speed streak. For the flow controlled by the transverse

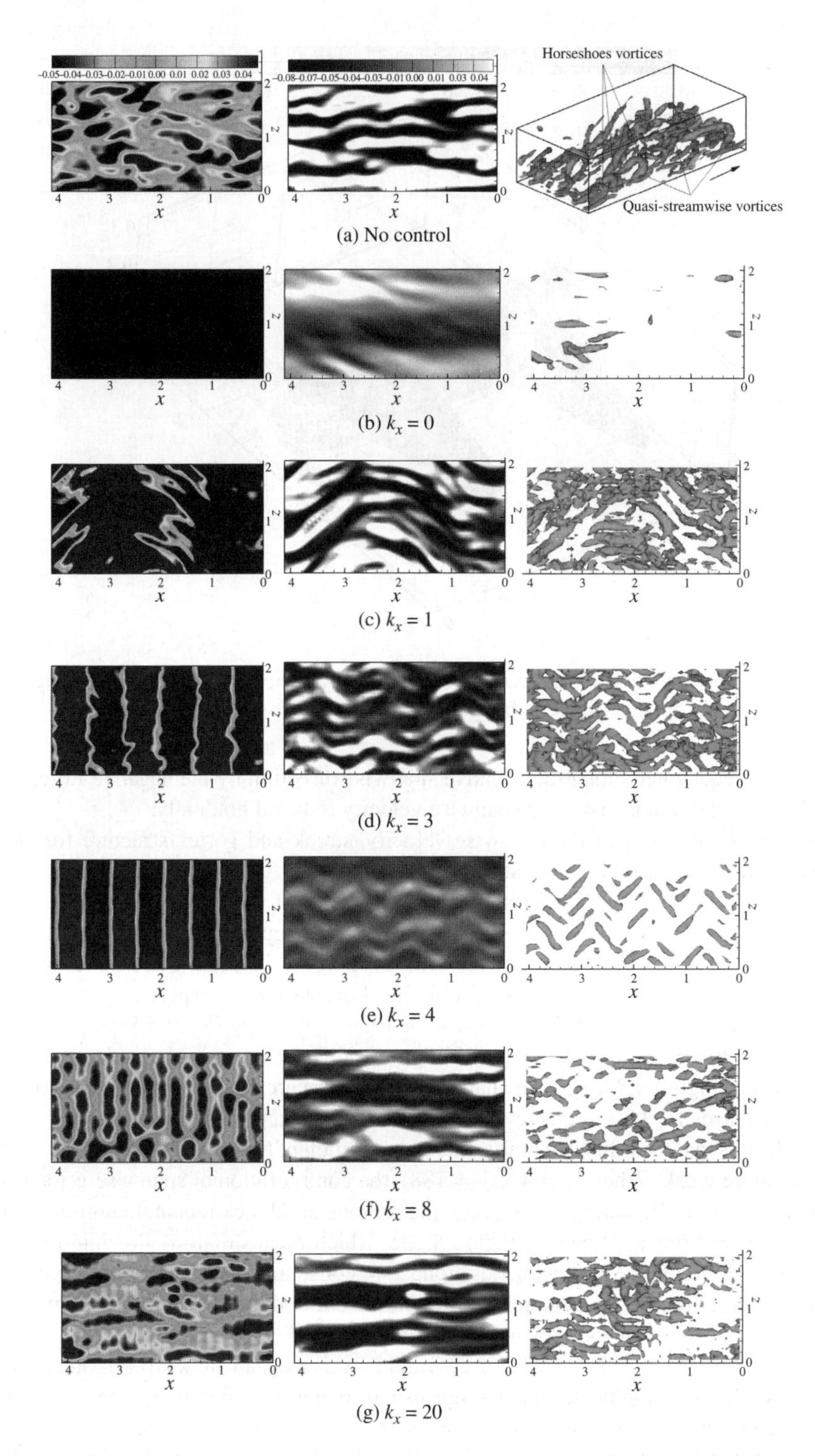

Figure 4.73 Snapshot distributions of the spanwise velocity, streak, and vortex structure near the wall for the controlled flows by the transverse wavy Lorentz force.[35] *Source*: Huang L P 2010. Reproduced with permission from Cambridge University Press

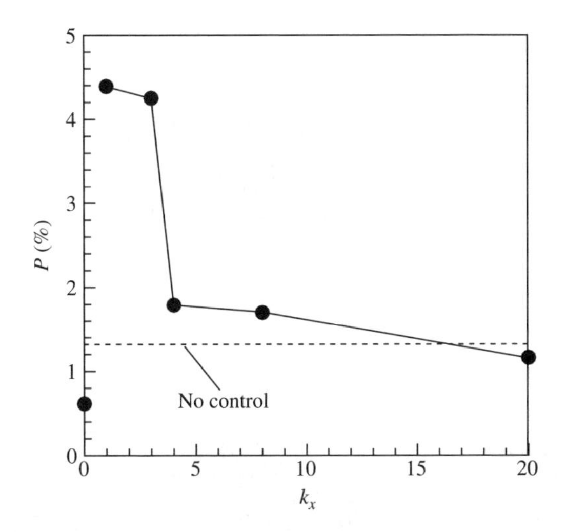

Figure 4.74 Variations of the averaged bursting frequency with k_x at $y^+ = 20$.[35] *Source*: Huang L P 2010. Reproduced with permission from Cambridge University Press

wavy Lorentz force (i.e., $k_x > 0$), the regularity and intensity of streaks are dependent on the wave number k_x. For the smaller wave number, the streaks are significantly sinusoidal with the increasing frequency and the decreasing amplitude as the wave number k_x increases, and its intensity decreases with the increase in k_x as shown in Figure 4.73c–e. Then for $k_x > 4$ ($\lambda_x^+ < 188$), the large-amplitude oscillation of streaks disappears gradually and the intensity of the streaks increases with the increase in k_x as shown in Figure 4.73f and g.

The third column in Figure 4.73 shows the snapshots of vortex structures near the wall. Figure 4.73c–g corresponds to the controlled cases of transverse wavy Lorentz forces. For the low-frequency waves, the longitudinal vortices meander significantly and arrange in order; the number of the vortices decreases as the wave number increases, as shown in Figure 4.73c–e. When $k_x > 4$ ($\lambda_x^+ < 188$), the longitudinal vortices are produced again, as shown in Figure 4.73f and g.

The variation of the bursting frequency P with the wave number k_x at $y^+ = 20$ is shown in Figure 4.74. Compared with the uncontrolled case represented by the dashed line in Figure 4.74, the bursting frequency is significantly reduced by the control of the spanwise oscillating Lorentz force ($k_x = 0$). As the transverse wavy Lorentz force is imposed to modify the near-wall flow, the bursting frequency decreases as k_x increases, and is maintained at a level higher than that for the uncontrolled case until $k_x > 17$ ($\lambda_x^+ < 44$).

The meandering of the near-wall low-speed streak is responsible for the instability of longitudinal vortices and the increase in the number of turbulent bursts; thus, the shapes of the streaks have a close dynamical relationship with the frequency of the turbulent bursts. The transverse wavy Lorentz force leads to the streak sinusoidal meandering and the instability of longitudinal vortices, so that the bursting frequency increases significantly when $k_x < 4$ ($\lambda_x^+ > 188$).

The distributions of the conditionally averaged fluctuating streamwise velocity $\langle u'(\xi, z)\rangle$ and its corresponding shaded contours in the (ξ, z) plane at $y^+ = 20$ are shown in Figure 4.75, where Figure 4.75a, for the uncontrolled case, has been discussed previously. From Figure 4.75b–g,

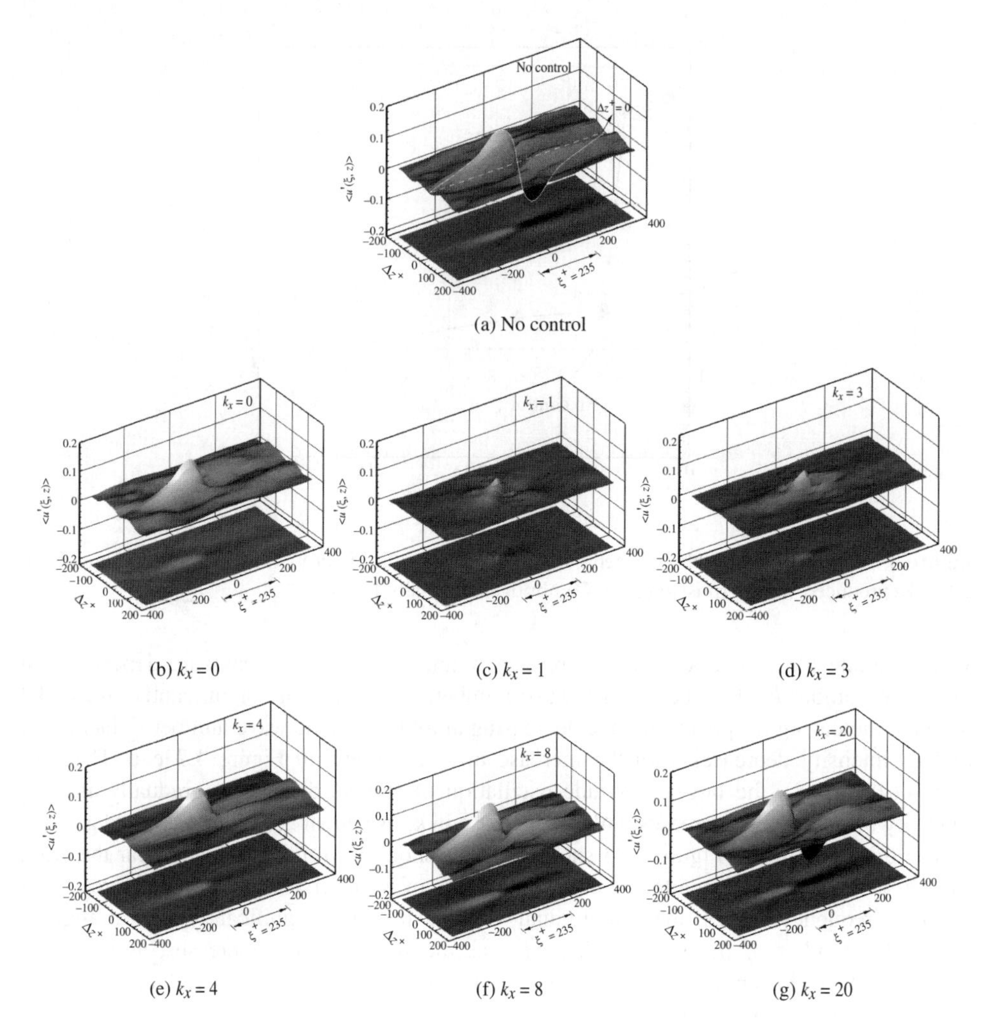

(a) No control

(b) $k_x = 0$ (c) $k_x = 1$ (d) $k_x = 3$

(e) $k_x = 4$ (f) $k_x = 8$ (g) $k_x = 20$

Figure 4.75 Conditional average of fluctuating streamwise velocities with a contour map in the (ξ, z) plane.[35] *Source*: Huang L P 2010. Reproduced with permission from Cambridge University Press

corresponding to the spanwise oscillating and transverse wavy Lorentz force controls, respectively, it is observed that the strength of the abrupt change of $\langle u'(\xi, z)\rangle$ near the location of detection reduces significantly both for the oscillating case and the low-frequency wavy case, but the strength increases with the increase in k_x for the wavy Lorentz force control.

The variation of the intensity with the wave number k_x is shown in Figure 4.76. Compared with the uncontrolled case represented by the dashed line in Figure 4.76, the bursting intensity is reduced by the control of the spanwise oscillating Lorentz force ($k_x = 0$). When the transverse wavy Lorentz force is imposed to modify the near-wall flow, the intensity of turbulent bursts is reduced greatly at the smaller k_x and then increases with the increase in k_x.

The variation of the drag reduction D_r, defined by Eq. (3.13), with the wave number k_x is shown in Figure 4.77. It is indicated that the largest amount of the drag reduction occurred in $k_x = 4$ ($\lambda_x^+ = 188$) is 42%. Compared to the no control case, under control by the transverse

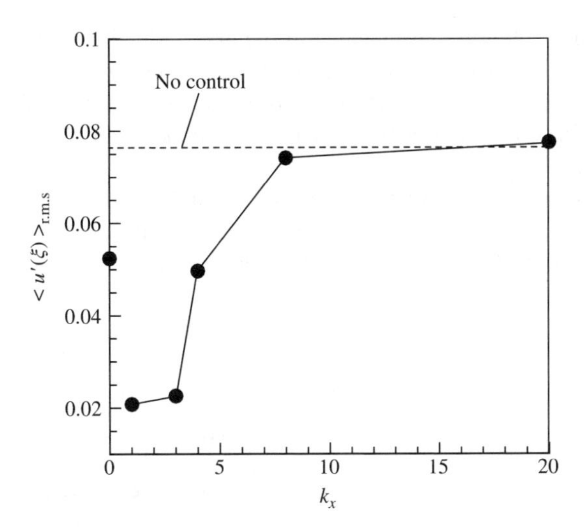

Figure 4.76 Variation of the intensity of bursts with the wave number k_x at $y^+ = 20$.[35] *Source*: Huang L P 2010. Reproduced with permission from Cambridge University Press

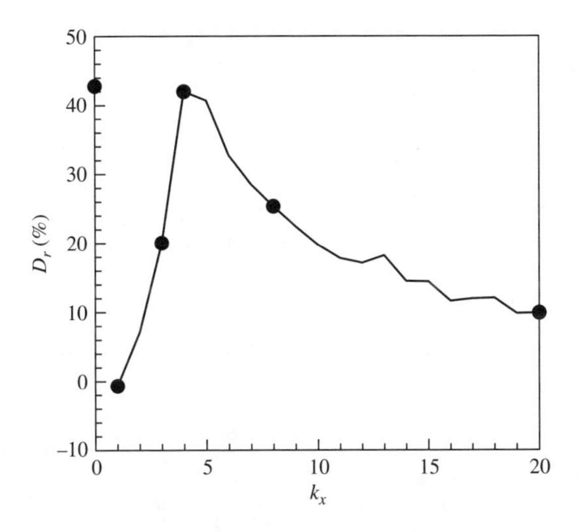

Figure 4.77 Variation of drag reduction with the wave number k_x.[35] *Source*: Huang L P 2010. Reproduced with permission from Cambridge University Press

wavy Lorentz force with $k_x = 4$ ($\lambda_x^+ = 188$), the burst frequency increases due to the streak meandering, which has a negative effect on the drag reduction; at the same time, the burst intensity decreases due to the modulation of the induced flow, which has a positive effect on the drag reduction. However, the decrease in the burst intensity is a dominant effect, which leads the drag reduction, as shown in Figure 4.77.

For control by the spanwise oscillating Lorentz force ($k_x = 0$), the decrease in both frequency and intensity of the burst events, as shown in Figures 4.74 and 4.76, leads to the drag reduction, which means that the mechanism of the drag reduction by the oscillating Lorentz force is different from that for the wavy Lorentz force control. In spite of the increase in the burst frequency for $k_x = 4$ ($\lambda_x^+ = 188$), the average drag reduction rate is close to that for $k_x = 0$, since the effect on decreasing the burst intensity for $k_x = 4$ ($\lambda_x^+ = 188$) is much stronger than that for $k_x = 0$.

The effects of the oscillation parameters, St and T, on the maximum drag reduction $D_{r\max}$, kept fixed in the parametric study of the wavy parameter k_x, are shown in Figures 4.78 and 4.79, respectively. Figure 4.78 shows that the optimal wave number $k_{x\ \mathrm{op}}$, corresponding to $D_{r\max}$,

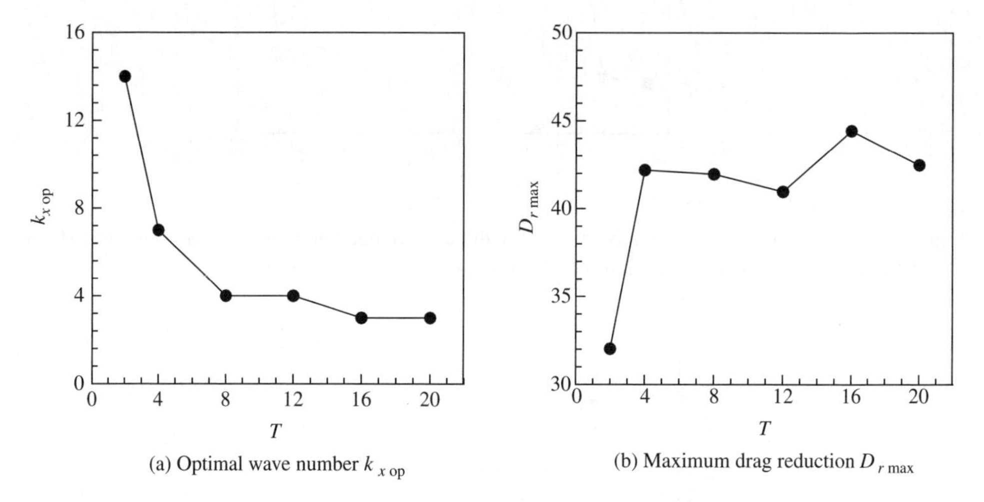

Figure 4.78 Effect of the time period T on (a) the optimal wave number $k_{x\ \mathrm{op}}$ and (b) the maximum drag reduction $D_{r\max}$.[35] *Source*: Huang L P 2010. Reproduced with permission from Cambridge University Press

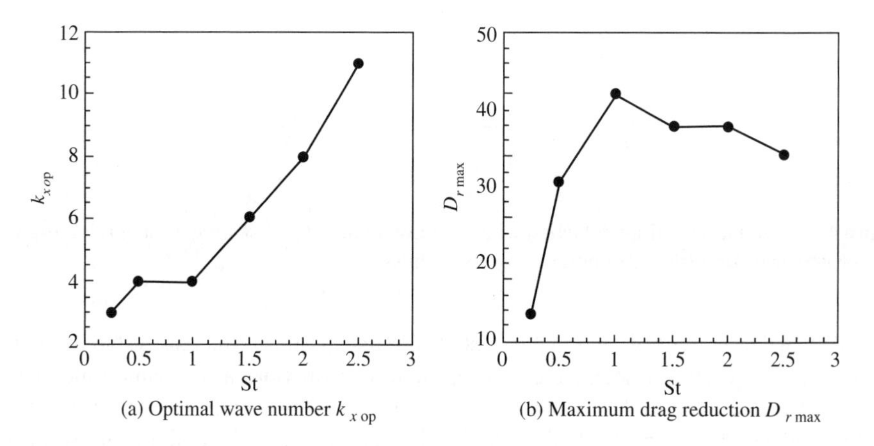

Figure 4.79 Effect of amplitude St on (a) the optimal wave number $k_{x\ \mathrm{op}}$ and (b) the maximum drag reduction $D_{r\max}$.[35] *Source*: Huang L P 2010. Reproduced with permission from Cambridge University Press

decreases monotonically with the increase in T at $\Delta = 0.02$ and $St = 1.0$, and meanwhile the maximum drag reduction grows and then is maintained at a higher level.

Figure 4.79 shows that the optimal wave number $k_{x\ \mathrm{op}}$ increases monotonically with the increase in St at $\Delta = 0.02$ and $T = 8.0$. Meanwhile, the maximum drag reduction $D_{r\ \max}$ increases markedly at smaller St and then decreases gradually with the increase in St.

When $St = 1.0$, $T = 8.0$, $k_x = 4$, in which a better drag reduction of more than 40% can be achieved, the snapshot distributions of the streak and vortex structure near the wall are shown in Figure 4.80a and b, respectively, where the black area represents the low-speed streak, the gray area represents the high-speed streak in Figure 4.80a, while the dark are and light area represent, respectively, positive value and negative value of vorticities in Figure 4.80b. These near-wall flow patterns are similar to that shown in Figure 4.53, for the case controlled by the spatially oscillating Lorentz force. However, these patterns travel downward in the streamwise direction with the traveling force.

The vortex structures and shaded contours of the spanwise vorticity in the y–z planes at different instants are shown in Figure 4.81. It should be noted that the positive vortex is tilted with the negative tilting angle, while the negative vortex with the positive tilting angle at any instant, similar to that shown in Figure 4.53, for the case controlled by the spatially oscillating Lorentz force. These tilting vortices should induce the negative spanwise vorticity as shown in Figure 4.81, thereby leading to drag reduction.

The subtraction of the scatter plot between the uncontrolled and controlled (with $St = 1.0$, $T = 8.0$, $k_x = 4$) cases is shown in Figure 4.82. The white circle and black circle correspond, respectively, to data points only occurring in the uncontrolled case and controlled case. The action of the Lorentz force is to move the data points from the white circle locations to the black

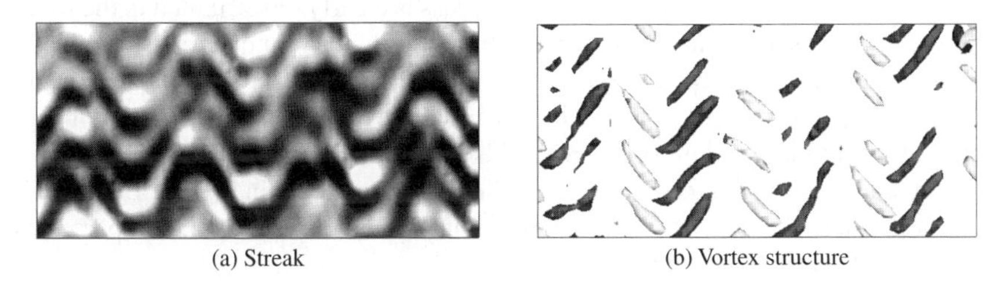

Figure 4.80 Distributions of the streak and vortex structure near the wall for the controlled flows by the transverse wavy Lorentz force

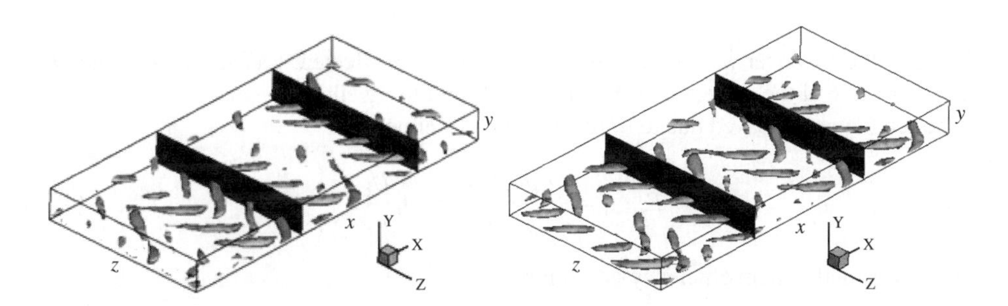

Figure 4.81 Vortex structures and shaded contours of spanwise vorticity on cross-flow planes in the wavy Lorentz force control

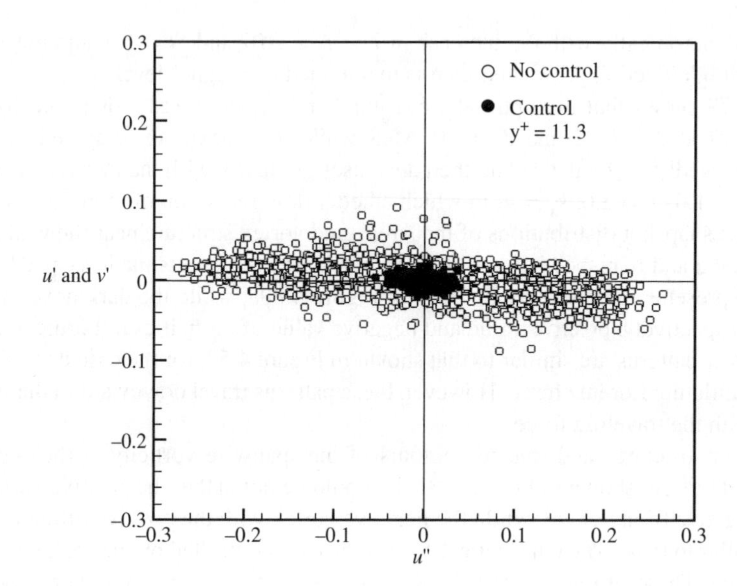

Figure 4.82 Subtraction of scatter plot of u' and v' at $y^+ = 11.3$ between controlled and uncontrolled flows

circle locations. Large numbers of black circles concentrating in the vicinity of the origin of coordinates indicate that the ejection and sweep activities induced by the streamwise vortices have been weakened, which results in the drag reduction.

For the control case with $k_x = 4$, the Reynolds stress is clearly redistributed in the four quadrants compared with the no control case. The probabilities of the four quadrants become 23.1%, 27.0%, 21.2%, 28.6%, respectively. Hence, the Q1 and Q3 events are significantly increased, while the Q2 and Q4 events are reduced,which leads to the drag reduction in terms of Eq. (3.4).

The instantaneous distributions of the Reynolds stress at the y–z plane and the corresponding skin-friction drag at the x–z plane for the controlled case are shown in Figure. 4.83. Obviously, Q4 events adjacent to the wall are significantly reduced due to the control, compared with the uncontrolled case, as shown in Figure 4.47a.

The power required to operate the control system is

$$P_{\text{used}} = \frac{VI_{\text{r.m.s.}}}{\rho u_\tau^3 A}, \tag{4.51}$$

where A is the area over which the Lorentz force control is affected, V is the voltages, and I_{rms} is the rms value of the current distribution over the control wall.

The scaling power saved due to drag reduction is defined as

$$P_{\text{saved}} = \frac{D_r \rho u_\tau^2 U_c A}{\rho u_\tau^3 A} = D_r \frac{U_c}{u_\tau}, \tag{4.52}$$

Thus, the total system efficiency is given by

$$\eta = \frac{P_{\text{saved}}}{P_{\text{used}}} \tag{4.53}$$

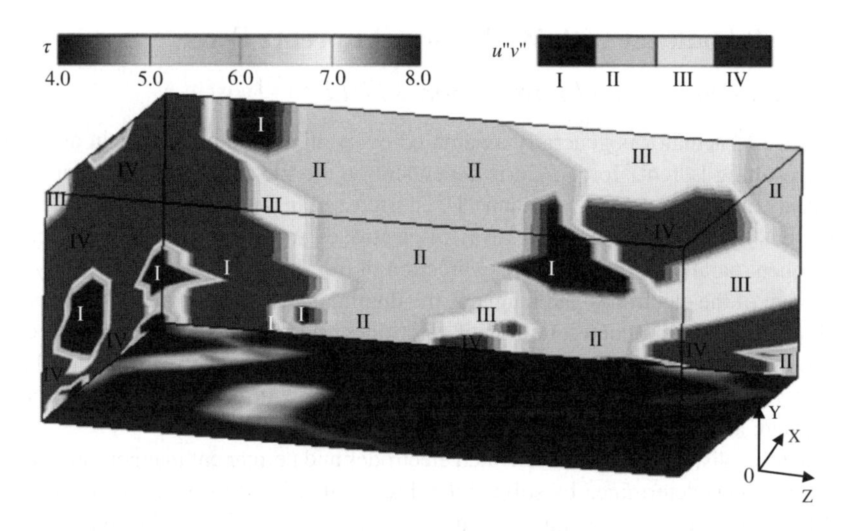

Figure 4.83 Instantaneous distributions of the Reynolds stress and wall drag

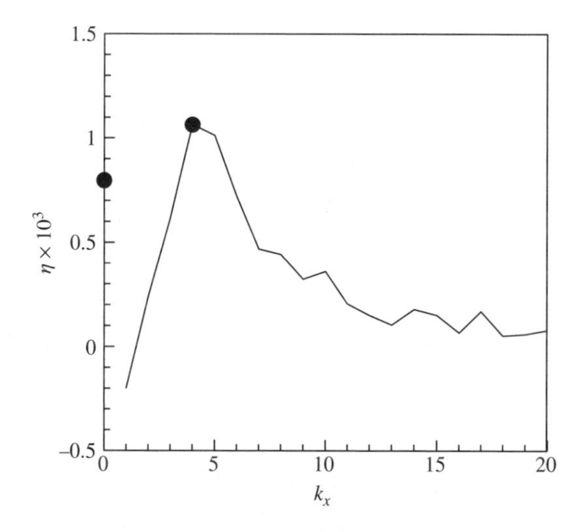

Figure 4.84 Effect of the wave number k_x on the total system efficiency.[35] *Source*: Huang L P 2010. Reproduced with permission of Springer

Figure 4.84 shows the total system efficiency values as a function of the wave number k_x for $St = 1.0$, $T = 8.0$, $\Delta = 0.02$. The efficiency η grows significantly at small k_x and then decrease with the increase in k_x. Finally, the efficiency does not change so markedly when k_x is high enough. The maximum value of η occurs at $k_x = 4$ ($\lambda_x^+ = 188$). Despite the poor efficiency, the maximum efficiency for the wavy Lorentz force control is higher than that for the spanwise oscillating Lorentz force control represented by an isolated black dot on the vertical axis in Figure 4.84.

4.4 Wall Turbulence with Wall-Normal Lorentz Force

4.4.1 Three-Dimensional Lorentz Force Field[36, 37]

A large area EM actuator is generally assembled by small EM actuator tiles in an array pattern. The resulting Lorentz force field is dependent on the tile configurations, that is, spatial arrangements of the electrodes and magnets. As discussed in the sections 4.2 and 4.3, the EM actuator tile is made by an alternating array of electrodes and magnets, which produced only one component parallel to the lengthwise direction of the electrodes and magnets. By varying the placement of the electrodes and magnets, the dominant near-wall force distribution can be changed, that is, different configurations of electrodes and magnets would produce different force fields. For a tile with the configuration, as shown in Figure 4.85, where the electrodes and magnets are placed perpendicular to one another, a three-dimensional Lorentz force field can be created, as shown in Figure 4.86.

The true force distribution when activated electrodes and permanent magnets are placed in a tile configuration is determined by solving the three-dimensional Maxwell's equations (4.20) and (4.18) for the electric and magnetic fields, respectively. In particular, the commercial finite element software package ANSYS is chosen to solve these equations. The calculated force field created by the tile shown in Figure 4.83 is shown in Figure 4.87. Figure 4.87a shows the force distributions on the wall, where the wall-normal components are represented by

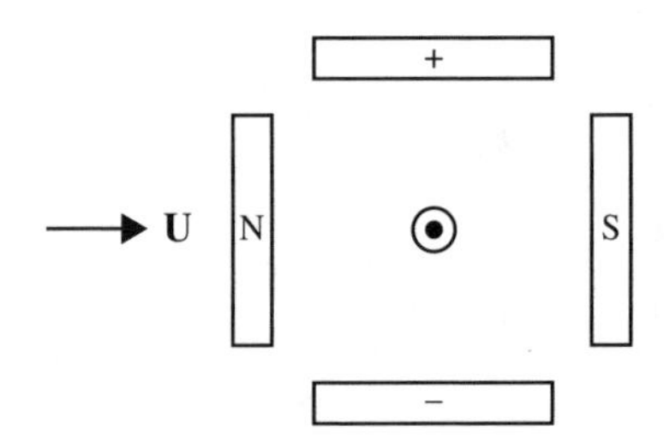

Figure 4.85 Wall normal EM actuator tile.[23] *Source*: Berger T W 2000. Reproduced with permission of Cambridge University Press

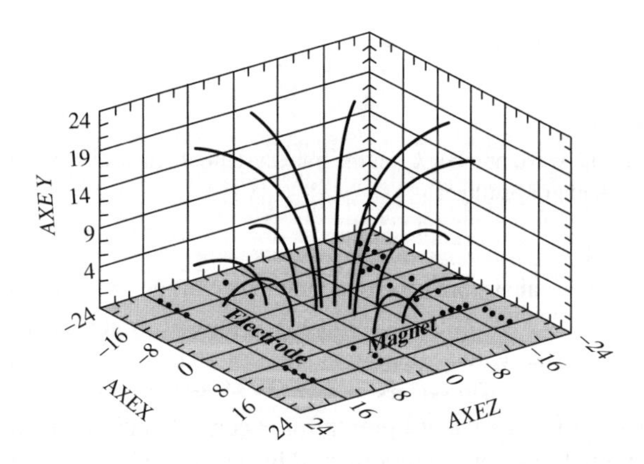

Figure 4.86 Distribution of forces lines three-dimensional Lorentz force field.[36] *Source*: Kral L D 1997. Reproduced with permission of Elsevier

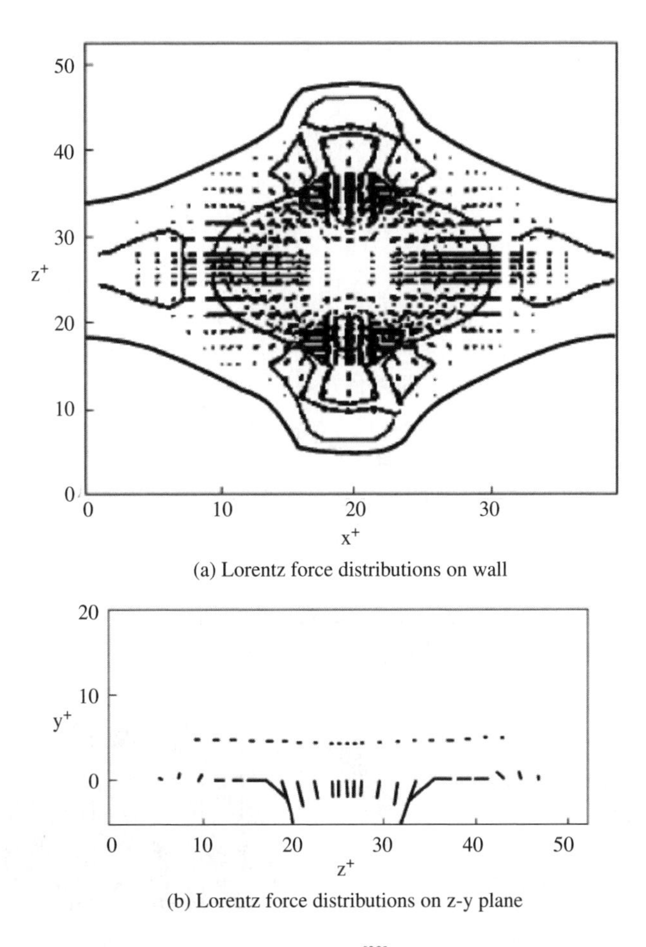

Figure 4.87 Wall-normal dominant force distribution.[23] *Source*: Berger T W 2000. Reproduced with permission of Cambridge University Press

contours, dashed lines indicate negative values. Vectors represent the streamwise and spanwise components. Figure 4.87b is a z–y plane view demonstrating the distributions of wall-normal and spanwise components by vectors. It is seen that the dominant wall-normal force component exists at the interior edges of each electrode. In the center of the tile, the magnitude of the wall-normal component is smaller but more uniform.

Figure 4.88 shows wall-normal force component distribution in y direction above the tile center. The wall-normal force directs to the wall (with negative value), and decays exponentially away from the wall.

4.4.2 Experiments on Wall Normal EM Actuator Tile

An EM actuator tile generating wall normal Lorentz force is shown in Figure 4.89, where a schematic and a photograph of the EM actuator are shown in Figure 4.89a and b, respectively. The permanent magnets made of rare earth material have a residual induction of 1.3 T. The electrode material used is platinum coated titanium.

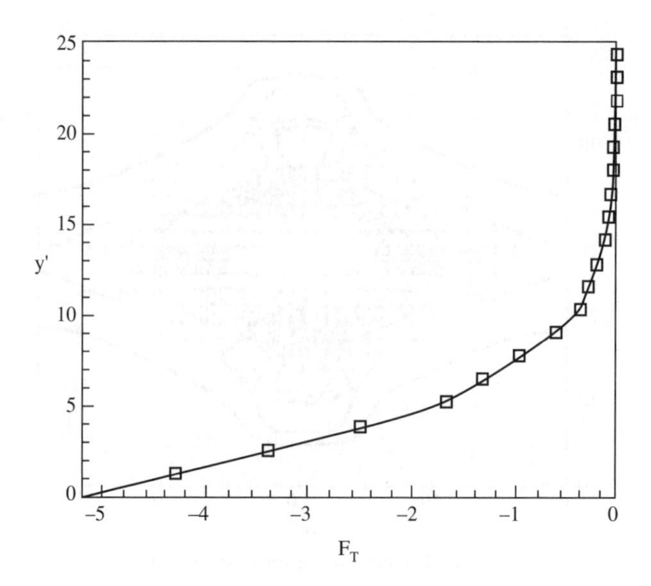

Figure 4.88 Wall-normal force component distribution above tile center.[23] *Source*: Berger T W 2000. Reproduced with permission of Cambridge University Press

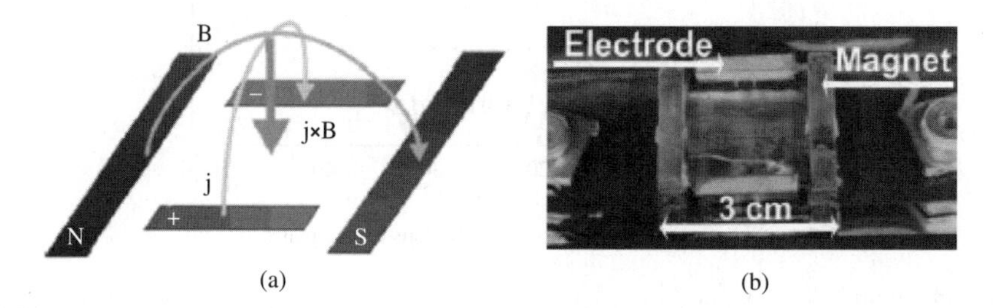

Figure 4.89 Wall normal EM actuator tile.[38] *Source*: Rossi L 2003. Reproduced with permission of Springer

The wall normal EM actuator placed within a 400 mm diameter is placed in a static tank filled with seawater (35 g l^{-1} NaCl). Electrode polarity is chosen so that the fluid would be pumped toward the wall. Figure 4.90 gives a cross-sectional view of the induced flow. In the two cases of continuous actuation (Figure 4.90a) and pulsed actuation (Figure 4.90b) (0.1 Hz with a time of actuation of 0.4 s), the resulting flow rotations confirm that the vorticity is one of the important mechanisms involved in electromagnetic flow control.

The induced vortex tube can be visualized by the fluid marked with fluorescein filled in a "pyramidal" volume above the EM actuator. As shown in Figure 4.91, there are two straight tubes parallel to the magnets (generated above each magnet) and two of horse shoe shape parallel to each electrode (generated in the region of the electrodes). These vortex tubes are connected in the corners via a "mushroom like" flow.

In order to discuss EM actuator actions on a laminar boundary layer, the experiments lay on preliminary visualization in a small seawater tunnel. The single actuator is mounted on

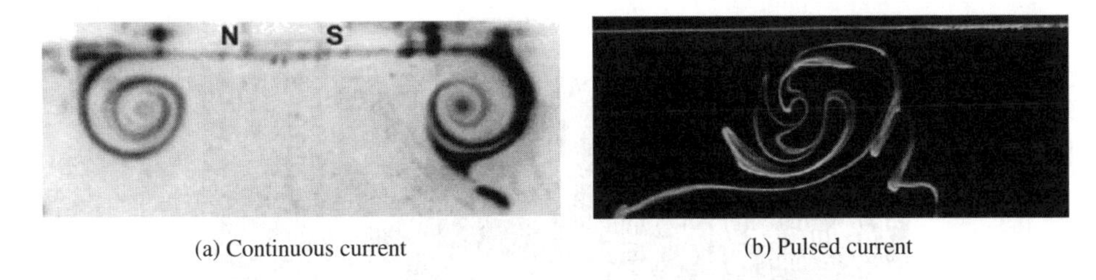

(a) Continuous current (b) Pulsed current

Figure 4.90 Side views of induced flow rotations.[38] *Source*: Rossi L 2003. Reproduced with permission of Springer

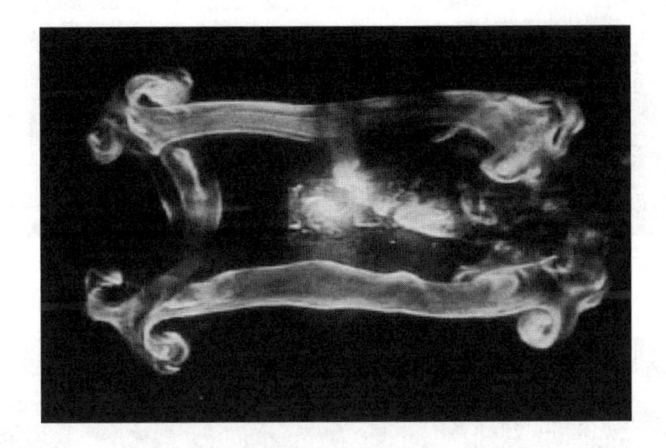

Figure 4.91 Visualization of induced vortex tube, top view.[38] *Source*: Rossi L 2003. Reproduced with permission of Springer

the upper wall of the tunnel about 20 cm downstream from the test section entrance. A wall half-sphere is used about 12 cm upstream of the actuator to generate the marking hairpin structures with the laminar boundary. The boundary layer is either pressed down toward the wall or blown away from the wall depending on the direction of the Lorentz forces, so that these hairpin structures are also turned away from or toward the wall again depending on the sign of the Lorentz force. In addition, the direction of the Lorentz force is dependent on the choice of electrode polarity. The effects of the EM force on the hairpin structures in the boundary are shown in Figure 4.92. The vortex structure is pulled into the wall for the negative force as shown in Figure 4.92a, whereas the vortex structure is blown away from the wall for the positive force as shown in Figure 4.92c.

4.4.3 Numerical Simulation of Wall Turbulence with Normal Lorentz Force

A near-wall turbulence structure, such as a streamwise vortex shown in Figure 4.93, not only induces the sweep and ejection events but also creates the spanwise velocity underneath the vortex. Therefore, the optimistically designed EM tile is that which can create a wall-normal distribution acting in opposition to the sweep and ejection events of a near-wall vortex while creating a spanwise force distribution that will counteract the induced spanwise velocity.

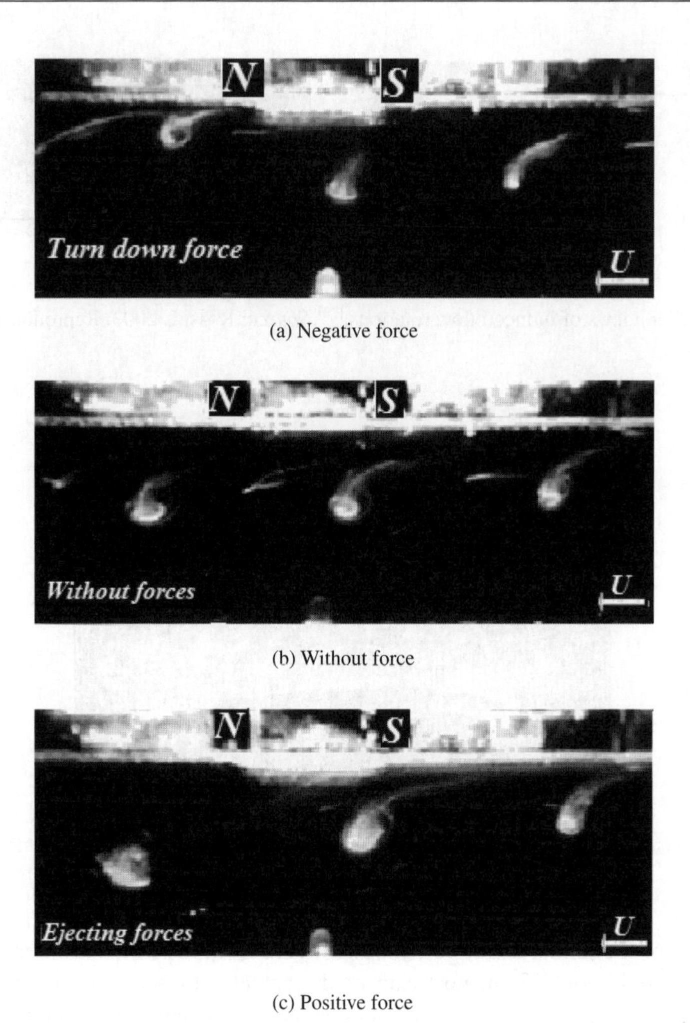

(a) Negative force

(b) Without force

(c) Positive force

Figure 4.92 Action of the wall normal Lorentz force on hairpin structures.[38] *Source*: Rossi L 2003. Reproduced with permission of Springer

An employed two-dimensional "idealized" force distribution with wall-normal and spanwise components is expressed as

$$f_y^+ = St\, e^{-\frac{y^+}{\Delta^+}} \sin(k_z^+ z^+) \cos(k_x^+ x^+) \tag{4.54}$$

$$f_z^+ = -\frac{St}{k_z^+ \Delta^+} e^{-\frac{y^+}{\Delta^+}} \cos(k_z^+ z^+) \cos(k_x^+ x^+) \tag{4.55}$$

where the streamwise Lorentz force component is ignored, and the force is now specified to vary sinusoidally in the streamwise direction for a tile with finitely long. k_x^+, defined as $k_x^+ = \frac{2\pi}{\lambda_x^+}$, is the streamwise wave number based on the wall unit.

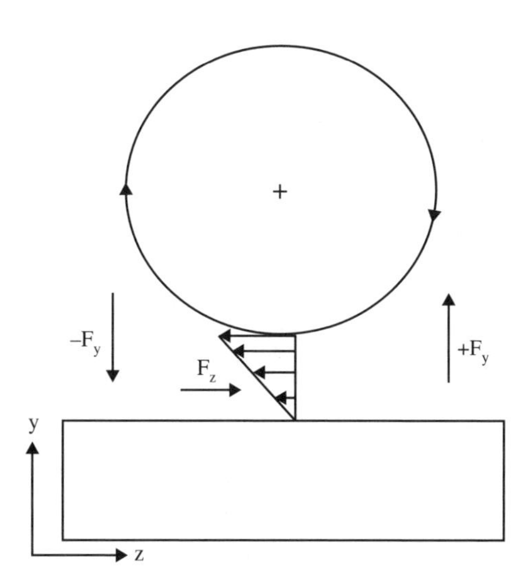

Figure 4.93 Optimal force distribution for the suppression of the streamwise vortex.[23] *Source*: Berger T W 2000. Reproduced with permission of Cambridge University Press

Figure 4.94a and b shows the effects of applying the force on the drag reduction for $Re_\tau = 100$ and $\Delta^+ = 10$ and $\Delta^+ = 20$, respectively. For each wavelength, λ_x^+, there appears to be an optimum *St*, that is inversely proportional to λ_x^+, resulting in the maximum drag reduction.

Different tile designs creating normal Lorentz force have been investigated numerically for wall turbulence control. In many cases, despite the fact that the control has a significant effect on the turbulent flow field and on the local skin friction in the vicinity of the actuators, the mean skin friction is only weakly affected compared with the uncontrolled simulations.

4.5 Closed Remarks

An EM actuator made of electrodes and magnets can produce a Lorentz force field in a conducting fluid. The force distribution is dependent on the EM actuator configuration. For example, one-dimensional force can be created by an alternating array of electrodes and magnets, that is, placing electrodes and magnets side by side, parallel to one another. In this case, the force direction can be changed by changing the orientation of the electrodes and magnets. A spatially oscillating force is produced by an alternating array of electrodes and magnets and a temporally oscillating force is produced by alternating the polarity the electrodes as well. Furthermore, a traveling wavy force can be generated by multiphase excitation. When the electrodes and magnets are placed perpendicular to one another in an EM actuator, three-dimensional force field will be created.

The Lorentz force generated when an electrically conducting fluid flows through an electromagnetic field is capable to modifying the near-wall flow. Among the three direction forces, the spanwise Lorentz force seems to be the most effective one. Four types of spanwise force, that is, spanwise force oscillating temporally or spatially and temporally oscillating spanwise force traveling in streamwise or spanwise directions, are discussed in this chapter. The control

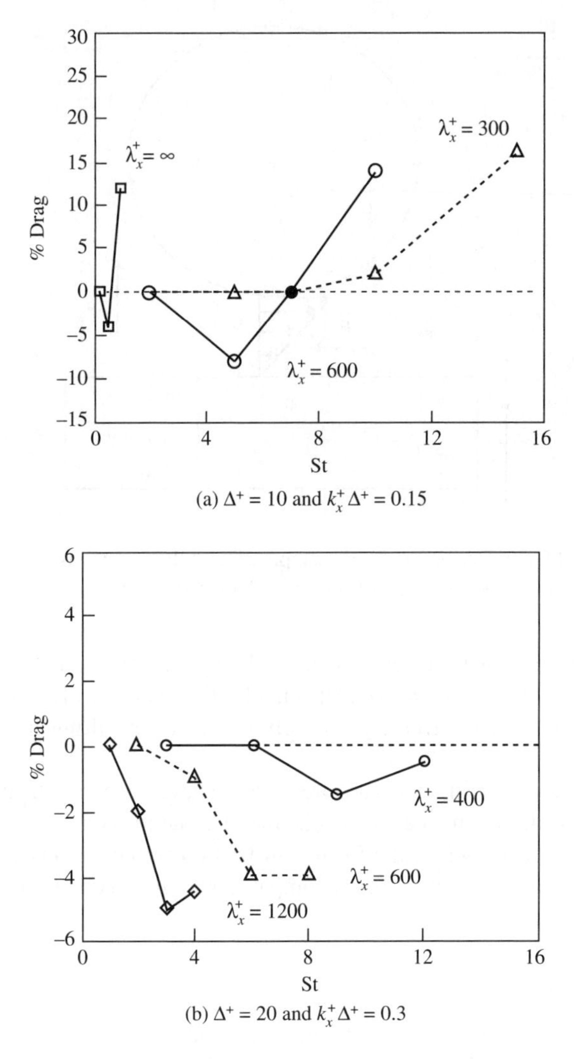

Figure 4.94 Variation of drag reduction with *St* number.[23] *Source*: Berger T W 2000. Reproduced with permission of Cambridge University Press

mechanism can be viewed as modulation of the near-wall turbulence (intrinsic flow) by the induced flow created by imposing the force into a corresponding laminar flow. The turbulence can be suppressed effectively, then resulting in a considerable drag reduction, when these forces applied with the specific waveform parameters can induce the perfect waves (induced flow) in the Stokes layer.

In case 1, that is, temporally oscillating force, no vortex structure and spanwise velocity distribution in the homogeneous directions exist in the induced flow field. After modulation, in the controlled flow field, the near-wall structures varies periodically, where the dominant vortex structures are tilted upward and downward alternatively, corresponding to the force direction. Then the weakened structures are almost eliminated gradually with increasing its lateral spacing. Subsequently, more and more tilted near-wall structures appear again, similar to that happening in the oscillating wall control. Both the special tilt of vortex structures, that is, the positive vortex tilts upward and the negative one tilts downward, and the elimination of near-wall structures can lead to the drag reduction.

In cases 2 and 3, that is, spatially oscillating force and transverse wavy Lorentz force, no vortex structure can be found in the induced flow field, but the spanwise velocities induced distribute sinuously in the streamwise direction. After modulation, the controlled flow field is dominated by stabilized, well-organized quasi-streamwise vortices, whereas few of hairpins vortices can be found. The vortices are inclined in the vertical plane and tilted in the horizontal plane, overlap with alternating sign in the streamwise direction as staggered arrays. Meanwhile, the positive vortex tilts upward and the negative one tilts downward and the same-signed vortices are arrayed, along the spanwise direction as columns. These tilted and stabilized near-wall structures lead to the drag reduction.

In case 4, that is, longitudinal wavy Lorentz force, the longitudinal vortex structures and streaks parallel to the wall, stretching through the considered area and moving with constant velocity and spanwise spacing, are created in the induced flow. After modulation, the controlled flow field is dominated by the set of streamwise vortices stretching in the streamwise direction, through the entire flow field and sweeping over the near-wall flow field in the spanwise direction, which is possessed of the basic characteristics of corresponding induced flow. This stabilized and relaminarized near-wall turbulent flow leads to the drag reduction.

For wall-normal force control, the mean skin friction is only weakly affected compared with the uncontrolled simulations unless the optimal closed controls are considered, to be discussed in Section 6.4.4.

References

[1] Lee C, Kim J. Control of the viscous sublayer for drag reduction. Phys. Fluids, 2002, 14(7):2523–2529.

[2] Gailitis A K and Lielausis O A. On the possibility of drag reduction of a flat plate in an electrolyte. Appl. Magnetohydrodyn. Trudy Inst. Fisiky AN Latvia SSR, 1961, 12:143–146.

[3] Nosenchuck D M and Brown G L. Discrete spatial control of wall shear stress in a turbulent boundary layer. In Near-Wall Turbulent Flows, edited by So R M C, Speziale C G and Launder B E, Elsevier, New York, 1993, pp. 689–698.

[4] Nosenchuck D M. Spatial and temporal characteristics of boundary layers controlled with the Lorentz force, 12th Australasian Fluid Mechanics Conference, University of Sydney Press, Sydney, 1955, pp. 93–96.

[5] Chen Y H, Fan B C, Chen Z H and Li H Z. Flow pattern and lift evolution of hydrofoil with control of electro-magnetic forces. Sci. China Ser. G-Phys. Mech. Astron., 2009, 52(9):1364–1374.

[6] Chen Y H, Fan B C, Chen Z H and Li H Z. Influences of Lorentz force on the hydrofoil lift. Acta Mechanica Sinica, 2009, 25:589–595.

[7] Zhang H, Fan B C, Chen Z H, Li H Z. Numerical study of the suppression mechanism of vortex-induced vibration by symmetric Lorentz force. J. Fluids Struc., 2014, 48:62–80.

[8] Henoch C and Stace J. Experimental investigation of a salt water turbulent boundary layer modified by an applied streamwise magnetohydrodynamic body force. Phys. Fluids, 1995, 7(6):1371.

[9] Crawford C H and Kamiadakis G E. Reynolds stress analysis of EMHD-controlled wall turbulence, Part I. streamwise forcing. Phys. Fluid, 1997, 9(3):788–806.

[10] Moreau R. Magnetohydrodynamics, Kluwer, Dordrecht, 1991.

[11] Sutton G W and Sherman A. Engineering Magnetohydrodynamics, McGraw-Hill, New York, 1965.

[12] Crawford C H. Direct numerical simulation of near-wall turbulence: Passive and active control, Ph.D. Thesis, Princeton University, 1998.

[13] Pang J, Choi K S, Aessopos A and Yoshida H. Control of near-wall turbulence for drag reduction by spanwise oscillating Lorentz force. AIAA 2004–2117, 2nd AIAA Flow Control Conference, June 28–July 1 2004, Portland, Oregon.

[14] Park J, Henock C and Breuz K. Drag reduction in turbulent flow using Lorentz force actuation. Fluid Mech. Appl., 2004, 74:315–318.
[15] Pang J and Choi K S. Turbulent drag reduction by Lorentz force oscillation. Phys. Fluids, 2004, 16(5):L35–L38.
[16] Mei D J, Fan B C, Chen Y H and Ye J F. Experimental investigation on turbulent channel flow utilizing spanwise oscillating Lorentz force. Acta Phys. Sin., 2010, 59(12):8335–8342.
[17] Huang L P, Fan B C and Mei D J. Mechanism of drag reduction by spanwise oscillating Lorentz force in turbulent channel flow. Theor. Appl. Mech. Lett., 2012, 2(1):012005.
[18] Breuer K S, Park J and Henoch C. Actuation and control of turbulent channel flow using Lorentz force. Phys. Fluids, 2004, 16(4):897–907.
[19] Choi K S, DeBisschop J R and Clayton B R. Turbulent boundary-layer control by means of spanwise-wall oscillation. AIAA J., 1998, 36:1157.
[20] Xu P, Choi, K S. Boundary layer control for drag reduction by Lorentz forcing. In: Flow Control and MEMS, Springer, Berlin, 2007, pp. 259–265.
[21] Xu P. Turbulent flow control using spanwise travelling wave via Lorentz force. Ph.D. Thesis, University of Nottingham, 2009.
[22] Huang L P, Choi K S, Fan BC and Chen Y H. Drag reduction in turbulent channel flow using bidirectional wavy Lorentz force. Sci. China-Phys. Mech. Astron., 2014, 57(11):2133–2140.
[23] Berger T W, Kim J, Lee C and Lim J. Turbulent boundary layer control utilizing the Lorentz force. Phys. Fluids, 2000, 12(3):631–649.
[24] Park J, Henoch C, McCamley M and Breuer, K S. Lorentz force control of turbulent channel flow. AIAA Paper 2003–4157.
[25] Berger T W. Turbulent boundary layer control utilizing the Lorentz force. Ph.D. Thesis, University of California, 2001.
[26] Mei D J, Fan B C, Huang L P, Dong G. Drag reduction in turbulent channel flow by spanwise oscillating Lorentz force. Acta Phys. Sin., 2010, 59(10):6786–6789.
[27] Mei D J. Drag reduction in turbulent channel flow utilizing spanwise oscillating Lorentz force. Ph.D. Thesis, Nanjing University of Science and Technology, 2011.
[28] Wu W T, Hong Y J and Fan B C. Vortex structures in turbulent channel flow modulated by spanwise Lorentz force with steady distributions. Acta Phys. Sin., 2014, 63(5):054702.
[29] Guo C F and Fan B C. Drag reduction in turbulent channel flow via spanwise Lorentz force with steady spatial distribution. J. Ship Mech., 2013, 17(4):336–345.
[30] Du Y, Symeonidis V and Karniadkis G E. Drag reduction in wall-bounded turbulence via a travelling wave. J. Fluid Mech., 2002, 457:1–34.
[31] Du Y and Karniadakis G E., Suppressing wall-turbulence via a transverse traveling wave. Science 2000, 288:1230–1234.
[32] Huang L P. Drag reduction in turbulent channel flows utilizing spanwise motions. Ph.D. Thesis, Nanjing University of Science and Technology, 2010.
[33] Huang L P, Fan B C. Effects of wave number of spanwise travelling wavy Lorentz force on wall turbulence control. J. Astronaut., 2012, 33(3):305–310.
[34] Huang L P, Choi K, Fan B C. Formation of low-speed ribbons in turbulent channel flow subject to a spanwise travelling wave. *Springer Proceedings in Physics*: Advances in Turbulence XIII. Proceedings of the 13th European Turbulence Conference, September 12–15, 2011, Warsaw, Poland.
[35] Huang L P, Fan B C, Dong G. Turbulent drag reduction via a transverse wave traveling along streamwise direction induced by Lorentz force. Phys. Fluids, 2010, 22:015103.
[36] Kral L D, Donovan J F and Cary A W. Numerical simulation and analysis of flow control using electromagnetic forcing. AIAA 1997–1797.
[37] Donovan J F, Kral L D and Cary A W. Characterization of a Lorentz force actuator. AIAA 1997–1918.
[38] Rossi L and Thibault J P. Electromagnetic forcing in turbulence and flow control: analytical definitions of EM parameters and hydrodynamic characterizations of flow. Phys. Fluids, 2003.

Part Three

Optimal Flow Control

5

Linear Optimal Flow Control

Flow control means an attempt to manipulate a flow, by altering the character of the flow field such as the mechanical state (e.g. velocity) and/or the thermodynamic state (e.g., temperature) to achieve a desired objective, including drag reduction, separation control, enhanced mixing, noise suppression, and so on.

As early as 1904, Prandtl[1] conducted a famous experiment to reduce the drag induced by a flow past a circular cylinder as shown in Figure 5.1, in which flow separation is delayed or suppressed by sucking fluid through a slit on the top side of the cylinder. This work marked the start of the experimental science of boundary layer control.

The investigation and development of flow control are stimulated effectively due to the military needs in World War II and the subsequent Cold War and the requirement of energy conservation in civilian sector by reason of the energy crises during the period 1970–1990. The recent important progress of both the numerical fluid mechanics and experimental fluid mechanics, which makes it possible to study complex flow and to understand its dynamic mechanism in depth, should promote scientifically the further development of flow control, since it is based on the understanding of controlled flow fields.[3–5] For example, the discovery of an organized vortical structure of turbulent boundary layers, opening up the possibility of wall-turbulence control and eventually of viscous drag, has attracted much attention from many fluid dynamicists to pursue drag reduction techniques by turbulence control. Therefore, flow control has been one of the frontiers in fluid mechanics.

Flow control involves passive and active devices based on energy expenditure. In passive control, flow is controlled without external energy expenditure, such as geometric shaping, large eddy break-up devices, placement of longitudinal grooves, or riblets on a surface. The passive control is inherently an open-loop control.

For active control, such as oscillating wall, synthetic jets, alternative suction and blowing, and electromagnetic force, the external energy is introduced into a flow in two ways, either in a predetermined manner or in a closed-loop form. In the predetermined method, energy input is set at the design stage in advance and remains fixed without regard to the particular state of the flow as discussed in Chapters 3 and 4; thus the control loop is open. In closed-loop control, also known as feedback control, energy input is continuously adjusted based on instantaneous state of the flow field.

Principles of Turbulence Control, First Edition. Baochun Fan and Gang Dong.

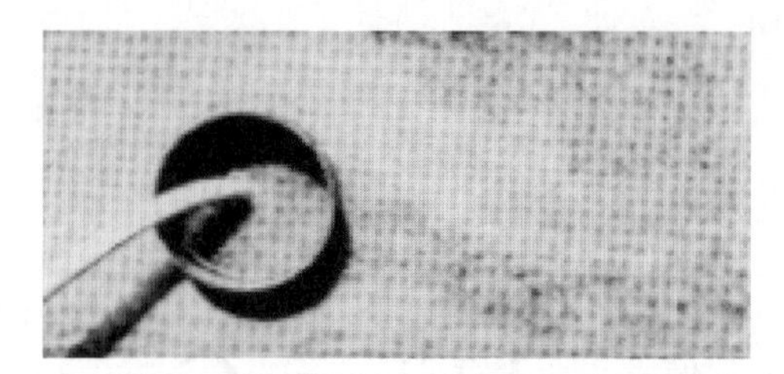

Figure 5.1 Depiction of Prandtl's 1904 experiment.[2] *Source*: Gunzburger M D 2003. Reproduced with permission of Elsevier

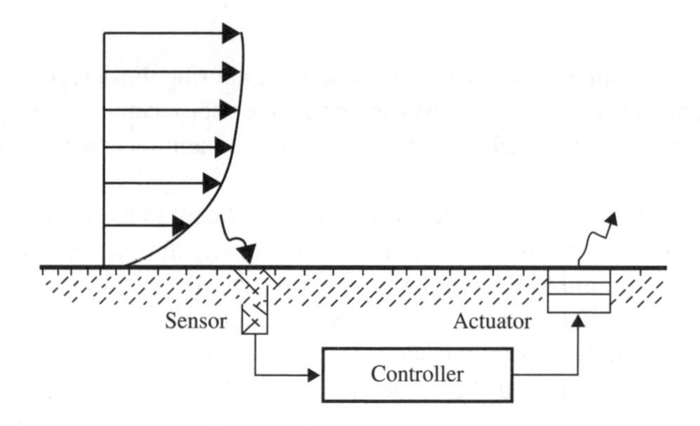

Figure 5.2 Schematic representation of a closed-loop control system

One of the main aims of flow control is to reduce the energy consumption; therefore the energy consumed in the realization of an active flow control must be considered. For a particular goal of flow control, the amount of consumed energy is dependent on the control method utilized. The best possible control method with lower consumed energy can be designed by solving the optimal feedback control problem.

The three essential components of optimal feedback control technique are sensors, controller, and actuators, as shown in Figure 5.2. The information about instantaneous flow fields, or usually called the plant, measured by means of sensors is inputted into the controller. Then the controller gives desired control variables to the actuators by using a feedback control law based on an approach of the optimal flow control; finally the actuators actualize the plant control, leading the variations of the plant. The new measured information will be extracted for changed plant, and the input control variables are adjusted inevitably. Therefore, in the actual operation of the feedback control system, the input to the plant varies with the instantaneous flow fields continuously. It is obvious that when considering an optimal feedback control, one needs some control theories to controller design, and also requires that both actuators and sensors be desired and utilized in an effective way, which is related to two key concepts, controllability and observability. Controllability is a property of both the actuator system and the state that determines whether all the state modes can be arbitrarily influenced by the actuator. Observability is related to the ability of a particular sensor system to reliably measure changes in the state.

Flow control is a particular interdisciplinary branch, incorporating the traditional fields of fluid mechanics and control theory.[6] Fluid mechanics and control theory have been developed individually as traditional scientific disciplines; hence it is essential to understand a consideration of both the fundamental flow physics and the requirements and limitations of control algorithms, and to know how these fields fit together.

Usually, there are two ways of balancing control theory and fluid mechanism. First, one reduces the order of governing flow equations, that is, transforming Navier–Stokes equations into linear equations, to fit with the linear optimal control theory so that the mature control theory can be used to solve the flow control problem directly. Alternatively, one develops the control algorithms to fit with the relevant flow physics. These will be discussed in Chapters 5 and 6, respectively.

In this chapter, we will first discuss linear optimal flow control. Section 5.1 represents some basic problems and fundamental conceptions in optimal control, and introduces conditional extremum of a function or a functional constraints by the ordinary or partial differential equations. In Section 5.2, we discuss the linear quadratic optimal control, which is mature correspondingly in optimal control. Sections 5.3–5.5 are devoted to the turbulence control with linear optimal control theory. While the reliability of turbulent control with a linear control theory is provided in Section 5.3, subsequently the linear optimal controls applied in two- and three-dimensional flows are discussed in Sections 5.4 and 5.5, respectively.

5.1 Optimal Control

5.1.1 Introduction[2, 5]

5.1.1.1 Optimization Problem

Energy consumption must be considered in the proceeding of the flow control. To raise the control efficiency and to decrease the control cost, the control variable should be varied optimistically according to the instantaneous flow fields, which is called optimal feedback control. The ingredients in an optimal feedback control problem are as follows:

1. State variable $\mathbf{x}(t)$
   In the fluid mechanics, state variables, such as velocity, pressure, and temperature, are the mechanical and thermodynamic variables that describe the flow.
2. Control variables $\mathbf{u}(t)$
   The control variables could be the heat flux or temperature at a wall, the inflow mass flow rate, or the body force, which influences the flow and can be manipulated by an actuator system.
3. Observations $\mathbf{y}(t)$
   In more realistic case, a complete state cannot be accurately measured at all times, only the observations are available for measurement by sensors.
4. State equation
   The candidate state and control variables are required to satisfy the constraint equation $F(\mathbf{x}(t), \mathbf{u}(t)) = 0$, called the state equation, which usually are the governing flow equations for flow control.

5. Objective set
   The goal of the flow control is to maintain the flow field at a desired configuration, which can be represented by an objective set $\Psi(\mathbf{x}(t_f), t_f) = 0$, where t_f is the terminal time.
6. Performance index
   The performance index (also called cost functional), which measures the energy consumption during the control, is introduced to formulate the optimal control problem, then the optimization problem is to find state and control variables that minimize the performance index subject to the requirement that the constraints are satisfied.

5.1.1.2 Conditional Extremum

The optimal flow control problem can be expressed as follows: Find an optimal relation between the control variable **u** and the system state variable $\mathbf{x}(t)$, called control law, minimizing the performance index $J(\mathbf{x}, \mathbf{u})$ constrained by the governing flow equation $\mathbf{F}(\mathbf{x}, \mathbf{u}) = 0$ and the corresponding boundary conditions. Therefore that is a conditional extremum problem.

Introduce the Lagrange multiplier $\boldsymbol{\lambda}$, or adjoint variable; then the Lagrange functional is

$$L = J(\mathbf{x}, \mathbf{u}) + \boldsymbol{\lambda}^{\mathrm{T}} \cdot \mathbf{F}(\mathbf{x}, \mathbf{u}) \tag{5.1}$$

where superscript T denotes the transpose of the vector.

Therefore, the optimal problem is as follows: Find control variable **u**, system state variable **x**, and adjoint variable $\boldsymbol{\lambda}$ such that $L(\mathbf{x}, \mathbf{u}, \boldsymbol{\lambda})$ is rendered stationary, that is, satisfying the necessary following conditions:

$$\frac{\delta L}{\delta \mathbf{x}} = 0 \tag{5.2}$$

$$\frac{\delta L}{\delta \mathbf{u}} = 0 \tag{5.3}$$

$$\frac{\delta L}{\delta \boldsymbol{\lambda}} = 0 \tag{5.4}$$

Equation (5.4) is equivalent to the condition

$$\lim_{\varepsilon \to 0} = \left(\frac{L\left(\mathbf{x}, \mathbf{u}, \boldsymbol{\lambda} + \varepsilon \widetilde{\boldsymbol{\lambda}}\right) - L(\mathbf{x}, \mathbf{u}, \boldsymbol{\lambda})}{\varepsilon} \right) = 0 \tag{5.5}$$

where $\widetilde{\boldsymbol{\lambda}}$ denotes a perturbation in the adjoint variable $\boldsymbol{\lambda}$. Then, substituting for L from Eq. (5.1), we have

$$\widetilde{\boldsymbol{\lambda}}^{T} \cdot \mathbf{F}(\mathbf{x}, \mathbf{u}) = 0 \tag{5.6}$$

Since $\widetilde{\boldsymbol{\lambda}}^{T}$ is arbitrary, we recover the constraint equation, that is, the state equation

$$\mathbf{F}(\mathbf{x}, \mathbf{u}) = 0 \tag{5.7}$$

Similarly, from Eq. (5.2)

$$\lim_{\varepsilon\to 0} = \left(\frac{L\left(\mathbf{x}+\varepsilon\widetilde{\mathbf{x}},\mathbf{u},\boldsymbol{\lambda}\right) - L(\mathbf{x},\mathbf{u},\boldsymbol{\lambda})}{\varepsilon} \right) = 0$$

$$\lim_{\varepsilon\to 0} = \frac{1}{\varepsilon}[(\mathrm{J}(\mathbf{x}+\varepsilon\widetilde{\mathbf{x}},\mathbf{u}) - \lambda^T\mathbf{F}(\mathbf{x}+\varepsilon\widetilde{\mathbf{x}},\mathbf{u})) - J(L(\mathbf{x},\mathbf{u}) - \lambda^T\mathbf{F}(\mathbf{x},\mathbf{u}))] = 0$$

or

$$\lim_{\varepsilon\to 0} = \left(\frac{J\left(\mathbf{x}+\varepsilon\widetilde{\mathbf{x}},\mathbf{u}\right) - J(\mathbf{x},\mathbf{u})}{\varepsilon} - \frac{\lambda^T(\mathbf{F}(\mathbf{x}+\varepsilon\widetilde{\mathbf{x}},\mathbf{u}) - \mathbf{F}(\mathbf{x},\mathbf{u}))}{\varepsilon} \right) = 0 \tag{5.8}$$

Introducing Taylor series, we have

$$J(\mathbf{x}+\varepsilon\widetilde{\mathbf{x}},\mathbf{u}) = J(\mathbf{x},\mathbf{u}) + \varepsilon\left(\frac{\partial J}{\partial \mathbf{x}}\right)\widetilde{\mathbf{x}}$$

Then, we have from Eq. (5.8) that

$$\widetilde{\mathbf{x}}^T\left(\left(\frac{\partial J}{\partial \mathbf{x}}\right)^* - \left(\frac{\partial \mathbf{F}}{\partial \mathbf{x}}\right)^*\boldsymbol{\lambda}\right) = 0 \tag{5.9}$$

where superscript "*" denotes the matrix conjugate transpose. Since $\widetilde{\mathbf{x}}^{\mathrm{T}}$ is arbitrary, we obtain the adjoint equation

$$\left(\frac{\partial J}{\partial \mathbf{x}}\right)^* = \left(\frac{\partial \mathbf{F}}{\partial \mathbf{x}}\right)^*\boldsymbol{\lambda} \tag{5.10}$$

Proceeding, as we did, for the adjoint equation yields the optimality condition

$$\left(\frac{\partial J}{\partial \mathbf{u}}\right)^* = \left(\frac{\partial \mathbf{F}}{\partial \mathbf{u}}\right)^*\boldsymbol{\lambda} \tag{5.11}$$

Collecting the above results yields the optimality system:
State equation $\mathbf{F}(\mathbf{x},\mathbf{u}) = 0$ (5.7)
Adjoint equation $\left(\frac{\partial J}{\partial \mathbf{x}}\right)^* = \left(\frac{\partial \mathbf{F}}{\partial \mathbf{x}}\right)^*\boldsymbol{\lambda}$ (5.10)
Optimal control law $\left(\frac{\partial J}{\partial \mathbf{u}}\right)^* = \left(\frac{\partial \mathbf{F}}{\partial \mathbf{u}}\right)^*\boldsymbol{\lambda}$ (5.11)
This coupled system is generally the formidable system to solve.

5.1.1.3 Extremum of Functional

For all functions $x(t)$ in a function set, J is called functional of $x(t)$, if it is completely dependent on $x(t)$, and denoted by

$$J = J[x(t)] \tag{5.12}$$

For a continuous functional $J(x)$ defined on the normed linear space R^n, its increment is expressed as

$$\Delta J(x) = J(x + \delta x) - J(x) = L(x, \delta x) + r(x, \delta x) \tag{5.13}$$

where $L(x, \delta x)$ is a linear continuous functional of δx, $r(x, \delta x)$ is a higher order infinitesimal of δx, then $\delta J = L(x, \delta x)$ is called the variation of functional $J(x)$, which is a linear main part of the increment $\Delta J(x)$.

The optimal control problem can be described as finding a control function so that an extremum of a performance index under confined conditions is given. When the controlled plant is described by a set of partial differential equations, the performance index is expressed by a functional. The control function can be determined by solving the extremum problem of a functional under the constraint of differential equations. The extremum problem of a functional can be solved by using the variational method, where the variational formula is almost identical with the corresponding differential formula.

Consider a simple example, that is, a problem with the following boundaries:

$$\mathbf{x}(t_0) = \mathbf{x}_0, \quad \mathbf{x}(t_f) = \mathbf{x}_f \tag{5.14}$$

The unconditional extremum of a functional $J(\mathbf{x})$ is expressed as

$$\min_{\mathbf{x}} J(\mathbf{x}) = \int_{t_0}^{t_f} L(\mathbf{x}, \dot{\mathbf{x}}, t)dt \tag{5.15}$$

where $\dot{\mathbf{x}} = \frac{d\mathbf{x}}{dt}$, $L(\mathbf{x}, \dot{\mathbf{x}}, t)$, and $\mathbf{x}(t)$ are continuously differentiable on the time interval $[t_0, t_f]$. Then, the extremum satisfies the following Euler equation:

$$\frac{\partial L}{\partial \mathbf{x}} - \frac{d}{dt}\frac{\partial L}{\partial \dot{\mathbf{x}}} = 0 \tag{5.16}$$

Under the constraint of equation $\mathbf{f}(\mathbf{x}, \dot{\mathbf{x}}, t) = 0$, the unconditional extremum is expressed as

$$\min_{\mathbf{x}} J(\mathbf{x}) = \int_{t_0}^{t_f} g(\mathbf{x}, \dot{\mathbf{x}}, t)dt \tag{5.17}$$

where $g(\mathbf{x}, \dot{\mathbf{x}}, t)$ and $\mathbf{x}(t)$ are continuously differentiable on the time interval $[t_0, t_f]$. Lagrangian function for the conditional extremum is

$$L(\mathbf{x}, \dot{\mathbf{x}}, \boldsymbol{\lambda}, t) = g(\mathbf{x}, \dot{\mathbf{x}}, t) + \boldsymbol{\lambda}^{\mathrm{T}}(t) \cdot \mathbf{f}(\mathbf{x}, \dot{\mathbf{x}}, t) \tag{5.18}$$

where $\boldsymbol{\lambda}$ is an adjoint variable.

Constructing a generalized functional

$$J_a = \int_{t_0}^{t_f} L(\mathbf{x}, \dot{\mathbf{x}}, \boldsymbol{\lambda}, t)dt \tag{5.19}$$

consequently the conditional extremum of the functional is equivalent to the unconditional extremum of the generalized functional. We have the Euler equation

$$\left[\frac{\partial g}{\partial \mathbf{x}} + \left(\frac{\partial \mathbf{f}^{\mathrm{T}}}{\partial \mathbf{x}}\right) \cdot \boldsymbol{\lambda}\right] - \frac{d}{dt}\left[\frac{\partial g}{\partial \dot{\mathbf{x}}} + \left(\frac{\partial \mathbf{f}^{\mathrm{T}}}{\partial \dot{\mathbf{x}}}\right) \cdot \boldsymbol{\lambda}\right] = 0 \tag{5.20}$$

with the confined condition

$$\mathbf{f}(\mathbf{x}, t) = 0 \tag{5.21}$$

which can be simplified as

$$\left[\frac{\partial g}{\partial \mathbf{x}} + \frac{\partial \mathbf{f}^{\mathrm{T}}}{\partial \mathbf{x}} \cdot \boldsymbol{\lambda}\right] - \frac{d}{dt}\frac{\partial g}{\partial \dot{\mathbf{x}}} = 0 \tag{5.22}$$

Thus, Eqs. (5.21) and (5.22) are the governing equations of the optimal system.

5.1.2 Optimal Control for Ordinary Differential Equations[7, 8]

The state-space representation of many control systems can be described by a set of ordinary differential equations and its initial conditions in the following form:

$$\frac{d\mathbf{x}}{dt} = \mathbf{f}(\mathbf{x}(t), \mathbf{u}(t), t) \tag{5.23}$$

$$\mathbf{x}(t_0) = \mathbf{x}_0 \tag{5.24}$$

where t is the time variable, $\mathbf{x}$ is a time-varying column vector, denoting the state of the system, and $\mathbf{u}$ is a column vector, indicating the input variable or control variable, which is unconfined and continuous in the time interval $[t_0, t_f]$.

The optimal performance index of the control system is written in the form

$$J = \int_{t_0}^{t_f} L(\mathbf{x}, \mathbf{u}, t)dt + \varphi(\mathbf{x}(t_f), t_f) \tag{5.25}$$

where both L and φ are nonnegative and continuously differentiable functions of $\mathbf{x}$ and $\mathbf{u}$. The terminal time t_f may be fixed or free.

5.1.2.1 Fixed Terminal Time

For fixed terminal time, where the terminal time t_f is constant, the terminal state of system $\mathbf{x}(t_f)$ might be confined, free, or fixed.

When the terminal state is confined, the objective set can be expressed as

$$\boldsymbol{\psi}(\mathbf{x}(t_f), t_f) = 0 \tag{5.26}$$

Hence, the optimal control problem can be regarded as a task of finding an optimal control $\mathbf{u}^*(t)$ and optimal trajectory $\mathbf{x}^*(t)$ to bring the system described by Eq. (5.23) from the initial state indicated by Eq. (5.24) to a desired state satisfying Eq. (5.26), by minimizing the performance index defined in Eq. (5.25). It is expressed mathematically as

$$\min_{u(t)} J = \varphi[\mathbf{x}(t_f)] + \int_{t_0}^{t_f} L(\mathbf{x}, \mathbf{u}, t)dt \tag{5.27}$$

with confined conditions $f(\mathbf{x}, \mathbf{u}, t) - \dot{\mathbf{x}}(t) = 0, \quad \mathbf{x}(t_0) = \mathbf{x}_0$,
and $\boldsymbol{\psi}(\mathbf{x}(t_f)) = 0$ which is a problem associated with a conditional extremum of functional confined by two equations.

Introducing the adjoint vectors $\boldsymbol{\lambda}$ and $\boldsymbol{\gamma}$, generalized functional can be constructed as

$$J_a = \varphi[\mathbf{x}(t_f)] + \boldsymbol{\gamma}^{\mathrm{T}}(t)\boldsymbol{\psi}[x(t_f)] + \int_{t_0}^{t_f} \{L(\mathbf{x}, \mathbf{u}, t) + \boldsymbol{\lambda}^{\mathrm{T}}(t)[\mathbf{f}(\mathbf{x}, \mathbf{u}, t) - \dot{\mathbf{x}}(t)]\}dt$$

From the variational point of view, J_a and J are equivalent. Therefore, the conditional extremum of J can be obtained from the determined unconditional extremum of J_a.

Introducing Hamilton function

$$H(\mathbf{x}, \mathbf{u}, \boldsymbol{\lambda}, t) = L(\mathbf{x}, \mathbf{u}, t) + \boldsymbol{\lambda}^{\mathrm{T}}(t)\mathbf{f}(\mathbf{x}, \mathbf{u}, t) \tag{5.28}$$

we have

$$J_a = \varphi[\mathbf{x}(t_f)] + \boldsymbol{\gamma}^{\mathrm{T}}(t)\boldsymbol{\psi}[\mathbf{x}(t_f)] + \int_{t_0}^{t_f} [H\mathbf{x}, \mathbf{u}, \boldsymbol{\lambda}, t) - \boldsymbol{\lambda}^{\mathrm{T}}(t)\dot{\mathbf{x}}(t)]dt \tag{5.29}$$

Applying integral by part

$$-\int_{t_0}^{t_f} \boldsymbol{\lambda}^{\mathrm{T}}(t)\dot{\mathbf{x}}(t)dt = -\boldsymbol{\lambda}^{\mathrm{T}}(t)\mathbf{x}(t)|_{t_0}^{t_f} + \int_{t_0}^{t_f} \dot{\boldsymbol{\lambda}}^{T}(t)\mathbf{x}(t)dt$$

Eq. (5.29) can be written as

$$\begin{aligned} J_a =& \varphi[\mathbf{x}(t_f)] + \boldsymbol{\gamma}^{\mathrm{T}}(t)\boldsymbol{\psi}[\mathbf{x}(t_f)] - \boldsymbol{\lambda}^{\mathrm{T}}(t_f)\mathbf{x}(t_f) + \boldsymbol{\lambda}^{\mathrm{T}}(t_0)\mathbf{x}(t_0) \\ &+ \int_{t_0}^{t_f} [H\mathbf{x}, \mathbf{u}, \boldsymbol{\lambda}, t) + \dot{\boldsymbol{\lambda}}^{\mathrm{T}}(t)\mathbf{x}(t)]dt \end{aligned}$$

The variation in this equation is

$$\begin{aligned} \delta J_a =& \delta\mathbf{x}^{\mathrm{T}}(t_f)\left[\frac{\partial\phi}{\partial\mathbf{x}\left(t_f\right)} + \frac{\partial\boldsymbol{\psi}^{\mathrm{T}}}{\partial\mathbf{x}(t_f)}\boldsymbol{\gamma}(t_f) - \boldsymbol{\lambda}(t_f)\right] \\ &+ \int_{t_0}^{t_f} \left[\left(\frac{\partial H}{\partial\mathbf{x}} + \dot{\boldsymbol{\lambda}}\right)^{\mathrm{T}}\delta\mathbf{x} + \left(\frac{\partial H}{\partial\mathbf{u}}\right)^{\mathrm{T}}\delta\mathbf{u}\right]dt \end{aligned}$$

δJ_a should be zero, when minimizing the performance index J_a. Note that $\delta \mathbf{x}(t)$, $\delta \mathbf{u}(t)$, and $\delta \mathbf{x}(t_f)$ are all arbitrary, then we have

$$\dot{\boldsymbol{\lambda}}(t) = -\frac{\partial H}{\partial \mathbf{x}} \tag{5.30}$$

$$\frac{\partial H}{\partial \mathbf{u}} = 0 \tag{5.31}$$

with the transversality condition

$$\boldsymbol{\lambda}(t_f) = \frac{\partial \varphi}{\partial \mathbf{x}(t_f)} + \frac{\partial \boldsymbol{\psi}^{\mathrm{T}}}{\partial \mathbf{x}(t_f)}\boldsymbol{\gamma}(t_f) \tag{5.32}$$

From Eq. (5.28), we have

$$\dot{\mathbf{x}}(t) = \frac{\partial H}{\partial \boldsymbol{\lambda}} = \mathbf{f}(\mathbf{x}, \mathbf{u}, t) \tag{5.33}$$

Since the right-hand side of the above equations are the partial derivative of the Hamilton function, Eqs. (5.30) and (5.33) are canonical equations composed of $2n$ first-order ordinary differential equations. The initial condition (5.24) and transversality condition (5.32) contain $2n$ boundary condition equations.

Equation (5.33) is a state-space equation, Eq. (5.30) is an adjoint equation, and $\boldsymbol{\lambda}$ is an adjoint vector. The extremal condition (or control equation) (5.31) is a mth algebraic equation, which determines the relation among optimal control $\mathbf{u}^*(t)$, optimal trajectory $\mathbf{x}^*(t)$, and adjoint vector $\boldsymbol{\lambda}^*(t)$.

If terminal states are free, no terminal condition exists, and the boundary conditions of the canonical equations will be

$$\mathbf{x}(t_0) = \mathbf{x}_0 \quad \boldsymbol{\lambda}(t) = \frac{\partial \varphi}{\partial \mathbf{x}(t_f)} \tag{5.34}$$

If terminal states are fixed, that is $\mathbf{x}(t_f) = \mathbf{x}_f$, the boundary conditions of the canonical equations will be

$$\mathbf{x}(t_0) = \mathbf{x}_0 \quad \mathbf{x}(t_f) = \mathbf{x}_f \tag{5.35}$$

It is obvious that a performance index satisfying the canonical equations is the optimal performance index and determines an optimal control.

5.1.2.2 Free Terminal Time

For free terminal time, the terminal state of system $\mathbf{x}(t_f)$ may be confined, free, or fixed, similar with the cases in fixed terminal time.

When the terminal state is confined, the optimal control problem can be expressed mathematically as

$$\min_{u(t)} J = \varphi[\mathbf{x}(t_f), t_f] + \int_{t_0}^{t_f} L(\mathbf{x}, \mathbf{u}, t)dt$$

$$f(\mathbf{x}, \mathbf{u}, t) - \dot{\mathbf{x}}(t) = 0, \quad \mathbf{x}(t_0) = \mathbf{x}_0$$

$$\boldsymbol{\psi}(\mathbf{x}(t_f), t_f) = 0$$

where t_f is free. Then, we have canonical equations

$$\dot{\mathbf{x}}(t) = \frac{\partial H}{\partial \boldsymbol{\lambda}}$$
$$\dot{\boldsymbol{\lambda}}(t) = -\frac{\partial H}{\partial \mathbf{x}} \tag{5.36}$$

and control equation

$$\frac{\partial H}{\partial \mathbf{u}} = 0 \tag{5.37}$$

with boundary conditions

$$\mathbf{x}(t_0) = \mathbf{x}_0$$
$$\boldsymbol{\lambda}(t_f) = \frac{\partial \phi}{\partial \mathbf{x}(t_f)} + \frac{\partial \boldsymbol{\psi}^{\mathrm{T}}}{\partial \mathbf{x}(t_f)}\boldsymbol{\gamma}(t_f)$$
$$\boldsymbol{\psi}(\mathbf{x}(t_f), t_f) = 0$$

If terminal states are free, no terminal condition exists, and the boundary conditions of the canonical equations are

$$\mathbf{x}(t_0) = \mathbf{x}_0 \quad \boldsymbol{\lambda}(t_f) = \frac{\partial \varphi}{\partial \mathbf{x}(t_f)}$$

If terminal states are fixed, that is $\mathbf{x}(t_f) = \mathbf{x}_f$, the boundary conditions of the canonical equations are

$$\mathbf{x}(t_0) = \mathbf{x}_0 \quad \mathbf{x}(t_f) = \mathbf{x}_f$$

5.2 Optimal Control of Linear Quadratic Systems[9, 10]

5.2.1 Linear Quadratic Optimal Control

Linear optimal control is a special sort of optimal control with computable solutions. The controlled system is assumed linear, and the optimal system input is supposed to depend linearly on its output. Hence, a linear and time-varying system in state-space form can be written as

$$\frac{d\mathbf{x}(t)}{dt} = A(t)\mathbf{x}(t) + B(t)\mathbf{u}(t)$$
$$\mathbf{x}(t_0) = \mathbf{x}_0 \tag{5.38}$$

where $\mathbf{x}(t)$ is a n-dimensional time-varying column vector, which denotes the system state, $\mathbf{u}(t)$ is a k-dimensional column vector, indicating the system input or system control, and A is the time-varying coefficient matrix with $(n \times n)$ dimension and $(n \times m)$ matrix B is the input matrix. If matrices A and B are constant, the system is time invariant.

Since $\mathbf{u}(t)$ is required to be an instantaneous function of $\mathbf{x}(t)$, we can write down a control law as

$$\mathbf{u}(t) = f(\mathbf{x}(t), t)$$

in the case of the linear control, given by

$$\mathbf{u}(t) = K'(t)\mathbf{x}(t)$$

In a more realistic case, a complete state cannot be accurately measured at all times; thus there is an observed variable of the form

$$\mathbf{y}(t) = C(t)\mathbf{x}(t) \tag{5.39}$$

which is available for measurement. $\mathbf{y}(t)$ is a p-dimensional column vector, called output of the system, which is usually measured by means of sensors, and the $(p \times n)$ matrix C is the output matrix function. Then, for a completely observable system, the entries of the state vector can be constructed from linear combinations of the system input $\mathbf{u}(t)$, output $\mathbf{y}(t)$, and their derivatives. Equation (5.38) is called the state differential equation, and Eq. (5.39) is called the output equation of the system. Then Eqs. (5.38) and (5.39) together are called the system equations, in which the controllability of a system is a property of the matrices A and B, while the observability is a property of the matrices A and C.

If $\mathbf{u}_0(t)$ is a given input to a system described by Eq. (5.33), $\mathbf{x}_0(t)$ is a known solution, that is,

$$\dot{\mathbf{x}}_0(t) = f[\mathbf{x}_0(t), \mathbf{u}_0(t), t]$$

We refer to $\mathbf{u}_0(t)$ and $\mathbf{x}_0(t)$ as nominal conditions, and assume that the system is operated close to nominal conditions. Therefore,

$$\mathbf{u}(t) = \mathbf{u}_0(t) + \widetilde{\mathbf{u}}(t)$$

$$\mathbf{x}(t) = \mathbf{x}_0(t) + \widetilde{\mathbf{x}}(t)$$

where $\widetilde{\mathbf{u}}(t)$ and $\widetilde{\mathbf{x}}(t)$ are small perturbations. Then, $\widetilde{\mathbf{u}}(t)$ and $\widetilde{\mathbf{x}}(t)$ approximately satisfy the following linear equation:

$$\dot{\widetilde{\mathbf{x}}}(t) = A(t)\widetilde{\mathbf{x}}(t) + B(t)\widetilde{\mathbf{u}}(t)$$

which is called the linearized state differential equation. The linearization procedure is a very common practice.

In terms of the control aims desired to achieve, the linear optimal control is classified into two types, that is, the regulator problem and the tracking problem. For the regulator problem, the initial output of the system, or any of its derivatives, is nonzero, and the problem is to apply a control to take the system from a nonzero state to the zero state. For the tracking problem, the system output, or a derivative, is required to track some prescribed functions.

5.2.1.1 Linear Quadratic Regulator (LQR)

Consider the linear time-varying system expressed by Eq. (5.38), defined in a finite-time interval $[0, T]$:

$$\begin{aligned} \frac{d\mathbf{x}(t)}{dt} &= A(t)\mathbf{x}(t) + B(t)\mathbf{u}(t) \\ \mathbf{x}(t_0) &= \mathbf{x}_0 \end{aligned} \tag{5.38}$$

The regulator problem is to apply control to take the system from a nonzero state to the zero state. Considering the consumption of control energy, it is necessary to keep the energy consumption bounded during the control course, where the energy consumption might be expressed by

$$\int_{t_0}^{T} \mathbf{u}^{\mathrm{T}}(t)R(t)\mathbf{u}(t)dt \tag{5.40}$$

where the superscript T denotes matrix transpose and R is the positive definite matrix to ensure $\mathbf{u}^{\mathrm{T}}(t)R\mathbf{u}(t) > 0$. In addition, to relax the aim that the system should actually achieve the zero state, we might ask that $\mathbf{x}^{\mathrm{T}}(T)Q(t)\mathbf{x}(t)$, with Q being a positive definite matrix, be made small, where $\mathbf{x}^{\mathrm{T}}(T)Q(t)\mathbf{x}(t)$ is a measure of the deviation from the zero state and the weighting matrix Q determines how much weight is attached to each of the components of the state.

If we suppose that the complete state can be accurately measured at all times, the quadratic performance index is written as

$$J = \frac{1}{2}\mathbf{x}^{\mathrm{T}}(T)M\mathbf{x}(T) + \frac{1}{2}\int_{t_0}^{T} [\mathbf{x}^{\mathrm{T}}(t)Q(t)\mathbf{x}(t) + \mathbf{u}^{\mathrm{T}}R(t)\mathbf{u}(t)]dt \tag{5.41}$$

This control problem is referred to as the "linear quadratic optimal control problem."

The optimal control $\mathbf{u}^*(t)$ can be obtained by minimizing the performance index, which represents the energy consumption during the control. It can be proved that the control law, indicating the dependence of the optimal control $\mathbf{u}^*(t)$ on both $\mathbf{x}(t)$ and t, is given by

$$\mathbf{u}^*(t) = -K(t)\mathbf{x}(t) \tag{5.42}$$

where the feedback gain matrix $K(t)$ has the form

$$K(t) = R^{-1}(t)B^{\mathrm{T}}(t)P(t)$$

and $P(t)$ is a symmetric matrix, satisfying a matrix Riccati equation in the form

$$-\frac{dP(t)}{dt} = A^{\mathrm{T}}(t)P(t) + P(t)A(t) + Q(t) - P(t)B(t)R^{-1}(t)B^{\mathrm{T}}(t)P(t) \tag{5.43}$$

with the boundary condition

$$P(T) = M \tag{5.44}$$

Then the optimal performance index is

$$J^* = \frac{1}{2}\mathbf{x}^{\mathrm{T}}(t_0)P(t_0)\mathbf{x}(t_0)$$

and optimal trajectory is

$$\begin{aligned} \dot{\mathbf{x}}(t) &= [A(t) - B(t)R^{-1}(t)B^{\mathrm{T}}(t)P(t)]\mathbf{x}(t) \\ \mathbf{x}(t_0) &= \mathbf{x}_0 \end{aligned} \tag{5.45}$$

PROOF

Introduce the Hamilton function to solve the optimal control problem; here it is

$$H = \frac{1}{2}\mathbf{x}^{\mathrm{T}}(t)Q(t)\mathbf{x}(t) + \frac{1}{2}\mathbf{u}^{\mathrm{T}}(t)R(t)\mathbf{u}(t) + \boldsymbol{\lambda}^{\mathrm{T}}(t)A(t)\mathbf{x}(t) + \boldsymbol{\lambda}^{\mathrm{T}}(t)B(t)\mathbf{u}(t) \tag{5.46}$$

Since $\mathbf{u}(t)$ is unspecified, the extremal condition is $\frac{\partial H}{\partial \mathbf{u}} = 0$, that is

$$\frac{\partial H}{\partial \mathbf{u}} = R\mathbf{u}^*(t) + B^{\mathrm{T}}\boldsymbol{\lambda}(t) = 0$$

Hence,

$$\mathbf{u}^*(t) = -R^{-1}(t)B^{\mathrm{T}}(t)\boldsymbol{\lambda}(t) \tag{5.47}$$

Introducing

$$\boldsymbol{\lambda}(t) = P(t)\mathbf{x}(t) \tag{5.48}$$

and substituting into Eq. (5.47), we get seeking Eq. (5.42), and terminate the proof. ■

From Eq. (5.46) $\dot{\mathbf{x}}(t) = \frac{\partial H}{\partial \boldsymbol{\lambda}} = A(t)\mathbf{x}(t) - B(t)R^{-1}(t)B^{\mathrm{T}}(t)\boldsymbol{\lambda}(t)$

Combining with Eq. (5.48), we get Eq. (5.45), which describes the optimal trajectory.

Differentiating Eq. (5.48) gives

$$\dot{\boldsymbol{\lambda}}(t) = \dot{P}(t)\mathbf{x}(t) + P(t)\dot{\mathbf{x}}(t) \tag{5.49}$$

Substituting in Eq. (5.45) gives

$$\dot{\boldsymbol{\lambda}}(t) = [\dot{P}(t) + P(t)A(t) - P(t)B(t)R^{-1}(t)B^{\mathrm{T}}(t)P(t)]\mathbf{x}(t) \tag{5.50}$$

From Eq. (5.46)

$$\dot{\boldsymbol{\lambda}}(t) = -\frac{\partial H}{\partial \mathbf{x}} = -Q(t)\mathbf{x}(t) - A^{\mathrm{T}}(t)\boldsymbol{\lambda}(t)$$

Combining with Eq. (5.48)

$$\dot{\boldsymbol{\lambda}}(t) = -[Q(t) + A^{\mathrm{T}}(t)P(t)]\mathbf{x}(t)$$

Compared with Eq. (5.50), we have

$$-\frac{dP(t)}{dt} = A^{\mathrm{T}}(t)P(t) + P(t)A(t) + Q(t) - P(t)B(t)R^{-1}(t)B^{\mathrm{T}}(t)P(t)$$

which is the Riccati equation (5.43).

For free terminal time, the boundary condition is

$$\boldsymbol{\lambda}(T) = \frac{\partial}{\partial \mathbf{x}(T)}\left[\frac{1}{2}\mathbf{x}^{\mathrm{T}}(T)M\mathbf{x}(T)\right] = M\mathbf{x}(T)$$

at terminal time $t = T$.

From Eq. (5.48), $\lambda(T) = P(T)\mathbf{x}(T)$

Hence $P(T) = M$

It can also be proved from the Riccati equation and its boundary conditions that the matrix $P(t)$ is unique, symmetric, and positive semidefinite.

As $T \to \infty$, corresponding to a steady LQR problem, $P(t)$, denoted as $\overline{P}$ here, satisfies the equation called algebraic Riccati equation (ARE) in the following form:

$$A^{\mathrm{T}}\overline{P} + \overline{P}A + Q - \overline{P}BR^{-1}B^{\mathrm{T}}\overline{P} = 0 \tag{5.51}$$

Then, we have

$$\mathbf{u}^{*}(t) = -R^{-1}B^{\mathrm{T}}\overline{P}\mathbf{x}(t) \tag{5.52}$$

The feedback gain matrix K and solution $\overline{P}$ of the Riccati equation (5.51) can be obtained by means of commercial software, such as MATLAB.[11, 12]

Designing control law using this optimization approach discussed above is referred to as a LQR design.

In a more realistic case, the output $\mathbf{y}(t)$ is induced as

$$\mathbf{y}(t) = C(t)\mathbf{x}(t) \tag{5.39}$$

Then, the quadratic performance index is written as

$$J = \frac{1}{2}\mathbf{y}^{\mathrm{T}}(T)M\mathbf{y}(T) + \frac{1}{2}\int_{t_0}^{\mathrm{T}} [\mathbf{y}^{\mathrm{T}}(t)Q(t)\mathbf{y}(t) + \mathbf{u}^{\mathrm{T}}R(t)\mathbf{u}(t)]dt$$

Substituting in Eq. (5.52) gives

$$J = \frac{1}{2}\mathbf{x}^{\mathrm{T}}(T)M_1\mathbf{x}(T) + \frac{1}{2}\int_{t_0}^{\mathrm{T}} [\mathbf{x}^{\mathrm{T}}(t)Q_1(t)\mathbf{x}(t) + \mathbf{u}^{\mathrm{T}}(t)R(t)\mathbf{u}(t)]dt \tag{5.53}$$

where $M_1 = C^{\mathrm{T}}(T)MC(T)$, $Q_1(t) = C^{\mathrm{T}}(t)Q(t)C(t)$

Hence Eq. (5.40) is still available, only if the Riccati equation is written as

$$-\frac{dP(t)}{dt} = A^{\mathrm{T}}(t)P(t) + P(t)A(t) + C^{\mathrm{T}}(t)Q(t)C(t) - P(t)B(t)R^{-1}(t)B^{\mathrm{T}}(t)P(t) \tag{5.54}$$

with the boundary condition

$$P(T) = C^{\mathrm{T}}(T)MC(T)$$

5.2.1.2 Linear Quadratic Tracking System

In an optical control, the outputs of the system are required to follow or track a desired trajectory in an optimal sense. For the regulator discussed above, the desired trajectory is simply the zero state. If the desired trajectory is a particular prescribed function of time $\mathbf{y}_l(t)$, the problem is called a tracking problem. Then there should be a cost term in the performance index involving the error vector $\mathbf{e}(t)$ defined as

$$\mathbf{e}(t) = \mathbf{y}_l(t) - \mathbf{y}(t)$$

which indicates the difference between the actual output vector $y(t)$ and the prescribed output vector $\mathbf{y}_l(t)$.

The performance index is as follows:

$$J = \lim_{t_f \to \infty} \int_{t}^{t_f} [\mathbf{e}^{\mathrm{T}}(t) Q \mathbf{e}(t) + \mathbf{u}^{\mathrm{T}}(t) R \mathbf{u}(t)] dt \tag{5.40}$$

Consider a finite time interval $[0, T]$, then

$$J = \frac{1}{2}\mathbf{e}^{\mathrm{T}}(T) M \mathbf{e}(T) + \frac{1}{2}\int_{t_0}^{\mathrm{T}} [\mathbf{e}^{\mathrm{T}}(t) Q(t) \mathbf{e}(t) + \mathbf{u}^{\mathrm{T}}(t) R(t) \mathbf{u}(t)] dt \tag{5.41}$$

where $\mathbf{e}^{\mathrm{T}}(t) Q \mathbf{e}(t)$ is a tracing error in the control and the weighting matrix Q determines how much weight is attached to each of the components of $\mathbf{e}(t)$. Since Q is a positive semidefinite, thus $\mathbf{e}^{\mathrm{T}}(t) Q \mathbf{e}(t) \geq 0$, which represents the penalty at time t for output trajectories $\mathbf{y}(t)$ which deviate from the desired trajectory $\mathbf{y}_l(t)$. $\mathbf{u}^{\mathrm{T}}(t) R \mathbf{u}(t)$ denotes the energy consumption in the control in trying to regulate $\mathbf{e}(t)$ to zero, since R is a positive definite, $\mathbf{u}^{\mathrm{T}}(t) R \mathbf{u}(t) > 0$. $\mathbf{e}^{\mathrm{T}}(T) M \mathbf{e}(T)$ is the final control precision at time T, where M is a positive semidefinite. The entire performance index reflects the cumulative penalty incurred over the time interval.

Substituting $\mathbf{e}(t) = \mathbf{y}_l(t) - \mathbf{y}(t) = \mathbf{y}_l(t) - C(t)\mathbf{x}(t)$ into the performance index (Eq. (5.41)), we have

$$\begin{aligned} J = {} & \frac{1}{2}[y_l(T) - C(T)\mathbf{x}(T)]^{\mathrm{T}} M [y_l(T) - C(T)\mathbf{x}(T)] \\ & + \frac{1}{2}\int_{t_0}^{\mathrm{T}} \{[y_l(T) - C(T)\mathbf{x}(T)]^{\mathrm{T}}(t) Q_1(t) [y_l(T) - C(T)\mathbf{x}(T)] + \mathbf{u}^{\mathrm{T}} R(t) \mathbf{u}(t)\} dt \end{aligned} \tag{5.55}$$

It can be proved that

$$\mathbf{u}^*(t) = -R^{-1}(t) B^{\mathrm{T}}(t) [P(t)\mathbf{x}(t) - \mathbf{g}(t)] \tag{5.56}$$

where $P(t)$ satisfies the following Riccati equation:

$$-\frac{dP(t)}{dt} = A^{\mathrm{T}}(t)P(t) + P(t)A(t) + C(t)Q(t)C^{\mathrm{T}}(t) - P(t)B(t)R^{-1}(t)B^{\mathrm{T}}(t)P(t) \tag{5.57}$$

with the boundary condition

$$P(T) = C^{\mathrm{T}}(T) M C(T) \tag{5.58}$$

And $\mathbf{g}(t)$ satisfies the equation as

$$-\dot{\mathbf{g}}(t) = [A(t) - B(t)R^{-1}(t)B^{\mathrm{T}}(t)P(t)]^{\mathrm{T}}\mathbf{g}(t) + C^{\mathrm{T}}(t)Q(t)\mathbf{y}_l(t) \tag{5.59}$$

with the boundary condition

$$\mathbf{g}(T) = C^{\mathrm{T}}(T)M\mathbf{y}_l(T) \tag{5.60}$$

And optimal trajectory

$$\dot{\mathbf{x}}(t) = [A(t) - B(t)R^{-1}(t)B^{\mathrm{T}}(t)P(t)]^{\mathrm{T}}\mathbf{x}(t) + B(t)R^{-1}(t)B^{\mathrm{T}}(t)\mathbf{g}(t) \tag{5.61}$$

with an initial condition

$$\mathbf{x}(t_0) = \mathbf{x}_0 \tag{5.62}$$

PROOF

The Hamilton function for this problem is

$$H = \frac{1}{2}[\mathbf{y}_l(t) - C(t)\mathbf{x}(t)]^{\mathrm{T}}Q(t)[\mathbf{y}_l(t) - C(t)\mathbf{x}(t)] + \frac{1}{2}\mathbf{u}^{\mathrm{T}}(t)R(t)\mathbf{u}(t)$$
$$+\mathbf{x}^{\mathrm{T}}(t)A^{\mathrm{T}}(t)\boldsymbol{\lambda}(t) + \mathbf{u}^{\mathrm{T}}(t)B^{\mathrm{T}}(t)\boldsymbol{\lambda}(t)$$

In terms of the extremal condition

$$\frac{\partial H}{\partial \mathbf{u}} = \mathbf{R}(t)\mathbf{u}(t) + \mathbf{B}^{\mathrm{T}}(t)\boldsymbol{\lambda}(t) = 0$$

we have $\mathbf{u}(t) = -R^{-1}(t)B^{\mathrm{T}}(t)\boldsymbol{\lambda}(t)$ (5.63)

Since $\frac{\partial^2 H}{\partial \mathbf{u}^2} = R(t) > 0$, Eq. (5.63) results in the minimum H.

From the canonical equations

$$\dot{\mathbf{x}}(t) = \frac{\partial H}{\partial \boldsymbol{\lambda}} = A(t)\mathbf{x}(t) - B(t)R^{-1}(t)B^{\mathrm{T}}(t)\boldsymbol{\lambda}(t) \tag{5.64}$$

and

$$\dot{\boldsymbol{\lambda}}(t) = -\frac{\partial H}{\partial \mathbf{x}} = -C^{\mathrm{T}}(t)Q(t)C(t)\mathbf{x}(t) - A^{\mathrm{T}}(t)\boldsymbol{\lambda}(t) + C^{\mathrm{T}}(t)Q(t)\mathbf{y}_l(t) \tag{5.65}$$

Introducing

$$\boldsymbol{\lambda}(t) = P(t)\mathbf{x}(t) - \mathbf{g}(t) \tag{5.66}$$

$$\dot{\boldsymbol{\lambda}}(t) = \dot{P}(t)\mathbf{x}(t) + P(t)\dot{\mathbf{x}}(t) - \dot{\mathbf{g}}(t) \tag{5.67}$$

with the transversality condition

$$\boldsymbol{\lambda}(t_f) = \frac{\partial}{\partial \mathbf{x}(t_f)}\left\{\frac{1}{2}\mathbf{e}^{\mathrm{T}}(t_f)M\mathbf{e}(t_f)\right\} = C^{\mathrm{T}}(t_f)MC(t_f)\mathbf{x}(t_f) - C^{\mathrm{T}}(t_f)M\mathbf{y}_l(t_f) \tag{5.68}$$

Substituting Eq. (5.66) into Eq. (5.64), we get

$$\dot{\mathbf{x}}(t) = [A(t) - B(t)R^{-1}(t)B^{\mathrm{T}}(t)P(t)]\mathbf{x}(t) + B(t)R^{-1}(t)B^{\mathrm{T}}(t)\mathbf{g}(t)$$

Substituting in Eq. (5.67)

$$\begin{aligned}\dot{\boldsymbol{\lambda}}(t) = {} & [\dot{P}(t) + P(t)A(t) - P(t)B(t)R^{-1}(t)B^{\mathrm{T}}(t)P(t)]\mathbf{x}(t) \\ & + P(t)B(t)R^{-1}(t)B^{\mathrm{T}}(t)\mathbf{g}(t) - \dot{\mathbf{g}}(t)\end{aligned} \tag{5.69}$$

From Eq. (5.65) and Eq. (5.66)

$$\dot{\boldsymbol{\lambda}}(t) = [-C^{\mathrm{T}}(t)Q(t)C(t) - A^{\mathrm{T}}(t)P(t)]\mathbf{x}(t) + A^{\mathrm{T}}(t)\mathbf{g}(t) + C^{\mathrm{T}}(t)Q(t)\mathbf{y}_l(t) \tag{5.70}$$

Comparing Eqs. (5.69) and (5.70), we get seeking Eqs. (5.57) and (5.59), and terminate the proof. ■

When $t = t_f$, from Eq. (5.66), we have

$$\boldsymbol{\lambda}(t_f) = P(t_f)\mathbf{x}(t_f) - \mathbf{g}(t_f)$$

Compared with Eq. (5.68), the boundary conditions are expressed by Eqs. (5.58) and (5.60).

5.2.2 Discrete Linear Quadratic Systems

A linear and discrete time system in state-space form is written as

$$\mathbf{x}(k+1) = A(k)\mathbf{x}(k) + B(k)\mathbf{u}(k), \quad \mathbf{x}(0) = \mathbf{x}_0 \tag{5.71}$$

where $k = 0, 1, 2, \ldots, N-1$, and the performance index

$$J = \frac{1}{2}\mathbf{x}^{\mathrm{T}}(N)M\mathbf{x}(N) + \frac{1}{2}\sum_{k=o}^{N-1}[\mathbf{x}^{T}(k)Q(k)\mathbf{x}(k) + \mathbf{u}^{\mathrm{T}}(k)R(k)\mathbf{u}(k)]dt \tag{5.72}$$

It can be proved that there exists a unique set of linear optimal control laws

$$\mathbf{u}^*(k) = -K(k)\mathbf{x}(k), \quad k = 0, 1, 2, \ldots, N-1 \tag{5.73}$$

which leads to the optimal performance index in the form

$$J^*[x(0), 0] = \frac{1}{2}\mathbf{x}^{\mathrm{T}}(0)P(0)\mathbf{x}(0)$$

and the feedback gain matrix

$$K(k) = [B^{\mathrm{T}}(k)P(k+1)B(k) + R(k)]^{-1}B^{\mathrm{T}}(k)P(k+1)A(k) \tag{5.74}$$

where $P(k)$ is a certain nonnegative definite symmetric solution of the following Riccati equation:

$$P(k) = [A(k) - B(k)K(k)]^{\mathrm{T}} P(k+1)[A(k) - B(k)K(k)] + K^{\mathrm{T}}(k)R(k)K(k) + Q(k) \quad k = 0, 1, 2, \ldots, N-1 \tag{5.75}$$

with boundary condition $P(N) = M$

Considering the output $\mathbf{y}(t)$, we have

$$\mathbf{x}(k+1) = A(k)\mathbf{x}(k) + B(k)\mathbf{u}(k), \quad \mathbf{x}(0) = \mathbf{x}_0$$
$$\mathbf{y}(k) = C(k)\mathbf{x}(k) \qquad k = 0, 1, 2, \ldots, N-1 \tag{5.76}$$

and the performance index

$$J = \frac{1}{2}\mathbf{y}^{\mathrm{T}}(N)M\mathbf{y}(N) + \frac{1}{2}\sum_{k=0}^{N-1}[\mathbf{y}^{\mathrm{T}}(k)Q(k)\mathbf{y}(k) + \mathbf{u}^{\mathrm{T}}(k)R(k)\mathbf{u}(k)] \tag{5.77}$$

An existing unique set of linear optimal control laws

$$\mathbf{u}^*(k) = -K(k)\mathbf{x}(k), \qquad k = 0, 1, 2, \ldots, N-1 \tag{5.78}$$

leads to the optimal performance index in the form

$$J^*[x(0), 0] = \frac{1}{2}\mathbf{x}^{\mathrm{T}}(0)P(0)\mathbf{x}(0)$$

and the feedback gain matrix

$$K(k) = [B^{\mathrm{T}}(k)P(k+1)B(k) + R(k)]^{-1}B^{\mathrm{T}}(k)P(k+1)A(k)$$

where $P(k)$ satisfies the following Riccati equation:

$$P(k) = [A(k) - B(k)K(k)]^{\mathrm{T}} P(k+1)[A(k) - B(k)K(k)] + K^{\mathrm{T}}(k)R(k)K(k) + C^{\mathrm{T}}(k)Q(k)C(k) \quad k = 0, 1, 2, \ldots, N-1 \tag{5.79}$$

with the boundary condition $P(N) = C^{\mathrm{T}}(N)MC(N)$

5.2.3 Linear Quadratic Gaussian (LQG) Control in the Presence of Noise

In practice, the state equation is usually disturbed by a disturbance noise, which influences the plant in an unpredictable way. Meanwhile the observed variable is contaminated with observation noise so that the observed variable may not directly give information of the plant. The plants we shall consider are of the form

$$\frac{d\mathbf{x}(t)}{dt} = A\mathbf{x}(t) + B\mathbf{u}(t) + \mathbf{v}(t)$$
$$\mathbf{y}(t) = C\mathbf{x}(t) + \mathbf{w}(t) \tag{5.80}$$

where $\mathbf{y}(t)$ is the observed variable, $\mathbf{v}(t)$ and $\mathbf{w}(t)$ represent the state excitation and observation noise, respectively, assumed to be white, Gaussian, and to have zero mean, which implies that all probabilistic information about noise is summed up in the covariance of the noise. Therefore, in the mathematical terms

$$E(\mathbf{v}(t)\mathbf{v}^{\mathrm{T}}(\tau)) = Q_e(t)\delta(t-\tau) \quad E(\mathbf{v}(t)) = 0 \tag{5.81}$$

$$E(\mathbf{w}(t)\mathbf{w}^{\mathrm{T}}(\tau)) = \mathbf{w}_e(t)\delta(t-\tau) \quad E(w(t)) = 0 \tag{5.82}$$

where E is the expectation operator and $Q_e(t)$ and $\mathbf{w}_e(t)$ are covariance, where $Q_e(t)$ is symmetric and nonnegative definite and $\mathbf{w}_e(t)$ is the positive definite.

Since the state variables are random variables associated with the noises, the performance index is defined in the following form with the expectation operators:

$$J = E\left\{\frac{1}{2}\mathbf{x}^{\mathrm{T}}(T)\,\mathbf{M}\mathbf{x}(T) + \frac{1}{2}\int_{t_0}^{\mathrm{T}}(\mathbf{x}^{\mathrm{T}}Q\mathbf{x} + \mathbf{u}^{\mathrm{T}}R\mathbf{u})dt\right\}$$

To deal with the optimal problem concerning the noise, two steps can be made: first, designing an estimator to get state estimate $\hat{\mathbf{x}}$ by solving the optimal estimation or filtering problem, and second, designing a controller to get the control law gain matrix K, which is independent of the noise, by using state estimate $\hat{\mathbf{x}}$. It can be proved that the calculations of state estimate $\hat{\mathbf{x}}$ and gain matrix K are separate problems, which are tackled independently. In the following discussions, some proofs are omitted for simplicity.

First, we shall get "best possible" estimated state $\hat{\mathbf{x}}$ by solving an optimal filtering problem, known as the Kalman–Bucy filter, described by the following equation:

$$\frac{d\hat{\mathbf{x}}}{dt} = A\hat{\mathbf{x}} + B\mathbf{u}(t) + K_e(C\hat{\mathbf{x}}(t) - \mathbf{y}) \tag{5.83}$$

where Kalman–Bucy filter gains $K_e = -P_eC^{\mathrm{T}}\mathbf{w}_e^{-1}$ and P_e satisfies the following Riccati equation:

$$-\frac{dP_e}{dt} = A^TP_e + P_eA + Q_e - P_eC^Tw_e^{-1}CP_e \tag{5.84}$$

Hence, the Kalman–Bucy filter can produce an estimated $\hat{\mathbf{x}}$ that has minimal mean square error among all estimates, by processing the data $\mathbf{y}(t)$ and $\mathbf{u}(t)$.

Then, we have the optimal control law

$$\mathbf{u}^*(t) = -K\hat{\mathbf{x}}(t) = -R^{-1}BP\hat{\mathbf{x}}(t) \tag{5.85}$$

where, P satisfies the following Riccati equation:

$$-\frac{dP(t)}{dt} = A^{\mathrm{T}}(t)P(t) + P(t)A(t) + Q(t) - P(t)B(t)R^{-1}(t)B^{\mathrm{T}}(t)P(t) \tag{5.86}$$

It is obvious that the control law of the stochastic optimal linear control problem in the presence of noise described by Eq. (5.85) is the same as that for the optimal linear regulator problem described by Eq. (5.42), except that the state $\mathbf{x}(t)$ is replaced with its minimum mean square linear estimate $\hat{\mathbf{x}}(t)$. The control and estimation problems are independent, which is called the "separation principle."

5.3 Linear Process in Near-Wall Turbulent Flow[13–16]

Incompressible Navier–Stokes equations for a channel flow in the nondimensional form are

$$\begin{aligned}\frac{\partial u_i}{\partial t} &= -\frac{\partial}{\partial x_k}u_i u_k + \frac{1}{\mathrm{Re}}\frac{\partial^2 u_i}{\partial x_k \partial x_k} - \frac{\partial p}{\partial x_i}\\ \frac{\partial u_i}{\partial x_i} &= 0\end{aligned} \tag{5.87}$$

where the flow variables are normalized by the channel half-height δ and centerline velocity U_c.

To describe the importance of the linear mechanism in wall-bounded turbulent flows and the reliability of turbulent control with a linear control theory, we first decompose the Navier–Stokes equations into linear and nonlinear parts and then discuss the linear process in turbulence. The flow is statistically homogeneous in the streamwise (x) and the spanwise (z) directions. By representing the wall-normal velocity v and the wall-normal vorticity ω_y in terms of Fourier transform in the streamwise and spanwise directions, Eq. (5.87) can be written in an operator form

$$\frac{d}{dt}\begin{bmatrix}\widetilde{v}\\ \widetilde{\omega}_y\end{bmatrix} = A\begin{bmatrix}\widetilde{v}\\ \widetilde{\omega}_y\end{bmatrix} + \begin{bmatrix}N_v\\ N_{\omega_y}\end{bmatrix} \tag{5.88}$$

where all nonlinear parts are lumped into N_v and N_{ω_y}.

The linearized Navier–Stokes system operator is defined as

$$A = \begin{bmatrix} L_{\mathrm{os}} & 0\\ L_c & L_{\mathrm{sq}}\end{bmatrix}$$

where

$$\begin{aligned}L_{\mathrm{os}} &= \Delta^{-1}\left(-ik_x U\Delta + ik_x\frac{d^2U}{dy^2} + \frac{1}{\mathrm{Re}}\Delta^2\right)\\ L_{\mathrm{sq}} &= -ik_x U + \frac{1}{\mathrm{Re}}\Delta\\ L_c &= -ik_z\frac{dU}{dy}\end{aligned}$$

L_c is the coupling operator, coupling $\widetilde{v}$ to $\widetilde{\omega}_y$, k_x and k_z are the streamwise and spanwise wavenumbers, respectively, $\Delta = \frac{\partial^2}{\partial y^2} - k^2$, $k^2 = k_x^2 + k_z^2$, and U is the mean velocity about which the Navier–Stokes equations are linearized.

Dropping the coupling term, L_c, Eq. (5.88) is written as

$$\frac{d}{dt}\begin{bmatrix}\tilde{v}\\ \tilde{\omega}_y\end{bmatrix}=\begin{bmatrix}L_{\text{os}} & 0\\ 0 & L_{\text{sq}}\end{bmatrix}\begin{bmatrix}\tilde{v}\\ \tilde{\omega}_y\end{bmatrix}+\begin{bmatrix}N_v\\ N_{\omega_y}\end{bmatrix} \tag{5.89}$$

This modified system can be viewed as representing a synthetic turbulent flow without the coupling term, or a controlled turbulent flow in which the coupling term is suppressed.

On the other hand, removing the nonlinear terms, Eq. (5.88) is written as

$$\frac{d}{dt}\begin{bmatrix}\tilde{v}\\ \tilde{\omega}_y\end{bmatrix}=\begin{bmatrix}L_{\text{os}} & 0\\ L_c & L_{\text{sq}}\end{bmatrix}\begin{bmatrix}\tilde{v}\\ \tilde{\omega}_y\end{bmatrix} \tag{5.90}$$

which is called the linearized Navier–Stokes equation.

The effects of the coupling term and nonlinear terms on the generation and self-sustaining mechanism of near-wall turbulence can be discussed by solving the above three modified nonlinear equations: Case 1, regular Navier–Stokes equation (5.88); Case 2, modified equations without coupling term (5.89); and Case 3, modified equations without nonlinear terms (5.90). Considering an initial velocity field with the mean velocity and random disturbances, where there are no organized turbulence structures present, a spectral method is used to get the numerical solutions of the above equations. Several snapshot distributions of the streamwise vorticity in a $y-z$ plane at different times $t^+ = 20$, 40, and 80, are shown in Figure 5.3, from the first column to the third column, respectively. Figure 5.3a shows the velocity field in Case 1, while Figures 5.3b and c represent Cases 2 and 3, respectively. It is evident that organized structures are discernible in all three cases at $t^+ = 20$, but they look different from each other.

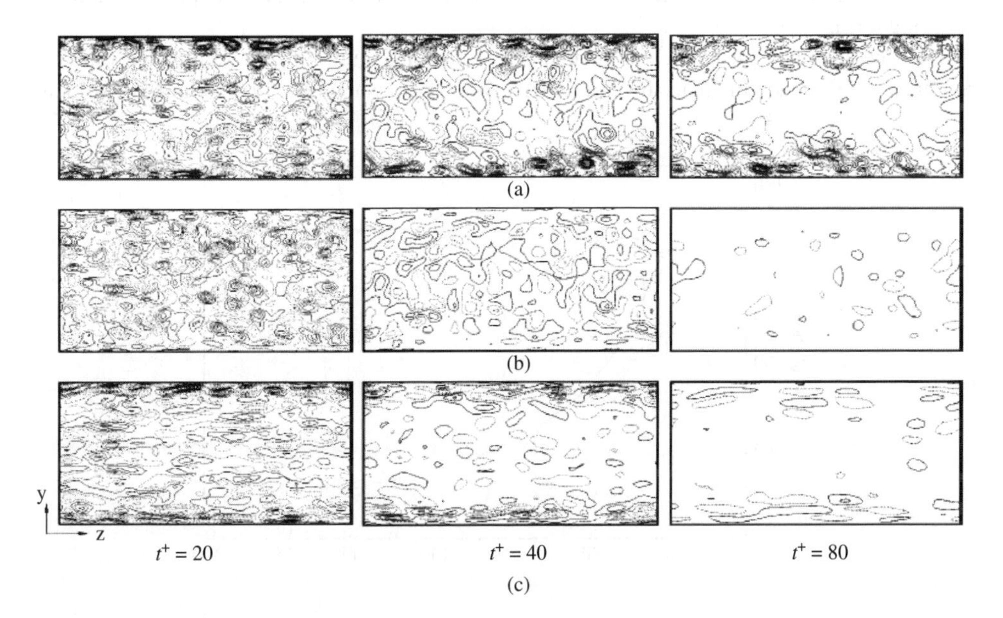

Figure 5.3 Contours of the streamwise vorticity in a $y-z$ plane for different cases.[13] *Source*: Kim J 2000. Reproduced with permission of Cambridge University Press

For Case 2 (without the linear coupling term), shown in Figure 5.3b, the vortical structures appearing at $t^+ = 20$ disappear quickly at $t^+ = 40$, and vanish completely in the wall region at $t^+ = 80$. These indicate that the linear process related to the linear coupling term plays an important role in nonlinear turbulent boundary layers. The vortical structures cannot be maintained without the linear coupling term, though it may be generated at an early time. It is also shown that the formation of streaky structure depends on $\overline{(v\omega_z)}$, which is related to the linear coupling term, where "–" indicates streamwise average; thus the streaky structure cannot be produced without the linear coupling term.

For Case 3 (without the nonlinear term), shown in Figure 5.3c, the vortical structures still exist near the wall at $t^+ = 40$, but the structures have larger spanwise scales than those in Case 1, which indicates that the instability of the streaks strengthens the streamwise vortices through a nonlinear process.

Finally, at $t^+ = 80$, the near-wall turbulence is self-sustained only for Case 1, shown in Figure. 5.3a. Hence, both the nonlinear terms and the linear coupling terms are necessary for the formation and maintaining of these structures in near-wall turbulence, otherwise turbulence will cease to exist in the absence of either mechanism.

A flow field of a channel turbulent flow with $Re_\tau = 100$ for Case 1 computed by the spectral method is used as an initial field, and starting from this initial field, the coupling term is dropped from the governing equations in the upper half of the channel. The calculated results are shown in Figures 5.4–5.6. Figure 5.4 shows time evolution of the mean shear at the both walls, where solid line is for the upper wall while dashed line is for the lower wall. Thick line is for Case 1 and thin line is for Case 2, that is, $L_c = 0$, in the upper half of the channel starting from $t^+ = 0$. It is shown that the wall shear is reduced drastically due to the absence of the coupling term.

Several snapshots of the velocity field described by contours of streamwise vorticity are shown in Figure 5.5. It is evident that streamwise vortices quickly disappear without the coupling term.

Root-mean-square turbulence intensities are shown in Figure 5.6, where the solid line represents the turbulence intensity for streamwise velocity, dashed line for wall-normal velocity,

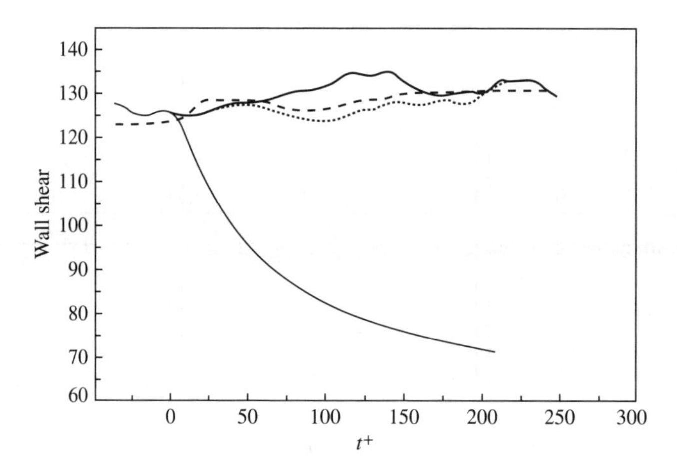

Figure 5.4 Time evolution of mean shear at wall.[13] *Source*: Kim J 2000. Reproduced with permission of Cambridge University Press

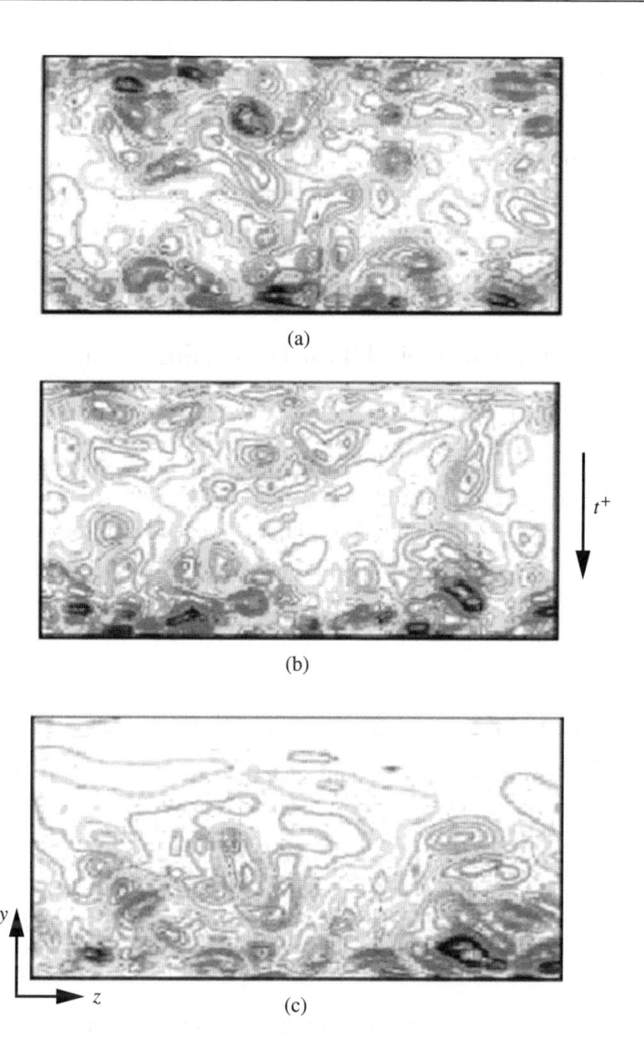

Figure 5.5 Contours of streamwise vorticity in $y - z$ plane at different times.[13] *Source*: Kim J 2000. Reproduced with permission of Cambridge University Press

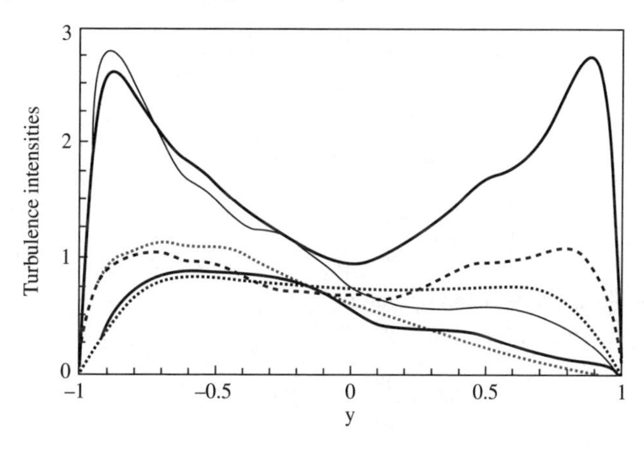

Figure 5.6 Root-mean-square turbulence intensities.[13] *Source*: Kim J 2000. Reproduced with permission of Cambridge University Press

and dash dotted line for spanswise velocity. The thick line is for $t^+ = 0$ while the thin line is for $t^+ = 180$. It is shown that the turbulence intensities are reduced drastically due to the absence of the coupling term.

Since the near-wall streamwise vortices, which have been found to be responsible for high skin-friction drag in turbulent boundary layer, cannot be sustained without the linear coupling term, these can be suppressed by reducing the effect of the coupling term in the wall region. Therefore, the near-wall turbulent flow can be controlled effectively by a linear control theory.

5.4 Linear Optimal Control of Two-Dimensional Flow[17–21]

In modern control theory, the linear optimal control is simple to implement physically and is mature in theory. All linear optimal control problems have readily computable solutions obtained by the commercial software. A crucial problem to discuss in the linear optimal flow control is how to transform the equations from the Navier–Stokes form in fluid mechanism into the state-space form in traditional control theory. For two-dimensional linearized systems described by Eq. (5.88), the $\widetilde{v}$ equation is completely decoupled with the $\widetilde{\omega}_y$ equation, since the coupling term $L_c\widetilde{v}$ vanishes due to $k_z = 0$. In many cases, only $\widetilde{v}$ equation is considered for the control design, in which stream function Ψ is induced replacing v, which will be discussed next.

5.4.1 Linearization of Navier–Stokes Equations

Consider small perturbations in a two-dimensional plane flow between two parallel stationary plates as shown in Figure 5.7. The flow variables can be written as

$$u(x, y, t) = U(y) + u'(x, y, t)$$

$$v(x, y, t) = v'(x, y, t)$$

$$p(x, y, t) = P(y) + p'(x, y, t)$$

where $U(y) = U_c(1 - y^2)$ is the base flow distributing in the y direction, $P(y)$ is the base pressure, and superscript "$'$" represents the perturbation state.

Substituting in Eq. (5.87), we have

$$\begin{aligned}
&\frac{\partial u'}{\partial t} + U\frac{\partial u'}{\partial x} + v'\frac{dU}{dy} = -\frac{\partial p'}{\partial x} + \frac{1}{\mathrm{Re}}\Delta u' \\
&\frac{\partial v'}{\partial t} + U\frac{\partial v'}{\partial x} = -\frac{\partial p'}{\partial y} + \frac{1}{\mathrm{Re}}\Delta v' \\
&\frac{\partial u'}{\partial x} + \frac{\partial v'}{\partial y} = 0
\end{aligned} \tag{5.91}$$

where $\Delta = \frac{\partial^2}{\partial x^2} + \frac{\partial^2}{\partial y^2}$ is a Laplacian operator.

By introducing a "stream function", Ψ,

$$u' = \frac{\partial \Psi}{\partial y}, \qquad v' = -\frac{\partial \Psi}{\partial x}$$

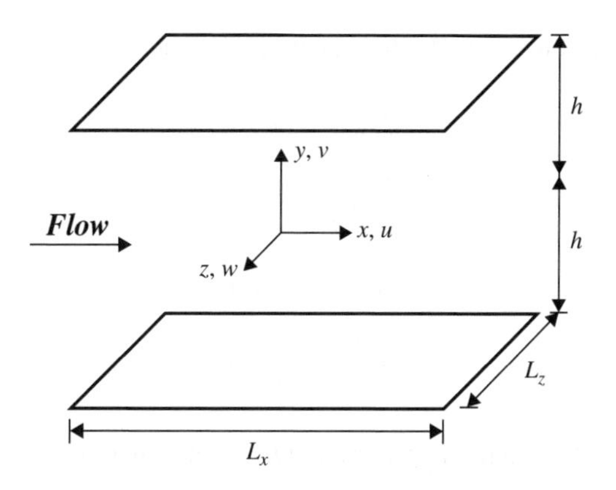

Figure 5.7 Incompressible channel flow

from Eq. (5.91), we have

$$\left(\frac{\partial}{\partial t} + U\frac{\partial}{\partial x}\right)\Delta\Psi - \frac{d^2U}{dy^2}\frac{\partial \Psi}{\partial x} = \mathrm{Re}^{-1}\Delta(\Delta\Psi) \tag{5.92}$$

Assume the periodic boundary condition in the streamwise (x) direction and no-slip boundary conditions in the rigid plates, then

$$\begin{aligned} &\Psi(x, y = -1, t) = 0 \\ &\frac{\partial \Psi}{\partial y}(x, y = -1, t) = 0 \\ &\Psi(x, y = 1, t) = 0 \\ &\frac{\partial \Psi}{\partial y}(x, y = 1, t) = 0 \end{aligned} \tag{5.93}$$

And an initial condition is

$$\Psi(x, y, t = 0) = g(x, y) \tag{5.94}$$

The boundary value problem is completely formed, which does not include any control terms.

While the flow is controlled by blowing/suction at the lower wall, the boundary conditions should now be modified to include boundary input. Represented as the known separable function $q(t)w(x)f(y)$, we have

$$\begin{aligned} &\Psi(x, y = -1, t) = q(t)w(x)f(y = -1) \\ &\frac{\partial \Psi}{\partial y}(x, y = -1, t) = q(t)w(x)\frac{\partial f(y = -1)}{\partial y} = 0 \\ &\Psi(x, y = 1, t) = 0 \\ &\frac{\partial \Psi}{\partial y}(x, y = 1, t) = q(t)w(x)\frac{\partial f(y = 1)}{\partial y} = 0 \end{aligned} \tag{5.95}$$

Note that these conditions constrain the function $f(y)$ such that

$$f(y=-1) \neq 0$$

$$\frac{\partial f(y=-1)}{\partial y} = 0$$

$$f(y=1) = 0$$

$$\frac{\partial f(y=1)}{\partial y} = 0$$

Many functions may be equally appropriate. One such function is

$$f(y) = \frac{1}{2}y^4 + \frac{1}{4}y^3 - y^2 - \frac{3}{4}y + 1$$

Equation (5.92), that is, an equation for a two-dimensional channel flow on the perturbation fields, is a homogeneous high-order partial differential equation with the inhomogeneous boundary condition (5.95).

Taking the normal velocity on the lower wall as a control variable, or an input variable produces

$$v(x, y=-1, t) = -q(t)\frac{\partial w(x)}{\partial x}f(y=-1) \tag{5.96}$$

5.4.2 *Spectral Decomposition of Linearized Flow*

To convert the above equations into the equations with the form being identical with Eq. (5.38), the equations should be modified as follows.

5.4.2.1 Inhomogeneous Equations and Homogeneous Boundary Conditions

Introducing

$$\Phi(x, y, t) = \Psi(x, y, t) - q(t)f(y)w(x) \tag{5.97}$$

and substituting into Eq. (5.92) lead to

$$\begin{aligned}
\frac{\partial}{\partial t}\frac{\partial^2\Phi}{\partial x^2} + \frac{\partial}{\partial t}\frac{\partial^2\Phi}{\partial y^2} &= -U(y)\frac{\partial^3\Phi}{\partial x^3} - U(y)\frac{\partial}{\partial x}\frac{\partial^2\Phi}{\partial y^2} + \frac{d^2U(y)}{dy^2}\frac{\partial\Phi}{\partial x} + \frac{1}{R_e}\frac{\partial^4\Phi}{\partial x^2} \\
&+ \frac{2}{R_e}\frac{\partial^2}{\partial x^2}\frac{\partial^2\Phi}{\partial y^2} + \frac{1}{R_e}\frac{\partial^4\Phi}{\partial y^4} - \frac{\partial q}{\partial t}\frac{\partial^2 w(x)}{\partial x^2}f(y) - \frac{\partial q}{\partial t}w(x)\frac{\partial^2 f(y)}{\partial y^2} \\
&- q(t)\frac{\partial^3 w(x)}{\partial x^3}U(y)f(y) - q(t)\frac{\partial w(x)}{\partial x}U(y)\frac{\partial^2 f(y)}{\partial y^2} + q(t)\frac{\partial w(x)}{\partial x}\frac{d^2U(y)}{dy^2}f(y) \\
&+ \frac{1}{R_e}q(t)\frac{\partial^4 w(x)}{\partial x^4}f(y) + \frac{2}{R_e}q(t)\frac{\partial^2 w(x)}{\partial x^2}\frac{\partial^2 f(y)}{\partial y^2} + \frac{1}{R_e}q(t)w(x)\frac{\partial^4 f(y)}{\partial y^4}
\end{aligned} \tag{5.98}$$

with boundary conditions as

$$\begin{aligned}
&\Phi(y=-1)=0\\
&\frac{\partial\Phi(y=-1)}{\partial y}=0\\
&\Phi(y=1)=0\\
&\frac{\partial\Phi(y=1)}{\partial y}=0
\end{aligned}\tag{5.99}$$

The streamwise component of shear at a single boundary point $(x_i, -1)$ is taken as an output of a control system, which is given by

$$z(x_i, y=-1, t)=\frac{\partial u(x_i, y=-1, t)}{\partial y}=\frac{\partial^2\Psi(x_i, y=-1, t)}{\partial y^2}$$

By observing Eq. (5.97)

$$z(x_i, y=-1, t)=\frac{\partial^2\Phi(x_i, y=-1, t)}{\partial y^2}+q(t)\frac{\partial^2 f(y=-1)}{\partial y^2}w(x_i)\tag{5.100}$$

5.4.2.2 Combined Fourier–Chebyshev Transform

An approximate solution of Eq. (5.98) can be expanded into the Fourier transform series in x direction and the Chebyshev polynomials in the y direction, and written in the truncated form as

$$\Phi(x, y, t)=\sum_{n=-N}^{N}\sum_{m=0}^{M}a_{nm}(t)P_n(x)T_m(y)\tag{5.101}$$

where Fourier function $P_n(x)=e^{in\alpha_0 x}$, a_{nm} is the spectral coefficient of solution Φ, α_0 is the fundamental wavenumber in the x direction, defined as $2\pi/L$, and L is the nondimensional length of the finite-length channel.

Substitution in Eq. (5.98) leads to

$$\begin{aligned}
&\sum_{n=-N}^{N}\sum_{m=0}^{M}\frac{da_{nm}(t)}{dt}\frac{d^2P_n(x)}{dx^2}T_m(y)+\sum_{n=-N}^{N}\sum_{m=0}^{M}\frac{da_{nm}(t)}{dt}P_n(x)\frac{d^2T_m(y)}{dy^2}\\
&+\sum_{n=-N}^{N}\sum_{m=0}^{M}a_{nm}(t)\frac{d^3P_n(x)}{dx^3}U(y)T_m(y)+\sum_{n=-N}^{N}\sum_{m=0}^{M}a_{nm}(t)\frac{dP_n(x)}{dx}U(y)\frac{d^2T_m(y)}{dy^2}\\
&-\sum_{n=-N}^{N}\sum_{m=0}^{M}a_{nm}(t)\frac{dP_n(x)}{dx}\frac{d^2U(y)}{dy^2}T_m(y)-\frac{1}{\mathrm{Re}}\sum_{n=-N}^{N}\sum_{m=0}^{M}a_{nm}(t)\frac{d^4P_n(x)}{dx^4}T_m(y)\\
&+\frac{2}{\mathrm{Re}}\sum_{n=-N}^{N}\sum_{m=0}^{M}a_{nm}(t)\frac{d^2P_n(x)}{dx^2}\frac{d^2T_m(y)}{dy^2}-\frac{1}{\mathrm{Re}}\sum_{n=-N}^{N}\sum_{m=0}^{M}a_{nm}(t)P_n(x)\frac{d^4T_m(y)}{dy^4}
\end{aligned}$$

$$= -q(t)\frac{\partial w(x)}{\partial x}U(y)\frac{\partial^2 f(y)}{\partial y^2} + q(t)\frac{\partial w(x)}{\partial x}\frac{d^2U(y)}{dy^2}f(y) + \frac{1}{R_e}q(t)\frac{\partial^4 w(x)}{\partial x^4}f(y)$$
$$+ \frac{2}{R_e}q(t)\frac{\partial^2 w(x)}{\partial x^2}\frac{\partial^2 f(y)}{\partial y^2} + \frac{1}{R_e}q(t)w(x)\frac{\partial^4 f(y)}{\partial y^4} \tag{5.102}$$

Multiplying by $P_l(x)T_k(y)$ and integrating yield

$$\sum_{m=0}^{M}\sum_{n=-N}^{N}\frac{da_{nm}(t)}{dt}\kappa_{nl}^2\beta_{mk}^0 + \sum_{m=0}^{M}\sum_{n=-N}^{N}\frac{da_{nm}(t)}{dt}\kappa_{nl}^0\beta_{mk}^2$$
$$= -\sum_{m=0}^{M}\sum_{n=-N}^{N}a_{nm}(t)\kappa_{nl}^3\beta_{mk}^{0+} - \sum_{m=0}^{M}\sum_{n=-N}^{N}a_{nm}(t)\kappa_{nl}^1\beta_{mk}^{2+}$$
$$+\sum_{m=0}^{M}\sum_{n=-N}^{N}a_{nm}(t)\kappa_{nl}^1\beta_{mk}^{0-} + \frac{1}{\mathrm{Re}}\sum_{m=0}^{M}\sum_{n=-N}^{N}a_{nm}(t)\kappa_{nl}^4\beta_{mk}^0$$
$$-\frac{2}{\mathrm{Re}}\sum_{m=0}^{M}\sum_{n=-N}^{N}a_{nm}(t)\kappa_{nl}^2\beta_{mk}^2 + \frac{1}{\mathrm{Re}}\sum_{m=0}^{M}\sum_{n=-N}^{N}a_{nm}(t)\kappa_{nl}^0\beta_{mk}^4 +$$
$$\frac{\partial q(t)}{\partial t}\{S_{lk}^1\} + q(t)\{S_{lk}^2\} \tag{5.103}$$

where $S_{lk}^1 = -\left[\frac{\partial^2 w(x)}{\partial x^2}, P_l(x)\right]_x \left[f(y), T_k(y)\right]_y - \left[w(x), P(x)\right]_x\left[\frac{\partial^2 f(y)}{\partial y^2}, T_k(y)\right]_y$

$$S_{lk}^2 = -\left[\frac{\partial^3 w(x)}{\partial x^3}, P_l(x)\right]_x\left[U(y)f(y), T_k(y)\right]_y - \left[\frac{\partial w(x)}{\partial x}, P_l(x)\right]_x\left[U(y)\frac{\partial^2 f(y)}{\partial y^2}, T_k(y)\right]_y$$
$$+\left[\frac{\partial w(x)}{\partial x}, P_l(x)\right]_x\left[\frac{d^2U(y)}{dy^2}, T_k(y)\right]_y + \frac{1}{R_e}\left[\frac{\partial^4 w(x)}{\partial x^4}, P_l(x)\right]_x\left[f(y), T_k(y)\right]_y$$
$$+\frac{2}{R_e}\left[\frac{\partial^2 w(x)}{\partial x^2}, P_l(x)\right]_x\left[\frac{\partial^2 f(y)}{\partial y^2}, T_k(y)\right] + \frac{1}{R_e}[w(x), P_l(x)]\left[\frac{\partial^4 f(y)}{\partial y^4}, T_k(y)\right]_y$$
$$l = -N, \ldots, N; \quad k = 0, \ldots, M$$

where [,] denotes the inner product,

$$[u(x), v(x)]_x = \int_{x_0}^{x_1} u(x)v(x)dx$$

$$\kappa_{nl}^j = \left[\frac{d^jP_n(x)}{dx^j}, P_l(x)\right]_x \quad j = 0, \ldots, 4$$

$$\beta_{mk}^{j} = \left[\frac{d^j T_m(y)}{dy^j}, T_k(y)\right]_y \qquad j = 0, \ldots, 4$$

$$\beta_{mk}^{0+} = [U(y)T_m(y), T_k(y)]_y$$

$$\beta_{mk}^{0-} = \left[\frac{d^2 U(y)}{dy^2} T_m(y), T_k(y)\right]_y$$

$$\beta_{mk}^{2+} = \left[U(y)\frac{d^2 T_m(y)}{dy^2}, T_k(y)\right]_y$$

The recurrence relationship of $\{T_k(y)\}$ (see Eq. (1.94)) is

$$T_{k+1}(y) = 2yT_k(y) - T_{k-1}(y)$$

with $T_0(y) = 1$ and $T_1(y) = y$.

Applying orthogonality property of the basis function leads to

$$\kappa_{nl}^{j} = \left[\frac{d^j P_n(x)}{dx^j}, P_l(x)\right]_x = \begin{cases} (in\alpha_0)^j, & \text{if} \quad n = l \quad \text{and} \quad j = 0, \ldots, 4 \\ 0, & \text{if} \quad n \neq l \quad \text{and} \quad j = 0, \ldots, 4 \end{cases}$$

Equation (5.103), a system of first-order differential equations constituted with $(2N + 1)(M + 1)$ equations, may be written in the form

$$\sum_{m=0}^{M} (\beta_{mk}^2 + l^2\alpha_0^2\beta_{mk}^0)\frac{da_{lm}(t)}{dt} = \sum_{m=0}^{M} (il^3\alpha_0^3\beta_{mk}^{0+} - il\alpha_0\beta_{mk}^{2+} + il\alpha_0\beta_{mk}^{0-})a_{lm}(t)$$

$$+ \frac{1}{\mathrm{Re}}\sum_{m=0}^{M} (l^4\alpha_0^4\beta_{mk}^0 - 2l^2\alpha_0^2\beta_{mk}^2 + \beta_{mk}^4)\,a_{lm}(t) + \frac{dq(t)}{dt}\{S_{l,k}^1\} + q(t)\{S_{l,k}^2\} \qquad (5.104)$$

5.4.3 Standard State-Space Representations of Linearized Flow

Equation (5.104) can be written in the block matrix form as

$$\begin{bmatrix} M_{-N} & 0 & \ldots & 0 \\ 0 & M_{-N+1} & 0 & 0 \\ 0 & 0 & \ddots & 0 \\ 0 & 0 & 0 & M_N \end{bmatrix} \begin{bmatrix} d\mathbf{a}_{l=-N}/dt \\ d\mathbf{a}_{l=-N+1}/dt \\ \vdots \\ d\mathbf{a}_{l=N}/dt \end{bmatrix}$$

$$= \begin{bmatrix} R_{-N} & 0 & \ldots & 0 \\ 0 & R_{-N+1} & 0 & 0 \\ 0 & 0 & \ddots & 0 \\ 0 & 0 & 0 & R_N \end{bmatrix} \begin{bmatrix} \mathbf{a}_{l=-N} \\ \mathbf{a}_{l=-N+1} \\ \vdots \\ \mathbf{a}_{l=N} \end{bmatrix} + [\Upsilon_1 \Upsilon_2] \begin{bmatrix} q(t) \\ \dfrac{dq(t)}{dt} \end{bmatrix} \qquad (5.105)$$

where $\mathbf{a}_{l=p}$ denotes pth column vector in the spectral coefficient matrix $\{a_{lk}\}$, M_p and R_p denote pth row vector in the matrixes $\{M_{mk}\}$ and $\{R_{mk}\}$ respectively, where

$$\{M_{mk}\} = \beta_{mk}^2 + l^2\alpha_0^2\beta_{mk}^0$$

and $\{R_{mk}\} = il^3\alpha_0^3\beta_{mk}^{0+} - il\alpha_0\beta_{mk}^{2+} + il\alpha_0\beta_{mk}^{0-} + \frac{1}{\mathrm{Re}}(l^4\alpha_0^4\beta_{mk}^0 - 2l^2\alpha_0^2\beta_{mk}^2 + \beta_{mk}^4)$

In a compact form,

$$M\frac{d\mathbf{a}}{dt} = R\mathbf{a} + [\Upsilon_1 \quad \Upsilon_2]\begin{bmatrix} q(t) \\ \dfrac{dq(t)}{dt} \end{bmatrix} \tag{5.106}$$

When M is a nonsingular matrix, we may invert to obtain

$$\frac{d\mathbf{a}}{dt} = [M^{-1}R]\mathbf{a} + [M^{-1}\Upsilon_1 \quad M^{-1}\Upsilon_2]\begin{bmatrix} q(t) \\ \dfrac{dq(t)}{dt} \end{bmatrix}$$

where $\mathbf{a}, R, \Upsilon_1$, and Υ_2 are all complex.

By expanding in terms of real and imaginary parts, we obtain

$$\frac{d\mathbf{a}_R}{dt} + i\frac{d\mathbf{a}_I}{dt} = M^{-1}R_R\mathbf{a}_R + iM^{-1}R_R\mathbf{a}_R + iM^{-1}R_I\mathbf{a}_I - M^{-1}R_I\mathbf{a}_I$$
$$+M^{-1}\Upsilon_{1R}q(t) + iM^{-1}\Upsilon_{1I}q(t) + M^{-1}\Upsilon_{2R}\frac{dq(t)}{dt} + iM^{-1}\Upsilon_{2I}\frac{dq(t)}{dt} \tag{5.107}$$

where the subscript R and I represent real and imaginary parts, respectively. Define the state-space variable as

$$\tilde{\mathbf{x}} = \begin{bmatrix} \mathbf{a}_R \\ \mathbf{a}_I \\ q(t) \end{bmatrix}$$

We have

$$\frac{d\tilde{\mathbf{x}}}{dt} = \begin{bmatrix} M^{-1}R_R & M^{-1}R_I & M^{-1}R_{1R} \\ M^{-1}R_I & M^{-1}R_R & M^{-1}\Upsilon_{1I} \\ 0 & 0 & 0 \end{bmatrix}\tilde{\mathbf{x}} + \begin{bmatrix} M^{-1}\Upsilon_{2R} \\ M^{-1}\Upsilon_{2I} \\ 1 \end{bmatrix}\frac{dq(t)}{dt} \tag{5.108}$$

It can be written further in the state-space form for a linear and time-varying system as

$$\frac{d\tilde{\mathbf{x}}}{dt} = \overline{A}\tilde{\mathbf{x}} + \overline{B}u(t) \tag{5.109}$$

where the control variable $u(t)$ is $\frac{dq(t)}{dt}$ and $q(t)$ is related to the blowing/suction on the lower wall.

Substituting Eq. (5.101) into Eq. (5.100) leads to

$$z(x_i, y = -1, t) = \sum_{n=-N}^{N} \sum_{m=0}^{M} a_{nm}(t) P_n(x_i) \frac{\partial^2 T_m(y = -1)}{\partial y^2} + q(t) \frac{\partial^2 f(y = -1)}{\partial y^2} w(x_i) \tag{5.110}$$

Let $D = \frac{\partial^2 f(y=-1)}{\partial y^2} w(x_i)$

By pulling out the complex spectral coefficients and denoting them as the vector **a**, we may construct a complex matrix, **O**

$$z(x_i, y = -1, t) = \mathbf{O}_{y=-1}\mathbf{a} + Dq(t) \tag{5.111}$$

By introducing the state-space variable $\widetilde{\mathbf{x}}$, stacking the real and imaginary parts of **a** as well as $q(t)$, we have

$$z_R(x_i, y = -1, t) = \mathbf{C}\widetilde{\mathbf{x}} \tag{5.112}$$

where $\mathbf{C} = [\mathbf{O}_R - \mathbf{O}_I\ D]$.

Equation (5.112) is the output equations, or the measurement equations for the control system.

The state differential equation (5.109) and the output equation (5.112) together construct the traditional control system equations. This is called a single-input single-output system (SISO), since only a single-input variable $u(t)$ related to the blowing/suction on the lower wall and a single-output variable $z(x_i, y = -1, t)$ related to the streamwise component of shear at a single boundary point $(x_i, -1)$ are involved in the control system.

The initial condition is set as

$$\Psi(x, y, t = 0) = g(x, y)$$

Then

$$\Phi(x, y, t = 0) = g(x, y) - q(t = 0)f(y)w(x)$$

$$= \sum_{n=-N}^{N} \sum_{m=0}^{M} b_{nm} P_n(x) T_m(y) \tag{5.113}$$

where b_{nm} are assumed to be known. Then

$$\mathbf{b} = \mathbf{a}(t = 0)$$

where **a** and **b** are both complex. Thus the initial condition on $\widetilde{\mathbf{x}}$ is

$$\widetilde{\mathbf{x}}(t = 0) = \begin{bmatrix} \mathbf{b}_R \\ \mathbf{b}_I \\ q\,(t = 0) \end{bmatrix} \tag{5.114}$$

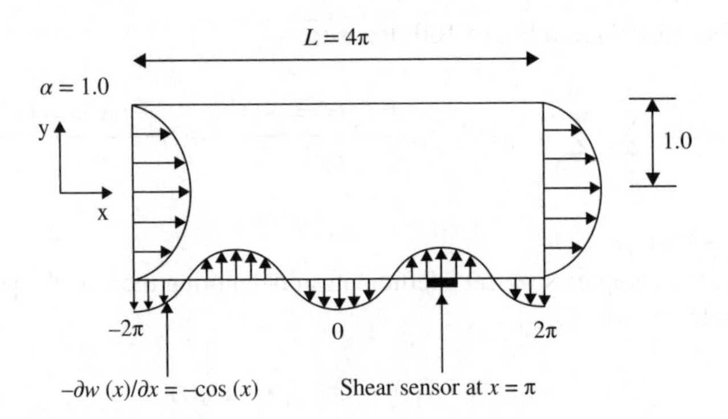

Figure 5.8 Linear optimal control of channel flow.[20] *Source*: Lee K H 1999. Reproduced with permission obtained from Elsevier

5.4.4 Linear Optimal Control of Channel Flow

Using the linear optimal flow control discussed above, we consider a channel flow with boundary blowing/suction and wall-shear measurement as shown in Figure 5.8.

The total nondimensional length of the channel, L, is 4π, and the channel half-height is $h = 1$. The fundamental wave number is $\alpha_0 = 0.5$. Only one wave number is included, corresponding to $l = 2$ and $\alpha = l\alpha_0 = 1.0$. The Reynolds number is chosen as $R_e = 10000$. In terms of the boundary input function in Eq. (5.96), $w(x)$ is set to $\sin(x)$. A single shear sensor is located at $x = \pi$.

For the LQR problem presented in Section 5.2.1, a controller can be designed with the following parameters:

$$Q_1(t) = C^{\mathrm{T}}(t)Q(t)C(t) = 0.001C^{\mathrm{T}}C \text{ and } R = I \tag{5.115}$$

where the matrix C is the output matrix function for the measurement taken at $x = \pi$. I is the identity matrix. With these parameters, the optimal feedback-gain matrix K is obtained from Riccati equation (5.53) via MATLAB.[12] Recalling that the input to Eq. (5.110) is in the form of $\frac{dq(t)}{dt}$, the optimal control input is integrated to compute the actual blowing/suction at the lower wall of channel

$$q(t) = -\int_0^t K\widetilde{\mathbf{x}}(\tau)d\tau \tag{5.116}$$

The outputs of system are required to follow a reference signal, which is simply the zero state for the LQR problem. Hence, the initial condition of Eq. (5.116) is given since the shear is driven to zero

$$q(0) = 0$$

Then, the physical blowing/suction defined by Eq. (5.96) takes the form

$$v(x, y = -1, t) = -q(t)\frac{\partial w(x)}{\partial x}f(y = -1) = K\cos(x)\int_0^t \widetilde{\mathbf{x}}(x_i, y = -1, \tau)d\tau \tag{5.117}$$

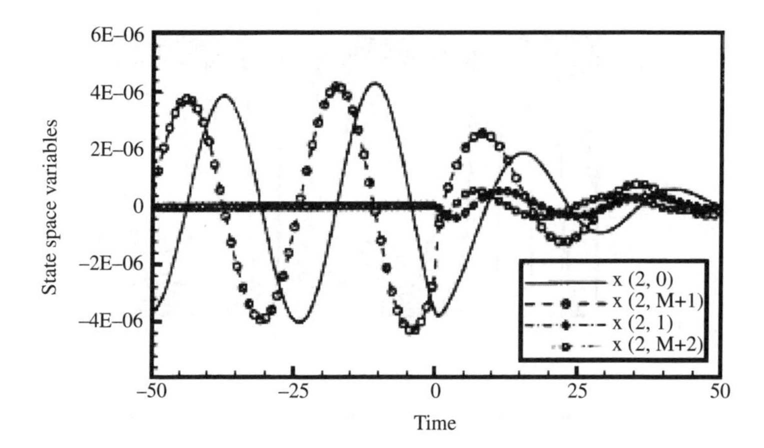

Figure 5.9 Time evolution of state-space variable with an LQR system.[20] *Source*: Lee K H 1999. Reproduced with permission obtained from Elsevier

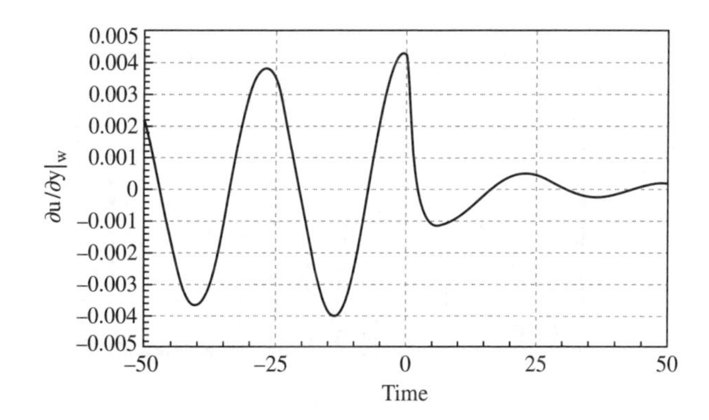

Figure 5.10 Time evolution of shear measured at $x = \pi$.[20] *Source*: Lee K H 1999. Reproduced with permission obtained from Elsevier

The state-space variable $a_{lm}(t)$ can be evaluated by solving Eq. (5.104) numerically. The evolution of the state-space variable is shown in Figure 5.9, assuming that the optimal controller is turned on after $t = 0$. In the figure, $x(2, 0)$ and $x(2, M + 1)$ correspond to the real and imaginary parts of $a_{10}(t)$, respectively. Likewise, $x(2, 1)$ and $x(2, M + 2)$ correspond to the real and imaginary parts of $a_{11}(t)$. Before turned on, even states ($x(2, 0), x(2, M + 1), \ldots$, etc.) have finite values while the odd states ($x(2, 1), x(2, M + 2), \ldots$, etc.) have zero values. Once the controller is turned on, these odd states are activated to grow, while the even states are suppressed. All states eventually decay.

The time history of wall-shear stress measured at $x = \pi$ is shown in Figure 5.10. After the optimal LQR controller is turned on, the shear is quickly suppressed.

While the flow control is implemented at time $t = 0$, the initial disturbance velocity field is shown in Figure 5.11, where only half of the channel in the streamwise direction is shown since the initial field is periodic. There exists a pair of vortices with a strong velocity gradient at the wall.

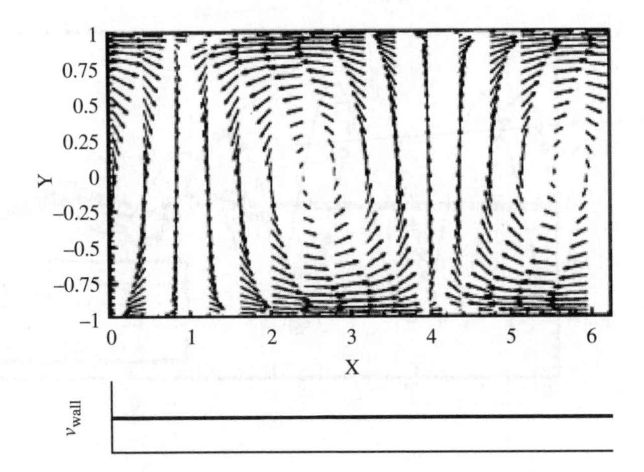

Figure 5.11 Disturbance velocity field and v_{wall} at $t = 0$.[20] *Source*: Lee K H 1999. Reproduced with permission obtained from Elsevier

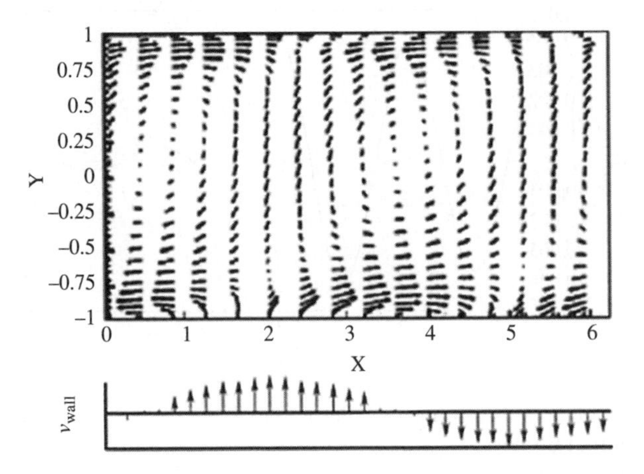

Figure 5.12 Disturbance velocity field and v_{wall} at $t = 7$.[20] *Source*: Lee K H 1999. Reproduced with permission obtained from Elsevier

The disturbance velocity field and the blowing/suction normal velocity at the wall, v_{wall} at $t = 7$, are shown in Figure 5.12. The whole flow field is affected by the control and the velocity gradient at the wall is much reduced. The distribution of normal velocity at the wall is sinusoidal.

The streamwise disturbance velocity profiles measured at $x = \pi$ at $t = 0$ and $t = 50$ are shown in Figure 5.13 for comparison, where the solid line is measured at $t = 50$ and dotted line at $t = 0$. It is apparent that the disturbance at the upper wall is also significantly reduced.

If the state excitation and observation noise are considered, the LQG controller presented in Section 5.2.3 is required. Additional parameters are needed along with those from Eq. (5.115), we use

$$Q_e = 10BB^{\text{T}} \tag{5.118}$$

$$W_e = 1 \tag{5.119}$$

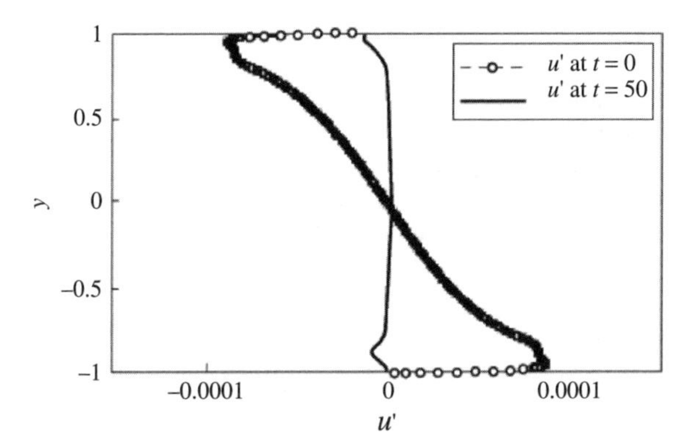

Figure 5.13 Stream disturbance velocity profile at $x = \pi$.[20] *Source*: Lee K H 1999. Reproduced with permission obtained from Elsevier

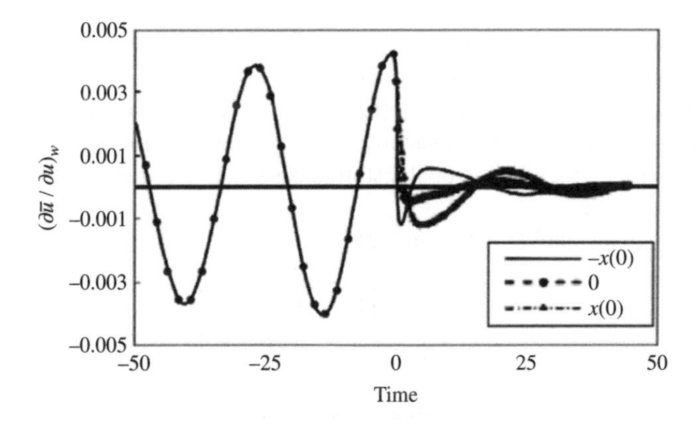

Figure 5.14 History of disturbance wall-shear stresses measured at $x = \pi$.[20] *Source*: Lee K H 1999. Reproduced with permission obtained from Elsevier

Kalman filter gain K_e is obtained from Riccati equation (5.84) via MATLAB. Based on the measured variable z, the state estimate $\hat{\mathbf{x}}_e$ can be obtained by solving Eq. (5.83), which is used to determine $\frac{dq(t)}{dt}$ by the optimal control law (5.85). The optimal control input, that is, the physical blowing/suction v_{wall} is determined by integrating $\frac{dq(t)}{dt}$.

For the LQG controller, three different initial estimated states are used: $x_e(0) = x(0)$, 0, and $-x(0)$, where $x_e(0)$ is an initial estimated state at $t = 0$ and $x(0)$ is an actual initial state at $t = 0$. The history of disturbance wall-shear stresses measured at $x = \pi$ for different estimated state at $t = 0$ is shown in Figure 5.14. It can be seen that the disturbance wall-shear stress is suppressed in the control. Figure 5.15 shows the error between the state-space variables and their estimates. It appears that the estimator takes a while to track the right state.

5.5 Linear Optimal Control of Three-Dimensional Flow[22–25]

In the previous section, the controller is designed for two-dimensional flow, which does not accommodate three-dimensional characteristics of a fully developed turbulent channel flow.

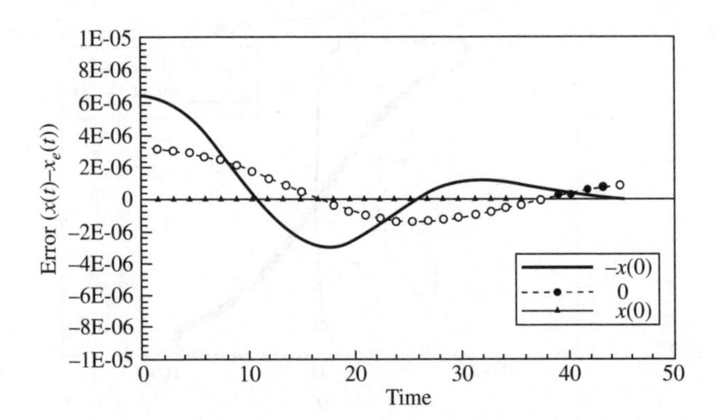

Figure 5.15 Time evolution of errors between actual state and its estimate.[20] *Source*: Lee K H 1999. Reproduced with permission obtained from Elsevier

Therefore, the linear optimal control for a three-dimensional flow will be discussed in this section to deal with the optimal turbulent control.

For three-dimensional disturbances described by Eq. (5.88), $\widetilde{\omega}_y$ is forced by $\widetilde{v}$ through the coupling term, while $\widetilde{v}$ evolves independently. Since the nonzero linear coupling term causes stronger nonnormality, three-dimensional disturbances have much larger transient growth than the two-dimensional ones.

The linearized Navier–Stokes equation (5.90)

$$\frac{d}{dt}\begin{bmatrix}\widetilde{v}\\ \widetilde{\omega}_y\end{bmatrix} = \begin{bmatrix}L_{os} & 0\\ L_c & L_{sq}\end{bmatrix}\begin{bmatrix}\widetilde{v}\\ \widetilde{\omega}_y\end{bmatrix} \tag{5.90}$$

is now discretized on a grid of $N + 1$ collocation points

$$y_j = \cos(\pi j/N) \quad \text{for} \quad 0 \le j \le N \tag{1.96}$$

Then, at the collocation points, we have (see Eq. (1.102))

$$\omega_y' = D\omega_y \quad \text{and} \quad \omega_y'' = D\omega_y'$$

where the hat accents (∼) have been dropped for notational convenience. The prime (′) indicates the derivative of ω_y with respect to y. The D is a first derivative matrix with $(N + 1) \times (N + 1)$ dimension. The homogeneous Neumann boundary condition on v is accomplished by modifying the first derivative matrix such that

$$\widetilde{D}_{lj} = \begin{cases} 0 & l = 0, \mathrm{N} \\ D_{lj} & 1 \le l \le \mathrm{N\text{-}1} \end{cases}$$

Differentiation of v with respect to y is then given by

$$v' = \widetilde{D}v, v'' = Dv', v''' = Dv'' \quad \text{and} \quad v'''' = Dv'''$$

With these derivative matrices, Eq. (5.90) can be written in the matrix form on all $N + 1$ collocation points such as

$$\begin{aligned}\frac{dv}{dt} &= Lv \\ \frac{d\omega_y}{dt} &= Cv + l\ \omega_y\end{aligned} \tag{5.120}$$

where the $(N+1)\times(N+1)$ matrices L, C and l represent the spatial discretization of L_{os}, L_c, and L_{sq}, respectively, which can be decomposed further such as

$$L = \begin{pmatrix} * & * & * \\ b_{11} & L_c & b_{12} \\ * & * & * \end{pmatrix},\ C = \begin{pmatrix} * & * & * \\ b_{21} & C_c & b_{22} \\ * & * & * \end{pmatrix},\ l = \begin{pmatrix} * & * & * \\ * & l_c & * \\ * & * & * \end{pmatrix}$$

where matrices L_c, C_c, and l_c are $(N-1)\times(N-1)$ and b_{11}, b_{12}, b_{21}, and b_{22} are $(N-1)\times 1$.

When the flow is controlled by blowing/suction at the upper and lower walls, Eq. (5.120) can be converted into a state-space equation

$$\frac{d\mathbf{x}(t)}{dt} = A\mathbf{x}(t) + B\mathbf{u}(t) \tag{5.121}$$

where the vector $\mathbf{x}$ contains the v_i and ω_i at the grid points on the interior of the channel, such as $\mathbf{x} = (v_1, \dots, v_{N-1}, \omega_{y,1}, \dots, \omega_{y,N-1})^{\mathrm{T}}$

A is $2(N-1)\times 2(N-1)$, B is $2(N-1)\times 2$, and defined as

$$A = \begin{bmatrix} L_c & 0 \\ CC_c & l_c \end{bmatrix}$$

$$B = \begin{bmatrix} -b_{11} & b_{12} \\ 0 & 0 \end{bmatrix}$$

$$\mathbf{u} = \begin{bmatrix} -v_0 \\ v_n \end{bmatrix}$$

The streamwise and spanwise components of shear at the wall are taken as the outputs of a control system, which can be expressed as

$$z_1 = -\frac{1}{\mathrm{Re}}\frac{\partial u}{\partial y}\bigg|_{\text{upper wall}},\quad z_2 = -\frac{1}{\mathrm{Re}}\frac{\partial u}{\partial y}\bigg|_{\text{lower wall}}$$

$$z_3 = -\frac{1}{\mathrm{Re}}\frac{\partial w}{\partial y}\bigg|_{\text{upper wall}},\quad z_4 = -\frac{1}{\mathrm{Re}}\frac{\partial w}{\partial y}\bigg|_{\text{lower wall}}$$

where $u = \frac{i}{k_x^2+k_z^2}\left(k_x\frac{\partial v}{\partial y} - k_z\omega_y\right)$

$$w = \frac{-i}{k_x^2+k_z^2}\left(k_z\frac{\partial v}{\partial y} - k_x\omega_y\right)$$

These measurements are expressed in terms of the discrete vectors v and ω_y

$$z_1 = (-aD\widetilde{D}v + bD\omega_y)|_{\text{upper wall}}, \quad z_2 = (aD\widetilde{D}v - bD\omega_y)|_{\text{lower wall}}$$

$$z_3 = (-bD\widetilde{D}v - aD\omega_y)|_{\text{upper wall}}, \quad z_4 = (bD\widetilde{D}v + aD\omega_y)|_{\text{lower wall}}$$

where $a = \frac{ik_x}{(k_x^2+k_z^2)R_e}$ and $b = \frac{ik_z}{(k_x^2+k_z^2)R_e}$

For the wall values, Decompose D and $D\widetilde{D}$ according to

$$D = \begin{pmatrix} * & c_3 & * \\ * & * & * \\ * & c_4 & * \end{pmatrix}, D\widetilde{D} = \begin{pmatrix} d_1 & c_1 & d_3 \\ * & * & * \\ d_2 & c_2 & d_4 \end{pmatrix}$$

where c_1, c_2, c_3, and c_4 are $1 \times (N-1)$, and d_1, d_2, d_3, and d_4 are 1×1.

Finally, the output equation or the measurement equation for the control system is constructed as

$$\mathbf{z} = \mathrm{C}\mathbf{x} + \mathrm{D}\mathbf{u} \tag{5.122}$$

where

$$\mathbf{z} = \begin{pmatrix} z_1 \\ z_2 \\ z_3 \\ z_4 \end{pmatrix}, \mathrm{C} = \begin{pmatrix} -ac_1 & bc_3 \\ ac_2 & -bc_4 \\ -bc_1 & -ac_3 \\ bc_2 & ac_4 \end{pmatrix}, \mathrm{D} = \begin{pmatrix} ad_1 & -ad_3 \\ -ad_2 & ad_4 \\ bd_1 & -bd_3 \\ -bd_2 & bd_4 \end{pmatrix}$$

C is $4 \times 2(N-1)$.

For the minimum requirement for the application of linear optimal control theory, a performance index is defined as

$$J = \lim_{T\to\infty} \frac{1}{T}\int_0^{T} [l\mathbf{x}^{\mathrm{T}}(t)Q\mathbf{x}(t) + \mathbf{u}^{\mathrm{T}}(t)R\mathbf{u}(t)]dt \tag{5.123}$$

where the weight parameter l represents the price of the control. By increasing or decreasing the value of l, emphasis is made on one or the other term. For example, if l is large, the emphasis is on the first term, so the control algorithm will find a control actuation to suppress the $\mathbf{x}^{\mathrm{T}}(t)Q\mathbf{x}(t)$ term more, at the expanse of larger control input.

For the LQR problem, the linear optimal control law has the feedback form

$$\mathbf{u}(t) = -K(t)\mathbf{x}(t)$$

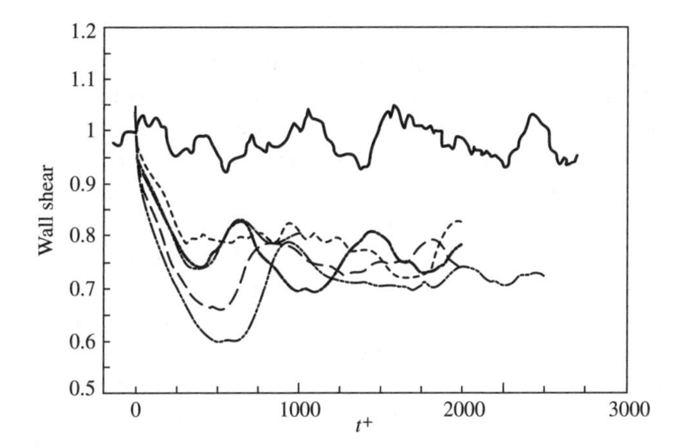

Figure 5.16 Time history of the mean wall-shear.[25] *Source*: Lim J 2003. Reproduced with permission of Cambridge University

where the feedback gain matrix $K(t)$ has the form

$$K(t) = R^{-1}(t)B^{\mathrm{T}}(t)P(t)$$

$P(t)$ satisfies the Riccati equation

$$A(t)P(t) + P(t)A(t) + lQ(t) - P(t)B(t)R^{-1}(t)B^{T}(t)P(t) = 0 \tag{5.124}$$

The eventual control objective here is to achieve the maximum mean drag reduction with the minimal amount of blowing and suction at the wall, found by minimizing the performance index. Many different choices are possible for the performance index, for example, different variables can be selected to form the performance index, such as the disturbance energy, the skin friction, and the linear coupling term. Figure 5.16 shows the mean drag histories under the control of different LQG controllers designed with the different performance indexes, where the solid line represents the uncontrolled case and the dotted line represents the case with the performance index defined by Eq. (5.123). It appears that these controls based on three-dimensional linearized Navier–Stokes equations can produce approximately 20–30% of statistically steady drag reduction.

5.6 Closed Remarks

The optimal control problem can be expressed as follows: Find an optimal relation between the control variable $\mathbf{u}$ and the system state variable $\mathbf{x}(t)$, called control law, minimizing the performance index $J(\mathbf{x}, \mathbf{u})$ constrained by the state equation $\mathbf{F}(\mathbf{x}, \mathbf{u}) = 0$ and the corresponding boundary conditions. It is a conditional extremum problem and can be transformed subsequently into an unconditional extremum problem by introducing an adjoint variable $\boldsymbol{\lambda}$. Finally, the closely coupled equations of the optimal system, consisting of the state equation, the adjoint equation, and the optimal control law, are obtained. When the state-space representation of the control system is described by a set of ordinary differential equations, that is, $\dot{\mathbf{x}} = f(\mathbf{x}, \mathbf{u}, t)$, the canonical equations are given by introducing Hamilton function.

For a linear and time-varying system, expressed by the state equation $\dot{\mathbf{x}} = A(t)\mathbf{x} + B(t)\mathbf{u}$, the quadratic integral is a very useful criterion to measure the control magnitude and the deviating extent. The linear optimal control with the quadratic performance index is called the linear quadratic optimal control. There exists a simple control law, $\mathbf{u}^* = -K(t)\mathbf{x}$, in this control, where the feedback gain matrix $K(t)$ with $K(t) = R^{-1}B^T P(t)$, $P(t)$ satisfying a matrix Riccati equation, is obtained by means of commercial software. In practice, the complete state cannot be measured, only the observations are available. Furthermore, the state equation and the observed variables are usually disturbed by noise. Hence, LQG control system is induced, where the estimator (filter) and the controller (regulator) are designed independently.

The importance of linear mechanism in nonlinear turbulent boundary layer has been shown numerically. Without the linear coupling term, the vortical structures cannot be maintained, as well as the streaky structure cannot be produced. Therefore, the near-wall turbulence can be controlled by using a linear control theory.

In the linear optimal flow control, a crucial problem is to transform the Navier–Stokes equations into the state equations. For two-dimensional flow approach, the Navier–Stokes equations are linearized by introducing the perturbation states, and then converted into a high-order partial differential equation about the stream function. Hereafter, a standard state equation for a linear and time-varying system in the spectral space could be derived by the combined Fourier–Chebyshev transform. Whereas, turbulence is three-dimensional, since the vortical structures cannot be maintained in absence of the linear coupling term in a two-dimensional flow, therefore, the linear optimal control for a three-dimensional flow should be applied to deal with the optimal turbulence control. Based on the linearized three-dimensional Navier–Stokes equations in terms of disturbances of the wall-normal velocity v and the wall-normal vorticity ω_y, a standard state equation could also be derived by the combined Fourier–Chebyshev transform.

References

[1] Prandtl L. Uber Flussigkeitsbewegung bei sehr kleiner Reibung. Proc. Third Int. Math. Cong, 484–491, Heidelberg, Germany, 1904.
[2] Gunzburger M D. Perspectives in Flow Control and Optimization. SIAM, Philadelphia, PA, 2003.
[3] Gad-el-Hak M, Pollard A and Bounet J P. Flow Control Fundamental and Practices. Springer, Berlin, 1998.
[4] Sritharan S. Optimal Control of Viscous Flow. SIAM, Philadelphia, PA, 1998.
[5] Abergel F and Teman R. On some control problems in fluid mechanics. Theoret. Comput. Fluid Dyn., 1990, 1:303–325.
[6] Bewley T B. Flow control: New challenges for a new renaissance. Prog. Aerosp. Sci., 2001, 37:21.
[7] Anderson B and Moore J B. Linear Optimal Control. Prentice-Hall, Inc., Upper Saddle River, NJ, 1971.
[8] Kwakernaak H and Sivan R. Linear Optimal Control System. John Wiley & Sons, Inc., Hoboken, NJ, 1972.
[9] Anderson B and Moore J B. Optimal Control, Linear Quadratic Methods. Prentice-Hall, Inc., Upper Saddle River, NJ, 1989.
[10] Dorato P. Linear-Quadratic Control. Prentice-Hall, Inc., Upper Saddle River, NJ, 1996.
[11] Shahian B and Hassul M. Control System Design using MATLAB. Prentice-Hall, Inc., Upper Saddle River, NJ, 1993.
[12] Grace A, Lamb A J, Little N J and Thompson C M. Control System Toolbox for Use with MATLA – User's Guide. The Math Works, Natick, MA, 1992.
[13] Kim J and Lim J. A linear process in wall bounded turbulent shear flows. Phys. Fluids, 2000, 12(8):1885–1888.
[14] Kim J and Lim J. A singular value analysis of boundary layer control. Phys. Fluids, 2004, 16(6):1980–1988.

[15] Kim J. Control of turbulence boundary layer. Phys. Fluids, 2003, 15(5):1093–1105.
[16] Kim J. Physics and control of wall turbulence for drag reduction. Phil. Trans. R. Soc. A, 2011, 369:1396–1411.
[17] Lee K H, Cortelezzi L, Kim J and Speyer J L. Applications of reduced-order controller to turbulent flows for drag reductions. Phys. Fluid, 2001, 13:1321.
[18] Hogberg M, Bewely T R and Henningson D S. Relaminarization of $Re_\tau = 100$ turbulence using gain scheduling and linear state-feed control. Phys. Fluids, 2003, 15:3572.
[19] Joshi S S, Speyer J L and Kim J J. A system theory approach to the feedback stabilization of infinitesimal and finite-amplitude disturbances in plane Poiseuille flow. J. Fluid Mech., 1997, 332:151–184.
[20] Lee K H. A Systems Theory Approach to Control of Transitional and Turbulent Flows. Ph.D. Thesis, Department of Mechanical Engineering, University of California, Los Angeles, CA, 1999.
[21] Baker J and Christonfides P D. Drag reduction in transitional linearized channel flow using distributed control. Int. J. Control., 2002, 75(15):1213–1218.
[22] Butler K M and Farrell B F. Three dimensional optimal perturbations in viscous shear flow. Phys. Fluids, 1992, A4:1637.
[23] Bewley T R. Optimal and Robust Control and Estimation of Transition, Convection, and Turbulence. Ph.D Thesis, Department of Mechanical Engineering, University of California, Los Angeles, CA, 1999.
[24] Lim J, Kim J, Kang S and Speyer J L. Linear controllers for turbulent boundary layers. Bull. Am. Phys. Soc., 2001, 46:156.
[25] Lim J. Control of Wall-bounded Turbulent Shear Flows using Modern Control Theory. Ph.D Thesis, Department of Mechanical Engineering, University of California, Los Angeles, CA, 2003.

6

Nonlinear Optimal Flow Control

The optimal flow control problem can be expressed as a conditional extremum problem of a performance index (or cost functional), that is, the determination of optimal values of control based on the Navier–Stokes (N-S) equations. This problem is very challenging due to its complexity and inherently strong nonlinear nature.

As discussed in Chapter 5, transforming the N-S equations into the linear state equations, then the optimal flow control problem can be designed for linear problems, so that the mature linear control theory is used directly. In practice, such linear control approaches are often successfully applied to the nonlinear problems even to the near-wall turbulence. However, for most engineering flows, it is not clear whether linear processes in other turbulent flows, such as jets and wakes, play an equally important role as doing in the near-wall turbulence and to what degree such inherently nonlinear process can, in general, be controlled using linear control theory. Since the linear optimal control approaches have limitations when used for turbulence control, therefore research and development of nonlinear optimal flow control theory is very necessary, even though that would be very difficult. Fortunately, the design of control systems directly for nonlinear flow system has been an active area of research, and considerable progress has been made in mathematical analysis and computation of controlling incompressible flows governed by the N-S equations.[1–10]

An optimal flow control problem can also be expressed further as a problem to find solutions of nonlinear sensitivity equations. There are two procedures to get the solutions, namely optimize-then-discretize approach and discretize-then-optimize approach. In the first one, the sensitivity equations are derived directly from the original equations through a formal minimization, and then solved numerically based on the discretized equations. The second one involves first discretizing the original equations and then performing the minimization procedure at the discrete level.

The suboptimal control refers to a control looking ahead only on one short time step into the future to minimize the cost functional, whereas the all possible developments of the flow over the entire time interval under consideration should be investigated in a truly optimal method. The two categories of suboptimal flow control are noticeable – spectrum-based suboptimal control and adjoint-based suboptimal control. For spectrum-based suboptimal control, a discretize-then-optimize approach, the derivations in the N-S equations are discretized first and then the minimization procedure is performed in the spectrum space by Fourier transform.

Principles of Turbulence Control, First Edition. Baochun Fan and Gang Dong.

While for adjoint-based suboptimal control, an optimize-then-discretize approach, the adjoint equations associated with the N-S equations are introduced first, and then the sensitivity equations are derived from the N-S and its adjoint equations.

In addition, the nonlinear optimal flow control method introduced worthily is neural network. Neural network attempts to develop models and controllers via some learning algorithm, performed independently of the N-S equations without regard to the details of the flow physics.

In this chapter, a nonlinear optimal flow control is stated. Some basic problems in optimal flow control are presented in Section 6.1. Section 6.2 introduces the spectrum-based suboptimal control. Based on this optimal control, the channel flow, backward-facing step flow, and cylinder flow controlled by blowing–suction on the wall are discussed as the typical examples. The adjoint-based optimal control is presented in Section 6.3, and both the channel flow controlled by blowing–suction on the wall and the cylinder flow controlled by Lorentz force are discussed. Section 6.4 concerns with neural network, the channel flows controlled by blowing–suction wall, deformed wall, and wall-normal Lorentz force individually based on neural networks are also discussed in the section.

6.1 Fundamentals of Optimal Flow Control[4]

6.1.1 Closed-Loop Flow Control

The flow can be manipulated by altering the governing equations via imposing body forces, or altering the boundary conditions via moving wall, riblet, and blowing–suction wall (see Chapters 3 and 4). Energy is introduced into the flow by all kinds of actuators for active control. The governing equations for an incompressible three-dimensional viscous flow with an active control via moving wall and externally imposed body force can be written as

$$\mathsf{N}\mathbf{q} = \begin{pmatrix} 0 \\ \mathrm{f} \end{pmatrix} \tag{6.1}$$

where $\mathbf{q} = \begin{pmatrix} p(\mathbf{x}, t) \\ \mathbf{u}(\mathbf{x}, t) \end{pmatrix}$

N denotes the N-S operator, then

$$\mathsf{N}\mathbf{q} = \begin{pmatrix} \dfrac{\partial u_j}{\partial x_j} \\ \dfrac{\partial u_i}{\partial t} + \dfrac{\partial u_j u_i}{\partial x_j} - \nu \dfrac{\partial^2 u_i}{\partial {x_j}^2} + \dfrac{\partial p}{\partial x_i} \end{pmatrix}$$

p is the pressure divided by the density, $\mathbf{u}$ is the velocity vector, ν is the kinematic viscosity, and f is the body force. As all variables are normalized with respect to the channel half-width δ and the free-stream velocity u_∞, the Reynolds number $\mathrm{Re} = u_\infty \delta / \nu$, and $\nu = 1/\mathrm{Re}$ in $\mathsf{N}\mathbf{q}$.

The boundary condition is

$$\mathbf{u}|_{\Gamma_k} = \mathbf{u}_{w,k} \quad \text{on} \quad \Gamma_k \tag{6.2}$$

where Γ_k denotes kth boundary and the moving velocity is represented by $\mathbf{u}_{w,k}$.

One of the main aims for control is to reduce the input energy; it can be achieved by the optimal feedback control (or closed-loop control), in which the control inputs, such as f and $\mathbf{u}|_{\Gamma_k}$, should vary with the transient flow-field $\mathbf{q}$ in terms of a control law. Thereby, it is required to design the control law to close the governing equations (6.1) and (6.2).

6.1.2 Cost Functional

For optimal flow control problems, one needs to choose what properties of the flow to target with the control. This choice is formulated as a cost functional (or performance index) such that the optimization problem could be represented as a cost functional to be minimized.

The performance of the control is quite sensitive to the definition of cost functional, thus, an adequate cost functional is of significant importance for flow control. In the optimization of the N-S control problem, an instantaneous cost functional may be written to represent a balance of the quantities that we want minimized and a measure of net control effort, which is usually taken to be a square value. For example, if a wall-normal velocity distributed over the wall section, denoted as ϕ, is to be applied to minimize the drag averaged over the representative wall section on intermediate time horizons $(0,\ T)$, the cost functional can be defined as follows:

1. *Minimization of drag*:

   A relevant cost functional for the minimization of drag on the upper and lower walls and over the control time horizon $(0, T)$ is

$$J_{DRAG}(\phi) = -d_1 \int_0^T \int_{\Gamma_2^{\pm}} \nu \frac{\partial u_1(\phi)}{\partial n} d\mathbf{x}dt + \frac{l^2}{2} \int_0^T \int_{\Gamma_2^{\pm}} \phi^2 d\mathbf{x}dt \tag{6.3}$$

   where the streamwise wall-shear stress $\frac{\partial u_1}{\partial n}$ is taken as a sensor signal input into the controller, then the first term in the right-hand side is a measure of the $\frac{\partial u_1}{\partial n}$ we would like to minimize and the negative sign is needed, since n is defined as an outward facing normal. d_1 is a dimensional constant for dimensional consistency. The second term is a measure of the magnitude of the control, where factor l represents the control price. It is small if the control is "cheap" and large if it is "expensive."

2. *Regulation of turbulent kinetic energy*:

   As discussed in Chapter 2, the turbulence causes the drag increase. It is thus reasonable for cost functional to target the turbulence (the "cause") over each time horizon rather than the drag increase due to the turbulence (the "effect"). Hence the turbulent kinetic energy (TKE) is taken as input, the cost functional for this formulation is

$$J_{\mathrm{TKE(reg)}}(\phi) = \frac{d_2}{2} \int_0^T \int_{\Omega} |\mathbf{u}(\phi)|^2 d\mathbf{x}dt + \frac{l^2}{2} \int_0^T \int_{\Gamma_2^{\pm}} \phi^2 d\mathbf{x}dt \tag{6.4}$$

3. *Regulation of enstrophy*:

   Regulation of the square of the vorticity (i.e., the enstrophy) is sometimes preferred over the regulation of the turbulent kinetic energy. The cost functional is

$$J_{\mathrm{ENS(reg)}}(\phi) = \frac{d_3}{2} \int_0^T \int_{\Omega} |\nabla \times \mathbf{u}(\phi)|^2 d\mathbf{x}dt + \frac{l^2}{2} \int_0^T \int_{\Gamma_2^{\pm}} \phi^2 d\mathbf{x}dt \tag{6.5}$$

4. *Terminal control of turbulent kinetic energy*:
   In sometimes, it is appropriate to sacrifice a piece to obtain long-term gain. Hence a cost functional may target to reduce values of the turbulent kinetic energy at the end of each time horizon, and is written as

$$J_{\mathrm{TKE(ter)}}(\phi) = \frac{d_4}{2}\int_{\Omega} |\mathbf{u}(\phi, T)|^2 d\mathbf{x} + \frac{l^2}{2}\int_0^T \int_{\Gamma_2^{\pm}} \phi^2 d\mathbf{x}dt \tag{6.6}$$

6.1.3 *Fréchet Differential*

An arbitrary perturbation ϕ' to the control ϕ may result in the perturbation $\mathbf{q}'$ to the flow state $\mathbf{q}$, which is expressed by a limiting process called the Fréchet differential

$$\mathbf{q}' = \lim_{\varepsilon\to 0} \frac{\mathbf{q}(\phi + \varepsilon\phi') - \mathbf{q}(\phi)}{\varepsilon} = \frac{D\mathbf{q}(\phi)}{D\phi}\phi' \tag{6.7}$$

where $\frac{D\mathbf{q}(\phi)}{D\phi}$ represents the sensitivity of the flow state $\mathbf{q}$ to small modification of the control ϕ. Subsequently, the perturbation $J'(\phi)$ may be defined as

$$J'(\phi, \phi') = \lim_{\varepsilon\to 0} \frac{J(\phi + \varepsilon\phi') - J(\phi)}{\varepsilon} = \int_0^T \int_{\Gamma_2^{\pm}} \frac{\mathrm{D}J(\phi)}{D\phi}\phi' d\mathbf{x}dt \tag{6.8}$$

The cost functional is in the extreme, if

$$\frac{\mathrm{D}J(\phi)}{D\phi} = 0 \tag{6.9}$$

Therefore the optimal control ϕ is a solution of Eq. (6.9), called the control sensitivity equation. Now, the optimal control problem could also be represented as a problem to find the solution of the sensitivity equation.

6.2 Spectrum-based Suboptimal Control

6.2.1 *Control of Channel Flow*

6.2.1.1 Analytic Solution of Fréchet Differential Equation

Three-dimensional turbulent channel flow is shown in Figure 6.1, where the interior of the domain is denoted by Ω and the boundaries of the domain in the x_i -direction are denoted by $\Gamma_i^{\pm}$. The governing equations, that is, the N-S and continuity equations, are written as

$$\frac{\partial u_i}{\partial t} + u_j\frac{\partial u_i}{\partial x_j} = -\frac{\partial p}{\partial x_i} + \frac{1}{\mathrm{Re}_\tau}\frac{\partial^2 u_i}{\partial x_j \partial x_j}$$

$$\frac{\partial u_j}{\partial x_j} = 0 \tag{6.10}$$

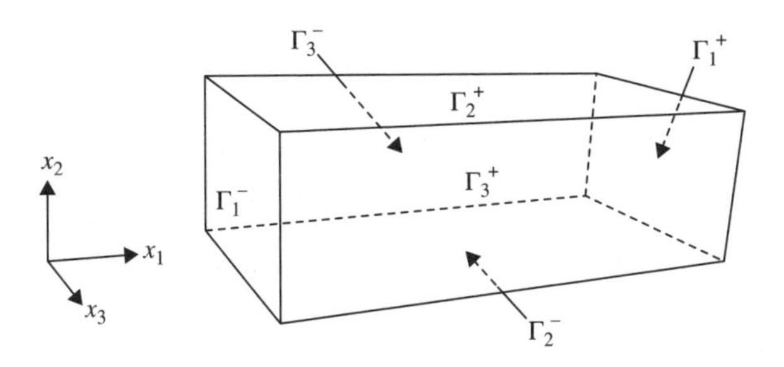

Figure 6.1 Three-dimensional channel flow

where t is the time, x_1, x_2, and x_3 are the streamwise, wall-normal, and spanwise directions, respectively, u_i are the corresponding velocity components, and p is the pressure. All variables are normalized by the channel half-width δ and the friction velocity u_τ. The u_τ is defined as $u_\tau = \sqrt{\overline{\tau}_w/\rho}$, where $\overline{\tau}_w$ is the mean skin friction on the wall for uncontrolled turbulent channel flow, defined as $\overline{\tau}_w/\rho = -\nu \partial u_1/\partial n|_{\Gamma_2^\pm}$ and ν is the kinematic viscosity. Then the Reynolds number is defined as $\text{Re}_\tau = \frac{u_\tau \delta}{\nu}$.

While the turbulent flow is controlled by blowing–suction on the walls $\Gamma_2^\pm$, the control input is the wall-normal velocity at the wall, ϕ, then the boundary condition is given such that

$$\mathbf{u}|_w = \phi \delta_{i2} \tag{6.11}$$

Considering the coherent structure of the near-wall turbulence, the periodic boundary conditions are applied both in x_1 and x_3 directions.

To discrete Eq. (6.10), the Runge–Kutta scheme and the Crank–Nicolson scheme are employed for the linear terms and the nonlinear terms, respectively, we have

$$u_i^{n+1} - \frac{\Delta t}{2R_{e\tau}} \frac{\partial^2 u_i^{n+1}}{\partial x_j \partial x_j} + \frac{\Delta t}{2} \frac{\partial p^{n+1}}{\partial x_i} + R^n = 0 \tag{6.12}$$

$$\frac{\partial u_j^{n+1}}{\partial x_j} = 0 \tag{6.13}$$

with

$$u_i^{n+1}|_w = \phi \delta_{i2} \tag{6.14}$$

where the superscripts n and $n+1$ denote the time step, and R^n includes the nonlinear terms and the explicit parts of the pressure gradient and viscous terms.

Fréchet differential of Eqs. (6.12)–(6.14) yield the governing equation for the perturbation states

$$u'^{n+1}_i - \frac{\Delta t}{2\text{Re}_\tau} \frac{\partial^2 u'^{n+1}_i}{\partial x_j \partial x_j} + \frac{\Delta t}{2} \frac{\partial p'^{n+1}}{\partial x_i} = 0 \tag{6.15}$$

$$\frac{\partial u'^{n+1}_j}{\partial x_j} = 0 \tag{6.16}$$

with

$$u'^{n+1}_i|_w = \phi'\delta_{i2} \tag{6.17}$$

where $(DR^n/D\phi)\phi' = 0$, since there is no effect of ϕ on the past flow fields. Note that there is no contribution from the nonlinear terms, thus making the equations linear. It may miss important flow dynamics; however it has been shown numerically that the contribution from the nonlinear terms is negligible in the boundary control with short optimization interval Δt.

Hereinafter, the superscript $n + 1$ is dropped, and all variables are understood to be at the $(n + 1)$ th time step.

By Fourier transform in a two-dimensional system $(x,\ z)$ (see Section 1.6.2), we have

$$u'_i = \sum_{k_x}\sum_{k_z} \widetilde{u}'_i(y)e^{ik_x x}e^{ik_z z}$$

$$p' = \sum_{k_x}\sum_{k_z} \widetilde{p}'(y)e^{ik_x x}e^{ik_z z}$$

where $x = x_1, y = x_2, z = x_3$; $\widetilde{u}_i(y)$ and $\widetilde{p}(y)$ are the Fourier coefficients of u'_i and p', respectively; and k_i is the wave number in the i direction. Then

$$\widetilde{u}'_1 - \frac{\Delta t}{2R_{e\tau}}\left(\frac{d^2}{dy^2} - k^2\right)\widetilde{u}'_1 + \frac{ik_x\Delta t}{2}\widetilde{p}' = 0 \tag{6.18}$$

$$\widetilde{u}'_2 - \frac{\Delta t}{2R_{e\tau}}\left(\frac{d^2}{dy^2} - k^2\right)\widetilde{u}'_2 + \frac{\Delta t}{2}\frac{d\widetilde{p}'}{dy} = 0 \tag{6.19}$$

$$\widetilde{u}'_3 - \frac{\Delta t}{2R_{e\tau}}\left(\frac{d^2}{dy^2} - k^2\right)\widetilde{u}'_3 + \frac{ik_z\Delta t}{2}\widetilde{p}' = 0 \tag{6.20}$$

$$ik_x\widetilde{u}'_1 + ik_z\widetilde{u}'_3 + \frac{d\widetilde{u}'_2}{dy} = 0 \tag{6.21}$$

with $\widetilde{u}'_i(0) = \widetilde{\phi}'\delta_{i2,}$ $\quad \widetilde{u}'_i(\infty) = 0$

where $k^2 = k_x^2 + k_z^2$ and $\widetilde{\phi}'$ is the Fourier coefficients of ϕ'.

From the above equations, we have

$$\frac{d^2\widetilde{p}'}{dy^2} - k^2\widetilde{p}' = 0 \tag{6.22}$$

which has a solution

$$\widetilde{p}' = \widetilde{p}'_w e^{-ky} \tag{6.23}$$

Substituting into Eqs. (6.18)–(6.20) yields

$$\widetilde{u}'_1(y) = \frac{\Delta t}{2}ik_x\widetilde{p}'_w(\exp[-(k^2 + 2\mathrm{Re}_\tau/\Delta t)^{1/2}y] - e^{-ky}) \tag{6.24}$$

$$\widetilde{u}'_3(y) = \frac{\Delta t}{2}ik_z\widetilde{p}'_w(\exp[-(k^2 + 2\mathrm{Re}_\tau/\Delta t)^{1/2}y] - e^{-ky}) \tag{6.25}$$

$$\widetilde{u}'_2(y) = \left(\widetilde{\phi}' - \frac{\Delta t}{2}k\widetilde{p}'_w\right)\exp[-(k^2 + 2\mathrm{Re}_\tau/\Delta t)^{1/2}y] + \frac{\Delta t}{2}k\widetilde{p}'_w e^{-ky} \tag{6.26}$$

With these, Eq. (6.21) reduces to

$$\left(-\left(k^2+\frac{2\mathrm{Re}_\tau}{\Delta t}\right)^{1/2}\left(\widetilde{\phi}'-\frac{\Delta t}{2}k\widetilde{p}'_w\right)-\frac{\Delta t}{2}k^2\widetilde{p}'_w\right)\exp[-(k^2+2\mathrm{Re}_\tau/\Delta t)^{1/2}y]=0$$

Since $2\mathrm{Re}_\tau \gg k^2$,

$$\widetilde{p}'_w=\frac{2}{\Delta tk}\widetilde{\phi}' \tag{6.27}$$

Substituting into Eqs. (6.23)–(6.26) yields the following solutions for Eqs. (6.18)–(6.21):

$$\widetilde{u}'_1(y)=\frac{ik_x}{k}\widetilde{\phi}'(\exp[-(2\mathrm{Re}_\tau/\Delta t)^{1/2}y]-e^{-ky}) \tag{6.28}$$

$$\widetilde{u}'_3(y)=\frac{ik_z}{k}\widetilde{\phi}'(\exp[-(2\mathrm{Re}_\tau/\Delta t)^{1/2}y]-e^{-ky}) \tag{6.29}$$

$$\widetilde{u}'_2(y)=\widetilde{\phi}'e^{-ky} \tag{6.30}$$

$$\widetilde{p}'(y)=\frac{2}{k\Delta t}\widetilde{\phi}'e^{-ky} \tag{6.31}$$

It is indicated from these solutions that the effect of the wall disturbance on the flow field decreases exponentially with the increase in the normal distance from the wall.

6.2.1.2 Control Law with Spanwise Pressure Gradient and Spanwise Shear Stress[11]

In the near-wall region, such as at $y^+ = 10$ shown in Figure 3.80, the locations on the different sides of a stremwise vortex, corresponding the high pressure and the low pressure are marked with + and −, respectively. The blowing and suction are applied at + and −, respectively, to control the turbulent flow via increasing the pressure gradient in the spanwise direction under the streamwise vortex near the wall. Thus, the cost functional is

$$J(\phi)=\frac{l}{2A\Delta t}\int_S\int_t^{t+\Delta t}\phi^2 dtdS-\frac{1}{2A\Delta t}\int_S\int_t^{t+\Delta t}\left(\frac{\partial p}{\partial z}\right)_w^2 dtdS \tag{6.32}$$

where control ϕ is the blowing–suction velocity at the wall, A is the area of the wall, and Δt is the interval of a time step. There is a minus sign in front of the second term on the right-hand side, since we want to maximize the pressure gradient.

Using a Fréchet differential,

$$\frac{DJ(\phi)}{D\phi}\phi'=\frac{l}{A\Delta t}\int_S\int_t^{t+\Delta t}\phi\phi' dtdS-\frac{1}{A\Delta t}\int_S\int_t^{t+\Delta t}\left.\frac{\partial p}{\partial z}\right|_w\left.\frac{\partial p'}{\partial z}\right|_w dtdS \tag{6.33}$$

The Fourier representation of the above equation is

$$\frac{\mathsf{D}\widetilde{J}(\phi)}{\mathsf{D}\phi}\widetilde{\phi'}^* = l\widetilde{\phi}\widetilde{\phi'}^* - \left.\frac{\partial\widetilde{p}}{\partial z}\right|_w \left.\frac{\partial\widetilde{p'}^*}{\partial z}\right|_w \tag{6.34}$$

where the superscript "*", denotes the complex conjugate.

From Eqs. (1.101) and (6.31),

$$\left.\frac{\partial\widetilde{p}}{\partial z}\right|_w = ik_z\widetilde{p}_w$$

$$\frac{\partial\widetilde{p'}^*}{\partial z}_w = -\frac{2ik_z}{k\Delta t}\widetilde{\phi'}^*$$

Equation (6.34) then reduces to

$$\frac{\mathsf{D}\widetilde{J}(\phi)}{\mathsf{D}\phi}\widetilde{\phi'}^* = l\widetilde{\phi}\widetilde{\phi'}^* - \frac{2k_z^2}{k\Delta t}\widetilde{p}_w\widetilde{\phi'}^*$$

For an arbitrary $\widetilde{\phi'}^*$, the above equation should be satisfied, yielding

$$\frac{\mathsf{D}\widetilde{J}(\phi)}{\mathsf{D}\phi} = l\widetilde{\phi} - \frac{2k_z^2}{k\Delta t}\widetilde{p}_w \tag{6.35}$$

If $\frac{\mathsf{D}\widetilde{J}(\phi)}{\mathsf{D}\phi} = 0$, then

$$\widetilde{\phi}(k_x, k_z) = C\frac{k_z^2}{k}\widetilde{p}_w = -C\frac{1}{k}\left.\frac{\partial^2\widetilde{p}}{\partial z^2}\right|_w \tag{6.36}$$

where C is a positive scale factor that determines the cost of the actuation. This equation indicates that the optimum wall actuation is negatively proportional to the second spanwise derivative of the wall pressure, with the high wave number components reduced by $1/k$.

The increase in the pressure gradient in the spanwise direction under the streamwise vortex near the wall will induce the spanwise flow, thus increasing the spanwise shear stress, $\partial w/\partial y$, at the wall. Thus another choice for the cost functional to be minimized is

$$J(\phi) = \frac{l}{2A\Delta t}\int_S\int_t^{t+\Delta t}\phi^2 dtdS - \frac{1}{2A\Delta t}\int_S\int_t^{t+\Delta t}\left(\frac{\partial w}{\partial y}\right)_w^2 dtdS \tag{6.37}$$

Following the procedure that leads to Eq. (6.36)

$$\widetilde{\phi}(k_x, k_z) = C\frac{ik_z}{k}\left.\frac{\partial\widetilde{w}}{\partial y}\right|_w = \frac{C}{k}\left(\frac{\partial}{\partial z}\left(\frac{\partial\widetilde{w}}{\partial y}\right)\right)_w \tag{6.38}$$

which indicates that the optimum wall actuation is proportional to the spanwise derivative of the spanwise shear at the wall, with the high wave number components reduced by $1/k$.

Hence, the control laws derived in the spectrum space, such as Eqs. (6.36) and (6.38), make the N-S control problems closed, which is called the suboptimal control with spectral method. Figure 6.2 shows contours of the streamwise vorticity in a cross-flow plane for the control with Eq. (6.38) based on wall-shear stress, where Figure 6.2a represents no control and Figure 6.2b represents control. It is evident that the strength of the streamwise vortices decreases significantly. Meanwhile, this control can lead to drag reduction as shown in Figure 6.3.

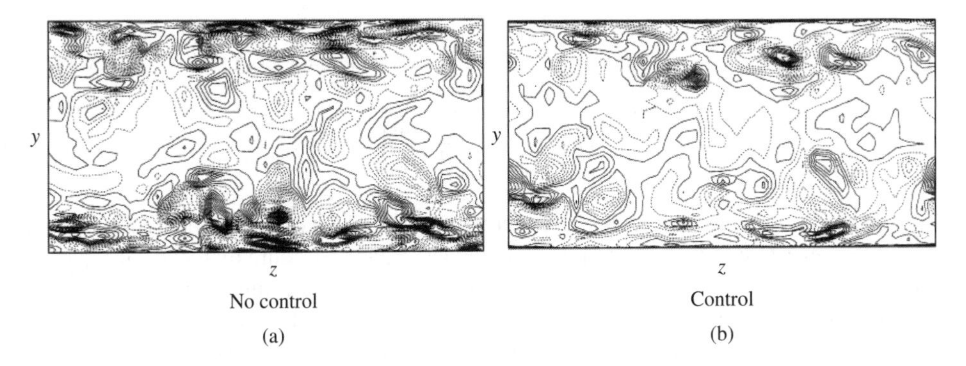

Figure 6.2 Contours of streamwise vorticity in a cross-flow plane: (a) no control and (b) control.[11] *Source*: Lee C 1994. Reproduced with permission from Cambridge University Press

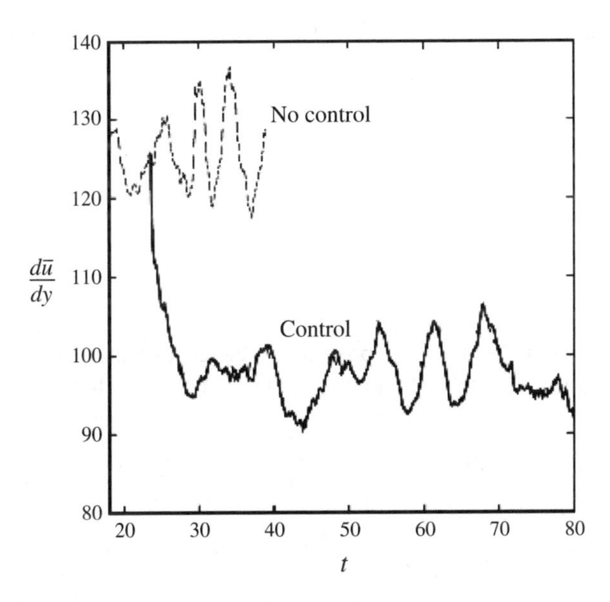

Figure 6.3 Mean streamwise wall shear stress.[11] *Source*: Lee C 1994. Reproduced with permission from Cambridge University Press

6.2.1.3 Control Law with Streamwise Shear Stress[12, 13]

From a practical point of view, it is desirable to use the streamwise wall-shear stress, $(\partial u/\partial y)_w$ as a sensor signal, thus the cost functional is given by

$$J(\phi) = \frac{l}{2A\Delta t}\int_S \int_t^{t+\Delta t} \phi^2 dt dS - \frac{1}{2A\Delta t}\int_S \int_t^{t+\Delta t} \left(\frac{\partial u}{\partial y}\right)_w^\alpha dt dS \tag{6.39}$$

The power to the wall shear, α, is chosen as 1 or 2. The use of $\alpha = 1$ leads to a trivial solution, $\phi = 0$. With $\alpha = 2$, following the procedure mentioned above, the control law is expressed as

$$\widetilde{\phi}(k_x, k_z) = C\frac{ik_x}{k}\frac{\partial \widetilde{u}}{\partial y}\bigg|_w \tag{6.40}$$

Unfortunately, however, the friction drag increases with Eq. (6.40), since the nonlinear terms, that is, the convective terms, are neglected in the derivative procedure that leads to Eq. (6.40). Hence some modifications should be considered to improve this control law.

As is well known, the friction drag on the wall is dominated by the Reynolds shear stress (i.e., $-uv'$) distributions near the wall, which is closely related with the sweep/ejection motions. The proposed cost functional is defined as

$$J(\phi) = \frac{l}{2A\Delta t}\int_s \int_t^{t+\Delta t} \phi^2 dt dS + \frac{l}{2A\Delta t}\int_s \int_t^{t+\Delta t} (-uv')_{y=Y} dt dS \tag{6.41}$$

where

$$u(y) = U(y) + u'(y) = (2y - y^2) + u'(y)$$

$$\frac{\partial u}{\partial y}\bigg|_w = 2 + \frac{\partial u'}{\partial y}\bigg|_w$$

The Reynolds shear stress at $y = Y$ appearing in the cost functional needs to be evaluated by Taylor series expansions of the streamwise and wall-normal velocities:

$$u'(y) = y\frac{\partial u'}{\partial y}\bigg|_w + \frac{y^2}{2}\frac{\partial^2 u'}{\partial y^2}\bigg|_w + O(y^3)$$

$$v'(y) = \phi + \frac{y^2}{2}\frac{\partial^2 v'}{\partial y^2}\bigg|_w + O(y^3)$$

Then the Reynolds shear stress at $y = Y$ is approximated as

$$-uv' = -\left(2Y + Y\frac{\partial u'}{\partial y}\bigg|_w\right)\phi = -Y\frac{\partial u}{\partial y}\bigg|_w \phi$$

Substitution of Eq. (6.42) into Eq. (6.41) yields

$$J(\phi) = \frac{l}{2A\Delta t}\int_{s}\int_{t}^{t+\Delta t} \phi^2 dtdS - \frac{Y}{2A\Delta t}\int_{s}\int_{t}^{t+\Delta t} \left.\frac{\partial u}{\partial y}\right|_w \phi dtdS$$

Applying the Fréchet differential,

$$\frac{DJ(\phi)}{D\phi}\phi' = \frac{l}{A\Delta t}\int_{S}\int_{t}^{t+\Delta t} \phi\phi' dtdS - \frac{Y}{2A\Delta t}\int_{S}\int_{t}^{t+\Delta t} \left(\phi' \left.\frac{\partial u}{\partial y}\right|_w + \phi \left.\frac{\partial u'}{\partial y}\right|_w\right) dtdS$$

By Fourier transform,

$$\frac{D\widetilde{J}}{D\phi}\widetilde{\phi'}^* = l\phi\widetilde{\phi'}^* - \frac{Y}{2}\left(\left.\frac{\partial \widetilde{u}}{\partial y}\right|_w \widetilde{\phi'}^* + \widetilde{\phi}\left.\frac{\partial \widetilde{u'}^*}{\partial y}\right|_w \right) \tag{6.42}$$

Differentiating Eq. (6.29) and assuming $2R_{e\tau} \gg k^2$, we get

$$\left.\frac{\partial \widetilde{u'}^*}{\partial y}\right|_w = m\frac{ik_x}{k}\widetilde{\phi'}^*$$

where

$$m = \sqrt{\frac{2\text{Re}}{\Delta t}} \tag{6.43}$$

Finally, by substituting Eq. (6.43) into Eq. (6.42), we find the control law

$$\widetilde{\phi} = \frac{\alpha}{1 - i\alpha m\frac{k_x}{k}} \left.\frac{\partial \widetilde{u}}{\partial y}\right|_w \tag{6.44}$$

where $\alpha = \frac{Y}{2l}$

6.2.2 *Control of Backward-Facing Step Flow*

6.2.2.1 Back-Facing Step Flow[14 15]

When a turbulent boundary-layer flow passes over a backward-facing step edge, separated flow behind the backward-facing step has a mixing layer induced by vortical structures. Generating streamwise vortices at or before a backward-facing step may be a possible control strategy for increasing mixing behind the step. As shown in Figure 6.4, the blowing and suction given from a spanwise slop Γ_c are applied to increase mixing. The actuation slop is located at the backward-facing step edge. The actuation varies in the spanwise direction and time based on pressure at Γ_s located near the reattachment position, where the wall-pressure fluctuations are generated.

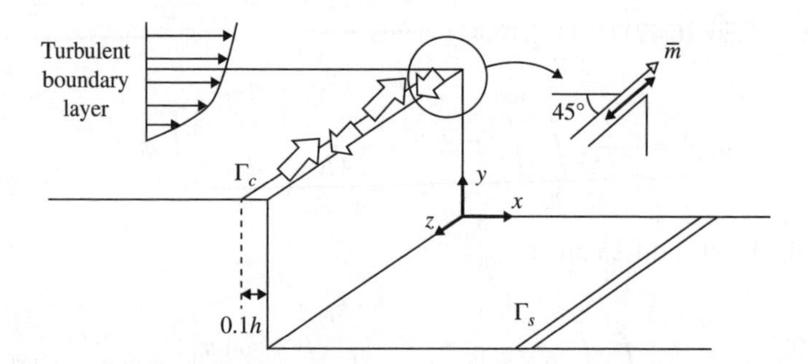

Figure 6.4 Control of turbulent flow over a backward-facing step.[14] *Source*: Kang S 2002. Reproduced with permission from Cambridge University Press

Therefore, the cost functional is defined as

$$J(\phi) = -\frac{1}{2}\int_{\Gamma_s} \left(\frac{\partial p'}{\partial z}\right)^2 dzdx + \frac{l^2}{2}\int_{\Gamma_c} \phi^2(z)dzdx \tag{6.45}$$

where ϕ is the control provided at Γ_c and p' is the pressure fluctuation sensed at Γ_s.

The governing equations for fluid flow and boundary conditions for the present problem are, respectively,

$$\frac{\partial u_i}{\partial t} + u_j\frac{\partial u_i}{\partial x_j} = -\frac{\partial p}{\partial x_i} + \frac{1}{\mathrm{Re}}\frac{\partial^2 u_i}{\partial x_j \partial x_j}$$

$$\frac{\partial u_j}{\partial x_j} = 0 \tag{6.46}$$

and

$$\begin{cases} u_i = \phi(z)\, m_i & \text{on } \Gamma_c \\ u_i = \text{given} & \text{otherwise} \end{cases} \tag{6.47}$$

where $\mathrm{Re} = \frac{u_\infty h}{\nu}$ is the Reynolds number, u_∞ is the free-stream velocity, ν is the kinematic viscosity, and m_i is the unit vector along the direction of blowing.

6.2.2.2 Control Law

We choose the Crank–Nicolson scheme for the pressure gradient and viscous diffusion terms and the Adams–Bashforth scheme for the convection terms. Then the discretized equations are

$$u_i^{n+1} - \frac{\Delta t}{2\mathrm{Re}}\frac{\partial^2 u_i^{n+1}}{\partial x_j \partial x_j} + \frac{\Delta t}{2}\frac{\partial p^{n+1}}{\partial x_i} + R^n = 0 \tag{6.48}$$

$$\frac{\partial u_j^{n+1}}{\partial x_j} = 0 \tag{6.49}$$

with

$$\begin{cases} u_i^{n+1} = \phi^{n+1}(z)\, m_i & \text{on}\,\Gamma_c \\ u_i^{n+1} = \text{given} & \text{otherwise} \end{cases} \tag{6.50}$$

where superscript n and $n+1$ denote the time step, and R^n includes the nonlinear terms and the explicit parts of the pressure gradient and viscous diffusion terms at the control time step n.

The Fréchet differential is applied to Eqs. (6.48)–(6.50), which yields

$$u_i' - \frac{\Delta t}{2R_e}\frac{\partial^2 u_i'}{\partial x_j \partial x_j} + \frac{\Delta t}{2}\frac{\partial p'}{\partial x_i} = 0 \tag{6.51}$$

$$\frac{\partial u_j'}{\partial x_j} = 0 \tag{6.52}$$

$$\begin{cases} u_i' = \phi'(z)\, m_i & \text{on}\,\Gamma_c \\ u_i' = 0 & \text{otherwise} \end{cases} \tag{6.53}$$

where $u_i' = \frac{Du_i^{n+1}}{D\phi^{n+1}}\phi'^{n+1}$

$$p' = \frac{Dp^{n+1}}{D\phi^{n+1}}\phi'^{n+1}$$

ϕ' is an arbitrary perturbation to ϕ.

The Fréchet differential system can be solved analytically with the aid of the Fourier transform in a simple geometry as mentioned in Section 6.2.1. However, when the flow geometry is complex, such as the present problem, it is not, in general, possible to solve it analytically. To present the problem, introduce a system as follows:

$$\eta_i - \frac{\Delta t}{2R_e}\frac{\partial^2 \eta_i}{\partial x_j \partial x_j} + \frac{\Delta t}{2}\frac{\partial \pi}{\partial x_i} = 0 \tag{6.54}$$

$$\frac{\partial \eta_j}{\partial x_j} = 0 \tag{6.55}$$

$$\begin{cases} \eta_i = \delta(z)\, m_i & \text{on}\,\Gamma_c \\ \eta_i = 0 & \text{otherwise} \end{cases} \tag{6.56}$$

where $\delta(z)$ denotes the Dirac delta function defined at $z = 0$, namely

$$\delta(z) = \begin{cases} 1 & z = 0 \\ 0 & z \neq 0 \end{cases}$$

The solution of Eqs. (6.51)–(6.53) is given as the following convolution integral form:

$$u_i'(x, y, z) = \int_0^{L_z} \eta_i(x, y, z - \varsigma)\phi'(\varsigma)d\varsigma \tag{6.57}$$

$$p'(x, y, z) = \int_0^{L_z} \pi(x, y, z - \varsigma)\phi'(\varsigma)d\varsigma \tag{6.58}$$

where L_z is the spanwise length of Γ_c.

By taking the Fréchet derivative of Eq. (6.45), and using Eqs. (6.57) and (6.58), as well as $\frac{DJ}{D\phi} = 0$, we can easily obtain

$$\phi(z) = \kappa \int_{\Gamma_s} \frac{\partial p'}{\partial \varsigma} \frac{\partial \pi(\varsigma - z)}{\partial \varsigma} d\varsigma dx \tag{6.59}$$

where κ is an ascent parameter determined such that root-mean-squared (RMS) of $\phi(z)$ is constant in time through the computation.

The solution π can be obtained by solving Eqs. (6.54)–(6.56) numerically. Then the control $\phi(z)$ is obtained through the convolution integral Eq. (6.59) with the measurement of p' at Γ_s at every control time step. Therefore, Eqs. (6.46), (6.47), and (6.59) constitute a complete system.

6.2.2.3 Control of Back-Facing Step Flow

The governing equations for the present problem are shown in Eqs. (6.46) and (6.47). The actuation slop is located at $-0.1h \le x \le 0$ and $y = h$, h is the step height. The angle of actuation is 45° with respect to the streamwise direction $-0.1h \le x \le 0$, L_s is located at $x_s = 6h$. The no-slip condition is used at the wall, and the periodic boundary condition is used in the spanwise direction. Assuming symmetry at the upper boundary, the no-stress condition is used here. To allow vortices to pass away smoothly, the boundary condition at the exit is the convective outflow condition

$$\frac{\partial u_i}{\partial t} + U_c \frac{\partial u_i}{\partial x} = 0$$

where U_c is the averaged streamwise velocity at the exit.

In numerical calculation, a variant of the fractional step method is employed to treat implicit coupling of the Navier–Stokes and continuity equations. The Crank–Nicolson method is used for the convection and diffusion terms in the wall-normal direction, and a third-order Runge–Kutta scheme is used for all other terms. The large eddy simulation (LES) technique is used with a dynamic subgrid model for the turbulent flow.

The time-average streamlines for the uncontrolled and suboptimal control cases are shown in Figure 6.5a and b, respectively. It is observed that the recirculating region becomes smaller and the reattachment length decreases with control.

Figure 6.6a and b shows the contours of the instantaneous spanwise and steamwise vorticity fluctuations, respectively. It is clear that the instantaneous spanwise and steamwise vorticity

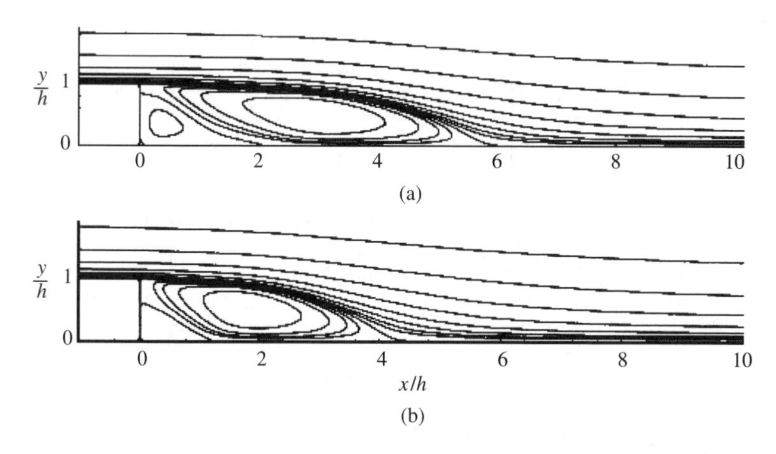

Figure 6.5 Time-averaged streamlines: (a) no control and (b) suboptimal control.[14] *Source*: Kang S 2002. Reproduced with permission from Cambridge University Press

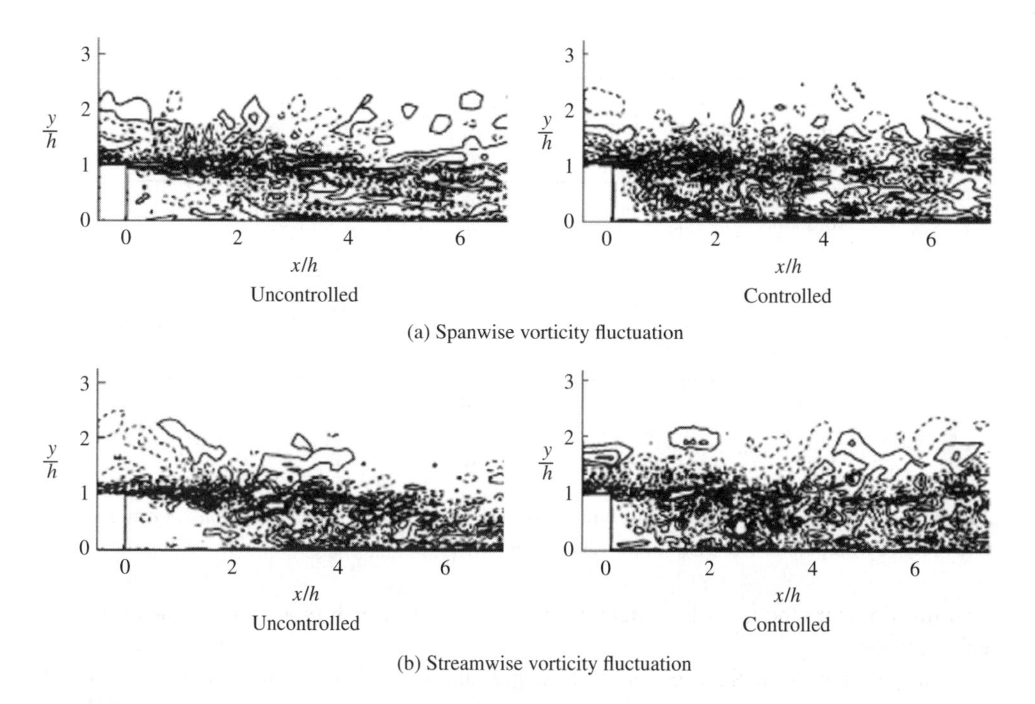

Figure 6.6 Contours of the instantaneous vorticity fluctuation.[14] *Source*: Kang S 2002. Reproduced with permission from Cambridge University Press

fluctuations are large even inside the recirculating region and at $1 < y/h < 2$ in the downstream locations owing to control, as compared to the uncontrolled case.

The distribution of control $\phi(z)$ at the spanwise direction, that is, the control profile, varies with the sensor location x_s, as shown in Figure 6.7, where (a)$x_s = 0$, $y_s = h$, (b)$x_s = 0$, $y_s = 0.8h$, (c)$x_s = y_s = 0$, (d)$x_s = 4h$, $y_s = 0$, (e)$x_s = 6h$, $y_s = 0$. It is obvious that the control profile becomes more sinusoidal as the increasing x_s, and can be modeled as a

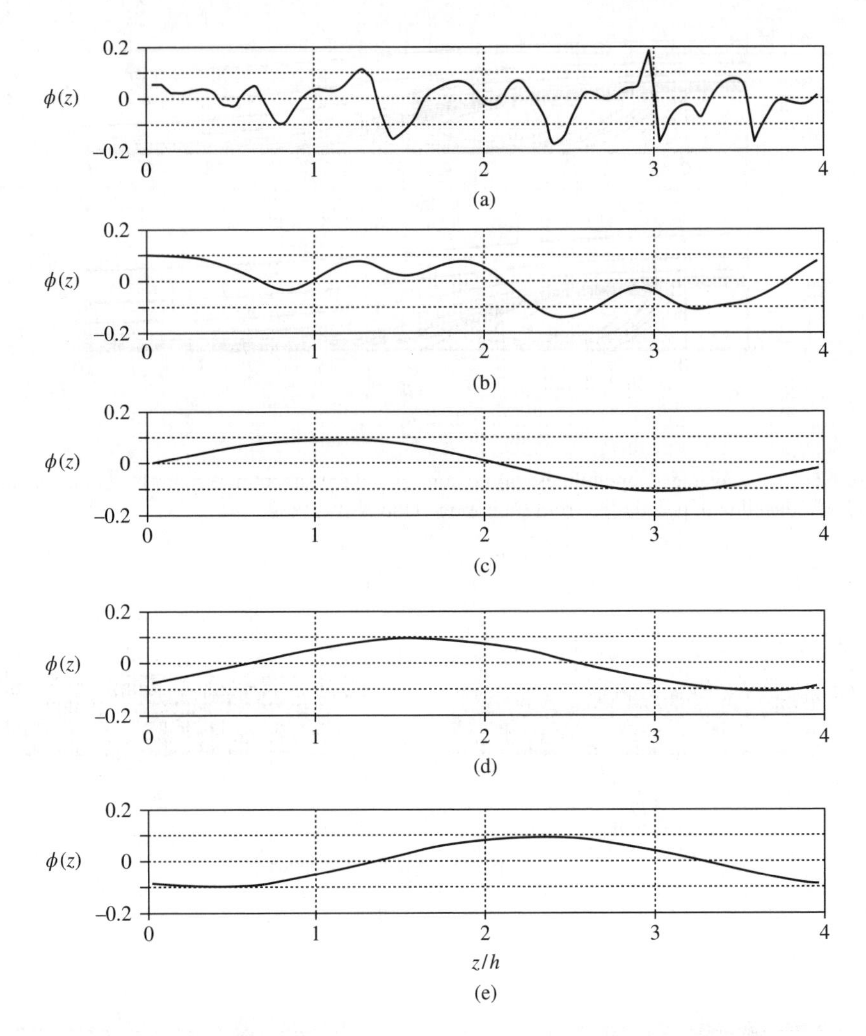

Figure 6.7 Variations of control profile with sensor locations in same instance.[14] *Source*: Kang S 2002. Reproduced with permission from Cambridge University Press

sine function with the time-dependent phase for $x_s = 6h$, which produces significant mixing enhancement.

Figure 6.8 shows the time sequence of the pressure isosurfaces for the suboptimal control, together with the blowing–suction profile at each instance. Here the arrow heading downstream denotes the blowing, and that heading upstream the suction. Each isosurface of the low pressure can be considered as a vortical structure, thus it is observed that inclined vortical structures are generated near the step. These inclined vortical structures have two components. One is the spanwise roll-up vortex generated by local blowing and suction, and the other is the streamwise-vorticity component generated by tilting of the spanwise vorticity due to the phase movement in the spanwise direction. These vortical structures become closer due to different convection velocities and interact among themselves in the downstream. This vortical interaction increases mixing and reduces the reattachment length.

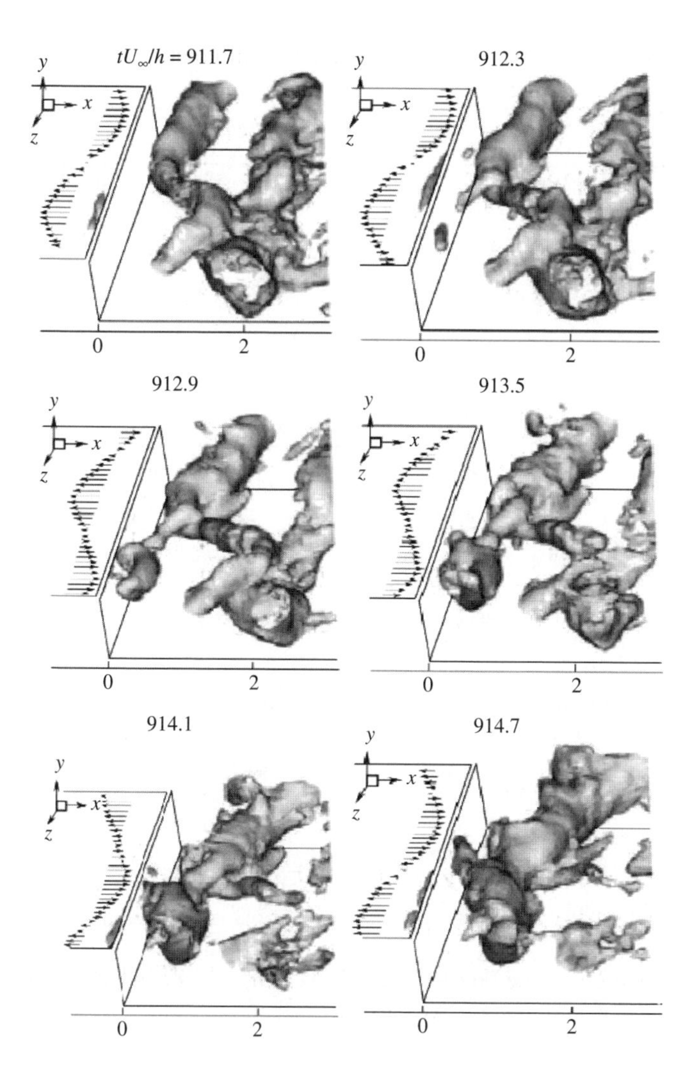

Figure 6.8 Time sequence of pressure isosurfaces for suboptimal control.[14] *Source*: Kang S 2002. Reproduced with permission from Cambridge University Press

6.2.3 Control of Cylinder Flow

6.2.3.1 Analytic Fréchet Differential Equation

Considering an incompressible flow over a circular cylinder, immersed in fluid vertically, controlled by blowing–suction on the wall, the governing equations are given in the polar coordinates (r, θ) with the boundary conditions

$$\left.\begin{aligned} &u_r|_{r=R} = \phi, \quad u_\theta|_{r=R} = 0 \\ &u_r|_{r\to\infty} = \cos\theta, \quad u_\theta|_{r\to\infty} = -\sin\theta \end{aligned}\right\}$$

After the discretization of the equations, the Fréchet differential yields

$$u'_r + \frac{\Delta t_c}{2}\frac{\partial p'}{\partial r} - \frac{\Delta t_c}{2\mathrm{Re}}\left(\frac{\partial^2 u'_r}{\partial r^2} + \frac{1}{r}\frac{\partial u'_r}{\partial r} + \frac{1}{r^2}\frac{\partial^2 u'_r}{\partial \theta^2} - \frac{1}{r^2}u'_r - \frac{2}{r^2}\frac{\partial u'_\theta}{\partial \theta}\right) = 0 \tag{6.60}$$

$$u'_\theta + \frac{\Delta t_c}{2}\frac{1}{r}\frac{\partial p'}{\partial \theta} - \frac{\Delta t_c}{2\mathrm{Re}}\left(\frac{\partial^2 u'_\theta}{\partial r^2} + \frac{1}{r}\frac{\partial u'_\theta}{\partial r} + \frac{1}{r^2}\frac{\partial^2 u'_\theta}{\partial \theta^2} - \frac{1}{r^2}u'_\theta - \frac{2}{r^2}\frac{\partial u'_r}{\partial \theta}\right) = 0 \tag{6.61}$$

$$\frac{1}{r}u'_r + \frac{\partial u'_r}{\partial r} + \frac{1}{r}\frac{\partial u'_\theta}{\partial \theta} = 0 \tag{6.62}$$

$$\frac{\partial^2 p'}{\partial r^2} + \frac{1}{r}\frac{\partial p'}{\partial r} + \frac{1}{r^2}\frac{\partial^2 p'}{\partial \theta^2} = 0 \tag{6.63}$$

with

$$\left.\begin{array}{ll} u'_r|_{r=R} = \phi', & u'_\theta|_{r=R} = 0 \\ u'_r|_{r\to\infty} = 0, & u'_\theta|_{r\to\infty} = 0 \end{array}\right\}$$

By Fourier transform in the θ direction (see Section 1.6.2), that is

$$u'_r = \sum_k \widetilde{q}_r e^{ik\theta}, \quad u'_\theta = \sum_k \widetilde{q}_\theta e^{ik\theta}$$

$$p' = \sum_k \widetilde{p} e^{ik\theta}, \quad \phi' = \sum_k \widetilde{\phi} e^{ik\theta}$$

where $\widetilde{q}_r$, $\widetilde{q}_\theta$, $\widetilde{p}$, and $\widetilde{\phi}'$ are the Fourier coefficients of u'_r, u'_θ, p', and ϕ', respectively, and k is the wave number in the θ direction, we have

$$\frac{d^2\widetilde{q}_r}{dr^2} + \frac{1}{r}\frac{d\widetilde{q}_r}{dr} - \left(\frac{2\mathrm{Re}}{\Delta t_c} + \frac{k^2+1}{r^2}\right)\widetilde{q}_r - \frac{2ik}{r^2}\widetilde{q}_\theta - \mathrm{Re}\frac{d\widetilde{p}}{dr} = 0 \tag{6.64}$$

$$\frac{d^2\widetilde{q}_\theta}{dr^2} + \frac{1}{r}\frac{d\widetilde{q}_\theta}{dr} - \left(\frac{2\mathrm{Re}}{\Delta t_c} + \frac{k^2+1}{r^2}\right)\widetilde{q}_\theta - \frac{2ik}{r^2}\widetilde{q}_r - \frac{ik}{r}\mathrm{Re}\frac{d\widetilde{p}}{dr} = 0 \tag{6.65}$$

$$\frac{1}{r}\widetilde{q}_r + \frac{d\widetilde{q}_r}{dr} + \frac{ik}{r}\widetilde{q}_\theta = 0 \tag{6.66}$$

$$\frac{d^2\widetilde{p}}{dr^2} + \frac{1}{r}\frac{d\widetilde{p}}{dr} - \frac{k^2}{r^2}\widetilde{p} = 0 \tag{6.67}$$

with

$$\widetilde{q}_r|_{r=R} = \widetilde{\phi}, \quad \widetilde{q}|_{r=R} = 0, \quad \widetilde{q}_r|_{r\to\infty} = 0 \quad \widetilde{q}_\theta|_{r\to\infty} = 0 \tag{6.68}$$

When $k = 0$,Eqs. (6.64)–(6.68) become

$$\frac{d^2\widetilde{q}_r}{dr^2} + \frac{1}{r}\frac{d\widetilde{q}_r}{dr} - \left(\frac{2\text{Re}}{\Delta t_c} + \frac{1}{r^2}\right)\widetilde{q}_r - \text{Re}\frac{d\widetilde{\rho}}{dr} = 0 \tag{6.69}$$

$$\frac{d^2\widetilde{q}_\theta}{dr^2} + \frac{1}{r}\frac{d\widetilde{q}_\theta}{dr} - \left(\frac{2\text{Re}}{\Delta t_c} + \frac{1}{r^2}\right)\widetilde{q}_\theta = 0 \tag{6.70}$$

$$\frac{1}{r}\widetilde{q}_r + \frac{d\widetilde{q}_r}{dr} = 0 \tag{6.71}$$

$$\frac{d^2\widetilde{\rho}}{dr^2} + \frac{1}{r}\frac{d\widetilde{\rho}}{dr} = 0 \tag{6.72}$$

with

$$\widetilde{q}_r|_{r=R,k=0} = \widetilde{\phi}_{k=0},\ \widetilde{q}_\theta|_{r=R,k=0} = 0,\ \widetilde{q}_r|_{r\to\infty,k=0} = 0,\ \widetilde{q}_\theta|_{r\to\infty,k=0} = 0 \tag{6.73}$$

The general solution of Eq. (6.72) is

$$\widetilde{\rho}_{k=0} = X_1 + X_2 \ln r$$

When $r \to \infty$, $\widetilde{\rho}$ should have a finite value, and thus $X_2 = 0$. Therefore

$$\widetilde{\rho}_{k=0} = \widetilde{\rho}_{r=R,k=0}(\text{constant}) \tag{6.74}$$

That is $\frac{d\widetilde{\rho}}{dt_{k=0}} = 0$.
Substitution into Eq. (6.69) yields

$$\frac{d^2\widetilde{q}}{dr^2} + \frac{1}{r}\frac{d\widetilde{q}}{dr} - \left(\frac{2\text{Re}}{\Delta t_c} + \frac{1}{r^2}\right)\widetilde{q}_r = 0 \tag{6.75}$$

Note that Eqs. (6.70) and (6.75) are modified Bessel equations, with solutions consisting of the modified Bessel functions:

$$\widetilde{q}_{r,k=0} = X_3 I_1(mr) + X_4 K_1(mr) \tag{6.76}$$

$$\widetilde{q}_{\theta,k=0} = X_5 I_1(mr) + X_6 K_1(mr) \tag{6.77}$$

where I_1 and K_1 are the modified Bessel equations of the first kind, of order 1, and the second kind, of order 1, respectively, and

$$m = \sqrt{\frac{2\text{Re}}{\Delta t_c}} \tag{6.78}$$

From the boundary condition (6.73), the coefficients of Eqs. (6.76) and (6.77) can be determined as

$$\widetilde{q}_{r,k=0} = \frac{\widetilde{\phi}_{k=0}}{K_1(mR)} K_1(mr) \tag{6.79}$$

$$\widetilde{q}_{\theta,k=0} = 0 \tag{6.80}$$

When $k \neq 0$, the general solution of Eq. (6.72) is

$$\widetilde{\rho} = X_7 r^k + X_8 r^k$$

When $r \to \infty$, $\widetilde{\rho}$ should have a finite value, and thus

$$X_8 = 0 \quad \text{for} \quad k < 0$$

or $X_7 = 0$ for $k > 0$

Therefore, the solution of Eq. (6.67) is

$$\widetilde{\rho}_{k\neq 0} = \widetilde{\rho}_{r=R,k\neq 0}\left(\frac{R}{r}\right)^{|k|} \tag{6.81}$$

where a new coefficient $\widetilde{\rho}_{r=R,k\neq 0}$ will be determined later.

Differentiation of Eq. (6.81) gives

$$\frac{d\widetilde{\rho}}{dr}_{k\neq 0} = -\frac{|k|}{r}\widetilde{\rho}_{r=R,k\neq 0}\left(\frac{R}{r}\right)^{|k|} \tag{6.82}$$

Substituting Eqs. (6.81) and (6.82) into Eqs. (6.64) and (6.65), respectively, yields

$$\frac{d^2\widetilde{q}_r}{dr^2} + \frac{1}{r}\frac{d\widetilde{q}_r}{dr} - \left(\frac{2\text{Re}}{\Delta t_c} + \frac{k^2+1}{r^2}\right)\widetilde{q}_r - \frac{2ik}{r^2}\widetilde{q}_\theta = -\frac{|k|}{r}\text{Re}\widetilde{\rho}_{r=R,k\neq 0}\left(\frac{R}{r}\right)^{|k|} \tag{6.83}$$

$$\frac{d^2\widetilde{q}_\theta}{dr^2} + \frac{1}{r}\frac{d\widetilde{q}_\theta}{dr} - \left(\frac{2\text{Re}}{\Delta t_c} + \frac{k^2+1}{r^2}\right)\widetilde{q}_\theta - \frac{2ik}{r^2}\widetilde{q}_r = \frac{ik}{r}\text{Re}\widetilde{\rho}_{r=R,k\neq 0}\left(\frac{R}{r}\right)^{|k|} \tag{6.84}$$

Using Eqs. (6.66) and (6.83) becomes

$$\frac{d^2\widetilde{q}_r}{dr^2} + \frac{3}{r}\frac{d\widetilde{q}_r}{dr} - \left(\frac{2\text{Re}}{\Delta t_c} + \frac{k^2-1}{r^2}\right)\widetilde{q}_r = -\frac{|k|}{r}\text{Re}\widetilde{\rho}_{r=R,k\neq 0}\left(\frac{R}{r}\right)^{|k|} \tag{6.85}$$

When the right-hand side of Eq. (6.85) is zero, Eq. (6.85) is also a modified Bessel equation. Therefore, the homogenous solution of Eq. (6.85) is

$$\widetilde{q}_{r,\text{ho}} = \frac{1}{r}\{X_8 I_{|k|}(mr) + X_9 K_{|k|}(mr)\}$$

where $I_{|k|}$ and $K_{|k|}$ are the modified Bessel functions of the first kind, of order $|k|$, and the second kind, of order $|k|$, respectively.

By assuming $\widetilde{q}_{r,\text{pa}} = X_{10} r^{\lambda}$, the particular solution of Eq. (6.85) is

$$\widetilde{q}_{r,\text{pa}} = \frac{\Delta t}{2}\frac{|k|}{r}\widetilde{\rho}_{r=R,k\neq 0}\left(\frac{R}{r}\right)^{|k|}$$

Thus, the general solution of Eq. (6.85) is

$$\widetilde{q}_{r,k\neq 0} = X_8\frac{1}{r}I_{|k|}(mr) + X_9\frac{1}{r}K_{|k|}(mr) + \frac{\Delta t_c}{2}\frac{|k|}{r}\widetilde{\rho}_{r=R,k\neq 0}\left(\frac{R}{r}\right)^{|k|} \tag{6.86}$$

From the boundary condition (6.68), we have

$$\widetilde{q}_{r,k\neq 0} = \frac{\Delta t_c}{2}\frac{|k|}{r}\widetilde{\rho}_{r=R,k\neq 0}\left(\frac{R}{r}\right)^{|k|} + C\frac{1}{r}K_{|k|}(mr) \tag{6.87}$$

where

$$C = \frac{R\widetilde{\phi}_{k\neq 0} - \frac{1}{2}\Delta t_c|k|\widetilde{\rho}_{r=R,k\neq 0}}{K_{|k|}(mR)} \tag{6.88}$$

Differentiation of Eq. (6.87) gives

$$\begin{aligned}\frac{dq_{r,k\neq 0}}{dr} &= -\frac{\Delta t_c}{2}\frac{|k|(|k|+1)}{r^2}\widetilde{\rho}_{r=R,k\neq 0}\left(\frac{R}{r}\right)^{|k|} \\ &+C\left\{-\frac{m}{r}K_{|k|+1}(mr) + \frac{|k|-1}{r^2}K_{|k|}(mr)\right\}\end{aligned} \tag{6.89}$$

where the differential relation between the modified Bessel functions K of orders n and $n+1$ is used,

$$\frac{d}{dr}K_n(mr) = -mK_{n+1}(mr) + \frac{n}{r}K_n(mr)$$

By substituting Eqs. (6.87) and (6.89) into Eq. (6.66), we get the general solution

$$\widetilde{q}_{\theta,k\neq 0} = \frac{i}{k}\left\{-\frac{\Delta t_c}{2}\frac{|k|^2}{r}\widetilde{\rho}_{r=R,k\neq 0}\left(\frac{R}{r}\right)^{|k|} + \frac{|k|}{r}CK_{|k|}(mr) - mCK_{|K|+1}(mr)\right\} \tag{6.90}$$

This solution satisfies the boundary condition (6.68).

Now, let us determine the value $\widetilde{\rho}_{r=R,k\neq 0}$. From the boundary condition (6.68), we obtain

$$C\left\{\frac{|k|}{R}K_{|k|}(mR) - mK_{|k|+1}(mR)\right\} = \frac{\Delta t_c}{2}\frac{|k|^2}{R}\widetilde{\rho}_{r=R,k\neq 0}$$

Substitution into Eq. (6.88) yields

$$\widetilde{\rho}_{r=R,k\neq 0} = \frac{2}{\Delta t_c}\frac{|k|K_{|k|}(mR) - mRK_{|k|+1}(mR)}{|k|\{(2|k|/R)K_{|k|}(mR) - mK_{|k|+1}(mR)\}}\widetilde{\phi}'_{k\neq 0}$$

Using Eq. (6.88), we get

$$C = \frac{R|k|}{2|k|K_{|k|}(mR) - mRK_{|k|+1}(mR)}\widetilde{\phi}'_{k\neq 0}$$

Substitution of Eqs. (6.83) and (6.84) into Eqs. (6.73), (6.80), and (6.81) yields

$$\widetilde{p}_{k\neq0}=\frac{2}{\Delta t_c}\frac{1}{|k|}\widetilde{\phi}_{k\neq0}\frac{A}{B}\left(\frac{R}{r}\right)^{|k|} \tag{6.91}$$

$$\widetilde{q}_{r,k\neq0}=\widetilde{\phi}_{k\neq0}\frac{A(R/r)^{|k|}+R|k|K_{|k|}(mr)}{Br} \tag{6.92}$$

$$\widetilde{q}_{\theta,k\neq0}=\frac{i|k|}{k}\widetilde{\phi}_{k\neq0}\frac{-A(R/r)^{|k|}+\{R|k|K_{|k|}(mr)-mRrK_{|k|+1}(mr)\}}{Br} \tag{6.93}$$

where

$$A=R|k|K_{|k|}(mR)-mR^2K_{|k|+1}(mR) \tag{6.94}$$

$$B=2|k|K_{|k|}(mR)-mRK_{|k|+1}(mR) \tag{6.95}$$

In summary, the analytic solutions of Eqs. (6.64)–(6.68) are

$$\widetilde{q}_{r,k=0}=\frac{\widetilde{\phi}'_{k=0}}{K_1(mR)}K_1(mr)$$

$$\widetilde{q}_{\theta,k=0}=0 \quad (6.80)$$

$$\widetilde{p}_{k=0}=\widetilde{p}_{r=R,k=0}(=\text{constant})$$

$$\widetilde{q}_{r,k\neq0}=\widetilde{\phi}'_{k\neq0}\frac{A(R/r)^{|k|}+R|k|K_{|k|}(mr)}{Br}$$

$$\widetilde{q}_{\theta,k\neq0}=\frac{i|k|}{k}\widetilde{\phi}'_{k\neq0}\frac{-A(R/r)^{|k|}+\{R|k|K_{|k|}(mr)-mRrK_{|k|+1}(mr)\}}{Br}$$

$$\widetilde{p}_{k\neq0}=\frac{2}{\Delta t_c}\frac{1}{|k|}\widetilde{\phi}'_{k\neq0}\frac{A}{B}\left(\frac{R}{r}\right)^{|k|}$$

where

$$m=\sqrt{\frac{2\mathrm{Re}}{\Delta t_c}}$$

$$A=R|k|K_{|k|}(mR)-mR^2K_{|k|+1}(mR)$$

$$B=2|k|K_{|k|}(mR)-mRK_{|k|+1}(mR)$$

Note that $\widetilde{\phi}'_{k=0}$ and $\widetilde{p}_{r=R,k=0}$ are the mean values of ϕ' and p' on the cylinder surface, respectively. From the continuity equation and boundary conditions, it is clear that $\widetilde{\phi}_{k=0}$(mean value of ϕ) $=\widetilde{\phi}'_{k=0}=0$. Then nonzero mean value can be easily implemented in the control procedure by assigning a different far-field boundary condition, such as

$$u|^{n+1}_{r=R_\infty}=\cos\theta+\phi R/R_\infty \tag{6.96}$$

6.2.3.2 Control Law

A flow past a cylinder may induce the vortex shedding behind the cylinder, resulting in the increase in the pressure drag. Therefore, three kinds of cost functional are defined with the actuation ϕ (blowing–suction on the cylinder surface here) to control vortex shedding which would be expected to reduce the drag.

$$J_1(\phi_1) = \int -p(\theta)|_{r=R} \cos\theta R d\theta + \frac{l^2}{2}\int {\phi_1}^2 R d\theta \tag{6.97}$$

$$J_2(\phi_2) = \frac{1}{2}\int (p_t - p(\theta)|_{r=R})^2 R d\theta + \frac{l^2}{2}\int {\phi_2}^2 R d\theta \tag{6.98}$$

$$J_3(\phi_3) = -\frac{1}{2}\int \left(\frac{\partial p(\theta)}{\partial\theta}|_{r=R}\right)^2 R d\theta + \frac{l^2}{2}\int {\phi_3}^2 R d\theta \tag{6.99}$$

where R is the cylinder radius and $p(\theta)|_{r=R}$ is the pressure on the cylinder surface. In $J_2(\phi_2)$, the square of difference between the target pressure p_t and the real pressure of the cylinder is taken as the sensor signal, where the inviscid flow pressure on the cylinder surface is considered as the target pressure, since a cylinder has no drag and lift on the cylinder surface for inviscid flow, that is

$$p_t = p_\infty + \frac{1}{2}\rho u_\infty^2(1 - 4\sin^2\theta)$$

In $J_3(\phi_3)$, the square of the pressure gradient on the cylinder surface is taken as the sensor signal, since the pressure gradient on the cylinder surface of the inviscid flow is much larger than that of the viscous flows.

Here, we restrict the sensing and actuation to local regions of the cylinder surface as shown in Figure 6.9, where the actuation ϕ is applied to a region of $\alpha \leq \theta \leq \beta$ and the sensing is restricted to a region of $\gamma \leq \theta \leq \delta$.

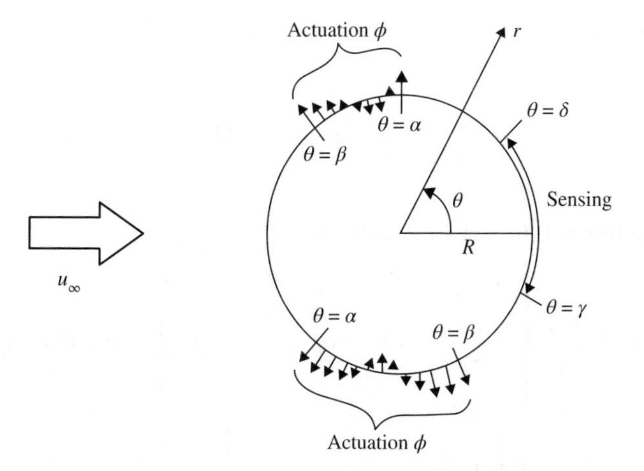

Figure 6.9 Schematic of sensing and actuation.[16] *Source*: Min C 1999. Reproduced with permission from Cambridge University Press

Rewrite Eq. (6.96) in the following form:

$$J_1(\phi_1) = \int_0^{2\pi} P_1 R d\theta + \frac{l^2}{2}\int_0^{2\pi} {\phi_1}^2 R d\theta \tag{6.100}$$

where

$$P_1 = \begin{cases} -p\,|_{r=R}\cos\theta & \gamma \le \theta \le \delta \\ 0 & \text{otherwise} \end{cases} \tag{6.101}$$

Taking the Fréchet derivative of Eqs. (6.100) and (6.101)

$$\frac{DJ_1(\phi_1)}{D\phi_1}\phi_1' = \int_0^{2\pi} \frac{DP_1}{D\phi_1}\phi_1' R d\theta + l^2 \int_0^{2\pi} \phi_1 \phi_1' R d\theta \tag{6.102}$$

with

$$\frac{DP_1}{D\phi_1}\phi' = \begin{cases} -\rho\,|_{r=R}\cos\theta & \gamma \le \theta \le \delta \\ 0 & \text{otherwise} \end{cases} \tag{6.103}$$

From Eqs. (6.74) and (6.91), we have

$$\widetilde{\rho}(k) = \widetilde{a}(k)\widetilde{\phi}_1(k) \tag{6.104}$$

with

$$\widetilde{a}(k) = \begin{cases} Y_0 = \text{constant} & k = 0 \\ \frac{2}{\Delta t_c}\frac{1}{|k|}Y_k & k \ne 0 \end{cases},\quad Y_0 = \frac{\widetilde{p}_{r=R,\,k=0}}{\widetilde{\phi}_{1,\,k=0}}, \quad Y_k = \frac{A}{R} \tag{6.105}$$

Using the convolution integral, we obtain $\rho(\theta)|_{r=R}$ from Eq. (6.104)

$$\rho(\theta)\bigg|_{r=R} = \frac{1}{2\pi}\int_0^{2\pi} (a(\theta-\tau)\phi_1'(\tau))d\tau \tag{6.106}$$

From Eqs. (6.103) and (6.106), Eq. (6.102) becomes

$$\begin{aligned} \frac{DJ_1(\phi_1)}{D\phi_1}\phi_1' &= \int_\gamma^\delta \left\{ -\frac{1}{2\pi}\int_0^{2\pi} \left(a(\theta-\tau)\,\phi_1'(\tau)\right) d\tau \cos\theta \right\} R d\theta + l^2 \int_0^{2\pi} \phi_1\phi_1' R d\theta \\ &= \int_0^{2\pi} \left\{ \frac{1}{2\pi}\int_\gamma^\delta -a(\tau-\theta)\cos\tau d\tau \right\} \phi_1'(\theta) R d\theta + l^2 \int_0^{2\pi} \phi_1\phi_1' R d\theta \end{aligned} \tag{6.107}$$

As $\frac{DJ_1(\phi_1)}{D\phi_1} = 0$, we obtain the following equation:

$$\phi_1 = \frac{\kappa}{2\pi}\int_{\gamma}^{\delta} a(\tau - \theta)\cos\tau d\tau$$

then

$$\phi_1(\theta) = \begin{cases} \frac{\kappa}{2\pi}\int_{\gamma}^{\delta} a(\tau - \theta)\cos\tau d\tau & \alpha \le \theta \le \beta \\ 0 & \text{otherwise} \end{cases} \tag{6.108}$$

Here the control ϕ_1 is applied to the region of $\alpha \le \theta \le \beta$ and is a function of θ only. Therefore the control ϕ_1 does not require any flow information.

If the sensing and actuation are applied over the cylinder surface, i.e., $(\alpha, \beta) = (\gamma, \delta) = (0, 2\pi)$, Eq. (6.108) becomes

$$\phi_1(\theta) = \frac{\kappa}{2\pi}\int_{0}^{2\pi} a(\tau - \theta)\cos\tau d\tau \tag{6.109}$$

Using Eq. (6.105) and the definition of the Fourier transformation $a(\theta) = \sum_k \widetilde{a}(k)e^{ik\theta}$, we obtain

$$a(\theta) = Y_0 + \frac{2}{\Delta t_c}\sum_{k\neq 0}\frac{Y_k}{|k|}\cos(k\theta)$$

and then $a(\tau - \theta) = Y_0 + \frac{2}{\Delta t_c}\sum_{k\neq 0}\frac{Y_k}{|k|}\{\cos(k\tau)\cos(k\theta) + \sin(k\tau)\sin(k\theta)\}$

Substitution into Eq. (6.109) yields

$$\begin{aligned}\phi_1(\theta) =& \frac{\kappa}{2\pi}\int_{0}^{2\pi}\{Y_0\cos\tau\}d\tau \\ &+ \frac{\kappa}{2\pi}\int_{0}^{2\pi}\left\{\frac{2}{\Delta t_c}\sum_{k\neq 0}\frac{Y_k}{|k|}\{\cos(k\tau)\cos(k\theta) + \sin(k\tau)\sin(k\theta)\}\cos\tau\right\}d\tau\end{aligned}$$

The first term of the right-hand side of this equation is identically zero. Application of the orthogonality of trigonometric function to the second term of the right-hand side yields

$$\phi_1(\theta) = \kappa\frac{2}{\Delta t_c}Y_1\cos\theta \tag{6.110}$$

with $Y_1 = \frac{RK_1(mR) - mR^2 K_2(mR)}{2K_1(mR) - mRK_2(mR)}$
where Y_1 is constant and thus $\phi_1(\theta)$ is a cosine function.

Rewrite Eq. (6.97) in the following form:

$$J_2(\phi_2) = \frac{1}{2}\int_0^{2\pi} P_2^2 R d\theta + \frac{l^2}{2}\int \phi_2{}^2 R d\theta$$

where

$$P_2 = \begin{cases} p_t - p\,|_{r=R} & \gamma \le \theta \le \delta \\ 0 & \text{otherwise} \end{cases}$$

Following the same procedure leading to Eq. (6.108), we have

$$\phi_2(\theta) = \begin{cases} \frac{\kappa}{2\pi}\int_0^{2\pi} P_2(\tau)\, a(\tau - \theta) d\tau & \alpha \le \theta \le \beta \\ 0 & \text{otherwise} \end{cases} \tag{6.111}$$

Here, $\phi_2(\phi)$ is related with the sensing of the instantaneous wall pressure on $\gamma \le \theta \le \delta$.
Rewrite Eq. (6.99) in the following form:

$$J_3(\phi_3) = -\frac{1}{2}\int_0^{2\pi} P_3^2 R d\theta + \frac{1}{2}\int \phi_3{}^2 R d\theta$$

where $P_3 = \begin{cases} \partial p/\partial\theta\,|_{r=R} & \gamma \le \theta \le \delta \\ 0 & \text{otherwise} \end{cases}$

It is easy to obtain

$$\phi_3(\theta) = \begin{cases} \frac{\kappa}{2\pi}\int_0^{2\pi} P_3(\tau)\, b(\tau - \theta) d\tau & \alpha \le \theta \le \beta \\ 0 & \text{otherwise} \end{cases} \tag{6.112}$$

where the Fourier coefficient $b(\theta)$ is

$$\widetilde{b}(k) = ik\widetilde{a}(k) = \begin{cases} 0 & k = 0 \\ \frac{2}{\Delta t_c}\frac{ik}{|k|}Y_k & k \ne 0 \end{cases} \tag{6.113}$$

6.2.3.3 Control of Cylinder Flow

The computational domain is shown in Figure 6.10, where the center of the cylinder is located at $x = 0, y = 0$ in the coordinate system. A periodic boundary condition is used at the branch cut and a convective outflow condition, $\frac{\partial u_i}{\partial t} + c\frac{\partial u_i}{\partial x} = 0$, is used for the outflow boundary condition, where c is the space-averaged streamwise velocity at the exit. This boundary condition allows vortices to smoothly pass away out of computational domain. Dirichlet boundary conditions are used at far-field boundaries, that is, $u = u_\infty, v = 0$, and also at the cylinder surface, that is, $u_r = \phi, u_\theta = 0$. A C-type grid system is employed, and a nonuniform mesh of 321×121 points is created using a hyperbolic grid-generation technique. The computational time step is $\Delta t = 0.015$, the control time interval is $\Delta t_c = 0.06$, and Reynolds number is Re = 100. The maximum blowing–suction value relative to the free stream velocity, $\phi_{\max}$, is kept constant during the control.

The time histories of the cost functional with blowing–suction applied all over the cylinder surface, that is, $(\alpha, \beta) = (\gamma, \delta) = (0, 2\pi)$, are shown in Figure 6.11, where dotted line, solid line, dash-dotted line, and dashed line represent $\phi_{\max} = 0.1$, $\phi_{\max} = 0.2$, $\phi_{\max} = 0.3$, and $\phi_{\max} = 0.4$, respectively. As the controls are applied at $t \geq 30$, J_1 and J_2 decrease and J_3 increases at a given $\phi_{\max}$, and with increasing $\phi_{\max}$, J_1 and J_2 further decrease and J_3 increases.

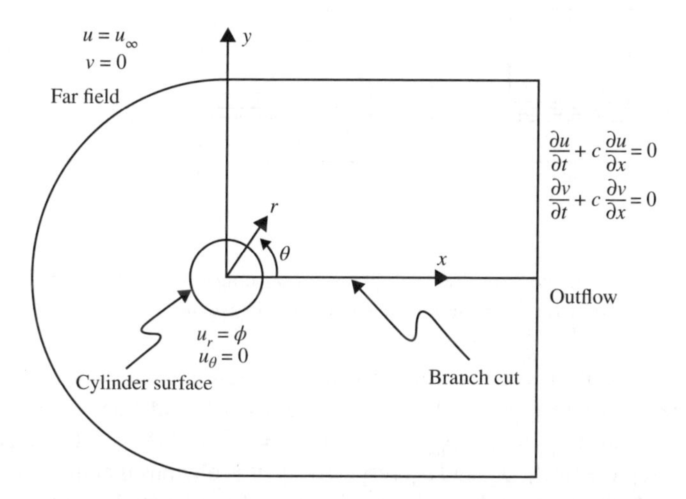

Figure 6.10 Computational domain and boundary conditions.[16] *Source*: Min C 1999. Reproduced with permission from Cambridge University Press

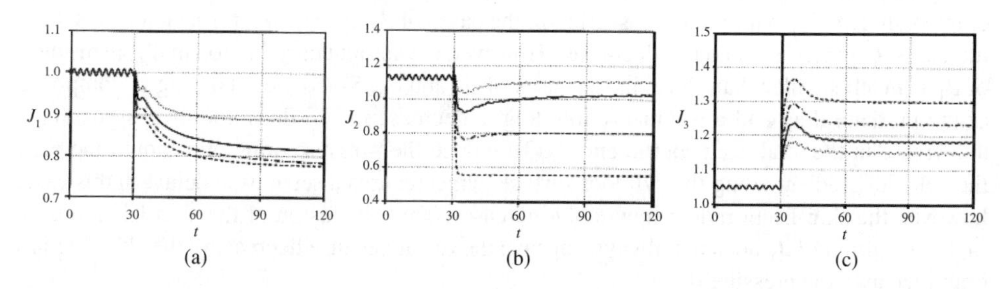

Figure 6.11 Variation in performance index with control.[16] *Source*: Min C 1999. Reproduced with permission from Cambridge University Press

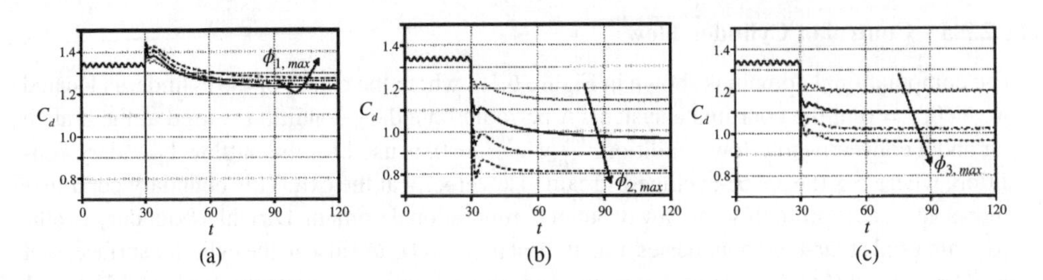

Figure 6.12 Variation in drag coefficient with control.[16] *Source*: Min C 1999. Reproduced with permission from Cambridge University Press

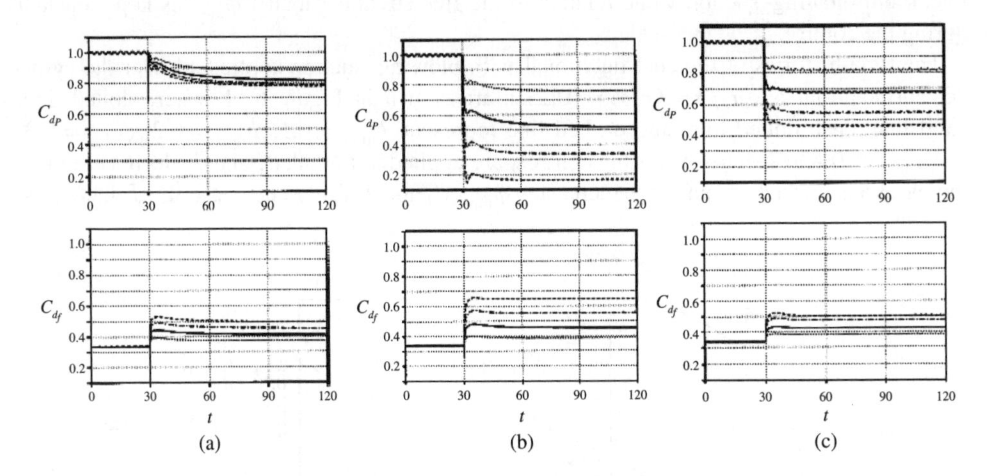

Figure 6.13 Variation in pressure drag and skin-friction drag coefficients with control.[16] *Source*: Min C 1999. Reproduced with permission from Cambridge University Press

The time histories of the drag coefficient with blowing–suction covered all over the cylinder surface are shown in Figure 6.12, where dotted line, solid line, dash-dotted line, and dashed line represent $\phi_{\max} = 0.1$, $\phi_{\max} = 0.2$, $\phi_{\max} = 0.3$, and $\phi_{\max} = 0.4$, respectively, and Figure 6.12a–c represents J_1, J_2, and J_3, respectively. It is obvious that the drag C_d is reduced with the control. In the cases of J_2 and J_3, C_d decreases further with increasing $\phi_{\max}$; however in the case of J_1, C_d first decreases and then increases as $\phi_{\max}$ increases.

Figure 6.13a–c shows the time histories of the pressure drag coefficient C_{dp} and skin-friction drag coefficient C_{df} with blowing–suction in the cases of J_1, J_2, and J_3, respectively. As $\phi_{\max}$ increases, C_{df} increases and C_{dp} decreases. However, the amount of reduction in C_{dp} in the case of J_1 is much smaller than those in the cases of J_2 and J_3. Since the total drag is composed of the pressure drag and the skin-friction drag, a decrease in C_{dp} does not always guarantee a decrease in the total drag coefficient C_d. To reduce the pressure drag, one should increase the velocity gradient along the cylinder surface, which causes a separation delay. In this case, however, the skin-friction drag inevitably increases. Thus, inclusion of the skin-friction drag in the definition of J_1 does not always improve the result because decreasing the skin-friction drag increases the pressure drag.

Figure 6.14 shows the instantaneous vorticity contours at Re = 100 in the case of J_2, where Figure 6.14a, b, c, and d represent no control, $\phi_{2,\max} = 0.1$, $\phi_{2,\max} = 0.2$, and $\phi_{2,\max} = 0.3$,

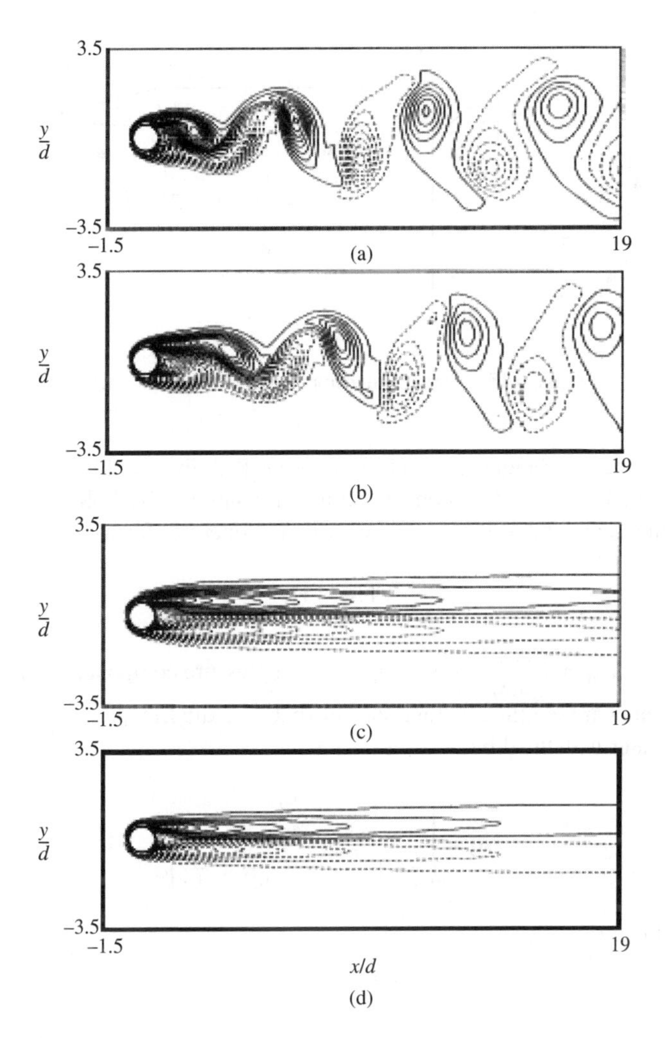

Figure 6.14 Instantaneous vorticity contours.[16] *Source*: Min C 1999. Reproduced with permission from Cambridge University Press

respectively. Positive contours are dashed. With $\phi_{2,\max} = 0.2$, vortex shedding becomes significantly weak but still exists. With $\phi_{2,\max} = 0.3$, however, vortex shedding completely disappears.

6.3 Adjoint-based Suboptimal Control[17–20]

6.3.1 Fundamentals of Adjoint-based Suboptimal Control

The optimal flow control based on the N-S equations can be expressed as a problem to find a conditional extremum confined by the nonlinear equations. Introducing adjoint variables and cost functional, the extremum can be achieved by solving the equations about the adjoint variables and the sensitivity of the cost functional, which is called adjoint-based optimal control.

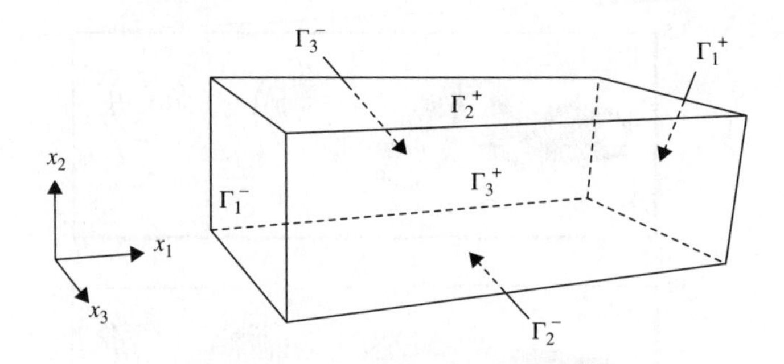

Figure 6.15 Three-dimensional channel flow geometry

A fully developed turbulent channel flow controlled by the blowing–suction on the walls as shown in Figure 6.15 is considered as an example; the N-S equations inside the three-dimensional rectangular domain Ω are written in operator form as

$$\mathbf{Nq} = \mathbf{F} \tag{6.114}$$

where, the flow state $\mathbf{q} = \begin{pmatrix} p(\mathbf{x}, t) \\ \mathbf{u}(\mathbf{x}, t) \end{pmatrix}$ is composed of a pressure component and a velocity component, all of which are continuous functions of space, **x**, and time, t.

The N-S operator is defined by

$$\mathbf{Nq} = \begin{pmatrix} \dfrac{\partial u_j}{\partial x_j} \\ \dfrac{\partial u_i}{\partial t} + \dfrac{\partial u_j u_i}{\partial x_j} - \nu \dfrac{\partial^2 u_i}{\partial x_j{}^2} + \dfrac{\partial p}{\partial x_i} \end{pmatrix} \tag{6.115}$$

and **F** is given by

$$\mathbf{F} = \begin{pmatrix} 0 \\ -\mathrm{r}P_x\,(x, t) \end{pmatrix}$$

An external pressure gradient $P_x(x, t)$ is applied to introduce a mean flow in the streamwise direction and **r** is a unit vector in the streamwise direction.

The boundary conditions on the state **q** are assumed to be periodic in the streamwise and spanwise directions, and a wall-normal control velocity is distributed over the wall such that

$$\mathbf{u} = -\phi\mathbf{n} \quad \text{on} \quad \Gamma_2^{\pm} \tag{6.116}$$

where **n** is the unit outward normal to the boundary.

Initial conditions on the velocity are prescribed such that

$$\mathbf{u} = \mathbf{u}_0 \quad \text{at} \quad t = 0 \tag{6.117}$$

A cost functional defined with the control ϕ (blowing–suction on the cylinder surface here) is treated in a fairly general form as

$$J(\phi) = \frac{1}{2}\int_0^T \int_\Omega |\mathsf{C}_1\mathbf{u}|^2 d\mathbf{x}dt + \frac{1}{2}\int_\Omega |\mathsf{C}_2\mathbf{u}(\mathbf{x},T)|^2 d\mathbf{x}$$

$$-\int_0^T \int_{\Gamma_2^\pm} \mathsf{C}_3 \nu \frac{\partial \mathbf{u}(\phi)}{\partial n} \cdot \mathbf{r} d\mathbf{x}dt + \frac{l^2}{2}\int_0^T \int_{\Gamma_2^\pm} \phi^2 d\mathbf{x}dt \tag{6.118}$$

where four cases of physical interest are included:

1. $\mathsf{C}_1 = d_1 I$ and $\mathsf{C}_2 = \mathsf{C}_3 = 0$ denote the case of regulation of turbulent kinetic energy;
2. $\mathsf{C}_1 = d_2 \nabla \times$ and $\mathsf{C}_2 = \mathsf{C}_3 = 0$ denote the regulation of the square of the vorticity;
3. $\mathsf{C}_2 = d_3 I$ and $\mathsf{C}_1 = \mathsf{C}_3 = 0$ represent the terminal control of turbulent kinetic energy;
4. $\mathsf{C}_3 = d_4 I$ and $\mathsf{C}_1 = \mathsf{C}_2 = 0$ represent the minimization of the time-average skin-friction in the direction of the mean flow.

The dimensional constants d_i are included to account for the relative weight of each individual term and $\sum_i d_i = 1$.

Taking the Fréchet differential of Eq. (6.114), the result is

$$N'(\mathbf{q})\mathbf{q}' = 0 \quad \text{on} \quad \Omega \tag{6.119}$$

$$\mathbf{u}' = -\phi'\mathbf{n} \quad \text{on} \quad \Gamma_2^\pm \tag{6.120}$$

$$\mathbf{u}' = 0 \quad \text{at} \quad t = 0 \tag{6.121}$$

where the flow perturbation state $\mathbf{q}' = \begin{pmatrix} p'(\mathbf{x},t) \\ \mathbf{u}'(\mathbf{x},t) \end{pmatrix}$, which is defined as the small perturbation to the state $\mathbf{q}$ arising from a small control perturbation ϕ' to the control ϕ. The linearized N-S operation $N'(\mathbf{q})\mathbf{q}'$ is given by

$$N'(\mathbf{q})\mathbf{q}' = \begin{pmatrix} \dfrac{\partial u'_j}{\partial x_j} \\ \dfrac{\partial u'_i}{\partial t} + \dfrac{\partial\left(u_j u'_i + u'_j u_i\right)}{\partial x_j} - \nu \dfrac{\partial^2 u'_i}{\partial x_j{}^2} + \dfrac{\partial p'}{\partial x_i} \end{pmatrix} \tag{6.122}$$

The operation $N'(\mathbf{q})\mathbf{q}'$ is a linear operation on the perturbation field $\mathbf{q}'$, though the operator $N'(\mathbf{q})$ is itself a function of the flow state $\mathbf{q}$.

The perturbation J' may be defined by a limiting process as the Fréchet differential of the cost functional J with respect to ϕ such that

$$J'(\phi,\phi') = \lim_{\varepsilon\to 0}\frac{J(\phi+\varepsilon\phi') - J(\phi)}{\varepsilon} = \int_0^T\int_{\Gamma_2^\pm}\frac{DJ(\phi)}{D\phi}\phi' d\mathbf{x}dt$$

From Eq. (6.118)

$$J'(\phi,\phi') = \int_0^T\int_\Omega C_1^* C_2\mathbf{u}\cdot\mathbf{u}' d\mathbf{x}dt + \int_\Omega (C_2^* C_2\mathbf{u}\cdot\mathbf{u}')_{t=T} d\mathbf{x}$$
$$-\int_0^T\int_{\Gamma_2^\pm} C_3^*\nu\frac{\partial\mathbf{u}'}{\partial n}\cdot\mathbf{r}d\mathbf{x}dt + l^2\int_0^T\int_{\Gamma_2^\pm}\phi\phi' d\mathbf{x}dt \tag{6.123}$$

The cost functional is in the extreme, if

$$J'(\phi,\phi') = 0 \quad \text{or} \quad \frac{DJ(\phi)}{D\phi} = 0 \tag{6.124}$$

Therefore, the optimal control ϕ is a solution of the sensitivity Eq. (6.124).

6.3.2 *Adjoint-based Suboptimal Control*

In adjoint-based suboptimal control, an adjoint flow field is introduced, defined judiciously with appropriate forcing in an interior equation, as well as appropriate boundary conditions and initial conditions. Then, the solution of the sensitivity equation may be determined by solving the adjoint flow field.

6.3.2.1 Adjoint Flow Field

Define an inner product of two vectors, $\mathbf{a}(t)$ and $\mathbf{a}^*(t)$, over the domain in consideration such that

$$\langle\mathbf{a}(t),\mathbf{a}^*(t)\rangle = \int_0^T \mathbf{a}(t)\cdot\mathbf{a}^*(t)dt$$

From the integration by parts

$$\int_0^T\left(\frac{\partial\mathbf{a}}{\partial t}\right)\cdot\mathbf{a}^* dt = \int_0^T \mathbf{a}\cdot\left(-\frac{\partial\mathbf{a}^*}{\partial t}\right)dt + \mathbf{a}\cdot\mathbf{a}^*|_{t=T} - \mathbf{a}\cdot\mathbf{a}^*|_{t=0}$$

or

$$\langle N\mathbf{a}, \mathbf{a}^* \rangle = \langle \mathbf{a}, N^*\mathbf{a}^* \rangle + b \tag{6.125}$$

where $N = \frac{\partial}{\partial t}$, $N^* = -\frac{\partial}{\partial t}$, $b = \mathbf{a} \cdot \mathbf{a}^*|_{t=T} - \mathbf{a} \cdot \mathbf{a}^*|_{t=0}$.

Therefore, considering a perturbation field $\mathbf{q}'$ and an adjoint field $\mathbf{q}^*$, which is defined as $\mathbf{q}^* = \begin{pmatrix} p^*(\mathbf{x}, t) \\ \mathbf{u} * (\mathbf{x}, t) \end{pmatrix}$, we have the identity

$$\langle \mathbf{q}^*, N'(\mathbf{q})\mathbf{q}' \rangle = \langle N'(\mathbf{q})^*\mathbf{q}^*, \mathbf{q}' \rangle + b \tag{6.126}$$

where the adjoint operator is given by

$$N'(\mathbf{q})^*\mathbf{q}^* = \begin{pmatrix} -\dfrac{\partial u_j^*}{\partial x_j} \\ -\dfrac{\partial u_i^*}{\partial t} - u_j\left(\dfrac{\partial u_i^*}{\partial x_j} + \dfrac{\partial u_j^*}{\partial x_i}\right) - \nu\dfrac{\partial^2 u_i^*}{\partial x_j{}^2} - \dfrac{\partial p^*}{\partial x_i} \end{pmatrix} \tag{6.127}$$

The operation $N'(\mathbf{q})^*\mathbf{q}^*$ is a linear operation on the adjoint field $\mathbf{q}^*$, and the operator $N'(\mathbf{q})^*$ is itself a function of the flow state $\mathbf{q}$. Equation (6.126) is the key to expressing J' in the desired form.

We also get several boundary terms

$$b = \int_\Omega (u_j^* u_j')|_{t=0}^{t=T} d\mathbf{x}$$

$$+\int_0^T \int_{\Gamma_2^\pm} n_j \left[p^* u_j' + u_i^*\left(u_j u_i' + u_j' u_i\right) - \nu\left(u_i^* \frac{\partial u_i'}{\partial x_j} - u_i' \frac{\partial u_i^*}{\partial x_j}\right) + u_j^* p'\right] d\mathbf{x}dt \tag{6.128}$$

where $\mathbf{n}$ denotes a unit outward normal to the surface and T is control interval. Therefore, an adjoint field can be defined using the operator $N'(\mathbf{q})^*$ together with appropriate forcing in an interior equation with appropriate boundary conditions and initial conditions such that

$$N'(\mathbf{q})^*\mathbf{q}^* = \begin{pmatrix} 0 \\ C_1^* C_1 \mathbf{u} \end{pmatrix} \quad \text{in } \Omega \tag{6.129}$$

$$\mathbf{u}^* = C_3^* \mathbf{r} \quad \text{on} \quad \Gamma_2^\pm \tag{6.130}$$

$$\mathbf{u}^* = C_2^* C_2 \mathbf{u} \quad \text{at} \quad t = T \tag{6.131}$$

where the adjoint operation $N'(\mathbf{q})^*\mathbf{q}^*$ is given in Eq. (6.127) and $\mathbf{r}$ is a unit vector in the streamwise direction. Note Eq. (6.129), though linear, has complexity similar to that of Eq. (6.114), and may be solved with similar numerical methods. Note also that the "initial" condition in Eq. (6.131) are defined at $t = T$.

Due to the judicious choice of the right-hand-side in Eq. (6.129) as well as boundary conditions and initial conditions (6.130) and (6.131), Eq. (6.124) reduces to

$$\int_0^T\int_\Omega \mathsf{C}_1^*\mathsf{C}_1\mathbf{u}\cdot\mathbf{u}'d\mathbf{x}dt + \int_\Omega (\mathsf{C}_2^*\mathsf{C}_2\mathbf{u}\cdot\mathbf{u}')_{t=T}dx - \int_0^T\int_{\Gamma_2^\pm} \nu\mathsf{C}_3^*\mathbf{r}\cdot\frac{\partial\mathbf{u}'}{\partial t}d\mathbf{x}dt = \int_0^T\int_{\Gamma_2^\pm} p^*\phi' d\mathbf{x}dt \tag{6.132}$$

Comparing with Eq. (6.123), we have

$$J'(\phi,\phi') = \int_0^T\int_{\Gamma_2^\pm}(p^* + l^2\phi)\phi' d\mathbf{x}dt = \int_0^T\int_{\Gamma_2^\pm}\frac{\mathsf{D}J(\phi)}{\mathsf{D}\phi}\phi' d\mathbf{x}dt$$

As ϕ' is arbitrary, this implies that

$$\frac{\mathsf{D}J(\phi)}{\mathsf{D}\phi} = p^* + l^2\phi \tag{6.133}$$

The cost functional is thus found to be a simple function of the adjoint problem. Then, the control sensitivity equation, that is, the control law, is expressed as

$$\frac{\mathsf{D}J(\phi)}{\mathsf{D}\phi} = p^* + l^2\phi = 0 \tag{6.134}$$

where p^* is the pressure in the adjoint flow field.

6.3.2.2 Receding-Horizon Predictive Control

There exists a complex functional relation between p^* and ϕ, involved in Eqs. (6.114) and (6.129), which may be solved with numerical methods described as follows.

Considering a flow governed by N-S equation (6.114), the control with a given input $\phi =$ const is initiated at $t = T_0$ and performed over the interval $(T_0, T_0 + T)$, where T is the optimization horizon. Then the evolutions of the flow-field $\mathbf{q}$ from $t = T_0$ to $t = T_0 + T$ can be obtained by integrations of Eq. (6.114) numerically. Subsequently, the developments of the adjoint flow field, $\mathbf{q}^*$, can also be obtained by the integration of Eq. (6.129) and marching backward from $t = T_0 + T$ to $t = T_0$, called receding integration, which is shown in Figure 6.16. Note that the computation of the adjoint field requires the storage of the flow-field $\mathbf{q}$. Therefore, the adjoint flow-field $\mathbf{q}^*$, such as p^*, at $t = T_0$ is finally established from a known ϕ. Based on p^* and ϕ, then the gradient information is determined by Eq. (6.133).

The purpose of optimal flow control is to determine the optimal control input function $\phi(t)$, which is achieved through the control law expressed by Eq. (6.134). With the gradient information, Eq. (6.134) can be solved by the iteration method such that

$$\phi^k = \phi^{k-1} - \alpha^k\frac{\mathsf{D}J(\phi^{k-1})}{\mathsf{D}\phi} \tag{6.135}$$

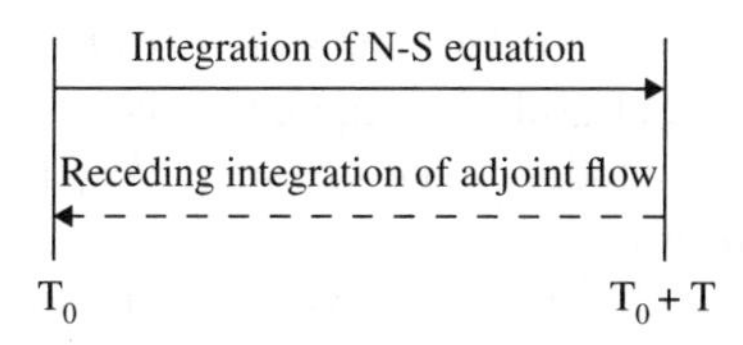

Figure 6.16 Receding integration

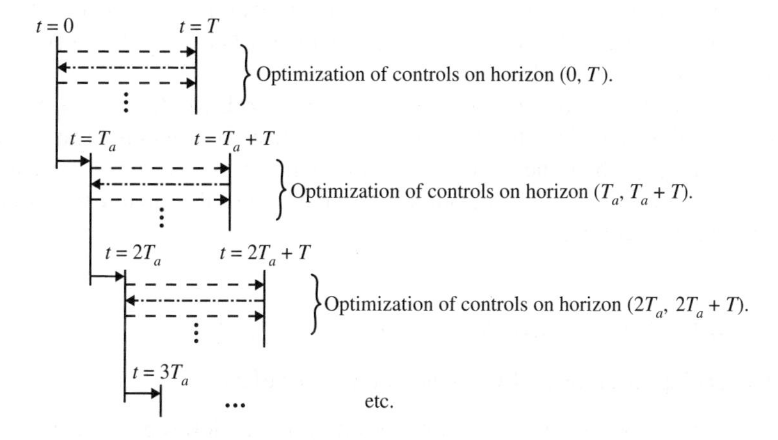

Figure 6.17 Sequence of events in receding-horizon predictive control.[17] *Source*: Bewley T R 2001. Reproduced with permission of Cambridge University

where k indicates the iteration number and α^k is a parameter of descent which governs the magnitude for update.

The receding-horizon predictive control setting is employed to discuss the adjoint-based optimal control problems as shown in Figure 6.17. When the flow field at $t = T_a$ is determined and the optimization horizon T is set, then the optimized control may be updated by the iteration as described earlier. The iteration is performed by the forward and backward integrations shown as the dashed lines and dot-dashed lines in the Figure 6.17 till the convergence. Once the optimal control ϕ is determined, the flow is advanced by some portion T_a, and the optimization process begins again as shown by the heavy solid arrows in Figure 6.17.

6.3.3 Near-Wall Turbulence Controlled by Blowing–Suction Wall

As all velocities are normalized by the wall shear velocity u_τ and all lengths are normalized by the channel half-width δ, with this normalization, the Reynolds number $\mathrm{Re}_\tau = u_\tau \delta / \nu$, and $\nu = 1/\mathrm{Re}_\tau$ in Eq. (6.115) for $\mathbf{Nq}$. The near-wall turbulence controlled by the blowing–suction wall is described by Eqs. (6.114)–(6.116) in this chapter. Fourier transform techniques are used to compute spatial derivatives in the homogeneous direction with 3/2 dealiasing of the nonlinear terms, and an exactly energy-conserving second-order finite difference scheme is used to compute spatial derivatives in the wall-normal direction. The flow is advanced in time using an explicit third-order Runge–Kutta method for all terms involving x_1 - and x_3 -derivatives and an implicit Crank–Nicholson method at each Runge–Kutta substep for all terms involving x_2 -derivatives. The computational grid is staggered in the wall-normal direction to prevent decoupling of the even and odd modes of the pressure. Very fine grid resolution is required near the wall to resolve the shear layer.

The turbulent channel flow with $\mathrm{Re}_\tau = 100$, controlled by the blowing–suction wall based on the terminal control of TKE in the receding-horizon framework (i.e., $C_1 = d_1 I$ and $C_2 = C_3 = 0$) is simulated with the dealiased collocation points $64 \times 65 \times 64$. The variations of drag and turbulent kinetic energy in the control process are shown in Figure 6.18a and b individually. It has been shown from the figure that the optimization horizon T over which the flow is optimized must be chosen carefully, and the flow relaminarizes when $T^+ \geq 25$, where superscript + represents time is normalized by ν / u_τ^2.

Figure 6.19 shows the performance of the optimized controls for three of the most promising optimal control formulations for $T^+ = 100$, that is, based on the minimization of J_{DRAG}, $J_{\mathrm{TKE(reg)}}$, and $J_{\mathrm{TKE(ter)}}$, where the opposition control strategy is also shown for comparison. Over the long term, the $J_{\mathrm{TKE(ter)}}$ formulation is the only one of the three formulations which relaminarizes this particular flow.

6.3.4 *Cylinder Flow Controlled by Lorentz Force*[21–24]

6.3.4.1 Governing Equations of Controlled Cylinder Flow

As we know, the viscous flow past a free bluff body may induce flow separations and vibrations of the body, accompanied by a large fluctuation in drag and lift forces. The above phenomena may be suppressed through flow control technologies, such as Lorentz force. Considering a circular cylinder, immersed in an incompressible and electrically low-conducting fluid vertically, the cylinder surface consists of two half cylinders mounted with alternating streamwise electrodes and magnets along the span of the cylinder in order of N, +, S, −, as shown in Figure 6.20. In this way, the Lorentz force generated is directed parallel to the cylinder surface.

The governing equations describing the incompressible electrically conducting viscous flow with an applied Lorentz force in a nondimensional form are given by

$$\frac{\partial \mathbf{u}}{\partial t} + (\mathbf{u} \cdot \nabla)\mathbf{u} = -\nabla p + \frac{2}{\mathrm{Re}} \nabla^2 \mathbf{u} + N\mathbf{F} \tag{6.136}$$

$$\nabla \cdot \mathbf{u} = 0 \tag{6.137}$$

where $\mathrm{Re} = \frac{2u_\infty a}{\nu}$ (Reynolds number), u_∞ is the free-stream velocity, ν is the kinematic viscosity, and a is the cylinder radius. The interaction parameter is defined as $N = \frac{j_0 B_0 a}{\rho u_\infty^2}$ with $j_0 = \sigma E_0$ being the current density, σ is the electric conductivity, E_0 is the electric field, and B_0 is the magnetic field. F is the distribution function of Lorentz force, given by

$$|F_\theta| = e^{-\alpha(r-1)} \tag{6.138}$$

$$F_r = 0 \tag{6.139}$$

where r and θ are polar coordinates, subscripts r and θ represent the components in r and θ directions, respectively, and α is a constant, representing the effective depth of Lorentz force in the fluid.

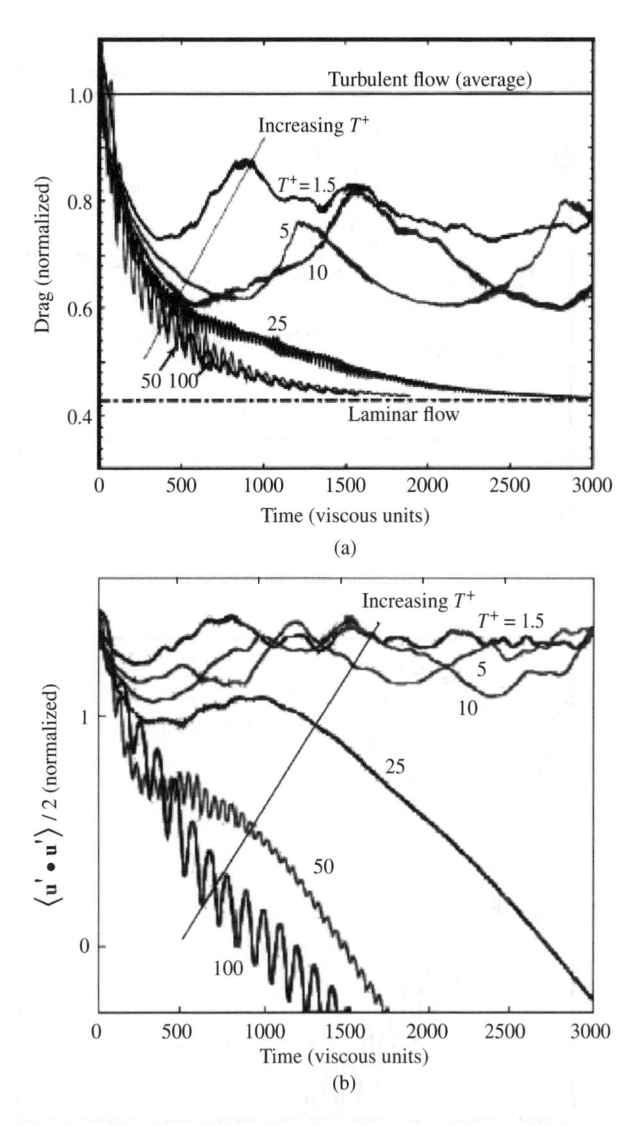

Figure 6.18 Variations of drag and turbulent kinetic energy under optimized control for formulations based on $J_{\text{TKE(ter)}}$ at $\text{Re}_\tau = 100$.[17] *Source*: Bewley T R 2001. Reproduced with permission of Cambridge University

For a laminar wake of a cylinder at low Reynolds numbers, the flow will be treated as a two-dimensional phenomenon. Then the stream function ψ and the vorticity Ω are introduced as follows in the exponential–polar coordinate system (ξ, η), where $r = e^{2\pi\xi}$ and $\theta = 2\pi\eta$.

$$\frac{\partial\psi}{\partial\eta} = U_r = H^{\frac{1}{2}}u_r, \quad -\frac{\partial\psi}{\partial\xi} = U_\theta = H^{\frac{1}{2}}u_\theta$$

and $\Omega = \frac{1}{r}\frac{\partial(ru_\theta)}{\partial r} - \frac{1}{r}\frac{\partial u_r}{\partial\theta}$
where u_r and u_θ are the velocity components in r and θ directions, respectively, $H = 4\pi^2 e^{2\pi\xi}$.

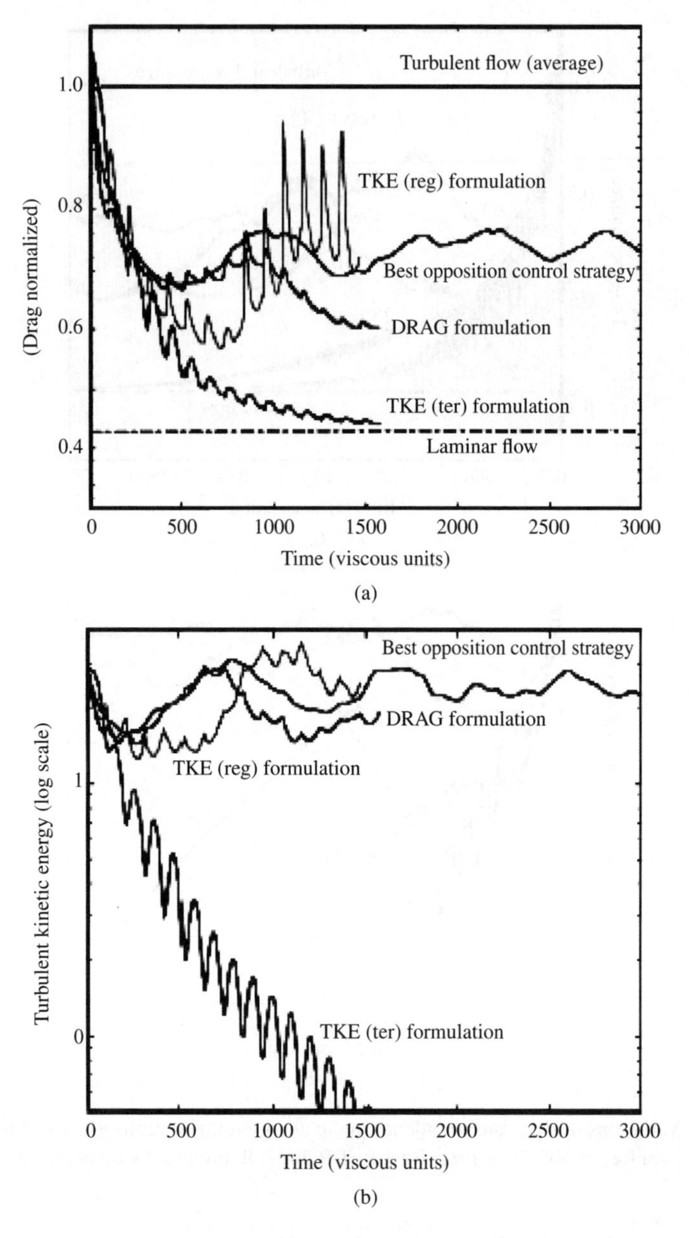

Figure 6.19 Variations in drag and turbulent kinetic energy under optimized control for different formulations (J_{DRAG}, $J_{\text{TKE(reg)}}$, and $J_{\text{TKE(ter)}}$), at $\text{Re}_\tau = 100$ and $T^+ = 100$.[17] *Source*: Bewley T R 2001. Reproduced with permission of Cambridge University

The stream-vorticity equations become

$$\frac{\partial^2 \psi}{\partial \xi^2} + \frac{\partial^2 \psi}{\partial \eta^2} = -H\Omega \tag{6.140}$$

$$H\frac{\partial \Omega}{\partial t} + \frac{\partial (U_r \Omega)}{\partial \xi} + \frac{\partial (U_\theta \Omega)}{\partial \eta} = \frac{2}{R_e}\left(\frac{\partial^2 \Omega}{\partial \xi^2} + \frac{\partial^2 \Omega}{\partial \eta^2}\right) + N\mathsf{F} \tag{6.141}$$

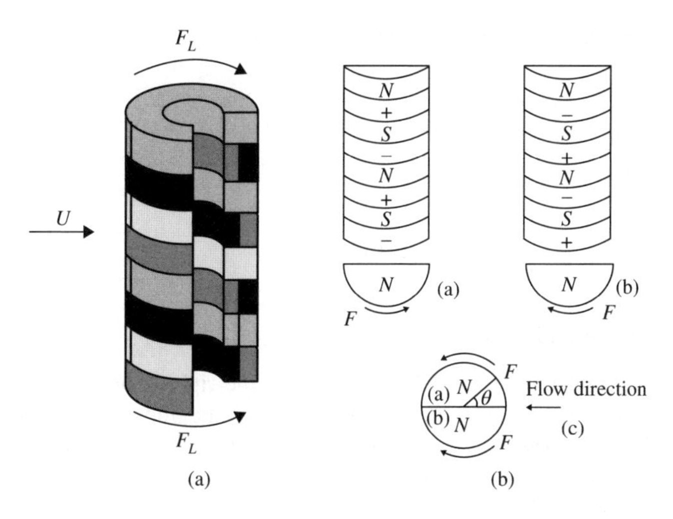

Figure 6.20 A circular cylinder for flow control: (a) cylinder equipped with electrodes and magnets and (b) Lorentz force on the cylinder surface.[23] *Source*: Zhang H 2010. Reproduced with permission of Cambridge University

where $\mathsf{F} = H^{\frac{1}{2}}\left(\frac{\partial F_\theta}{\partial \xi} + 2\pi F_\theta\right)$.

Except on the cylinder surface where the no-slip condition is imposed, the flow is considered to be potential initially and therefore

at $t = 0, \psi = 0, \Omega = -\frac{1}{H}\frac{\partial^2\psi}{\partial \xi^2}$ on $\xi = 0$

$$\psi = -2sh(2\pi\xi)\sin(2\pi\eta), \Omega = 0, \text{on} \quad \xi > 0$$

$$\text{at } t > 0, \psi = 0, \Omega = -\frac{1}{H}\frac{\partial^2\psi}{\partial \xi^2}, \text{on}, \quad \xi = 0$$

$$\psi = -2sh(2\pi\xi)\sin(2\pi\eta), \quad \Omega = 0, \text{on} \quad \xi = \xi_\infty$$

The total drag is composed of the pressure drag and the friction drag. The pressure drag coefficient $C_{d\rho}$ and the friction drag coefficient $C_{d\tau}$ are defined as

$$C_{d\rho} = \frac{F_{px}}{\frac{1}{2}\rho u_\infty^2 d}, C_{d\tau} = \frac{F_{\tau x}}{\frac{1}{2}\rho u_\infty^2 d}$$

where F_{px} and $F_{\tau x}$ denote the pressure drag and the friction drag, respectively. By derivation, we have

$$C_{d\rho} = \frac{2}{R_e}\int_0^1 -\frac{\partial \Omega}{\partial \xi}\sin(2\pi\eta)d\eta - 2\pi F_\theta(0)\int_0^1 N\sin(2\pi\eta)d\eta \tag{6.142}$$

$$C_{d\tau} = \frac{2}{R_e}\int_0^1 2\pi\,\Omega\sin(2\pi\eta)d\eta \tag{6.143}$$

Similarly, the lift force can be expressed as

$$C_{lp} = \frac{2}{R_e} \int_0^1 -\frac{\partial \Omega}{\partial \xi} \cos(2\pi\eta) d\eta - 2\pi F_\theta(0) \int_0^1 N \cos(2\pi\eta) d\eta \tag{6.144}$$

$$C_{l\tau} = \frac{2}{R_e} \int_0^1 2\pi\Omega \cos(2\pi\eta) d\eta \tag{6.145}$$

6.3.4.2 Optimal Control Law

The purpose of optimal flow control is to determine the right control input that effectively tailors the flow. Hence, the control input should vary with the transient flow field and depends on the state variables of the flow field. The key element of an optimal flow control is the minimization of a cost functional. Therefore, the cost functional should contain three kinds of important physical variables, that is, the control input, the state variables of the flow field, and the price of the control. In the present work, the interaction parameter, N, a dimensionless magnitude of Lorentz force, is taken as a control input. The state variable, a quantity of physical interest regulated or minimized in control such as the enstrophy Ω^2, the skin-friction drag $\Omega \sin(2\pi\eta)$, $\Omega \cos(2\pi\eta)$, and the skin-pressure drag $\frac{\partial \Omega}{\partial \xi} \sin(2\pi\eta)$, $\frac{\partial \Omega}{\partial \xi} \cos(2\pi\eta)$ can be chosen optionally. Then we have several kinds of cost functional such as

$$J_1 = -\frac{H}{2} \int_0^T \int_\Sigma \Omega^2 dv dt + \frac{l^2}{2} \int_0^T \int_0^1 N^2 d\eta dt \tag{6.146}$$

$$J_2 = -\frac{2}{R_e} \int_0^T \int_0^1 \frac{\partial \Omega}{\partial \xi} \sin(2\pi\eta) d\eta dt + \frac{l^2}{2} \int_0^T \int_0^1 N^2 d\eta dt \tag{6.147}$$

$$J_3 = -\frac{2}{R_e} \int_0^T \int_0^1 \frac{\partial \Omega}{\partial \xi} \cos(2\pi\eta) d\eta dt + \frac{l^2}{2} \int_0^T \int_0^1 N^2 d\eta dt \tag{6.148}$$

$$J_4 = -\frac{2}{R_e} \int_0^T \int_0^1 \Omega \sin(2\pi\eta) d\eta dt + \frac{l^2}{2} \int_0^T \int_0^1 N^2 d\eta dt \tag{6.149}$$

$$J_5 = -\frac{2}{R_e} \int_0^T \int_0^1 \Omega \cos(2\pi\eta) d\eta dt + \frac{l^2}{2} \int_0^T \int_0^1 N^2 d\eta dt \tag{6.150}$$

The Fréchet differential yields

$$J_1{}' = -H \int_0^T \int_\Sigma \Omega\Omega' dv dt + l^2 \int_0^T \int_0^1 NN' d\eta dt \tag{6.151}$$

$$J_2{}' = -\frac{2}{R_e} \int_0^T \int_0^1 \frac{\partial \Omega'}{\partial \xi} \sin(2\pi\eta) d\eta dt + l^2 \int_0^T \int_0^1 NN' d\eta dt \tag{6.152}$$

$$J_3{}' = -\frac{2}{R_e}\int_0^T\int_0^1 \frac{\partial\Omega'}{\partial\xi}\cos(2\pi\eta)d\eta dt + l^2\int_0^T\int_0^1 NN'd\eta dt \tag{6.153}$$

$$J_4{}' = -\frac{2}{R_e}\int_0^T\int_0^1 \Omega'\sin(2\pi\eta)d\eta dt + l^2\int_0^T\int_0^1 NN'd\eta dt \tag{6.154}$$

$$J_5{}' = -\frac{2}{R_e}\int_0^T\int_0^1 \Omega'\cos(2\pi\eta)d\eta dt + l^2\int_0^T\int_0^1 NN'd\eta dt \tag{6.155}$$

The purpose of optimal flow control is to determine the optimal control input function $N(t)$, which is achieved through the control sensitivity equation

$$J'(\phi,\phi') = 0 \quad \text{or} \quad \frac{\mathsf{D}J(\phi)}{D\phi} = 0$$

This equation implicates a complex relations among the optimal control input function $N(t)$ and the state variables of the flow field. Therefore, a special method should be developed to solve this equation.

N-S equations (6.161), (6.162) can be written in the operator form as

$$N_{(\mathbf{q})}\mathbf{q} = \begin{pmatrix} -H\Omega \\ N\mathsf{F}(\xi) \end{pmatrix} \tag{6.156}$$

where the flow state $q = \begin{pmatrix} \Omega \\ \psi \end{pmatrix}$, $N_{(\mathbf{q})}\mathbf{q}$ is the N-S operator.

Taking the Fréchet differential, the linearized perturbation equations are given by

$$N'_{(\mathbf{q})}\mathbf{q}' = \begin{pmatrix} -H\Omega' \\ N'\mathsf{F}(\xi) \end{pmatrix} \tag{6.157}$$

with initial and boundary conditions

$$\text{at}\, t = 0,\ \psi' = 0, \Omega' = 0, \frac{\partial\psi'}{\partial\xi} = 0, \frac{\partial\psi'}{\partial\eta} = 0$$

$$\text{at}\ t > 0, \psi' = 0, \frac{\partial\psi'}{\partial\xi} = -U'_\theta = 0, U'_r = 0 \quad \text{on} \quad \xi = 0$$

$$\psi' = 0, \frac{\partial\psi'}{\partial\xi} = 0, \Omega' = 0, \frac{\partial\Omega'}{\partial\xi} = 0 \quad \text{on} \quad \xi = \xi_\infty$$

where the flow perturbation state $q' = \begin{pmatrix} \Omega' \\ \psi' \end{pmatrix}$, $N'_{(\mathbf{q})}\mathbf{q}'$ is given by

$$N'_{(\mathbf{q})}\mathbf{q}' = \begin{pmatrix} \frac{\partial^2\psi'}{\partial\xi^2} + \frac{\partial^2\psi'}{\partial\eta^2} \\ H\frac{\partial\Omega'}{\partial t} + \frac{\partial(U_r\Omega' + U'_r\Omega)}{\partial\xi} + \frac{\partial(U_\theta\Omega' + U'_\theta\Omega)}{\partial\eta} - \frac{2}{R_e}\left(\frac{\partial^2\Omega'}{\partial\xi^2} + \frac{\partial^2\Omega'}{\partial\eta^2}\right) \end{pmatrix}$$

Here, the adjoint operation $N'(\mathbf{q})^*\mathbf{q}^*$ is given by

$$N'_{(\mathbf{q})^*}\mathbf{q}^* = \begin{pmatrix} \frac{\partial^2\psi^*}{\partial\xi^2} + \frac{\partial^2\psi^*}{\partial\eta^2} \\ H\frac{\partial\Omega^*}{\partial t} + \frac{\partial(U_r\Omega^* + U^*_r\Omega)}{\partial\xi} + \frac{\partial(U_\theta\Omega^* + U^*_\theta\Omega)}{\partial\eta} - \frac{2}{R_e}\left(\frac{\partial^2\Omega^*}{\partial\xi^2} + \frac{\partial^2\Omega^*}{\partial\eta^2}\right) \end{pmatrix}$$

with $\mathbf{q}^* = \begin{pmatrix} \Omega^* \\ \psi^* \end{pmatrix}$.

Let $b = \langle N'_{(\mathbf{q})} \mathbf{q}' \mathbf{q}^* \rangle$

$$= \int_{\Sigma} H(\Omega' \psi^*) \Big|_T dv - \frac{2}{R_e} \int_0^T \left(\int_0^1 A_\infty d\eta - \int_0^1 A_1 d\eta \right) dt \tag{6.158}$$

where $A_\infty = \left(\Omega' \frac{\partial \psi^*}{\partial \xi} + \psi^* \frac{\partial \Omega'}{\partial \xi} \right)_{\xi=\infty}$

$$A_1 = \left(\Omega' \frac{\partial \psi^*}{\partial \xi} + \psi^* \frac{\partial \Omega'}{\partial \xi} \right)_{\xi=1}$$

According to the integration by parts, we get

$$b = \langle N'_{(q)} q', q^* \rangle - \langle q', N'_{(q)*} q^* \rangle \tag{6.159}$$

From perturbation equation (6.157),

$$\langle N'_{(q)} q', q^* \rangle = \int_0^T \int_{\Sigma} (-H\Omega' \Omega^* + N' F \psi^*) dv dt \tag{6.160}$$

From Eqs. (6.158), (6.159), (6.160),

$$\int_{\Sigma} H(\Omega' \psi^*) \Big|_T dv - \frac{2}{R_e} \int_0^T \left(\int_0^1 A_\infty d\eta - \int_0^1 A_1 d\eta \right) dt$$

$$= \int_0^T \int_{\Sigma} (-H\Omega' \Omega^* + N' F \psi^*) dv dt - \langle q', N'_{(q)*} q^* \rangle \tag{6.161}$$

If the enstrophy is chosen as the quantity of physical interest which is directly related to the vortical structures in the flow field, then the governing equations of the adjoint flow field are

$$N'_{(q)*} q^* = \begin{pmatrix} -H(\Omega^* - \Omega) \\ 0 \end{pmatrix}$$

$$\text{at } t = T, \quad \psi^* = 0, \quad \Omega^* = 0$$

$$\text{at } t < T, \quad \psi^* = 0, \quad \frac{\partial \psi^*}{\partial \xi} = 0 \quad \text{on} \quad \xi = 0$$

$$\psi^* = 0, \quad \Omega^* = 0 \quad \text{on} \quad \xi = \infty.$$

Therefore, Eq. (6.146) can be written as

$$\int_0^T \int_{\Sigma} N' \mathsf{F} \psi^* dv dt = \int_0^T \int_{\Sigma} H \Omega \Omega' dv dt$$

Substitution into Eq. (6.151) yields

$$J_1' = \int_0^T \frac{DJ_1}{DN} N' dt = -\int_0^T \int_\Sigma N' \mathsf{F}\psi^* dv dt + l^2 \int_0^T \int_0^1 NN' d\eta dt$$

Then

$$\frac{DJ_1}{DN} = -\int_\Sigma \mathsf{F}\psi^* dv + l^2 N \tag{6.162}$$

The cost functional is in the extreme, as

$$\frac{DJ_1}{DN} = 0$$

Then the optimal control law is given by

$$\int_\Sigma \mathsf{F}\psi^* dv = l^2 N \tag{6.163}$$

When the several cases of physical interest are considered in a cost functional of the generic form

$$J = -\frac{Hd_1}{2} \int_0^T \int_\Sigma \Omega^2 dv dt - \frac{2d_2}{R_e} \int_0^T \int_0^1 \frac{\partial \Omega}{\partial \xi} K_1 d\eta dt - \frac{2d_3}{R_e} \int_0^T \int_0^1 \frac{\partial \Omega}{\partial \xi} K_2 d\eta dt$$
$$- \frac{2d_4}{R_e} \int_0^T \int_0^1 \Omega K_1 d\eta dt - \frac{2d_5}{R_e} \int_0^T \int_0^1 \Omega K_2 d\eta dt + \frac{l^2}{2} \int_0^T \int_0^1 N^2 d\eta dt \tag{6.164}$$

where $K_1 = \sin(2\pi\eta)$ and $K_2 = \cos(2\pi\eta)$. The constants d_i account for the relative weight of each individual term and $\sum_i d_i = 1$.

The Fréchet differential yields

$$J' = -H \int_0^T \int_\Sigma d_1 \Omega \Omega' dv dt - \frac{2}{R_e} \int_0^T \int_0^1 \left(d_2 \frac{\partial \Omega'}{\partial \xi} K_1 + d_3 \frac{\partial \Omega'}{\partial \xi} K_2 + d_4 \Omega' K_1 + d_5 \Omega' K_2 \right) d\eta dt$$
$$+ l^2 \int_0^T \int_0^1 NN' d\eta dt$$

Here the adjoint flow field is governed by

$$N'_{(q)^*} q^* = \begin{pmatrix} -H \left(d_1 + d_2 + d_3 + d_4 + d_5 \right) \Omega^* - d_1 \Omega \\ 0 \end{pmatrix} \tag{6.165}$$

With initial and boundary conditions

$$\text{at}\, t = T, \quad \psi^* = 0, \quad \Omega^* = 0$$

$$\text{at}\, t < T, \quad \psi^* = d_2K_1 + d_3K_2, \quad \frac{\partial \psi^*}{\partial \xi} = d_4K_1 + d_5K_2 \quad \text{on} \quad \xi = 0$$

$$\psi^* = 0, \quad \Omega^* = 0 \quad \text{on} \quad \xi = \infty$$

Hence Eq. (6.161) is written as

$$\int_0^T \int_\Sigma N' \mathsf{F} \psi^* dv dt = \int_0^T \int_\Sigma d_1 H \Omega \Omega' dv dt$$

$$+ \frac{2}{R_e} \int_0^T \int_0^1 \left\{ (d_2K_1 + d_3K_2) \frac{\partial \Omega'}{\partial \xi} + (d_4K_1 + d_5K_2)\Omega' \right\} d\eta dt$$

Substitution into Eq. (6.162) yields

$$\frac{DJ}{DN} = -\int_\Sigma F\psi^* dv + l^2 N$$

Then the optimal control law is

$$\int_\Sigma \mathsf{F} \psi^* dv = l^2 N$$

which is identical with Eq. (6.163). The desired optimal control input N is thus found to be a simple function of the stream function ψ^* in the adjoint problem proposed in Eq. (6.165), which is independent of the chosen physical interest.

6.3.4.3 Lorentz Force Control of Cylinder Wake

Equations (6.156), (6.163), and (6.165) form a complete system describing a nonlinear adjoint-based optimal control approach of cylinder wake flow by using electromagnetic forcing. These equations can be solved numerically, where the equation of vorticity transport is solved by using the alternative-direction implicit (ADI) algorithm, and the equation of stream function is integrated by means of a fast Fourier transform (FFT) algorithm, which have the accuracy of second order in space and first order in time. During the optimal control, the control input $N(t)$ varying with the transient flow field is determined by Eq. (6.163), in which the right-hand side term is in proportion to N and the left-hand side term is related to an integration of ψ^* over the adjoint flow field, which is dependent upon N. Therefore, the optimal $N(t)$ is determined at the intersection point of the two curves shown in Figure 6.21, where the solid line 1 denotes the integral curve related with the adjoint flow and the dashed line 2 denotes the line in proportion to N.

When the control input N and the period of the control time T are set, the developments of the flow-field $\mathbf{q}$ from $t = 0$ to $t = T$ can be obtained by integrations of Eq. (6.156) numerically.

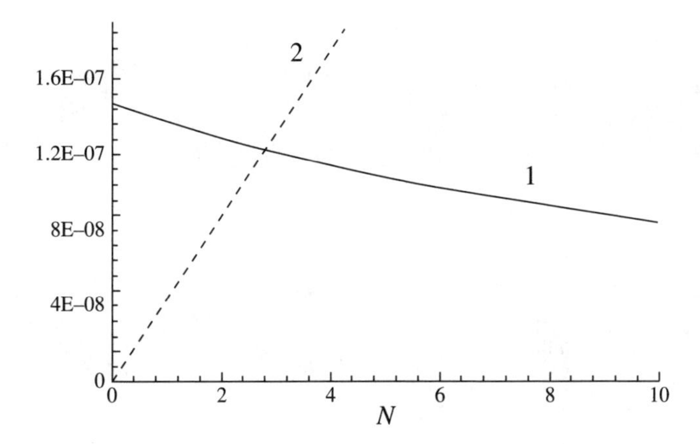

Figure 6.21 Optimal control input N determined by iteration.[23] *Source*: Zhang H 2010. Reproduced with permission of Cambridge University

Then the developments of the adjoint flow-field $\mathbf{q}^*$ with marching backward from $t = T$ to $t = 0$ are also obtained by integration of Eq. (6.165) numerically. Finally, $\int_{\Sigma} \mathsf{F}\psi^* dv$ is obtained by integrating the ψ^* over the adjoint flow field at $t = 0$. The iteration processes are performed to get the intersection point.

Optimal Control Input

Based on calculated results, the variations of the optimal input $N(t)$ with time t are shown in Figure 6.22. For $d_1 = 1$, that is, enstrophy regulation, the value of N increases rapidly at the triggered time $t = 500$ and after a short time interval, it tends to remain constant, due to the tendency for the flow field to steady. For $d_2 = 1$, that is, friction drag regulation, the control input N increases dramatically at the beginning of the control, then decreases and oscillates, and finally steadies on a constant level. For $d_i = 0.2$ ($I = 1.5$), that is, the regulations with several cases of physical interest, the variations in optimal input N are similar with that of

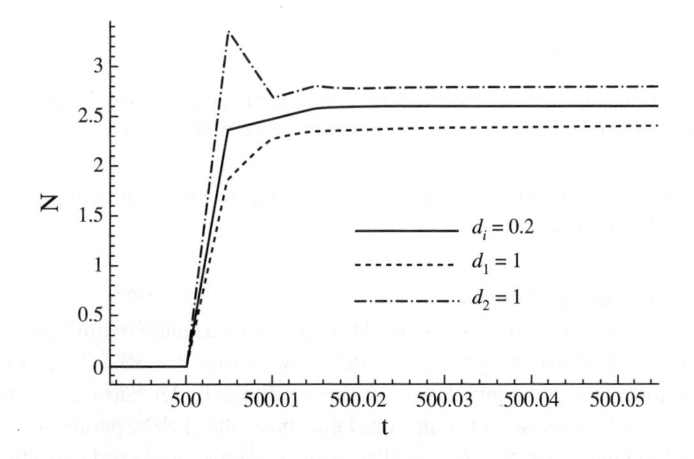

Figure 6.22 Variations in optimal input during the optimal control process.[24] *Source*: Zhang H 2010. Reproduced with permission of Cambridge University

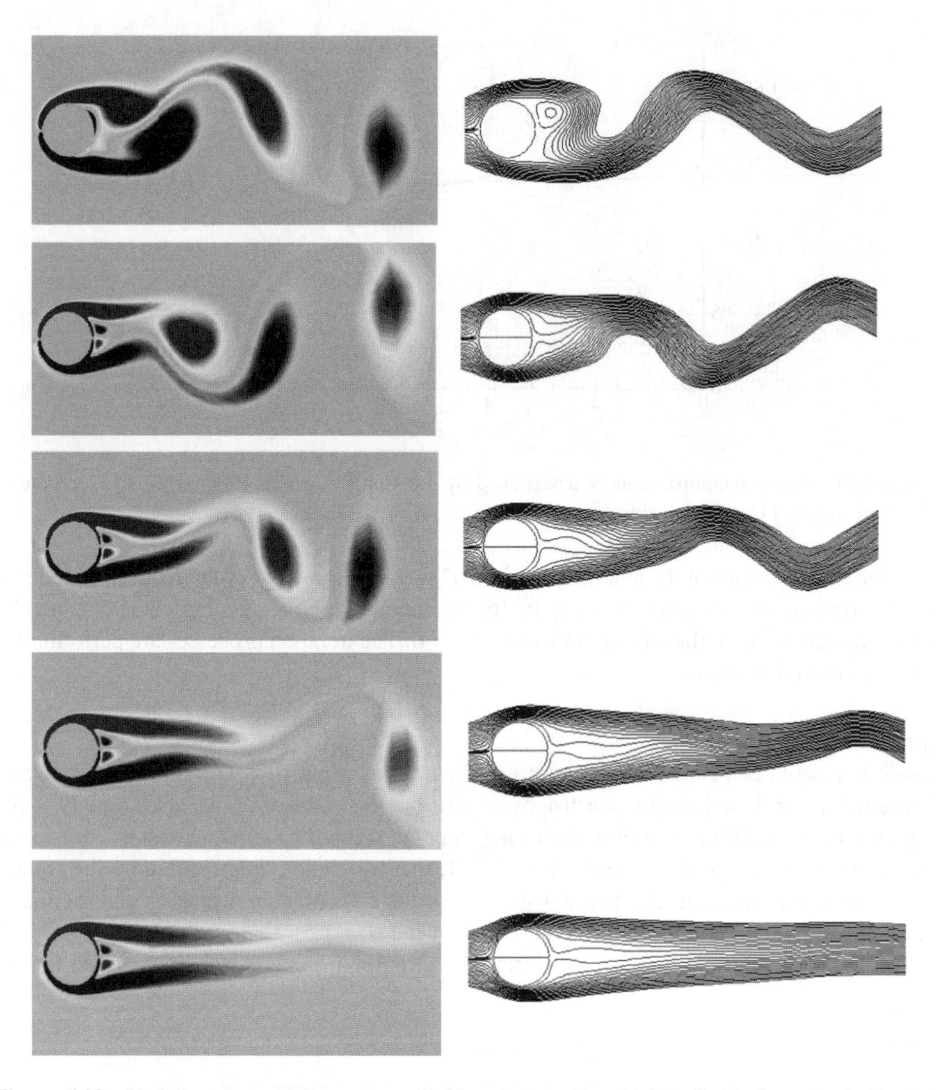

Figure 6.23 Variations in vorticity contours (left) and streamline (right) of cylinder wake in the flow control process.[24] *Source*: Zhang H 2010. Reproduced with permission of Cambridge University

$d_1 = 1$ during the optimal control process. However, there is no significant difference for the final input N of different regulations.

Evolution of Flow Field in Control

Figure 6.23 shows the evolution process of the vorticity (left) and streamline (right) contours with the optimal Lorentz force ($d_i = 0.2$) switched on at time $t = 500$. When the flow passes through the cylinder, the boundary layer appears on the cylinder surface due to the viscous effect. The effects of "adverse" pressure gradient make the flow separate from the cylinder surface, just behind the separation points. The vortexes appear and shed periodically from the cylinder, forming the vortex street. With the application of optimal Lorentz force, the flow in

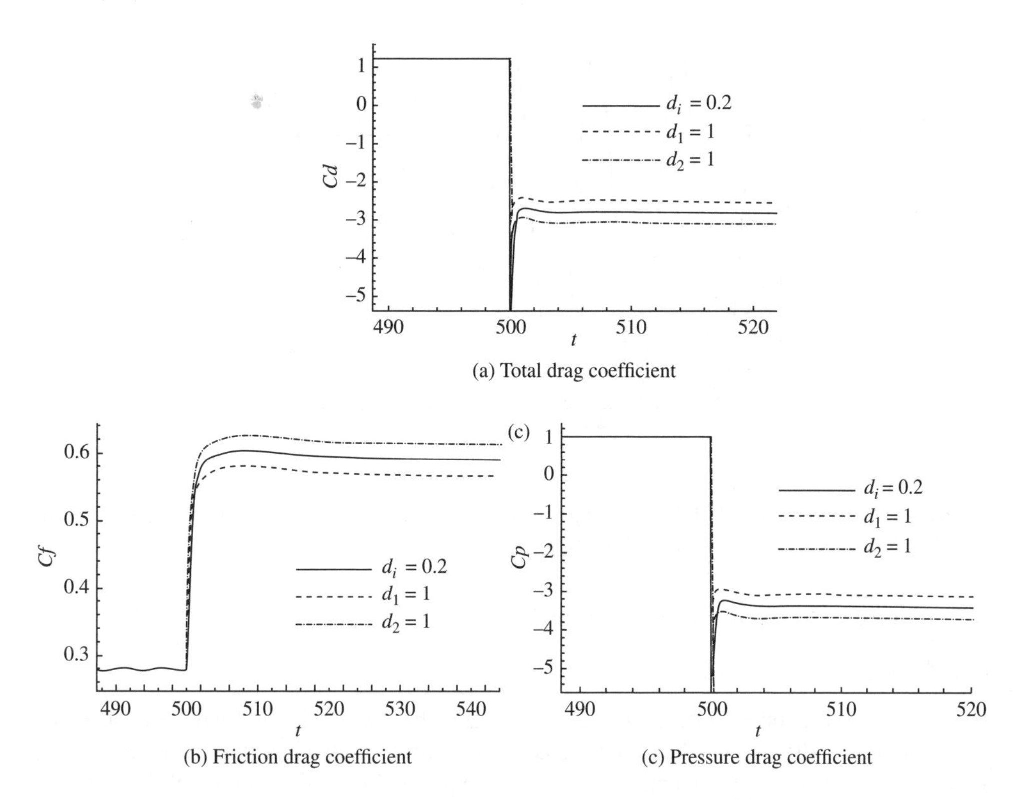

Figure 6.24 Time history of drag force during optimal control process.[24] *Source*: Zhang H 2010. Reproduced with permission of Cambridge University

the boundary layer is accelerated by the Lorentz force which causes the flow separation point to move downstream along the cylinder surface and be suppressed fully. Finally, the vortex street behind the cylinder is mitigated and the flow field turns to stable.

Drag and Lift Coefficients

The time history of the total drag in control, denoted as C_d, is shown in Figure 6.24a. It has been shown that the drag oscillates due to the shedding of vortex on the cylinder surface before the control and decreases dramatically at the beginning of the control, and finally becomes steady at a negtive level. The total drag coefficient is made up of the pressure drag and friction drag coefficients, denoted as C_p and C_f, respectively. The time histories of the friction drag and pressure drag coefficients in control are shown in Figure 6.24b and c, respectively. The increase in vorticity on the wall due to the control leads to the raise in the friction drag consequently. However, the pressure drag coefficient decreases in control, which is dominant in the total drag force and makes the total drag force decrease. For $d_1 = 1$, the total drag coefficient and the pressure drag coefficient are larger and the friction drag coefficient is smaller than that for $d_2 = 1$ and $d_i = 0.2$ $(I = 1 - 5)$ due to the smallest value of control input N in the three cases.

The time history of the lift coefficient in control is shown in Figure 6.25. After the control, as vortex shedding is suppressed, the oscillation of lift coefficient decreases and disappears finally and at that time the lift is kept at zero level. For $d_1 = 1$, the lift coefficient tends to be steady more quickly because the control input varies smoothly.

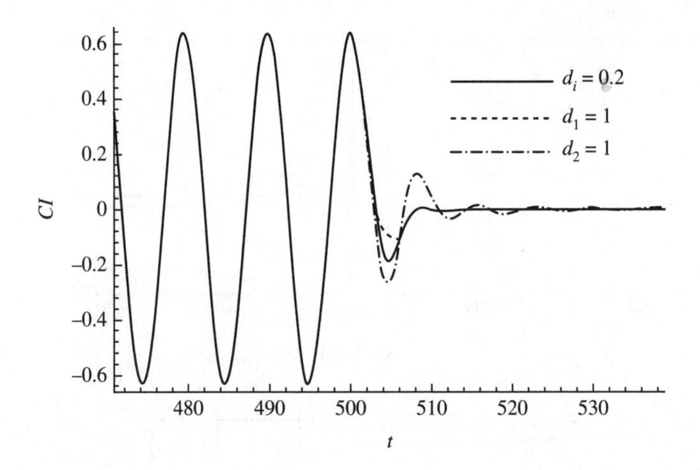

Figure 6.25 Time history of lift force during optimal control process.[24] *Source*: Zhang H 2010. Reproduced with permission of Cambridge University

6.4 Neural Network in Flow Control

6.4.1 *Neural Network*[25, 26]

For flow control applications, neural networks offer the possibility of adaptive controllers, which automatically outputs the optimum parameters after suitable learning. A standard feed-forward neural network architecture is shown in Figure 6.26, which consists of three layers, that is, input layer, hidden layer, and output layer. The data of instantaneous flow fields (or the plant) are inputted into the neural network from the input layer, then the desired control variables generated by the network are given to the actuator from the output layer. In the middle hidden layer, a kernel of the neural network, consisting of a set of neurons, each neuron is related with the input layer and the output layer by special transfer functions. The job of each neuron is to evaluate each of the input signals from the input layer to that particular neuron,

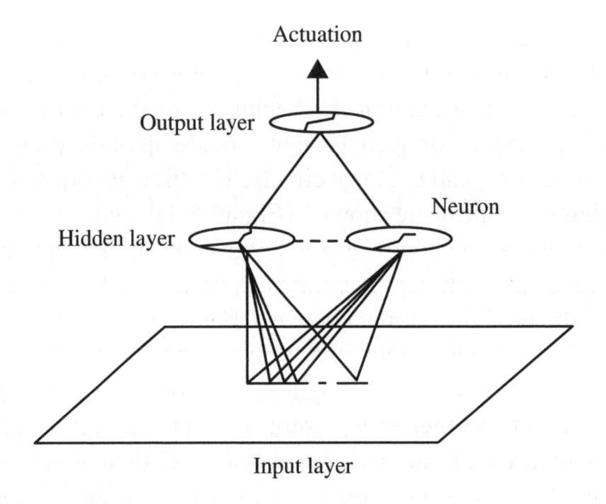

Figure 6.26 Neural network architecture.[26] *Source*: Lee C 1997. Reproduced with permission from Cambridge University Press

calculate the input for each neuron, and finally determine what the output of each neuron should be.

For instance, when a linear function is used as a transfer function between the input layer and the hidden layer, then the input for each neuron can be written as

$$I_{Hj} = \sum_i W_{ij} I_i - W_b \tag{6.166}$$

where I_{Hj} is the input of jth neuron, I_i denotes ith data point of the plant, and W_{ij} is weight. The network has a single set of weights (a template) convolved over the entire input space, that is, we use the same set of weights for each neuron. Taking a hyperbolic tangent as a transform function further, the output of each neuron is written as

$$O_{Hj} = W_a \tanh \quad I_{Hj} - W_c \tag{6.167}$$

where O_{Hj} is the output of jth neuron and W_a and W_c are constant for gain and bias weights, respectively. Furthermore, as a linear function is also used as a transfer function between the output layer and the hidden layer, the final control variables can be obtained via the sum of the input data of the plant with shared constant weights.

6.4.2 Near-Wall Turbulence Controlled by Blowing–Suction Wall

6.4.2.1 Neural Network

The objective here is to seek a control law, that is, the correlation between the wall-shear stresses, $\left.\frac{\partial w}{\partial y}\right|_W$, and the wall actuation, in the form of blowing and suction at the wall, by a neural network to achieve a skin-friction reduction. Based on the standard feed-forward network with hyperbolic tangent hidden units and a linear output unit, the functional form of the neural network for numerical approach is

$$v_{jk} = W_a \tanh\left(\sum_{i=-(N-1)/2}^{(N-1)/2} W_i \left.\frac{\partial w}{\partial y}\right|_{j,k+i} - W_b \right) - W_c$$

$$1 \le j \le N_x, \qquad 1 \le k \le N_z \tag{6.168}$$

where the subscripts j and k denote the numerical grid point at the wall in the streamwise and spanwise directions, respectively. N_x and N_z are the number of computational domain gird points in each direction, $v_{j\,k}$ is the blowing/suction velocity at the wall, and W_i denote input weights. The size of the template is initially chosen to include information about a single streak and streamwise vortex, and then is varied to find an optimal size. Here, N is the total number of input weights in the spanwise direction. The summation is only done over the spanwise direction.

Using an algorithm, such as a scaled conjugate-gradient learning algorithm, for given pairs of $\left(v_{jk}^{des}, \left.\frac{\partial w}{\partial y}\right|_{jk} \right)$, network is trained to minimize the sum of a weighted-squared error given by

$$\text{error} = \frac{1}{2} \sum_j \sum_k e^{\lambda |v_{jk}^{\text{des}}|} (v_{jk}^{\text{des}} - v_{jk}^{\text{net}})^2 \tag{6.169}$$

where v^{des} is the desired output value and v^{net} is the network output value given by Eq. (6.168) and λ is the error scale of the training error. The weights are given by some random values in the initial phase of training, and then are modified in the training process by the algorithm to minimize the error. As the error reaches its asymptotic limit after some training cycles, both the distribution of input weights and the input parameters are determined.

6.4.2.2 Off-line Training and Control

The training data for a neural network off-line are obtained from a numerical simulation of channel flow under the opposition control (see Section 3.6.3), in which blowing and suction at the wall is equal and opposite to the wall-normal component of velocity at $y^+ = 10$. The flow regime is turbulent channel flow with $\mathrm{Re}_\tau = 100$, where Re_τ is the Reynolds number based on the wall-shear velocity u_τ and the channel half-width h. The spectral method (see Section 1.6.2) is employed for calculations with the computational domain $(4\pi, 2, 4\pi/3)h$, and a grid resolution of (32, 65, 32) in the (x, y, z) directions, respectively. Each time step contains a 32×32 array of input values $\partial w/\partial y|_W$ and the corresponding actuations, $-v$ at $y^+ = 10$ The several networks are trained, with one hidden unit and different sized input templates: $7 \times 1, 7 \times 3, 7 \times 5, 9 \times 1, 9 \times 3, 11 \times 1$, and 11×3 (the number of input points in the spanwise direction by the number of input points in the streamwise direction). The weight distributions, obtained from the off-line training, in the spanwise direction at the same streamwise location for different input templates are shown in Figure 6.27. The same pattern for all seven input template sizes is observed. Increasing the number of neurons in the hidden layer only marginally improves performance, whereas increasing the template size significantly reduces the final training error.

Two cases of input template are considered to control a channel flow: a control scheme based on seven weights in the spanwise direction and another based on the same seven weights plus three more in the immediate downstream location. In Figure 6.28, the mean shear-stress variations at the wall obtained with these two controls are plotted along with the no-control case, where thick solid line, dotted line, and thin solid line denote 7 weights, 10 weights, and no control, respectively. Nearly 18% drag reduction is achieved by the control.

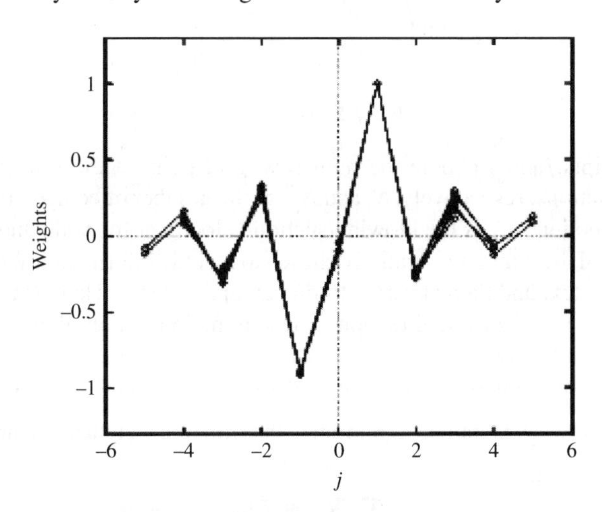

Figure 6.27 Weight distributions from off-line training for various sizes of input template.[26] *Source*: Lee C 1997. Reproduced with permission from Cambridge University Press

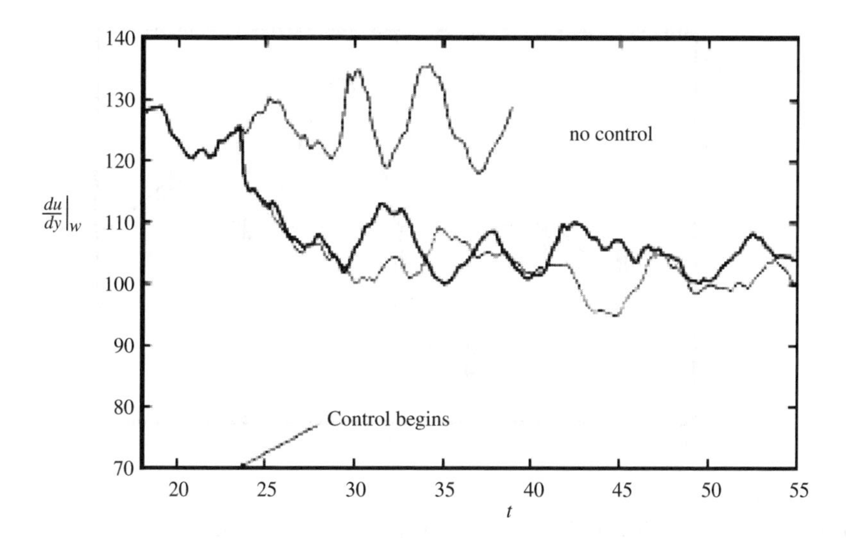

Figure 6.28 Mean wall-shear stress histories for two different off-line controls.[26] *Source*: Lee C 1997. Reproduced with permission from Cambridge University Press

6.4.2.3 On-line Control

In the previous section, the weights are obtained by off-line training, before control. However, since the system to be controlled is time-varying and nonlinear, this approach is not likely to generalize well. Continuous on-line training allows a controller to adapt to the evolution of the system. There are various schemes for on-line neural network control. The most direct scheme is adaptive-inverse model control. In this scheme, the wall actuations and the weights for new time step are obtained from input values $\partial w/\partial y|_W$ in previous step by iterations. This iteration is described briefly as follows: the outputs (wall actuations) and weights for ith interation can be obtained from the ith input $\partial w/\partial y|_W$ by the network, then the new inputs for $(i+1)$th iteration are obtained from the inverse plant mapping from wall-shear stress to wall actuations by the numerical solve of the N-S equations. This process starts again with the new inputs, till the error reaches its asymptotic limit. Finally, the wall actuations and weights for new time step are given by on-line training. The desired inputs to the controller to start the iteration are a fractional reduction in the shear stress from the previous step, that is,

$$\left(\frac{\partial w}{\partial y}\right)_{t+\Delta t}^{\mathrm{des}} = \eta \left(\frac{\partial w}{\partial y}\right)_t$$

where $0 < \eta < 1$. Good performance are achieved for the range of $\eta = 0.8 - 0.85$.

A turbulent channel flow at Reynolds number $\mathrm{Re}_\tau = 100$ is used to test the neural network, where the number of the neurons, the size of the input template, and the error scale of the training error are 1, 7×1, and 5, respectively. The converged weight distributions for 20 consecutive time steps after an asymptotic state is reached are shown in Figure 6.29, where ○ denotes the weight got from on-line training and solid line is plotted by $A(1 - \cos(\pi j)/j$. The pattern is only slightly different from the one obtained from the off-line training shown in Figure 6.16, and emerges immediately after the on-line control begins.

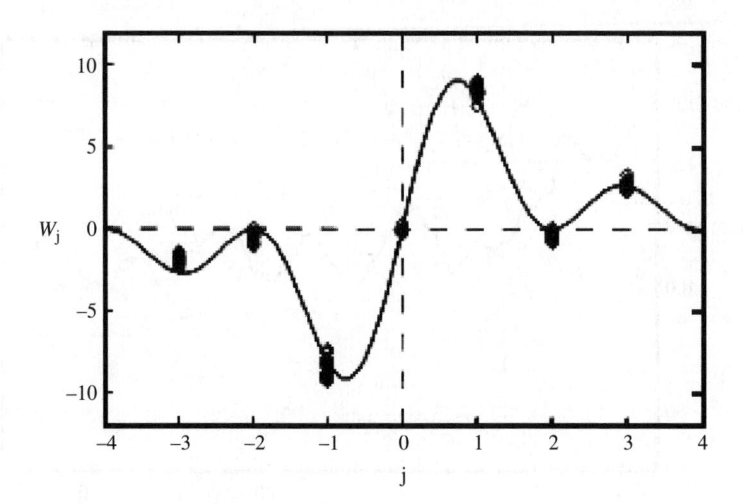

Figure 6.29 Weight distributions from on-line training.[26] *Source*: Lee C 1997. Reproduced with permission from Cambridge University Press

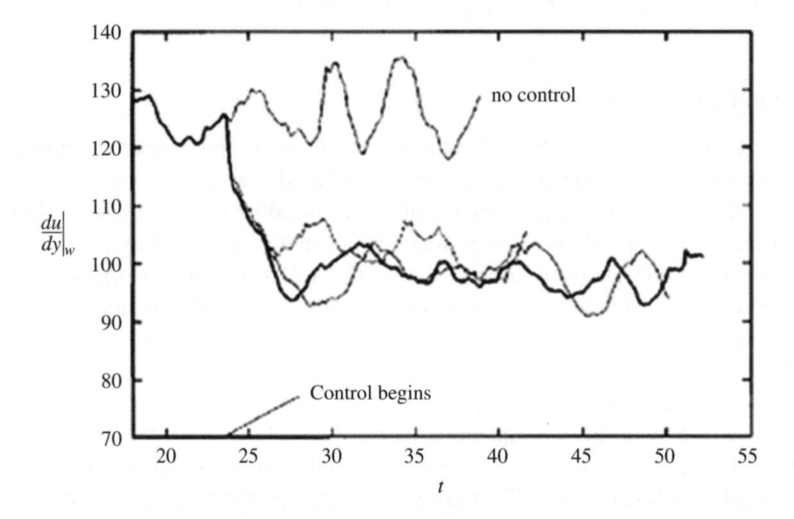

Figure 6.30 Mean wall-shear stress histories for two different on-line controls.[26] *Source*: Lee C 1997. Reproduced with permission from Cambridge University Press

Time histories of the wall-shear stress for on-line control with three different input template sizes, 5×1, 7×1, and 9×1, are shown in Figure 6.30, where thick and thin solid lines denote 7×1 and 9×1, respectively, and dotted line denotes 5×1. It is obvious that the drag drops to about 80% of that observed without control, for 7×1 and 9×1. The template size of 5×1, however, does not produce as much reduction.

The results of above controls indicate that at least seven spanwise points should be used for good performance. The simplified network with a single neuron and a template size of 7×1 has the following functional form:

$$v_{jk} = W_a \tanh(W_b g - W_c) - W_d \tag{6.170}$$

with

$$g = \sum_{i=-3}^{3} W_i \left.\frac{\partial w}{\partial y}\right|_{j,k+i} \tag{6.171}$$

where the bias weights (W_c and W_d) are negligibly small. The gain weights (W_a and W_b) changes in time significantly, although their product remains almost constant.

6.4.2.4 Simple Control Scheme

The previous discussion suggests that effective control can be achieved by a simple control scheme based only on the weighted sum of the wall-shear stress.

The distributions of weights can be approximated by

$$W_j = A\frac{1-\cos(\pi j)}{j} \tag{6.172}$$

where $j = 0$ corresponds to the point where the control is applied. This equation is the inverse Fourier transform of $ik_z/|k_z|$, that is,

$$\int_{-k_m}^{k_m} \frac{ik_z}{|k_z|}\exp(-ik_z z)dk_z = 2\frac{1-\cos(k_m z)}{z} = 2\frac{1-\cos(\pi j)}{j} \tag{6.173}$$

where $k_m = \pi/\Delta z$ is the maximum wave number and Δz is the numerical grid spacing, thus $z = j\Delta z$. From this result and the convolution theorem, one can suggest the following simple control law:

$$\widetilde{v}_w = C\frac{ik_z}{|k_z|}\left.\frac{\partial\widetilde{w}}{\partial y}\right|_w = C\frac{1}{|k_z|}\frac{\partial}{\partial z}\left(\frac{\partial\widetilde{w}}{\partial y}\right) \tag{6.174}$$

which is identical with Eq. (6.38), and implies that the optimum blowing and suction at the wall is proportional to $\partial(\partial w/\partial y)/\partial z$, which counteracts the up-and-down motion induced by a streamwise vortex. Equation (6.174) is equivalent to

$$v_{jk} = C\sum_{i=-(N-1)/2}^{(N-1)/2} W_i \left.\frac{\partial w}{\partial y}\right|_{j,k+i} \tag{6.175}$$

where W_i is given by Eq. (6.172). Magnitude of the weights decays with increasing distance from the center, which allows for good approximation using only a small number of weights. The constant C is chosen so that the root-mean-squared (rms) value of the actuator is kept at $0.15u_\tau$.

Time histories of the wall-shear stress for a no-control and a control based on Eqs. (6.172) and (6.175) with seven points are shown in Figure 6.31. It is clear that the drag decreases obviously after the control.

Contours of streamwise vorticity in a cross-plane after control with seven fixed weights, applied at both walls, are shown in Figure 6.32, where solid line denotes positive contours and negative contours are dotted. It has been shown that the near-wall streamwise vortices are suppressed successfully with the control, leading to a significant drag reduction.

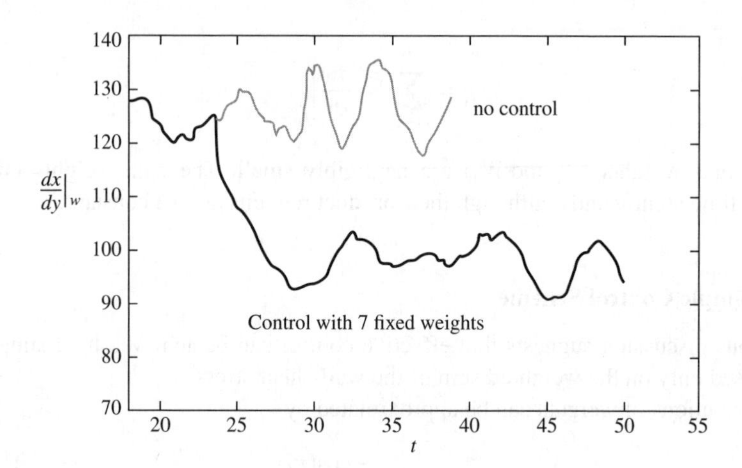

Figure 6.31 Mean wall-shear stress history for a simple control with seven fixed weights.[26] *Source*: Lee C 1997. Reproduced with permission from Cambridge University Press

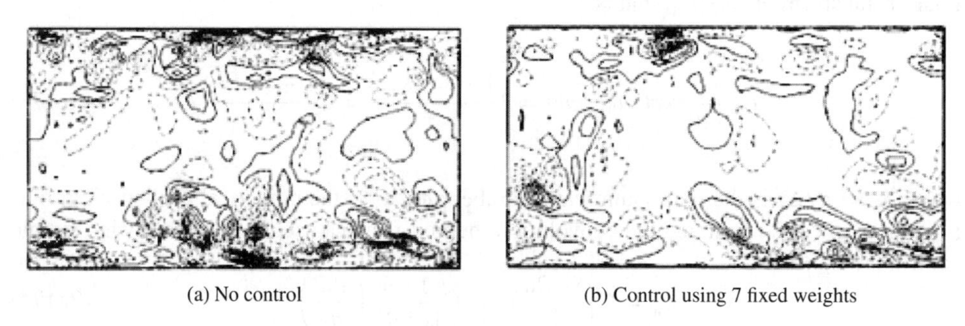

Figure 6.32 Contours of streamwise vorticity in a cross-flow plane: (a) no control and (b) control using seven fixed weights.[26] *Source*: Lee C 1997. Reproduced with permission from Cambridge University Press

The control laws expressed in terms of the Fourier coefficients presented in Section 6.2.1 are impractical to implement, since they require information over the entire spatial domain. Therefore, the inverse discrete Fourier transform of ik_z/k and k_z^2/k is sought, such as Eq. (6.174), so that the convolution integral can be used to express the control laws in physical space. The discrete representation of each law then becomes

$$v(x_j, z_k) = C\sum_{j'}\sum_{k'} W^w_{j'k'} \left.\frac{\partial w}{\partial y}\right|_w (x_{j+j'}, z_{k+k'}) \tag{6.176}$$

and

$$v(x_j, z_k) = C\sum_{j'}\sum_{k'} W^p_{j'k'} p_w(x_{j+j'}, z_{k+k'}) \tag{6.177}$$

respectively. The subscripts j and k denote the discretizing indices in the x- and z-directions, respectively. The weights $W^w_{j,k}$ and $W^p_{j,k}$ are given in Tables 6.1 and 6.2.

Table 6.1 Weight distribution of $W^{w}_{j,k}$ [11]

		$k(= z/\Delta z)$						
		0	1	2	3	4	5	6
$j(= x/\Delta x)$	0	**0.0000**	**1.0000**	**−0.1039**	**0.2679**	**−0.0852**	**0.1419**	−0.0671
	1	0.0086	0.0537	0.0503	0.0310	0.0340	0.0148	0.0237
	2	0.0001	−0.0104	0.0059	0.0051	0.0100	0.0074	0.0092

Source: Lee C 1994. Reproduced with permission of Cambridge University Press.

Table 6.2 Weight distribution of $W^{p}_{j,k}$ [11]

		$k(= z/\Delta z)$						
		0	1	2	3	4	5	6
$j(= x/\Delta x)$	0	**1.000**	**−0.4427**	**0.0031**	**−0.0440**	**0.0044**	**−0.0138**	**0.0032**
	1	**0.0413**	**−0.0007**	**−0.0040**	**−0.0037**	**−0.0029**	**−0.0021**	−0.0016
	2	**−0.0110**	0.0057	0.0019	0.0011	0.0002	0.0000	−0.0003

Source: Lee C 1994. Reproduced with permission of Cambridge University Press.

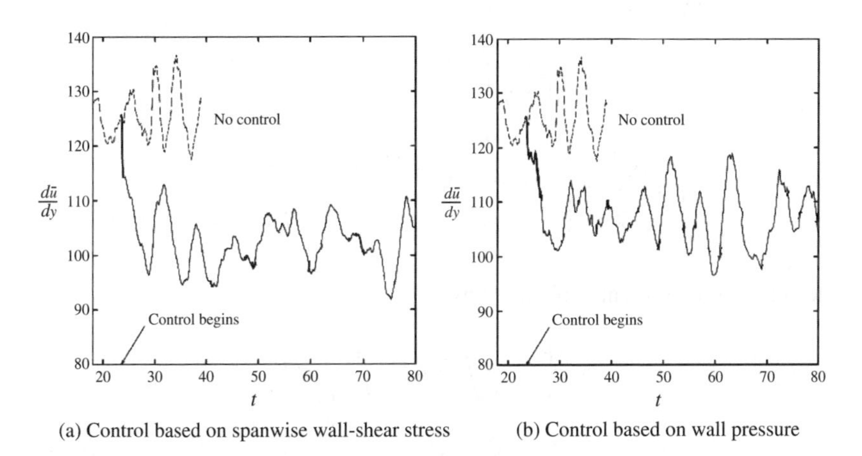

Figure 6.33 Mean streamwise wall-shear stress histories for various control laws.[11] *Source*: Lee C 1994. Reproduced with permission of Cambridge University Press

Time histories of the wall-shear stress for different control laws based on Eqs. (6.176) and (6.177), compared to the no-control case, are shown in Figure 6.33a and b, respectively. Drag reduction is achieved by the control laws.

6.4.3 Near-Wall Turbulence Controlled by Deformed Wall

To address the deformed walls, a new coordinate set (τ, ξ_i) is introduced as follows:

$$\tau = t, \quad \xi_1 = x_1, \quad \xi_2 = \frac{x_2 - \eta_0}{1 + \eta}, \quad \xi_3 = x_3$$

where $\eta_0 = (\eta_u + \eta_d)/2$ and $\eta = (\eta_u - \eta_d)/2$. η_u and η_d are, respectively, the displacements of the upper and lower walls with respect to the uncontrolled state and thus are only function of variables (t, x_1, x_3). Then, the governing equations are written as

$$\frac{\partial u_i}{\partial \tau} = -\frac{\partial u_i u_j}{\partial \xi_j} - \frac{\partial p}{\partial \xi_i} + \frac{1}{\mathrm{Re}} \frac{\partial^2 u_i}{\partial \xi_j \partial \xi_j} - \frac{dP}{d\xi_1}\delta_{i1} + S_i \tag{6.178}$$

$$\frac{\partial u_i}{\partial \xi_i} = -S \tag{6.179}$$

where

$$S_i = -\varphi_t \frac{\partial u_i}{\partial \xi_2} - \phi_j \frac{\partial u_i u_j}{\partial \xi_2} - \phi_j \frac{\partial P}{\partial \xi_2}\delta_{1j}$$

$$+\frac{1}{\mathrm{Re}}\left(2\phi_j \frac{\partial^2 u_i}{\partial \xi_j \partial \xi_2} + \phi_j \phi_j \frac{\partial^2 u_i}{\partial \xi_2^2} + \frac{1}{2}\frac{\partial \phi_j \phi_j}{\partial \xi_2}\frac{\partial u_i}{\partial \xi_2}\right)$$

$$S = \phi_j \frac{\partial u_j}{\partial \xi_2}$$

$$\phi_j = \varphi_j - \delta_{j2}$$

$$\varphi_t = \frac{\partial \xi_2}{\partial t} = -\frac{1}{1+\eta}\left(\xi_2 \frac{\partial \eta}{\partial \tau} + \frac{\partial \eta_0}{\partial \tau}\right)$$

$$\varphi_j = \frac{\partial \xi_2}{\partial x_j} = \begin{cases} -\frac{1}{1+\eta}\left(\xi_2 \frac{\partial \eta}{\partial \xi_j} + \frac{\partial \eta_0}{\partial \xi_j}\right) & j = 1, 3 \\ \frac{1}{1+\eta} & j = 2 \end{cases}$$

The mean pressure gradient is obtained by

$$-\frac{dP}{d\xi_1} = -\int b^* d\xi^3 / \int \frac{1}{\varphi_2} d\xi^3$$

where

$$b^* = u_1 \frac{\partial \eta}{\partial \tau} + \frac{1}{\varphi_2}\left(-\frac{\partial (u_1 u_j)}{\partial \xi_j} - \frac{\partial P}{\partial \xi_1} + \frac{1}{\mathrm{Re}}\frac{\partial^2 u_1}{\partial \xi_j \partial \xi_j} + S_1\right)$$

with boundary conditions

$$u_1 = u_3 = 0$$

Based on the suboptimal control law (6.38), the wall-normal velocity is proportional to the spanwise derivative of the spanwise velocity gradient at the wall, we have

$$\widetilde{u}_2 = \frac{\partial \widetilde{\eta}_d}{\partial \tau} = -C\frac{ik_3}{k}\varphi_2 \widetilde{\frac{\partial u_3}{\partial \xi_2}}\bigg|_{\xi_2=-1} \tag{6.180}$$

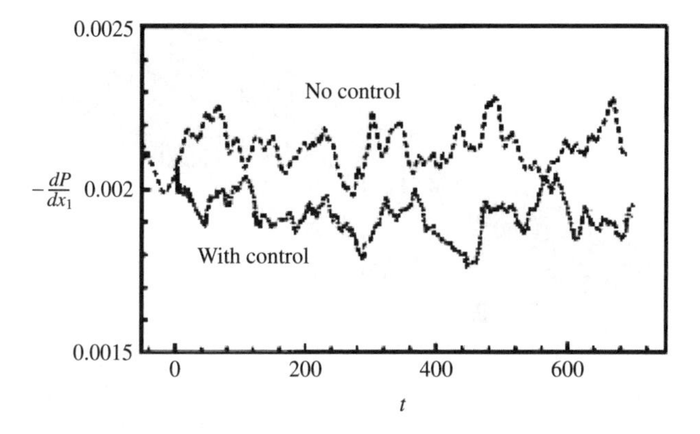

Figure 6.34 Mean streamwise pressure gradient history for control by deformed wall.[27] *Source*: Kang S 2000. Reproduced with permission of Cambridge University

From Eq. (6.176)

$$u_2 = C \sum_{k'=-5}^{5} W_{k'} \varphi_2(i, k + k') \left. \frac{\partial u_3}{\partial \xi_2} \right|_{(\xi_{1(i)}, \xi_2 = -1, \xi_3(k+k'))} \tag{6.181}$$

where the subscripts, i and k, denote the grid indices in the streamwise and spanwise directions, respectively.

The numerical method is based on a semi-implicit, fractional step method: the diffusion and nonlinear terms are advanced in time with the Crank–Nicolson method and a third-order Runge–Kutta method, respectively. All spatial derivatives are discretized with a second-order central-difference scheme. Time history of mean streamwise pressure gradient for the suboptimal control based on Eq. (6.181) is shown in Figure 6.34, which shows that drag reduction is obtained with the active wall motions based on Eq. (6.181).

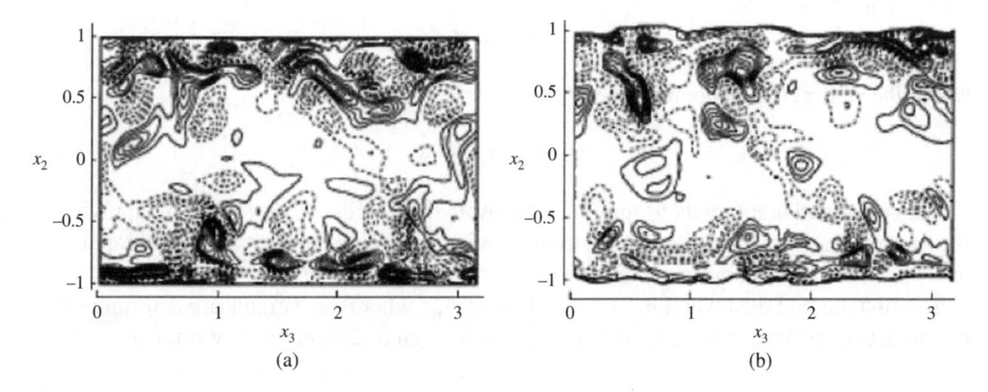

Figure 6.35 Contours of instantaneous streamwise vorticity.[27] *Source*: Kang S 2000. Reproduced with permission of Cambridge University

Figure 6.36 Instantaneous shape of wall deformation.[27] *Source*: Kang S 2000. Reproduced with permission of Cambridge University

Contours of the instantaneous streamwise vorticity for no control and optimal control are shown in Figure 6.35a and b, respectively. The active wall motions significantly weaken the strength of the streamwise vorticity near the wall.

Instantaneous wall shape for active wall motion is shown in Figure 6.36. It is interesting to note that the wall shape is elongated in thc streamwise direction and resemble riblets in appearance.

6.4.4 *Near-Wall Turbulence Controlled by Lorentz Force*

The governing equations for an incompressible electrically conducting viscous flow with an applied wall-normal Lorentz force are

$$\frac{\partial u_i}{\partial t} = -\frac{\partial}{\partial x_k} u_i u_k + \frac{1}{\mathrm{Re}} \frac{\partial^2 u_i}{\partial x_k \partial x_k} - \frac{\partial p}{\partial x_i} + f_i \delta_{i2}$$

$$\frac{\partial u_i}{\partial x_i} = 0 \tag{6.182}$$

with boundary condition

$$\mathbf{u}|_w = 0 \tag{6.183}$$

where the wall-normal Lorentz force is defined as

$$f_y = St(x, z) e^{-\frac{y}{\Delta}} \tag{6.184}$$

If a cost functional targets to minimize the wall-shear stresses, $\partial w/\partial y|_W$, by applying the wall-normal Lorentz force, where St, strength of Lorentz force, is taken as control input, there are two methods of suboptimal control to adjust the magnitude of St.

The first method uses weighted values of $\partial w/\partial y|_W$, where the weights are determined by a neural network. For instance, the control law with seven fixed weights is written as

$$St(x_k, z_l) = C \sum_{-3}^{3} W_j \left. \frac{\partial w}{\partial y} \right|_{k,l+j} \tag{6.185}$$

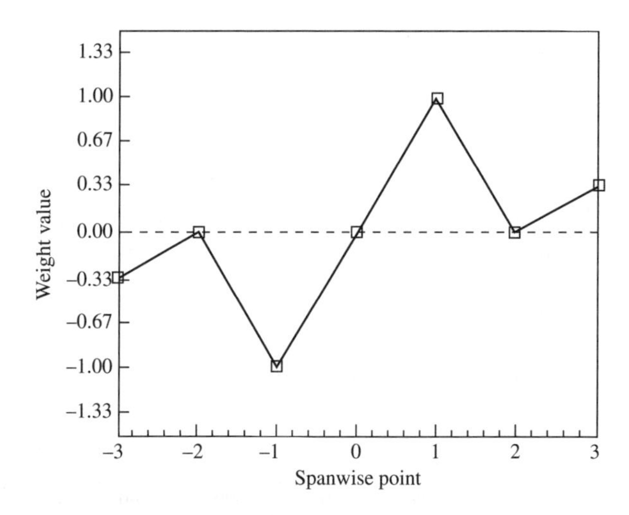

Figure 6.37 Weight distribution of seven-point closed-loop control scheme.[28] *Source*: Berger T W 2001. Reproduced with permission of AIP Publishing LLC

where C is chosen such that the resulting $St|_{\mathrm{RMS}}$, root-mean-square amplitude of force, matches a predetermined value; i and k denote the streamwise and spanwise indices, respectively. The values of weight at seven points are 1/3, 0, −1, 0, 1, 0, 1/3, respectively, or $W_j = \frac{1-(-1)^j}{2j}$, as shown in Figure 6.37.

This control method is performed in the following manner: from the specified $St|_{\mathrm{RMS}}$, the input signal $\left.\frac{\partial w}{\partial y}\right|_{i,k,W}$ and the weights W_j found from neural network, $\phi_{i,k}|_{\mathrm{wall}}$, defined as $\phi_{i,k}|_{\mathrm{wall}} = W_j \left.\frac{\partial w}{\partial y}\right|_{i,k+j,\mathrm{wall}}$, is obtained. Consequently, $\phi_{\mathrm{RMS}} = \sqrt{(\phi_{i,k}^2|_w)}$ is also obtained. Then we have constant C in terms of $C = St|_{\mathrm{RMS}}/\phi_{i,k}|_{\mathrm{wall}}$. Substituting into Eq. (6.185), the optimal St is found, and the Lorentz force for the next time step is gotten from Eq. (6.184) finally.

In the second method, the control law expressed in terms of the Fourier coefficients is referred as the "direct method" and given as

$$\widetilde{S}_t(k_x, k_z) = C\frac{ik_z}{k}\left.\frac{\partial \widetilde{w}}{\partial y}\right|_W \tag{6.186}$$

where $\frac{ik_z}{|k_z|}\left.\frac{\partial \widetilde{w}}{\partial y}\right|_W$, or $\frac{1}{|k_z|}\frac{\partial}{\partial z}\left(\widetilde{\frac{\partial w}{\partial y}}\right)_w$, represents a force counteracting the sweep and ejection motions near the wall.

For a turbulent channel flow, with $\mathrm{Re}_\tau = 100$, $St|_{\mathrm{RMS}} = 4$ and $\Delta^+ = 10$, time histories of drag for the two optimal control schemes based on Eqs. (6.185) and (6.186), respectively, are shown in Figure 6.38, where solid line represents no control, dashed line is seven-point control, and dotted line is direct method. Both methods are successful in reducing drag.

Contours of the streamwise vorticity at four different instances in time response to the application of the seven-point control method are shown in Figure 6.39, where dashed lines indicate negative values. It is demonstrated that the streamwise vortices are completely suppressed.

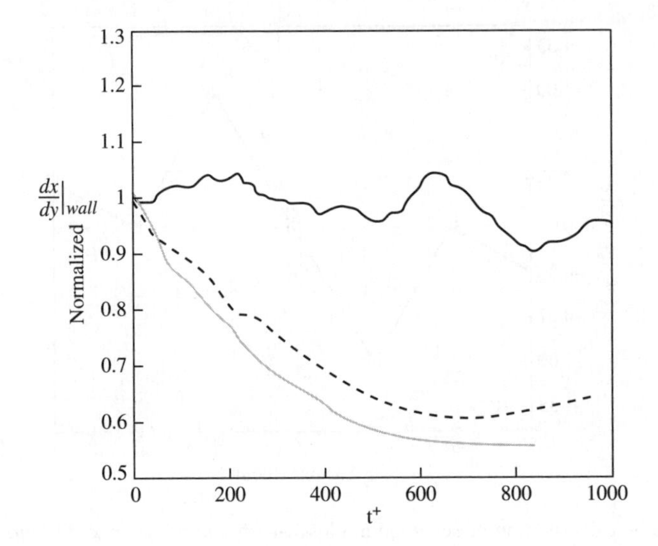

Figure 6.38 Drag histories for no control, seven-point weight, and direct methods.[28] *Source*: Berger T W 2001. Reproduced with permission of AIP Publishing LLC

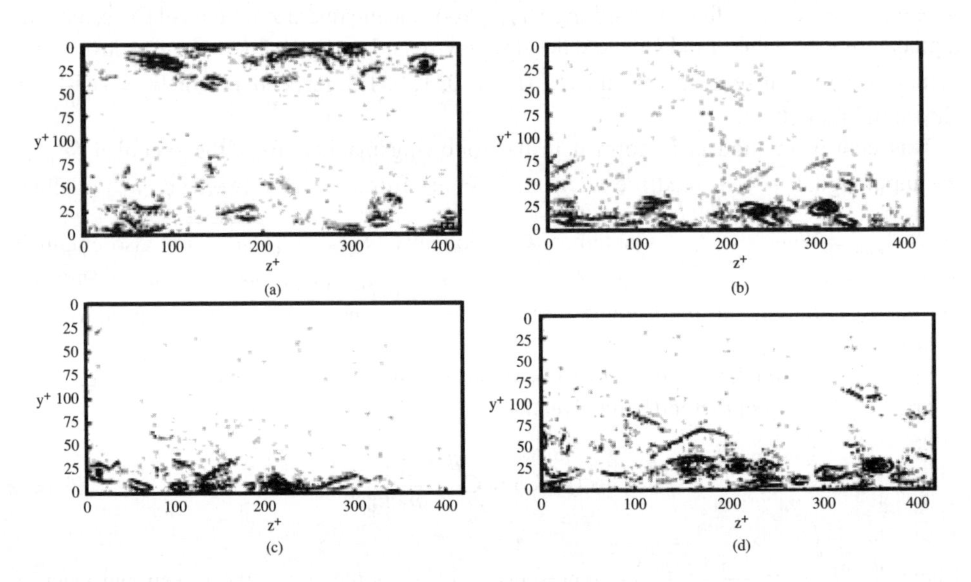

Figure 6.39 Contours of the streamwise vorticity at four different instances in time.[28] *Source*: Berger T W 2001. Reproduced with permission of AIP Publishing LLC

6.5 Closed Remarks

The optimal flow control problem is to determine an optimal control variable, ϕ, satisfying the control sensitivity equation, $\frac{DJ(\phi)}{D\phi} = 0$, where the cost functional $J(\phi)$ is written to represent the physical problem of interest. The control sensitivity equation can be derived directly from the original equations or the discretized equations through a formal minimization.

These problems are usually discussed numerically due to complexity. Hence, the flow control methods are classified as optimize-then-discretize approach and discretize-then-optimize approach based on the performing sequence.

For spectrum-based suboptimal control, a discretize-then-optimize approach, the derivation in N-S equations is discretized. Then the equations for the perturbation state are obtained by Fréchet differential of these discretized equations, where the nonlinear terms are dropped since the explicit schemes used for discretizing the nonlinear terms so that the analytic solutions in the spectrum space can be obtained for the case of simple flow geometry. Likewise, the control sensitivity equation in the spectrum space is given by the minimization procedure in the spectrum space. Finally, the control law in the spectrum space is derived from the analytic solutions and the sensitivity equation, which involves only wall measurements.

For adjoint-based suboptimal control, an optimize-then-discretize approach, the adjoint equations associated with N-S equations are introduced such that the Fréchet derivations of cost functional J' can be expressed by the adjoint state as well as the control law is found, which relates optimal control ϕ with the flow state and the adjoint state. Thus, the optimality system consists of N-S equations, adjoint equations, and control law. This system is solved with numerical methods. First, the N-S equations with an initial guess of ϕ are integrated numerically over an optimization horizon. From these solutions one can solve the corresponding adjoint equations by receding integration, such that the adjoint flow field at initial time is established. Then, the gradient information is determined, and the control law can be solved by the iteration method that gets the optimal control ϕ. Furthermore the flow is advanced by some time portion, and the optimization process begins again. It is noteworthy that the adjoint method requires full flow-field information, the measurements of which are not possible in practice.

Although significant barriers present in the practice applications of these numerical optimization methods, these strategies provide perhaps the most rigorous theoretical framework and can improve understanding of flow physics. In addition, the results of these optimal approaches can be taken as the training samples for neural networks.

Neural network makes no explicit reference to the flow dynamics governed by the N-S equations, and the optimum parameters are given by suitable learning. Therefore, it can provide convenient, fast algorithms to deal with the nonlinear dynamical problems.

References

[1] Gunzburger M D. Perspectives in Flow Control and Optimization. SIAM, Philadelphia, PA, 2003.

[2] Gad-el-Hak M. Modern developments in flow control. Appl. Mech. Rev., 1996, 49:365–379.

[3] Abergel F and Teman R. On some control problems in fluid mechanics. Theoret. Comput. Fluid Dyn., 1990, 1:303–325.

[4] Bewley T. Flow control: New challenges for a new Renaissance. Prog. Aerosp. Sci., 2001, 37:21.

[5] Moin P and Bewely T. Feedback control of turbulence. Appl. Mech. Rev., 1994, 47(6-2):S3–S12.

[6] Lumley J and Blossey P. Control of turbulence. Ann. Rev. Fluid Mech., 1998, 30:311–327.

[7] Hinze M. Optimal and instantaneous control of the instantaneous Navier–Stokes equations. Ph.D. Thesis, Technical University Berlin, 2000.

[8] Chevalier M. Feedback and adjoint based control of boundary layer flows. Technical Reports, Royal Institute of Technology Department of Mechanics, Sweden, 2004.

[9] Choi H, Temam R, Moin P and Kim J. Feedback control for unsteady flow and its application to the stochastic Burgers equation. J. Fluid Mech., 1993, 253:509.

[10] Chang Y. Approximate models for optimal control of turbulent channel flow. PhD Thesis, Rice University, Houston, TX, 2000.

[11] Lee C, Kim J and Choi H. Suboptimal control of turbulent channel flow for drag reduction. J. Fluid Mech., 1994, 358:245–258.
[12] Fukagata K and Kasagi N. Suboptimal control for drag reduction via suppression of near wall Reynolds shear stress. Int. J. Heat Fluid Flow, 2004, 25:341–350.
[13] Fukagata K, Iwamoto K and Kasagi N. Contribution of Reynolds stress distribution to the skin friction in wall-bounded flows. Phys. Fluids, 2002; 14:L73–L76.
[14] Kang S and Choi H. Suboptimal feedback control of turbulent flow over a backward facing step. J. Fluid Mech., 2002, 463:201–227.
[15] Choi H, Hinze M and Kunisch K. Instantaneous control of backward-facing step flows. Appl. Numer. Maths, 1999, 31:133–158.
[16] Min C and Choi H. Suboptimal feedback control of vortex shedding at low Reynolds number. J. Fluid Mech., 1999, 401:123–156.
[17] Bewley T R, Moin P and Teman R. DNS-based predictive control of turbulence: An optimal benchmark for feedback algorithms. J. Fluid Mech., 2001, 447:179–225.
[18] Lee H C. Optimal control problems in fluid mechanics. Information Center for Mathematical Sciences, 1998, 1:215–221.
[19] Bewley T R. Optimal and robust control and estimation of transition, convection, and turbulence. Ph.D. Thesis, Stanford University, CA, 1999.
[20] Bewley T R, Temam R and Ziane M. A general framework for robust control in fluid mechanics. Physica D, 2000, 138:360–392.
[21] Zhang H, Fan B C, Chen Z H, Dong G and Zhou B M. Open-loop and optimal control of cylinder wake via electro-magnetic fields. Chinese Sci. Bull., 2008, 53(19):2946–2952.
[22] Zhang H, Fan B C and Chen Z H. Control approaches of cylinder wake by electromagnetic force. Fluid Dny. Res., 2009, 41(4):045507.
[23] Zhang H, Fan B C and Chen Z H. Optimal control of cylinder wake by electromagnetic force based on the adjonit flow field. Eur. J. Mech. B Fluids, 2010, 29(1):53–60.
[24] Zhang H, Fan B C and Chen Z H. Computations of optimal cylinder flow control in a weakly conductive fluids. Comput. Fluids, 2010, 39(8):1216–1266.
[25] Fujisawa N and Nakabayashi T. Neural network control of vortex shedding from a circular cylinder using rotational feedback oscillations. J. Fluids Struct., 2002, 16(1):113–119.
[26] Lee C, Kim J, Babcock D and Goodman R. Application of neural networks to turbulence control for drag reduction. Phys. Fluids, 1997, 9(6):1740–1747.
[27] Kang S and Choi H. Active wall motions for skin-friction drag reduction. Phys. Fluids, 2000, 12(12):3301–3304.
[28] Berger T W. Turbulent boundary layer control utilizing the Lorentz force. Ph.D. Dissertation, 2001.

Index

Principles of Turbulence Control, First Edition. Baochun Fan and Gang Dong.